Methods in Molecular Biology

Series Editor
John M. Walker
School of Life and Medical Sciences
University of Hertfordshire
Hatfield, Hertfordshire, UK

For further volumes:
http://www.springer.com/series/7651

For over 35 years, biological scientists have come to rely on the research protocols and methodologies in the critically acclaimed Methods in Molecular Biology series. The series was the first to introduce the step-by-step protocols approach that has become the standard in all biomedical protocol publishing. Each protocol is provided in readily-reproducible step-by step fashion, opening with an introductory overview, a list of the materials and reagents needed to complete the experiment, and followed by a detailed procedure that is supported with a helpful notes section offering tips and tricks of the trade as well as troubleshooting advice. These hallmark features were introduced by series editor Dr. John Walker and constitute the key ingredient in each and every volume of the Methods in Molecular Biology series. Tested and trusted, comprehensive and reliable, all protocols from the series are indexed in PubMed.

Manuscript Instructions https://drive.google.com/file/d/1K4Az5vpAFVzYadJ3jKx2-QSo7Dqa_SNA/view?usp=sharing

Immunophenotyping

Methods and Protocols

Second Edition

Edited by

Angélique Biancotto

Ecosystem & Scientific Innovation, Sanofi, Gentilly, France

Pradeep K. Dagur

Frederick, MD, USA

Humana Press

Editors
Angélique Biancotto
Ecosystem & Scientific Innovation
Sanofi
Gentilly, France

Pradeep K. Dagur
Frederick, MD, USA

ISSN 1064-3745 ISSN 1940-6029 (electronic)
Methods in Molecular Biology
ISBN 978-1-0716-5606-8 ISBN 978-1-0716-5607-5 (eBook)
https://doi.org/10.1007/978-1-0716-5607-5

This Humana imprint is published by the registered company Springer Science+Business Media, LLC, part of Springer Nature.
The registered company address is: 1 New York Plaza, New York, NY 10004, U.S.A.

Preface

Immunophenotyping continues to evolve at a remarkable pace, driven by technological innovations that have fundamentally transformed how we characterize and analyze immune cell populations. Since the previous edition of this volume, the field has witnessed significant advances in instrumentation, reagent development, and analytical approaches that have expanded both the depth and breadth of immunophenotypic analysis.

The emergence of full spectral flow cytometry has revolutionized multiparametric analysis, enabling researchers and clinicians to construct panels of unprecedented complexity while overcoming traditional limitations of fluorochrome spillover. Simultaneously, the integration of single-cell transcriptomics with surface protein detection through technologies such as CITE-seq has opened new dimensions in cellular characterization, bridging the gap between phenotype and function. Spatial biology approaches now permit immunophenotyping within the context of tissue architecture, providing critical insights into cellular interactions and microenvironmental influences that were previously inaccessible.

This edition reflects these transformative developments while maintaining a strong foundation in established methodologies essential for both research and clinical applications. The volume is organized into three main parts.

The first part addresses fundamental immune cell populations, presenting detailed protocols for the immunophenotyping of naïve, memory, and effector T cells; B lymphocytes; regulatory cell subsets, including Tregs and natural dendritic cells; NK cells and innate lymphoid cells; neutrophil subsets; and plasmacytoid dendritic cells. A dedicated chapter on nonhuman primate models addresses the growing need for translational immunophenotyping in preclinical research, while functional assessment through NK cell stimulation assays demonstrates the integration of phenotypic and functional analysis.

The second part focuses on clinical applications in hematological malignancies and cellular therapies. Comprehensive chapters cover immunophenotyping approaches for acute myeloid leukemia, acute lymphoblastic leukemia, and chronic lymphocytic leukemia. The critical role of flow cytometry in minimal/measurable residual disease detection is addressed, reflecting the increasing importance of sensitive disease monitoring in patient management. Recognizing the rapid expansion of adoptive cell therapies, this edition includes new chapters on CAR-T cell characterization and immunophenotyping of cellular therapy products targeting oncogenic antigens.

The third part presents methodological advances and analytical frameworks essential for modern immunophenotyping. The chapters on full spectral cytometry, lyophilized reagent systems, and single-cell multi-omic approaches, including CITE-seq and spatial transcriptomics, provide practical guidance for implementing these technologies. Recognizing that data analysis has become as critical as data acquisition, dedicated chapters address guidelines for standardized gating strategies, quality control in immunophenotyping, and advanced biostatistical approaches for unsupervised analysis and immunoprofiling data analytics.

It is no longer possible to present a comprehensive compendium of all immunophenotyping methods in a single volume. However, this collection aims to provide researchers and clinicians with representative, practical protocols spanning fundamental techniques to cutting-edge technologies. Whether establishing core immunophenotyping capabilities or implementing advanced analytical platforms, readers will find detailed methodologies contributed by experts actively working at the forefront of their respective fields.

We are deeply grateful to each of the authors who generously contributed their time and expertise to this volume. Their dedication to sharing detailed, reproducible protocols ensures that these methods can be successfully implemented in laboratories worldwide. We also wish to acknowledge the foundation established by the previous edition, especially the former editor Dr. Phil McCoy, who has guided countless researchers and clinicians in their immunophenotyping endeavors.

Paris, France
Frederick, MD, USA

Angélique Biancotto
Pradeep K. Dagur

Contents

Contributors

YASMINE ADIMI • *Lab Sciences, Translational Medicine Unit, Sanofi, Vitry, France*
SHADY BANYAMIN • *Rutgers Cancer Institute, New Brunswick, NJ, USA*
YVONNE BAUMER • *Social Determinants of Obesity and Cardiovascular Risk Laboratory, Cardiovascular Branch, Division of Intramural Research, National Heart, Lung, and Blood Institute, Bethesda, MD, USA; Department of Biomedical and Translational Sciences, Eastern Virginia Medical School at Old Dominion University, Norfolk, VA, USA*
WILSON DOS-SANTOS BELE • *Target Disease and Systems Biology, Research and Development, Sanofi, Vitry, France*
RAUL BRAYLAN • *Department of Laboratory Medicine, Clinical Center, National Institutes of Health (NIH), Bethesda, MD, USA*
JOLENE L. CARDINALI • *Department of Pathology & Laboratory Medicine, Hartford Hospital, Hartford, CT, USA*
CHIH-HSIUNG CHEN • *Rutgers, Cancer Institute of New Jersey, New Brunswick, NJ, USA; Rutgers Cancer Institute, New Brunswick, NJ, USA*
MINGYI CHEN • *Hematopathology, UT Southwestern Medical Center, Dallas, TX, USA*
SANDRA CURRAS-ALONSO • *Target Disease and Systems Biology, Research and Development, Sanofi, Vitry, France; Target, Disease & Systems Biology, Research and Development, Sanofi, Vitry, France*
PRADEEP K. DAGUR • *Frederick, MD, USA; Flow Cytometry Core, NHLBI, NIH, Bethesda, MD, USA*
ANUSHREE A. DATAR • *Center for Cancer and Immunology Research, Children's National Hospital, Washington, DC, USA*
FABIEN DELAHAYE • *Target Disease and Systems Biology, Research and Development, Sanofi, Vitry, France; Target, Disease & Systems Biology, Research and Development, Sanofi, Vitry, France*
DANA DELGADO • *Flow Cytometry Unit, Laboratory of Pathology, National Cancer Institute, National Institutes of Health, Bethesda, MD, USA*
NILESH DESHPANDE • *Department of Hematopathology, ACTREC, Tata Memorial Center, HBNI University, Navi Mumbai, India*
EMIKO DESVAUX • *Biomarker Data Analytics, Vitry-sur-seine, France*
JOSEPH A. DIGIUSEPPE • *Department of Pathology & Laboratory Medicine, Hartford Hospital, Hartford, CT, USA*
ANAGHA DIVEKAR • *Department for Cellular Analysis, Biolegend, San Diego, CA, USA*
SARAH FIDAI • *Rutgers Cancer Institute, New Brunswick, NJ, USA*
FRANKLIN FUDA • *Hematopathology, UT Southwestern Medical Center, Dallas, TX, USA*
SHARON GERMANS • *Hematopathology, UT Southwestern Medical Center, Dallas, TX, USA*
SITARAM GHOGALE • *Department of Hematopathology, ACTREC, Tata Memorial Center, HBNI University, Navi Mumbai, India*

Valentina Giudice • *Hematology and Transplant Center, University Hospital "San Giovanni di Dio e Ruggi d'Aragona", Salerno, Salerno, Italy; Department of Medicine, Surgery and Dentistry "Scuola Medica Salernitana", University of Salerno, Baronissi, Salerno, Italy*

Marisa Gorrese • *Hematology and Transplant Center, University Hospital "San Giovanni di Dio e Ruggi d'Aragona", Salerno, Salerno, Italy*

Laurent Gorvel • *Centre de Recherche en Cancérologie de Marseille, CRCM, Institut Paoli-Calmettes, Inserm, CNRS, Aix Marseille University, Marseille, France*

Samuel Granjeaud • *Centre de Recherche en Cancérologie de Marseille, CRCM, Institut Paoli-Calmettes, Inserm, CNRS, Aix Marseille University, Marseille, France*

Bruce Greig • *Department of Pathology, Microbiology, and Immunology, Hematopathology/Flow Cytometry, Vanderbilt University Medical Center, Vanderbilt Medical Laboratories, Nashville, TN, USA*

Kim Han • *Laboratory of Mitochondrial Biology and Metabolism, NHLBI, NIH, Bethesda, MD, USA*

Patrick J. Hanley • *Center for Cancer and Immunology Research, Children's National Hospital, Washington, DC, USA*

Taha B. Hayal • *Hematology Branch, National Heart, Lung, and Blood Institute, National Institutes of Health, Bethesda, MD, USA*

Steven C. Hockman • *Flow Cytometry Core, NHLBI, NIH, Bethesda, MD, USA*

Pallavi Kanwar Galera • *Department of Pathology and Laboratory Medicine, Memorial Sloan Kettering Cancer Center, New York, NY, USA*

Keyvan Keyvanfar • *Flow Cytometry Core, NHLBI, NIH, Bethesda, MD, USA*

Timothy Kirtek • *Hematopathology, UT Southwestern Medical Center, Dallas, TX, USA*

Christopher A. Lazarski • *Center for Cancer and Immunology Research, Children's National Hospital, Washington, DC, USA*

Clement Levin • *Lab Sciences, Translational Medicine Unit, Sanofi, Vitry, France*

Wan-Chi Lin • *Flow Cytometry Core, NHLBI, NIH, Bethesda, MD, USA*

Eugénie Lohmann • *Centre de Recherche en Cancérologie de Marseille, CRCM, Institut Paoli-Calmettes, Inserm, CNRS, Aix Marseille University, Marseille, France*

Maria Lopez-Ocasio • *Flow Cytometry Core, NHLBI, NIH, Bethesda, MD, USA*

Holden T. Maecker • *Institute for Immunity, Transplantation, and Infection, Stanford University School of Medicine, Stanford, CA, USA*

J. Philip McCoy • *Brunswick, GA, USA*

Laurel G. Mendelsohn • *Social Determinants of Obesity and Cardiovascular Risk Laboratory, Cardiovascular Branch, Division of Intramural Research, National Heart, Lung, and Blood Institute, Bethesda, MD, USA*

Amazigh Mokhtari • *Biomarker Data Analytics, Vitry-sur-seine, France*

Nadia Nasir • *Laboratory of Pathology, Center for Cancer Research, National Cancer Institute, National Institutes of Health, Bethesda, MD, USA*

Christelle Piperoglou • *APHM, Hôpital de la Timone, Marseille Immunopole Profiling Platform, Marseille, France*

Tiffany M. Powell-Wiley • *Social Determinants of Obesity and Cardiovascular Risk Laboratory, Cardiovascular Branch, Division of Intramural Research, National Heart, Lung, and Blood Institute, Bethesda, MD, USA; Department of Biomedical and Translational Sciences, Eastern Virginia Medical School at Old Dominion University,*

Norfolk, VA, USA; Intramural Research Program, National Institute on Minority and Health Disparities, Bethesda, MD, USA

ABHINAV SAURABH • *Social Determinants of Obesity and Cardiovascular Risk Laboratory, Cardiovascular Branch, Division of Intramural Research, National Heart, Lung, and Blood Institute, Bethesda, MD, USA*

ANKIT SAXENA • *Immune Monitoring Flow Cytometry Shared Resource, Rutgers, Cancer Institute of New Jersey, New Brunswick, NJ, USA; Rutgers Cancer Institute, New Brunswick, NJ, USA*

CARMINE SELLERI • *Hematology and Transplant Center, University Hospital "San Giovanni di Dio e Ruggi d'Aragona", Salerno, Salerno, Italy; Department of Medicine, Surgery and Dentistry "Scuola Medica Salernitana", University of Salerno, Baronissi, Salerno, Italy*

SANDEEP K. SRIVASTAVA • *Center for Cancer and Immunology Research, Children's National Hospital, Washington, DC, USA; Cellular Therapies Division, Hoxworth Blood Center, University of Cincinnati, Cincinnati, OH, USA*

JANET STAATS • *Cedar Grove, NC, USA*

MARYALICE STETLER-STEVENSON • *Laboratory of Pathology, Center for Cancer Research, National Cancer Institute, National Institutes of Health, Bethesda, MD, USA*

PRASHANT R. TEMBHARE • *Department of Hematopathology, ACTREC, Tata Memorial Center, HBNI University, Navi Mumbai, India*

SAHITHI THANIKANTI • *Rutgers Cancer Institute, New Brunswick, NJ, USA*

FRÉDÉRIC VÉLY • *APHM, Hôpital de la Timone, Marseille Immunopole Profiling Platform, Marseille, France; Aix Marseille Université, CNRS, INSERM, CIML, Marseille, France*

HAO-WEI WANG • *Laboratory of Pathology, Center for Cancer Research, National Cancer Institute, National Institutes of Health, Bethesda, MD, USA*

CHUANFENG WU • *Hematology Branch, National Heart, Lung, and Blood Institute, National Institutes of Health, Bethesda, MD, USA*

Part I

Cell Type-Focused Profiling

Chapter 1

Multiparametric Flow Cytometry Analysis of Naïve, Memory, and Effector T-Cells

Chih-Hsiung Chen, Pradeep K. Dagur and Ankit Saxena

Abstract

Polychromatic flow cytometry enables the detection and characterization of markers that are helpful in defining the phenotype of various cell subsets. Here, we described a flow cytometry-based method to characterize the phenotype of naïve, memory, and effector T cells. Being able to differentiate these cells is crucial in understanding immune response and immune profiling. Naïve T cells enable the body to fight off new, unrecognized infections and diseases, and Memory T cells are enriched for response to recall antigens. Furthermore, the antigen-experienced T-cell populations can be broadly divided into effector and memory cell compartments, which are needed to sustain a responsive immune system. Simplistically, the effector T cells require active antigenic stimulation to eliminate pathogens. On the other hand, Memory T cells are described as cells that remain present without antigenic stimulation and can expand rapidly upon secondary challenges. Recently, with the identification of central and effector memory T-cell subsets, tremendous effort has been devoted to characterize markers on the surfaces of these cells. Though various markers have been used to identify the subsets, no single marker that segregates one subset from the other has been described. Thus, multiple markers are needed to subset the cells in order to characterize them. Here we report the verification of a nine-color panel (CD3, CD4, CD8, CD45RO, CD28, CD95, CCR7, Live/Dead Aqua, Dump channel-CD19, CD14, CD56, CD16) that can successfully identify six distinct CD4 and CD8 T-cell populations within the naïve and effector cell subsets from human donors.

Keywords Immunophenotyping, Multicolor-flow cytometry, Whole blood, T cells, Memory T cells, Naïve T cells

1 Introduction

T cell phenotype has long been used as a means of functionally classifying T cell subsets [1, 2]. The development of techniques for protein detection at the single-cell level allowed detailed correlations between the functional properties of T cells and their phenotype [3, 4]. Following positive and negative selection in the thymus, CD4+ and CD8+ T cells are released in the periphery as mature naïve T cells. Mature CD4 and CD8 T cells in the extrathymic environment are long-lived cells that can remain in interphase for many weeks or

Angélique Biancotto and Pradeep K. Dagur (eds.), *Immunophenotyping*, Methods in Molecular Biology, vol. 3087, https://doi.org/10.1007/978-1-0716-5607-5_1,

months [5–7]. In response to antigenic challenge, naïve T (TN) cells with specificity for a certain antigenic epitope proliferate and differentiate and later die >95% after antigenic clearance, leaving behind a small pool of long-lived memory cells [8–10]. The process of conversion of a naïve T cell to memory is a sequential process where less differentiated cells give rise to more differentiated progeny in response to antigenic stimulation or homeostatic signaling. Often these subsets are associated with signature molecules and display phenotypic and gene-specific properties, which are useful to define them as a cluster of cells. Though the progression of the naïve and effector memory cell subsets can be broadly staged as different subsets, such as naïve T (TN), T-stem cell memory (TSCM), T-central memory (TCM), T-terminal memory (TTM), T-effector memory (TEM), and T-terminal effector (TTE) [9, 10]. While naïve T cells are homogeneous, diversity becomes extreme in the antigen-experienced memory compartment. The precise identification of these subsets is pivotal for both basic and clinical immune monitoring of patients undergoing immunotherapy for T cell-based therapeutic approaches [11, 12].

Since multiparameter flow cytometry approaches are becoming more popular, if not routine, in immunology laboratories, it would be logical that human T-cell differentiation should be delineated using a minimum set of canonical markers, i.e., CD45RO (or CD45RA), CCR7, CD28, and CD95. The differential expression of these markers allows the identification of six subsets in the peripheral blood of healthy humans: T_N, T_{SCM}, T_{CM}, T_{TM}, T_{EM}, and T_{TE} cells (Fig. 1).

Multi-color flow cytometric analysis of T cells has revealed that a single marker or a combination of two markers does not allow the identification of pure TN cells [2]. Ideally, the more markers are included to define TN cells in a polychromatic panel, the better. We describe a nine-color panel to identify the abovementioned subsets (Table 1).

2 Materials

Prepare all solutions using distilled water (dH_2O) and analytical-grade reagents. Prepare and store all reagents at room temperature (RT) (unless indicated otherwise). Diligently follow all waste disposal regulations when disposing of waste materials.

1. Sample material: human peripheral blood, bone marrow (in anticoagulant–heparin, EDTA, etc.), and lymphoid tissue.
2. RBC lysis buffer: ammonium chloride lyse (10× concentration) NH_4Cl (ammonium chloride) 8.02 g; $NaHCO_3$ (sodium bicarbonate) 0.84 g; EDTA (disodium) 0.37 g; dissolve in 100 mL with dH_2O. Store at 4°C for 6 months.

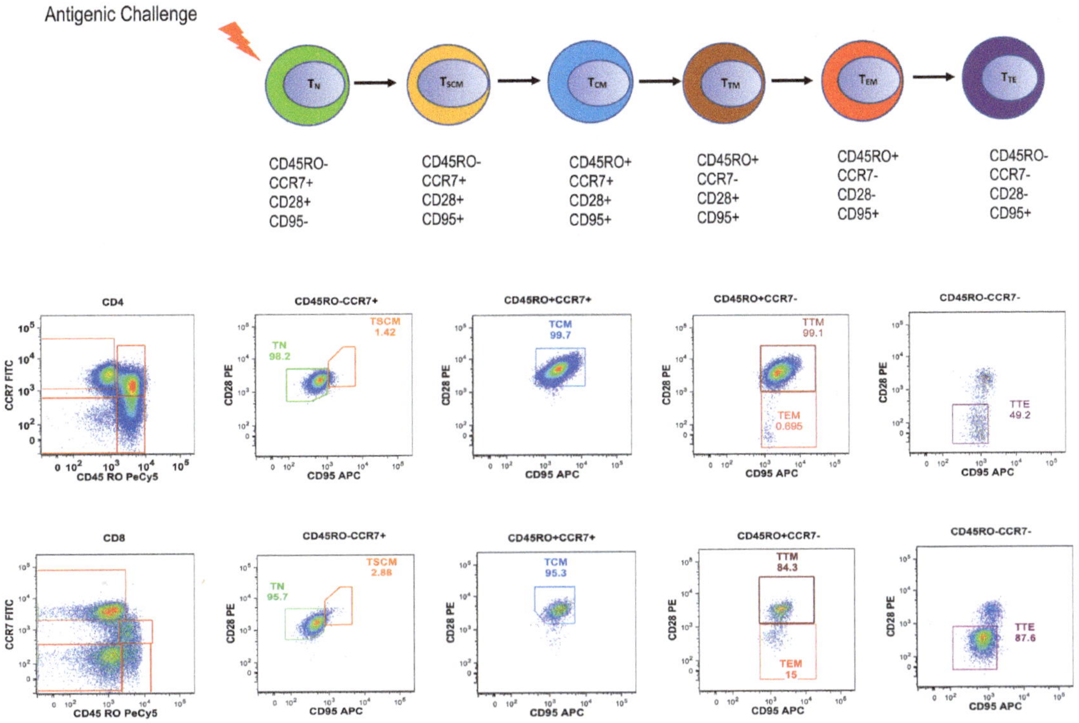

Fig. 1 T cell heterogeneity among naïve and memory subsets. CD45R0, CCR7, CD28, and CD95 help in identifying six major subsets of T cells. While differentiating (through activation) from T_{SCM} to T_{CM}, T_{TM}, T_{EM} and culminating in T_{TE} cells, memory T cells progressively lose or acquire specific surface protein markers. Following an encounter with an antigen, these naive T cells develop into effectors, for which the phenotype is highly dynamic and unpredictable. When the antigen is cleared, surviving effector T cells that survive return to memory state. Bottom-Multiparametric flow cytometry identification of multiple T cell subsets among naïve and memory compartments in the peripheral blood of healthy individuals according to expression of CD45R0, CCR7, CD28, and CD95

Working solution: dilute 10 mL 10× concentrate with 90 mL dH_2O. Refrigerate until use.

3. Staining buffer—1× phosphate buffered saline (PBS): dissolve 8 g NaCl, 0.2 g KCl, 1.15 g Na_2HPO_4, and 0.2 g, KH_2PO_4 in 800 mL distilled water (dH_2O), 10% FCS or 1% bovine serum albumin (BSA). Adjust the pH to 7.4 and the final volume to 1 L (*see* **Note 1**).
4. Fluorochrome-labeled antibody diluted to the appropriate concentration as determined by titration and viability dye (Table 2) (*see* **Note 2**).
5. 12 × 5 mm round-bottom polystyrene test tubes or 96-well round-bottom, V-bottom, or flat-bottom microtiter plates for high-throughput screening (HTS) (*see* **Note 3**).

Table 1
Rationale for panel development

Priority rating	Category	Rationale for priority level	Reagents concerned
1	Dump channel	Detector with the least number of good conjugates available	Monoclonal antibodies against CD16, CD14, CD56, and CD20 human antigen
2	Live/Dead	To gate out dead cells	AquaBlue Live/Dead Dye
3	T-cell subset markers	Identify CD4+ and CD8+ T-cell compartments	Monoclonal antibodies against CD3, CD4, CD8
4	Differentiation markers	Reason for the development of panel	CCR7 and CD45RO human antigen
5	Differentiation markers: additional	Identification of T_{TE} and T_{SCM} subsets within naïve-like cells	Monoclonal antibodies against CD95, CD28 human antigen

The table highlights important points to be considered in order to successfully design and execute a multicolor panel. Points to be considered are ordered under priority rating, followed by rationale and reagents concerned.

6. 40–100 μm nylon mesh for isolating cells from lymphoid tissues.
7. Centrifuge with adapters for 5 mL, 12 × 5 mm round-bottom test tubes, 50- and 15-mL Falcon tubes, or 96-well round-bottom microtiter plates.
8. Fixation buffer: 2% paraformaldehyde in 1× PBS.
9. Compensation beads (BD Biosciences).
10. Flow cytometer.
11. FlowJo analysis software (Tree Star Inc.).

3 Methods

3.1 Blood Sample Collection and Cell Preparation

1. Whole blood lysis: collect blood in sodium heparin, sodium citrate, or EDTA tube and keep it at RT on a nutator. Transfer 5–10 mL blood with a disposable Pasteur pipette into a 50 mL Falcon tube and add 40–45 mL of RBC lysis buffer. Place the 50 mL Falcon tube on a nutator for 2 min Centrifuge tube at 300 g for 5 min at 25°C. Carefully remove the lysis buffer and break the cell pellet by slow vortex or gentle tapping. Repeat the addition of 45–50 mL of RBC lysis buffer and keep cells

Table 2
Antibody panel details: antigen-specific mAb used for various cell surface antigens

Specificity	Fluorochrome	Ab clone	Vendor	Catalog number	Volume
CD3	BV450	UCHT1	BD Biosciences	560365	3.5 µl
CD4	APC-Cy7	RPA-T4	BD Biosciences	557871	3.5 µl
CD8	BV605	SK19	BD Biosciences	56416	3.5 µl
CD28	PE	CD28.2	BD Biosciences	555730	15.0 µl
Dead cells	AquaBlue	–	Life Technologies	L34957	100 µl[a]
CCR7	FITC	150503	R&D	FAB197F	10 µl
CD45RO	PE-Cy5	UCHL-1	BD Biosciences	555494	15 µl
CD95	APC	DX2	BD Biosciences	558814	15 µl
CD20	PE-Cy7	L27	BD Biosciences	335793	15 µl
CD56	PE-Cy7	NCAM16.2	BD Biosciences	335809	15 µl
CD14	PE-Cy7	M5E2	BD Biosciences	557742	3.5 µl
CD16	PE-Cy7	3G8	BD Biosciences	557742[b]	3.5 µl

APC allophycocyanin, *APC-Cy7* allophycocyanin–cyanine H7 tandem, *FITC* fluorescein isothiocyanate, *PE* phycoerythrin, *PE-Cy7* phycoerythrin–cyanine 7 tandem, *Pe-Cy5* peridinin–cyanine 5 tandem, *BV450* brilliant violet 450, *BV605* brilliant violet 605 violet [a]AquaBlue was pre-diluted 1:0.0035 in FACS buffer. The final dilution is thus 1:0.35

on a nutator for 1 min. Centrifuge tube at 300 × *g* for 5 min at 25°C, followed by lysis buffer removal and pellet breaking. Now add 10 mL of staining buffer (RT) to the cells and give a final wash to replace the remnant RBC lysis buffer with the staining buffer. Pellet the cells (5 min, 300 × *g*) and aspirate the supernatant. Centrifuge tube at 300 × *g* for 5 min at 25°C. to make a final cell concentration of 1–2 × 10^6 cells/mL (*see* **Note 4**).

2. Add fresh staining buffer and adjust cell suspension to a concentration of 1–2 × 10^6 cells/mL in ice-cold PBS, 10% FCS. If large clumps are seen, the sample may be passed through the 40–100 µm mesh to get rid of clumps and obtain a single-cell suspension. Cells are usually stained in polystyrene round-bottom 12 × 75 mm^2 Falcon tubes. However, they can be stained in any container for which you have an appropriate centrifuge, e.g., test tubes, microcentrifuge tubes, and 96-well round-bottomed microtiter plates (*see* **Note 5**).
3. Add a titrated amount of fluorochrome-conjugated antibodies and Live/Dead aqua stain along with FcR blocker or in the

presence of 2 µl of normal mouse serum to reduce background staining.

4. Incubate cells at 4°C for 15 min (*see* **Note 6**).
5. Add 2 mL of staining buffer and wash the cells (5 min, 300 × *g*). Resuspend the pellet in 0.4–0.5 mL of staining buffer.
6. If sample acquisition is to be done after 24 h, then fix cells using fixation buffer. Add 0.5 mL of fixation buffer to the cells and incubate at RT for 15–20 min in dark (*see* **Note 7**).
7. Pellet the cells (5 min, 500 × *g*) and decant the fixation buffer-containing supernatant (paraformaldehyde is toxic: handle with care; dispose of according to material safety regulations) (*see* **Note 8**).
8. Resuspend cells with 0.5 mL of staining buffer, keep it in the dark and cold till samples are acquired on the cytometer (It is advised to run processed samples as soon as possible or within 72 h of staining).

3.2 Data Collection on Flow Cytometer

Specific methods depend on the available make and model of the flow cytometer. It is important to appropriately establish forward- and side-scatter gates to exclude debris and cellular aggregates from analysis. If cells are stained with multiple fluorescent labels, attention must be paid to balance the signal amplification among photomultiplier tubes (PMTs) in a way that the particular fluorophore is brighter in its respective PMT in order to minimize potential spillover between fluorochromes and detectors.

1. Single-color controls for each fluorochrome are prepared and unstained cells are kept as a negative control. Single-color controls can be prepared both by using cells or compensation beads, as long as a bright signal is observed in each respective detector PMT (*see* **Note 9**).
2. Here, BD LSR-II was used to acquire stained samples (Table 3) using BD FACS DIVA v8.0 (BD) acquisition and analysis software.
3. Open DIVA software and connect to the instrument. Look for the green light appearing in the bottom-right corner of the software to ensure a proper connection. Once the instrument is connected, establish a new experiment in the browser window and add the appropriate number of tubes to acquire.
4. When the workstation is connected to the cytometer and an experiment is open, a gray pointer icon is displayed next to tubes in the browser. To activate acquisition controls, click the icon next to the tube you want to acquire to set as the current tube pointer. The icon turns green, and the tube becomes the active tube in the acquisition dashboard.

Table 3
Instrument configuration

Laser	Laser		Spectral range	Optical filters			
Wavelength [nm]	Power [mW]	Laser type	For detector [nm]	Dichroic #1 [nm]	Dichroic #2 [nm]	Band pass [nm]	Fluorochrome
633	200	DPSS	750–810	–	740	780/60	APC-Cy7
			650–670	685	–	660/20	APC
532	150	DPSS	760–800	–	740	780/40	PE-Cy7
			640–680	690	640	660/40	PE-Cy5
			562–587	600	–	575/25	PE
488	100	DPSS	505–525	685	505	515/20	FITC
405	100	DPSS					
			590–620	630	595	605/40	BV605
			564–606	595	630	660/40	BV650
			540–580	557	535	560/40	AquaBlue
			425–475	535	–	450/50	BV421

The panel was optimized for an LSR II with the listed optical elements. In our setup, the dichroic filters act as long-pass filters (Dichroic #2) for one detector, sending longer wavelengths through the respective band-pass filters and reflecting shorter wavelengths down the optical path to the next dichroic filter (Dichroic #1). *DPSS* diode-pumped solid-state

5. Under tab view select, browser, inspector, cytometer, acquisition, dashboard, worksheet panels.
6. Click the "Parameters" tab in the "Cytometer" window and delete all the parameters that you don't need. To delete parameters, click the selection button next to each parameter that you want to delete. Hold down the Ctrl key to select more than one. After you are finished selecting, click the "delete" button. If needed, parameters can be added or modified by clicking the "Add" button.
7. Use the experiment layout to create labels, enter values for keywords, and enter acquisition criteria for each tube in an experiment.
8. Add or change labels. Select the field(s) listing the fluorochromes to be labeled, and type to enter a label.
9. Once you have created your experiment, you are ready to create compensation controls. Select experiment > compensation setup > create compensation controls. Alternatively, right-click the experiment icon and select compensation setup > create compensation controls. A dialog appears where you

can add or delete controls, define label-specific controls, or change the order of the compensation controls.

10. Leave the "Include separate unstained control tube/well" checkbox selected when you are running an unstained sample as one of your compensation controls.
11. Record data for the unstained and single-color controls using appropriate signal amplification by PMTs to ensure that a bright signal is observed in each respective detector PMT (*see* **Note 10**; compensation controls can only be acquired on the normal worksheet, not on the global worksheet).
12. At the end, select experiment > compensation setup > calculate compensation.
13. Enter a compensation setup name, click Link and Save, and click OK. To apply compensation values to that experiment's cytometer settings, click "Apply Only."
14. Toggle back to global worksheet (in case the gating strategy changes in a different tube, then use "sheet" instead of "global work sheet").
15. Use menu commands or plot buttons to create dot plots and contour plots on the worksheet.
16. To create a plot using a plot button, click the appropriate button on the worksheet toolbar, and click once on the worksheet to draw a plot of default size.
17. Acquire cells at an event rate of 3000–5000 events per second.
18. Use a series of plots and establish a gating hierarchy to ensure recording of debris and doublet-free live T cells (at least 250,000 single live T cells to ensure a sufficient number of rare subsets to analyze in the recorded data file, if needed).

3.3 Data Analysis

1. On Flow-jo software, open a new workspace and add compensation FCS files and samples FCS files in separate group folders.
2. If required, compensation can be recalculated using compensation FCS files and newly calculated compensation can be applied to the sample FCS files for further data analysis (*see* **Note 11**).
3. Open the FCS files and start creating gates as follows.
4. Define the gated populations as shown in (Fig. 2).
5. Lymphocyte gate: Create a bivariant polychromatic plot and gate click FSC-A on X-axis and SSC-A on Y-axis. Make a polygonal gate around lymphocyte population as shown. Please ensure in the active gate check box the gated events label is checked in the flow-jo workspace.

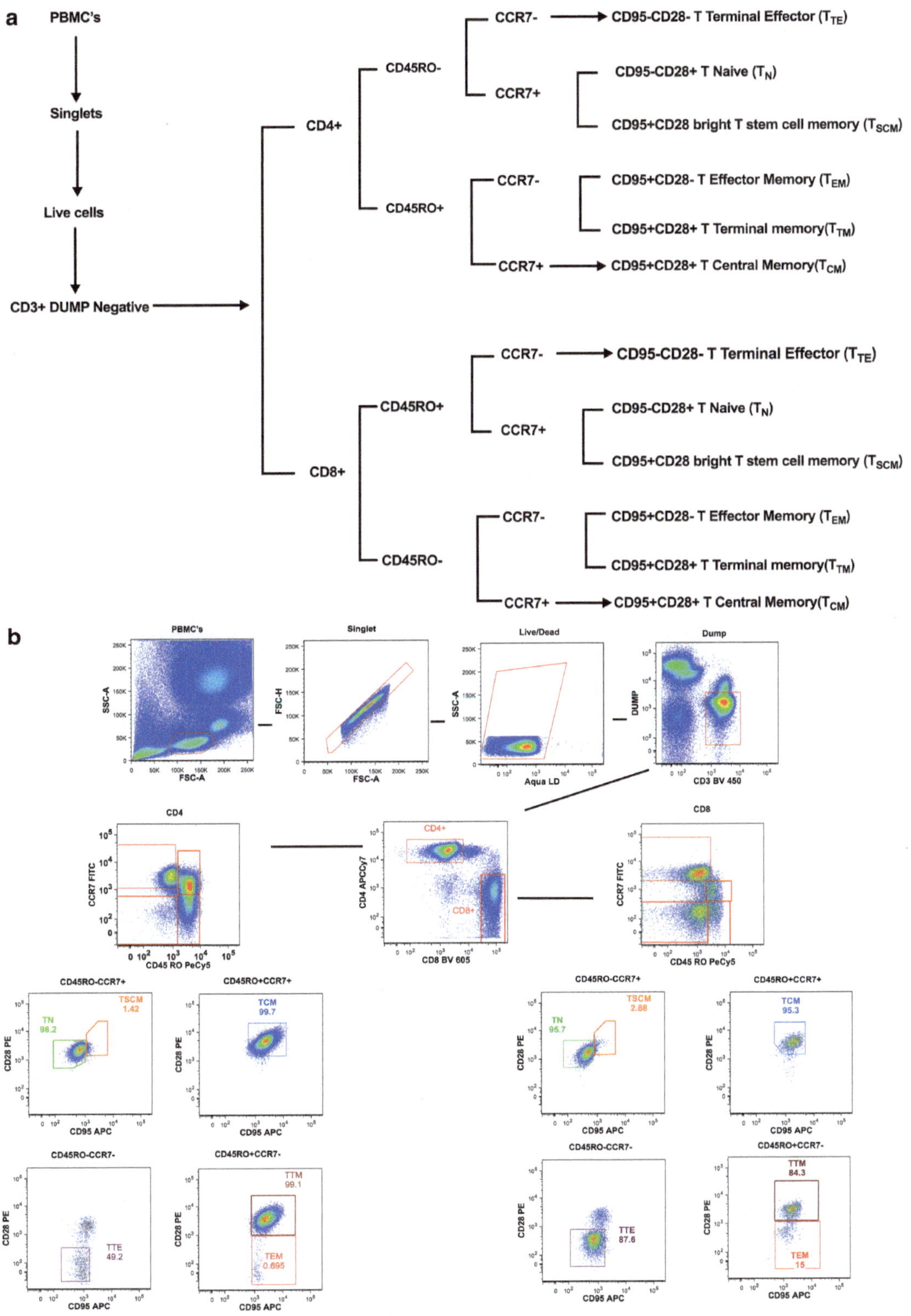
a
PBMC's
Singlets
Live cells
CD3+ DUMP Negative
CD4+
CD45RO-
CD45RO+
CCR7-
CCR7+
CD95-CD28- T Terminal Effector (TTE)
CD95-CD28+ T Naive (TN)
CD95+CD28 bright T stem cell memory (TSCM)
CD95+CD28- T Effector Memory (TEM)
CD95+CD28+ T Terminal memory(TTM)
CD95+CD28+ T Central Memory(TCM)
CD8+
b
PBMC's
Singlet
Live/Dead
Dump
SSC-A
FSC-A
FSC-H
Aqua LD
DUMP
CD3 BV 450
CD4
CD8
CCR7 FITC
CD45 RO PeCy5
CD4 APCCy7
CD8 BV 605
CD4+
CD8+
CD45RO-CCR7+
CD45RO+CCR7+
CD45RO-CCR7-
CD45RO+CCR7-
CD28 PE
CD95 APC
TN 98.2
TSCM 1.42
TCM 99.7
TTE 49.2
TTM 99.1
TEM 0.695
TN 95.7
TSCM 2.88
TCM 95.3
TTE 87.6
TTM 84.3
TEM 15

6. Singlet gating: select gate and right click, select drill down bi-variant polychromatic plot with FSC-A on X-axis and FSC-H on Y-axis for preselected lymphocytes. Make a gate around the singlets. If the area scaling for FSC detector and Blue LASER is properly adjusted while data recording, then singlets will be visible on a diagonal line when FSC-H and FSC-A parameters are plotted as X- and Y-axis of bivariate plot. Doublets will fall towards FSC-A but not FSC-H and can be gated out to identify single cells.
7. Live cell gating: Gate Aqua LD negative subsets as live cells.
8. Gate dump negative cells, excluding out B cells, NK cells, Monocytes.
9. Draw a histogram and select CD3 cells.
10. On gated CD3 cells, select CD4 and CD8 populations.
11. Select CCR7 on Y-axis and CD45RO on X-axis for both CD4 and CD8 subsets and analyze as follows.
12. Divide this population into four subsets based on the presence or absence of CD45RO and CCR7 as CD45RO+CCR7+ , CD45RO+CCR7−, CD45RO−CCR7+, CD45RO−CCR7− on CD4 and CD8 T cell subsets.
13. Drill down CD45RO−CCR7+ subset gate on bivariate plot for CD95 and CD28. Define CD28 bright CD95+ cells as T_{SCM} and CD95−CD28+ as T_N.
14. Drill down CD45RO+CCR7+ subsets and identify CD95+ CD28+ expressing cells as T_{CM}.
15. Drill down CD45RO+CCR7− subset and identify CD95+ CD28+ expressing cells as $T_{TM,}$ whereas CD95+CD28− as T_{EM} subset.
16. Drill down CD45RO−CCR7− subsets and identify CD28− and CD95− as T_{TE} cells.
17. Drag the gated population from the workspace to layout editor to prepare gating strategy schemes and to display data.

Fig. 2 Gating strategy for identifying naïve and memory T cell subsets. (**a**) The sequential gating strategy defines gated total lymphocytes from a whole blood lysed sample. (**b**) Lymphocytes were gated for singlets, live and dump channel, which excluded B, T, NK, and monocytes. CD3-gated cells were further gated for CD4, CD8, followed by CD45RO and CCR7. On the basis of CD45RO and CCR7, the memory and naïve T cells were defined as follows. From CD45RO−CCR7+ subset gate on CD95 and CD28. T_{SCM}—CD45RO−CCR7+ CD28^{bright} CD95+, T_N—CD45RO−CCR7+ CD95−CD28+, T_{CM}—CD45RO+CCR7− CD95+CD28+, T_{TM}—CD45RO+ CCR7− CD95+CD28+, T_{EM}—CD45RO+CCR7− CD95+CD28−, T_{TE}—CD45RO− CCR7−CD28− and CD95−

4 Notes

1. Use of azide in FACS buffer should be avoided, specially if cells are to be sorted and used later for functional studies. Azide inhibits various metabolic functioning of cells, which could hamper the downstream functional studies of sorted cells. Also, unfixed cells kept long in azide buffer may show more death than normal.
2. To avoid non-specific binding of antibodies/reagents, each fluorochrome-conjugated antibody/reagent should be optimally titrated before use.
3. High-throughput screen (HTS) should be set according to the manufacturer's instruction; please refer to the manual. It's important to note that there is a dead volume of 30 μl for plates of all kinds. The samples should preferably be run at a low flow rate between 1 and 2 μl/sec for up to 10^6 cells.
4. Avoid prolonged (>5 min) incubation with ACK lysis buffer, as it will significantly affect the lymphocyte viability.
5. Supplementing the medium with 20 μg/mL DNAse helps to dissociate cell aggregates that result from DNA released by dead cells. This is especially helpful if working with cryopreserved cells, as freeze–thaw may cause cell lysis.
6. Cell staining at 4°C in the dark is a preferred method, as it avoids antibody capping.
7. If there are any questions about the pathogenicity of the cells involved, then fixation of cells after staining and before running them through a flow cytometer is highly recommended for immunophenotyping purposes, not for live cell sorting.
8. Fixative solution should be removed after incubation by washing cells with FACS buffer. Prolonged fixation may alter fluorochrome properties and may lead to spillover in other channels.
9. Compensation beads should not be washed as cells, as it may lead to loss of beads.
10. Optimize PMT voltages using unstained and single colors in order to obtain minimal spectral overlap.
11. Compensation matrix may still require to be adjusted during analysis. After adjusting the compensation, electronically update the biexponential transformation. On flow jo- Select Platform < custom transformation < compensation matrix. Select an appropriate compensation matrix. And appropriately adjust biexponential scales in order to visualize the population clusters on scales.

References

1. Maecker HT, McCoy JP, Nussenblatt R (2012) Standardizing immunophenotyping for the Human Immunology Project. Nat Rev Immunol 12(3):191–200. https://doi.org/10.1038/nri3158
2. De Rosa SC, Herzenberg LA, Herzenberg LA, Roederer M (2001) 11-color, 13-parameter flow cytometry: identification of human naive T cells by phenotype, function, and T-cell receptor diversity. Nat Med 7(2):245–248. https://doi.org/10.1038/84701
3. Picker LJ, Singh MK, Zdraveski Z, Treer JR, Waldrop SL, Bergstresser PR, Maino VC (1995) Direct demonstration of cytokine synthesis heterogeneity among human memory/effector T cells by flow cytometry. Blood 86(4):1408–1419
4. An X, Varadarajan N (2018) Single-cell technologies for profiling T cells to enable monitoring of immunotherapies. Curr Opin Chem Eng 19:142–152. https://doi.org/10.1016/j.coche.2018.01.003
5. Sprent J, Cho JH, Boyman O, Surh CD (2008) T cell homeostasis. Immunol Cell Biol 86(4):312–319. https://doi.org/10.1038/icb.2008.12
6. Sprent J, Tough DF (1994) Lymphocyte life-span and memory. Science 265(5177):1395–1400
7. Laidlaw BJ, Craft JE, Kaech SM (2016) The multifaceted role of CD4(+) T cells in CD8(+) T cell memory. Nat Rev Immunol 16(2):102–111. https://doi.org/10.1038/nri.2015.10
8. Zhang N, Bevan MJ (2011) CD8(+) T cells: foot soldiers of the immune system. Immunity 35(2):161–168. https://doi.org/10.1016/j.immuni.2011.07.010
9. Mahnke YD, Brodie TM, Sallusto F, Roederer M, Lugli E (2013) The who's who of T-cell differentiation: human memory T-cell subsets. Eur J Immunol 43(11):2797–2809. https://doi.org/10.1002/eji.201343751
10. Bresser K, Kok L, Swain AC, King LA, Jacobs L, Weber TS, Perie L, Duffy KR, de Boer RJ, Scheeren FA, Schumacher TN (2022) Replicative history marks transcriptional and functional disparity in the CD8(+) T cell memory pool. Nat Immunol 23(5):791–801. https://doi.org/10.1038/s41590-022-01171-9
11. Golubovskaya V, Wu L (2016) Different subsets of T cells, memory, effector functions, and CAR-T immunotherapy. Cancers 8(3):36. https://doi.org/10.3390/cancers8030036
12. Maggi E, Munari E, Landolina N, Mariotti FR, Azzarone B, Moretta L (2024) T cell landscape in the microenvironment of human solid tumors. Immunol Lett 270:106942. https://doi.org/10.1016/j.imlet.2024.106942

Check for updates

Chapter 2

Enumeration of Benign and Malignant Plasmacytoid Dendritic Cells in Peripheral Blood and Bone Marrow by Flow-Cytometric Analysis

Timothy Kirtek, Franklin Fuda, Sharon Germans and Mingyi Chen

Abstract

Dendritic cells (DCs) are powerful antigen-presenting cells involved in regulating immune response. Plasmacytoid dendritic cells (pDCs) are a subtype of DCs present in small quantities in the bone marrow, peripheral blood, and lymph nodes. They are an important component of our immune system's regulation and response to diseases. They activate T cells and play a critical role in immune tolerance. In this chapter, we review the immunophenotypic features of pDCs and provide a practical protocol for pDCs enumeration in the peripheral blood and in bone marrow samples.

Keywords Flow cytometry, Dendritic cells, Plasmacytoid dendritic cells, Immunophenotyping

1 Introduction

Dendritic cells (DCs) are highly specialized cells that link the innate and adaptive immune responses [1]. DCs are a heterogeneous population of cells that regulate the immune response through their function as potent antigen-presenting cells. They have a special capability to present antigens to naive T cells, leading to T cell activation and further maturation. Additionally, depending on the inducing stimulus, they play a critical role in immune tolerance [2].

Two subtypes of DCs have been well described, conventional dendritic cells (cDCs) and plasmacytoid dendritic cells (pDCs), both derived from a common DC precursor arising from CD34+ hematopoietic stem cells (HSC). Plasmacytoid dendritic cell (pDC) precursors, also called type I IFN (alpha/beta/omega)-producing cells (IPCs), are the key effectors in the innate immune system by their extraordinary capacity to produce type I IFNs against microbial infection or other antigen stimulation. Precursor pDCs (pre-pDCs) differentiate into pDCs in the bone marrow and then are released

Angélique Biancotto and Pradeep K. Dagur (eds.), *Immunophenotyping*, Methods in Molecular Biology, vol. 3087,
https://doi.org/10.1007/978-1-0716-5607-5_2,

into peripheral blood, which represent 0.1%–0.8% of peripheral blood mononuclear cells in both humans and mice [3].

pDCs with plasmacytoid morphology are specialized in rapidly secreting massive amounts of type 1 interferon following viral stimulation, which promotes the function of natural killer cells, B cells, T cells, and myeloid DCs through type 1 interferons during an antiviral immune response. At a later stage of viral infection, pDCs differentiate into a unique type of mature dendritic cell to directly regulate the function of T cells; thus, pDCs may act as a link between innate and adaptive immunity. They then migrate to and reside in lymphoid tissues, including lymph nodes, tonsils, spleen, thymus, bone marrow, and Peyer's patches [4].

Mature pDCs stimulate distinct types of CD4 T helper (Th-1) cells in response to endogenous antigens. Likewise, pDCs can be stimulated by exogenous antigens like viruses [5]. Activated pDCs secrete type I interferon that activates and enhances T cells and natural killer (NK) cells [6–8]. Thus, pDCs play a critical role in a variety of immune-mediated pathophysiological processes, such as graft-versus-host disease (GvHD), allograft rejection, and autoimmune disorders [2, 9]. pDCs are present in small quantities in primary and secondary lymphoid organs, they accumulate in a variety of inflammatory conditions, including Kikuchi-Fujimoto lymphadenopathy, hyaline-vascular Castleman disease, and autoimmune diseases, and in certain malignancies such as classical Hodgkin lymphoma, carcinomas, and chronic myelomonocytic leukemia or myelodysplastic syndromes [10]. The counterpart of neoplastic transformation to highly aggressive malignancy is now known as blastic plasmacytoid dendritic cell neoplasm (BPDCN). Formerly called blastic natural killer cell lymphoma or CD4/CD56 hematodermic neoplasm, BPDCN [11].

Following hematopoietic stem cell transplantation (HSCT), pDCs facilitate HSC engraftment and induce tolerance to prevent GvHD [7]. Several studies highlight the roles pDCs play in immune reconstitution after HSCT for hematolymphoid malignancies and how they relate to disease relapse, patient survival, and risk of GvHD [1, 12]. Higher percentages of pDCs have been associated with successful bone marrow engraftment after allogeneic HSCT for leukemias. On the other hand, lower pDC counts may predict relapse, death, and acute graft-versus-host disease [1]. It has been reported that the percentage of pDCs following HSCT is an independent indicator for adverse clinical outcomes [1, 12].

pDC populations are present in low quantities in bone marrow, peripheral blood, and lymph nodes. They show plasmacytoid morphology and can be defined by their immunophenotype. In general, normal mature pDCs in the bone marrow express CD4, CD22, CD36, CD38, CD45, CD123, CD303, and HLA-DR. CD22 expression is best visualized when using the S-HCL-1 clone, other clones, including HIB22, are often negative in pDCs [13].

They are negative for CD11b, CD13, CD15, and CD64 and are predominantly negative for CD56 [12, 14]. Other markers that may be positive in subsets of normal pDCs include CD2, CD5, CD7, CD33, and CD56 [15, 16]. Immature pDCs may express CD34 and CD117 [16, 17]. Of note, subtle immunophenotypic variations have been described in pDCs in lymph nodes compared to bone marrow [14]. Neoplastic pDCs may show immunophenotypic aberrancy, such as increased CD56 expression, as seen in cases of blastic plasmacytoid dendritic cell neoplasm (BPDCN) [18]. Given the critical role of pDCs in the immune system, particularly in allogeneic HSCT, this chapter provides a practical approach to identify and quantify pDCs using multiparametric flow cytometry. Our protocol is designed to distinguish pDCs from other hematopoietic cells present in bone marrow.

2 Materials

2.1 Preparation of Bone Marrow Mononuclear Cells

Anticoagulated bone marrow or peripheral blood sample is usually collected in heparinized or ethylenediaminetetraacetic acid (EDTA) tubes. It is ideal to analyze the sample within 48 h of collection.

2.2 Cell Staining Buffer [PBS/Azide/BSA (1% PAB)]

500 mL Dulbecco's phosphate-buffered saline (DPBS) (10×); 22.75 g sodium azide; 50.0 g bovine serum albumin (BSA) (1%); adjust the pH to 7.2–7.4 with HCl and final volume to 5 L with additional deionized H_2O; stored at 2–8°C, brought to room temperature before use.

2.3 1% PFA Cell Fixation Solution

100 mL of paraformaldehyde (PFA) at 10% concentration into 900 mL of phosphate-buffered saline (PBS, 10×); stored at 2–8°C, brought to room temperature before use.

2.4 RBC Lysis Buffer

80 g of NH_4Cl (ammonium chloride); 10 g of $KHCO_3$ (potassium bicarbonate); 3.7 g of EDTA (ethylenediaminetetraacetic acid); adjust the pH to 7.2–7.4 with HCl and final volume to 1 L with additional deionized H_2O; Use a sterile filter, 0.22 μm polyethersulfone (PES) filter; stored at 2–8°C, brought to room temperature before use.

2.5 Monoclonal Antibodies

The listed antibody panels (Tables 1–3) are suggested monoclonal antibodies combinations [19] to allow the enumeration of the pDCs among other myeloid and lymphoid cells using four- or ten-color flow cytometry (*see* **Notes 1–3**).

Table 1
Suggested monoclonal antibodies for ten-color flow cytometry panels

Surface marker	Fluorochrome	Clone
CD2	FITC	S5.2
CD3	BV421	SK7 (Leu4)
CD4	PerCP-Cy5.5	SK3 (Leu3A)
CD5	PerCP-Cy5.5	L17F12
CD8	APC-H7	SK1
CD11b	APC	D12
CD13	PE	L138
CD15	FITC	MMA/HI98
CD22	APC	S-HCL-1
CD34	APC-R700	8G12
CD36	FITC	FA6.152
CD38	PE-Cy7	HIT2
CD45	V500-C	2D1
CD56	BV605	NCAM16.2
CD64	APC-R700	MD22
CD123	BV421	9F5/7G3
CD303	PE-Cy7	BDCA-2
HLA-DR	APC-H7	L243

Table 2
Antibody panel for four-color flow cytometry

Tube	FITC	PE	PerCP	APC
1	CD36	CD123	HLA-DR	CD22
2	CD4	CD123	HLA-DR	CD56
3	CD34	CD123	CD45	CD303

3 Methods

1. In general, it is advisable to stain 500,000 cells in each tube with an appropriate volume of monoclonal antibodies and acquire at least 100,000 cellular events for analysis.

Table 3
Antibody panel for ten-color flow cytometry

Tube	FITC	PE	PerCP-Cy5.5	PE-Cy7	APC	APC-R700	APC-H7	BV421	V500-C	BV605
1	CD15	CD13	CD4	CD38	CD11b	CD64	HLA-DR	CD123	CD45	–
2	CD36	CD34	CD4	CD303	CD22	–	HLA-DR	CD123	CD45	CD56

2. Several pretest factors that are critical and can affect the final result include bone marrow collection method, anticoagulation, and transportation.
3. Testing the sample as early as possible is encouraged, ideally within 24 h or in less than 48 h of collection.
4. All testing should be carried out in sterile conditions (*see* **Notes 4 and 5**).

3.1 Specimen Processing

1. Resuspend the anticoagulated specimen by inverting the sample several times (*see* **Note 6**).
2. Decant the specimen into a 50 mL conical tube.
3. Dissolve clots and remove particles.
4. Use a filter to remove undissolved particles (70 μm nylon mesh) if needed.
5. Add 3 mL NH_4Cl (ammonium chloride) lysis buffer into each tube and vortex.
6. Incubate for 10 min at room temperature (RT).
7. Centrifuge for 7 min 800 ×*g* at RT.
8. Discard the supernatant.
9. Resuspend the cell pellet in 50 mL 1% PAB (PBS/Azide/BSA) and mix gently.
10. Centrifuge for 7 min 800 ×*g* at RT.
11. Discard the supernatant.
12. Resuspend the cell pellet in 40 mL PAB (1%) and mix gently.
13. Centrifuge for 7 min 800 ×*g* at RT.
14. Discard the supernatant.
15. Resuspend the cell pellet in 40 mL PAB (1%) and mix gently.
16. Count cells of the filtered bone marrow using the Coulter AcT diff 2 Analyzer.

3.2 Monoclonal Antibody Staining

1. Pipette the appropriate quantity of patient sample to achieve 500,000 cells in each 12 × 75 mm tube.
2. Pipette 2–3 mL of PAB (1%) into each tube.
3. Add the precise amount of each monoclonal antiboy or antibody cocktail to the appropriate tube.
4. Vortex each tube gently.
5. Incubate for 15–20 min in the refrigerator (2–8°C) in the dark.
6. Pipette 2–3 mL of PAB (1%) into each tube.
7. Vortex each tube gently.
8. Add 250 μL of 1% PFA fixative and vortex.

9. Run the tubes or store them in the refrigerator (2–8°C) until they can be acquired on the flow cytometer, preferably within 1 h of staining.

3.3 Data Acquisition

1. Sequential gating strategies are recommended to identify and quantify the pDCs (*see* **Notes 1, 7, and 8**) (Fig. 1).
   (a) Gate out all doublets and nonviable cells (*see* **Notes 9 and 10**).
   (b) Gate all HLA-DR+/CD45+ and all CD4+ cell populations.
   (c) Exclude T lymphocytes (CD3+).
   (d) Exclude monocytes (CD14+/CD64+/CD36+).
   (e) Exclude all events that are CD123− and/or HLA-DR−.

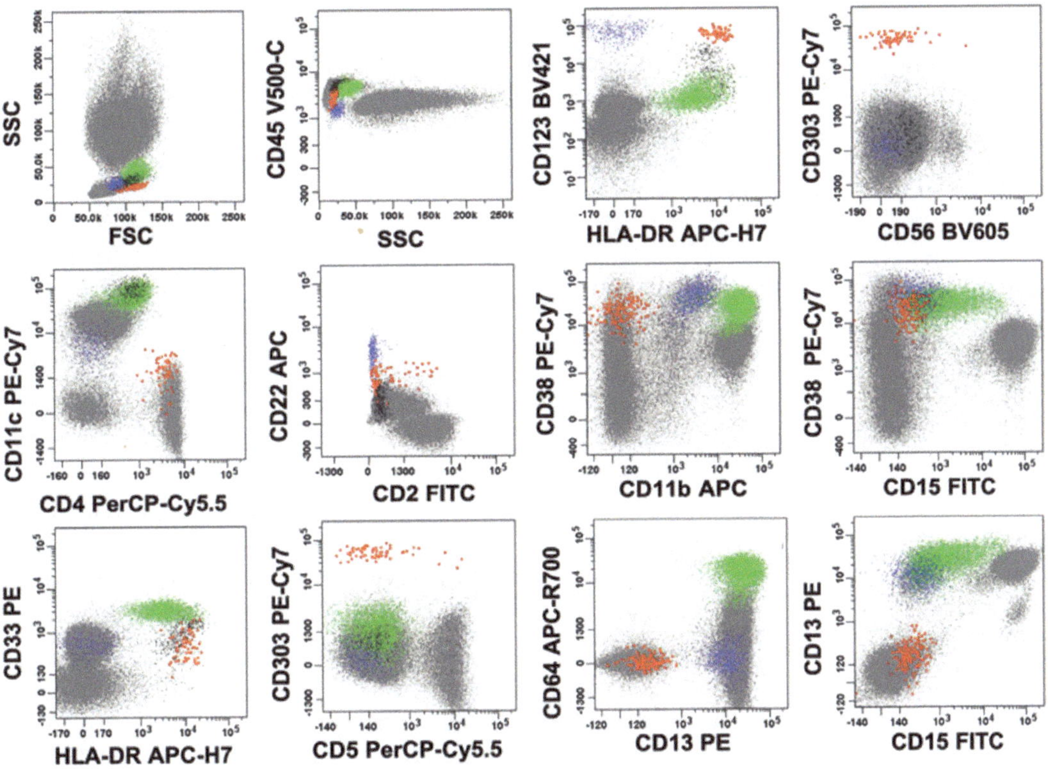

Fig. 1 Multiparametric flow cytometry analysis of a normal bone marrow specimen. The multicolor flow cytometry panel includes several monoclonal antibodies, such as CD2, CD4, CD5, CD11b, CD11c, CD13, CD15, CD22, CD33, CD38, CD45, CD56, CD64, CD123, CD303, HLA-DR. pDCs (red population) are found just below monocytes (green population), that is, the same size as monocytes but less SSC, in the FSC/SSC plot. Additionally, pDCs are slightly dimmer for CD45 with less SSC compared to monocytes in the CD45/SSC plot. In CD123/HLA-DR plot, pDCs are double positive for CD123/HLA-DR compared to basophils (blue population) that are CD123+/HLA-DR−. In this case, the nonneoplastic pDCs are characteristically positive for CD4, CD22 (S-HCL-1 clone), CD33, CD38, CD45, CD123, CD303, and HLA-DR, and largely negative for CD2, CD5, CD11c, CD13, CD15, CD56, and CD64

(f) Report the percentage of pDCs as pDC% of total viable events.

3.4 Data Interpretation

1. Nonneoplastic pDCs in the bone marrow and peripheral blood frequently express CD4, CD22dim, CD33, CD36, CD38, CD45, CD123, CD303, and HLA-DR [19, 20], and lack other myelomonocytic or lymphoid markers (*see* **Notes 2, 7, and 11–13**).
2. In the forward/side scatter (FSC vs. SSC) plot, pDCs are found just below monocytes in the two-dimensional scatter plot (i.e., same size as monocytes, but less SSC).
3. In the CD45 vs. SSC plot, pDCs show slightly dimmer for CD45 and less SSC compared to monocytes.
4. In CD3 vs. CD4 plots, the cluster of nonmonocytic CD4+ /CD3− is mainly pDCs, provided no aberrant T cells are present.
5. In the CD45 vs. CD34 plot, pDCs may show a maturation spectrum where early pre-pDCs (Stage I) are CD34+ and CD45 dimmer than mature pDCs, which are CD34− [21], (*see* **Note 14**).

4 Notes

1. A panel of monoclonal antibodies is used to better characterize the pDCs and discriminate these cells from other hematolymphoid cells that share some similar immunophenotypic findings.
2. It has been reported that small subsets of pDCs may express CD7 or CD2, particularly in bone marrow samples. However, the majority of pDCs should be negative for these markers [9]. Also, CD33 can be negative in a minority of cases.
3. The antibodies must be validated and titered for optimal signal-to-noise ratio under the settings to be used, particularly if the antibodies are pre-cocktailed.
4. Reagents should be used according to the manufacturer's instructions.
5. Daily QC, calibration controls, compensation procedure, and preventive maintenance must be performed according to the manufacturer's recommendations.
6. It is crucial to evaluate the specimen for hemodilution, which may affect the pDC count accuracy.
7. Clinical and morphologic correlation:

(a) Knowing the clinical presentation and the indications for testing is essential in order to address clinicians' concerns.

(b) In cases with suspected increased pDCs, an examination of cytospin morphology of the corresponding sample is recommended. Additionally, correlation with the bone marrow biopsy and clot section is required to provide an accurate assessment.

8. In normal conditions, pDCs show reproducible staining patterns. They usually present close to or within the "blast gate." While they can be detected using pre-gating strategies, we recommend the use of cluster analysis, which allows for identifying populations in ungated areas that may harbor pDC populations.

9. It is important to gate-out doublets ("doublet discrimination") during the pDC analysis by examining forward scatter area over height (FSC-A and FSC-H) and then excluding all events that do not represent single cells.

10. Excluding nonviable cells or debris can be performed by gating out events with very low forward scatter in FSC-A vs. SSC-A plot. The other option to remove the nonviable cells is to use a DNA-binding dye, such as 7-amino-actinomycin D (i.e., 7-AAD), which can provide a good assessment of sample integrity. However, it is essential to ensure that excluding nonviable cells does not interfere with the analysis of the cell of interest.

11. Our protocol for evaluating pDCs can be incorporated into the routine evaluation of all bone marrow and peripheral blood samples and is particularly useful if an increased non-monocytic CD4+/CD3− population is present.

12. The use of isotype controls is essential to exclude false-positive events (nonspecific staining).

13. It is advised to include fluorescence minus one (FMO) controls in multiparametric flow cytometry, which helps in identifying the proper gating boundaries.

14. Based on our experience, early immature pDCs may show decreased CD303 expression. This is mainly seen in bone marrow samples.

References

1. Reddy V et al (2004) Low dendritic cell count after allogeneic hematopoietic stem cell transplantation predicts relapse, death, and acute graft-versus-host disease. Blood 103(11):4330–4335

2. Abe M et al (2005) Plasmacytoid dendritic cell precursors induce allogeneic T-cell hyporesponsiveness and prolong heart graft survival. Am J Transplant 5(8):1808–1819

3. Soumelis V, Liu YJ (2006) From plasmacytoid to dendritic cell: morphological and functional switches during plasmacytoid pre-dendritic cell differentiation. Eur J Immunol 36(9):2286–2292
4. Liu YJ (2005) IPC: professional type 1 interferon-producing cells and plasmacytoid dendritic cell precursors. Annu Rev Immunol 23:275–306
5. McKenna K, Beignon AS, Bhardwaj N (2005) Plasmacytoid dendritic cells: linking innate and adaptive immunity. J Virol 79(1):17–27
6. Tomasello E et al (2018) Molecular dissection of plasmacytoid dendritic cell activation in vivo during a viral infection. Embo J 37:19
7. Auletta JJ, Devine SM, Waller EK (2016) Plasmacytoid dendritic cells in allogeneic hematopoietic cell transplantation: benefit or burden? Bone Marrow Transplant 51(3):333–343
8. Swiecki M, Colonna M (2015) The multifaceted biology of plasmacytoid dendritic cells. Nat Rev Immunol 15(8):471–485
9. Elze MC et al (2015) Dendritic cell reconstitution is associated with relapse-free survival and acute GVHD severity in children after allogeneic stem cell transplantation. Bone Marrow Transplant 50(2):266–273
10. Hamadeh F et al (2020) Flow cytometry identifies a spectrum of maturation in myeloid neoplasms having plasmacytoid dendritic cell differentiation. Cytom B Clin Cytom 98(1):43–51
11. Jegalian AG, Facchetti F, Jaffe ES (2009) Plasmacytoid dendritic cells: physiologic roles and pathologic states. Adv Anat Pathol 16(6):392–404
12. Su RJ, Green R, Chen M (2018) Enumeration of bone marrow plasmacytoid dendritic cells by multiparameter flow cytometry as a prognostic marker following allogeneic hematopoietic stem cell transplantation. Blood Cells Mol Dis 69:107–112
13. Reineks EZ et al (2009) CD22 expression on blastic plasmacytoid dendritic cell neoplasms and reactivity of anti-CD22 antibodies to peripheral blood dendritic cells. Cytom B Clin Cytom 76(4):237–248
14. Wang W et al (2021) Immunophenotypic characterization of reactive and neoplastic plasmacytoid dendritic cells permits establishment of a 10-color flow cytometric panel for initial workup and residual disease evaluation of blastic plasmacytoid dendritic cell neoplasm. Haematologica 106(4):1047–1055
15. Zhang H et al (2017) A distinct subset of plasmacytoid dendritic cells induces activation and differentiation of B and T lymphocytes. Proc Natl Acad Sci U S A 114(8):1988–1993
16. El Hussein S, Wang W (2024) Flow cytometry profiling of plasmacytoid dendritic cell neoplasms. Cancers 16:11
17. Martín-Martín L et al (2015) Classification and clinical behavior of blastic plasmacytoid dendritic cell neoplasms according to their maturation-associated immunophenotypic profile. Oncotarget 6(22):19204–19216
18. Khoury JD et al (2022) The 5th edition of the World Health Organization classification of haematolymphoid tumours: myeloid and histiocytic/dendritic neoplasms. Leukemia 36(7):1703–1719
19. Chen W, Luu HS (2017) Immunophenotyping by multiparameter flow cytometry. Methods Mol Biol 1633:51–73
20. Alculumbre S, Pattarini L (2016) Purification of human dendritic cell subsets from peripheral blood. Methods Mol Biol 1423:153–167
21. Martín-Martín L et al (2009) Immunophenotypical, morphologic, and functional characterization of maturation-associated plasmacytoid dendritic cell subsets in normal adult human bone marrow. Transfusion 49(8):1692–1708

Chapter 3

Neutrophil Subset Immunophenotyping

Marisa Gorrese, Carmine Selleri and Valentina Giudice

Abstract

Neutrophils, the most abundant circulating leukocyte population, are principally involved in pathogen clearance through three major mechanisms: phagocytosis, degranulation, and release of neutrophil extracellular traps. Along with these roles, neutrophils can promote or suppress tumor growth and metastatic processes. These different immunological functions reflect a great heterogeneity of maturation and activation status, as at least two main circulating populations can be identified: low-density granulocytes and normal-density neutrophils. In this chapter, we provide a detailed protocol for multiparametric flow cytometry immunophenotyping of circulating and bone marrow neutrophil subsets, as well as recommendations for protocol optimization, troubleshooting, and clinical sample analysis.

Keywords Flow cytometry, Immunophenotyping, Neutrophils, Multiparametric flow cytometry, Diagnostic flow cytometry

1 Introduction

Neutrophils are the most abundant circulating mononuclear cell population and are historically recognized for their role in pathogen clearance through three different physiological mechanisms: phagocytosis, degranulation, and release of neutrophil extracellular traps (NETs). However, neutrophils are currently considered an important immunomodulatory cell population, as they can also promote or suppress tumor growth and metastasis or inflammation [1]. These opposite functions reflect a great heterogeneity of their maturation and activation status, and two distinct circulating subsets have been historically identified, only based on different stratification abilities after Ficoll-Paque or dextran density gradient separation [2]. Based on this cytoplasmic complexity, low-density granulocytes (LDGs) and normal-density neutrophils (NDNs) have been described; however, within each population, several subsets with different biological activities and phenotypes have been characterized [3]. Immature NDNs share similarities with immature LDGs, as they both derive from differentiating metamyelocytes that are prematurely released

Angélique Biancotto and Pradeep K. Dagur (eds.), *Immunophenotyping*, Methods in Molecular Biology, vol. 3087, https://doi.org/10.1007/978-1-0716-5607-5_3,

into the bloodstream before terminal differentiation during an emergency granulopoiesis [1]. Conversely, activated neutrophils exhibit proinflammatory properties due to their ability to release reactive oxygen species, granules, and NETs [4]. Within LDG subsets, inflammatory LDGs comprise of more immature cells that are produced under emergency hematopoiesis, while immunosuppressive LDGs, also known as granulocyte-like myeloid-derived suppressor cells, can inhibit T cell proliferation and activation, as well as interferon (IFN)-γ production [5].

Neutrophil maturation and activation status and subset distribution are impaired in hematological malignancies and non-hematological conditions, especially autoimmune disorders and cardiovascular diseases [6–11]. For example, circulating activated LDGs spontaneously release NETs as a source of autoantigens and produce large amounts of proinflammatory cytokines, such as tumor necrosis factor(TNF)α, IFN-γ, and type I IFNs, in various autoimmune disorders, like systemic lupus erythematosus, rheumatoid arthritis, or autoimmune vasculitis, where they mediate endothelial cell dysfunction and vascular damage [6, 7]. Moreover, altered neutrophil maturation curve assessed by CD11b vs CD16 expression analysis and myeloperoxidase (MPO) cytoplasmic expression associated with neutrophil degranulation are hallmarks of myelodysplastic syndromes (MDS) [12–15]. Indeed, increased neutrophil MPO expression has been proposed as a highly specific neutrophilic marker of differential diagnosis between myelodysplasias and other myeloproliferative neoplasms [13,16].

In this chapter, we describe the procedure for neutrophil subset immunophenotyping in various biological samples, such as fresh whole peripheral blood mononuclear cells (PBMCs) and bone marrow mononuclear cells (BMMNCs), or different body fluids. This panel can be used for diagnostic and research purposes to identify alterations in neutrophil immunophenotype and frequency that could help in the differential diagnosis of hematological malignancies or monitoring hematopoietic reconstitution after chemotherapy and stem cell transplantation. In this protocol, neutrophil subset immunophenotyping was performed using whole peripheral blood (PB), bone marrow (BM), broncho-alveolar lavage (BAL), pleural fluid, or ascitic fluid specimens stained with various ten-color panels for neutrophils, T and B cells, monocytes, and leukemic cells identification and characterization. Finally, we provide manual gating strategy analysis, and notes for technical recommendations are also reported.

2 Materials

2.1 Buffers and Materials Required

1. IO test lysing solution 10× (Beckman Coulter). Freshly prepare a 1× dilution with distilled water (ddH_2O). For 50 mL solution, add 5 mL of IO Test Lysing Solution 10× in 45 mL of ddH_2O. Store at +4°C.
2. Phosphate-buffered saline (PBS) (IsoFlow Sheath Fluid, Beckman Coulter). Keep at room temperature.
3. Acquisition buffer: IsoFlow Sheath Fluid (Beckman Coulter) with 2% of fetal bovine serum (FBS; Gibco, Thermo Fisher Scientific, Waltham, MA, USA). For 50 mL solution, add 1 mL of FBS in 49 mL of IsoFlow Sheath Fluid. Add to samples at room temperature, while store at +4°C.
4. Ethylenediaminetetraacetic acid (EDTA) or heparin tubes for blood collection.
5. Sterile 100–150 mL graduated containers for BAL, pleural, and ascitic fluid collection.
6. 100 μm membrane filters for BAL filtration (*see* **Note 1**).
7. 50 mL Falcon tubes and 5 mL polypropylene tubes for flow cytometry assays.

2.2 Flow Cytometry Instrument and Setting

The cytometer used should be equipped with violet (405 nm), blue (488 nm), and red (628/640 nm) lasers. Filters should detect wavelengths for the following fluorochromes: FITC or AlexaFluor 488; PE; ECD or PE-Dazzle; PC5.5; PC7; APC; APC-AlexaFluor 700; APC-AlexaFluor 750; Pacific Blue; and Krome Orange. In this protocol, a Navios/EX or a DxFlex cytometer (Beckman Coulter, Brea, CA) is employed for sample acquisition.

For Navios/EX and DxFlex cytometer fluorochrome channel configurations are summarized in Table 1.

In our ten-color staining panels, channels Violet610, Violet660, and Violet780 are not used; however, they can be employed to deepen immunophenotypic analysis (*see* **Note 2**).

Daily quality control is performed using Flow-Check Pro Fluorospheres (Beckman Coulter), while compensation using Flow-Set and a compensation kit (Beckman Coulter). Data acquisition and analysis can be carried out using Navios EX Software v2.0 or Kaluza C Analysis Flow Cytometry Software v1.1 (Beckman Coulter).

2.3 Antibodies

Antibodies used for neutrophil immunophenotyping are as follows (Table 2).

2.4 Samples

Neutrophil immunophenotyping can be performed on different types of human specimens, including:

Table 1
Flow cytometer configurations

	Navios/EX		DxFlex	
Laser	**Filters**	**Channel name**	**Filters**	**Channel name**
Blue, 488 nm	525/40 BP	FITC	525/40 BP	FITC
	575/30 BP	PE	585/42 BP	PE
	614/20 BP	ECD	610/20 BP	ECD
	675/20 BP	PC5	690/50 BP	PC5.5
	755 LP	PE-Cy7	780/60 BP	PC7
Red, 638 nm	660/20 BP	APC	660/10 BP	APC
	725/20 BP	APC-A700	712/25 BP	APC-A700
	755 LP	APC-Cy7	780/60 BP	APC-A750
Violet, 405 nm	450/50 BP	PB	450/45 BP	PB450
	550/40 BP	KO	525/40 BP	KO525
			610/20 BP	Violet610
			660/10 BP	Violet660
			780/60 BP	Violet780

- whole fresh PB, obtained by drawing blood in heparinized or ethylenediaminetetraacetic acid (EDTA) tubes and processed within 12 h maximum;
- whole BM, obtained from bone marrow aspiration in EDTA tubes, and processed within 12 h maximum;
- fresh BAL, obtained from a diagnostic procedure where a bronchoscope is inserted into the airways, and sterile saline solution is instilled and then aspirated back, and fluids collected into a sterile 100 mL container; this type of sample should be processed as quickly as possible, within 8 h maximum, without freezing/thawing cycles;
- fresh pleural and ascitic fluids, obtained from thoracentesis and paracentesis diagnostic procedures, respectively, with the collection of fluids contained within the pleural or intrabdominal spaces. These types of samples should be processed as quickly as possible, within 8 h maximum, without freezing/thawing cycles.

Table 2
Antibodies used characteristics

MoAbs	Fluorochrome	Clone	Manufacturer	Purpose
CD45	Krome Orange	J33	Beckman Coulter	Pan-leukocyte marker
CD117	PE PC5.5	104D2D1	Beckman Coulter	HSPC marker
CD34	APC-A700	581	Beckman Coulter	HSC marker
CD33	APC APC-A750	D3HL60.251	Beckman Coulter	Neutrophil identification
CD33	PE-Dazzle 594	WM53	SONY	Neutrophil identification
CD13	PC5.5	Immu103.44	Beckman Coulter	Neutrophil maturation
CD64	ECD	22	Beckman Coulter	Neutrophil identification
CD66b	FITC	80H3	Beckman Coulter	Neutrophil phenotype
CD36	FITC	FA6.152	Beckman Coulter	Monocytes
CD24	PE	ALB9	Beckman Coulter	Neutrophil phenotype
CD16	PB	3G8	Beckman Coulter	Neutrophil maturation
CD15	PB PE PC5	80H5	Beckman Coulter	Neutrophil identification
CD14	PE APC-A750	RMO52	Beckman Coulter	Monocytes
CD11b	PC7	Bear1	Beckman Coulter	Neutrophil maturation
CD11c	PE	BU15	Beckman Coulter	DCs/neutrophils
CD11a	FITC	25.3	Beckman Coulter	DCs/neutrophils
CD56	PE ECD	N901	Beckman Coulter	NK
CD19	FITC PC5.5	J3-119	Beckman Coulter	B lymphocytes
CD3	APC	UCHT1	Beckman Coulter	T lymphocytes
CD5	APC-A750	BL1a	Beckman Coulter	B lymphocytes
CD7	PC7	8H8.1	Beckman Coulter	T lymphocytes
HLA-DR	FITC	Immu-357	Beckman Coulter	HPSCs and activation
CD157	PE	RF3	Beckman Coulter	GPI-anchored protein
FLAER	A488	–	CEDARLANE	GPI-anchored protein marker
Lineage[a]	APC	UCHT1, HIB19, 2H7, 5.1H11	BIOLEGEND	Lymphocyte exclusion

[a]Lineage comprises anti-CD3, anti-CD19, anti-CD20, and anti-CD56 MoAbs *MoAbs* monoclonal antibodies, *PE* phycoerythrin, *PC5.5* PE-cyanine 5.5, *APC-A700* allophycocyanin AlexaFluor700, *APC-A750* APC-AlexaFluor750, *FITC* fluorescein, *PB* Pacific Blue, *PC* PE-cyanine 7, *HSPC* hematopoietic stem and progenitor cell, *HSC* hematopoietic stem cell, *NK* natural killer cells, *DCs* dendritic cells

3 Methods

3.1 PB Neutrophil Subset Immunophenotyping

This ten-color staining panel allows the identification of principal circulating leukocyte populations (e.g., T and B lymphocytes, natural killer [NK] cells, NKT cells, monocytes, neutrophils, and hematopoietic stem and progenitor cells [HSPCs]), and neutrophil maturation stage (CD16 vs CD11b maturation curve).

1. Label 5 mL polypropylene tubes according to experimental schemes or patients' names.
2. Add the following antibodies at saturating concentrations to each tube (*see* **Note 3**):
   - 5 μL of CD19-FITC;
   - 5 μL of CD14-PE;
   - 5 μL of CD56-ECD;
   - 5 μL of CD117-PC5.5;
   - 5 μL of CD11b-PC7;
   - 5 μL of CD33-APC;
   - 5 μL of CD34-APC-AlexaFluor 700;
   - 5 μL of CD5-APC-AlexaFluor 750;
   - 5 μL of CD16-Pacific Blue;
   - 5 μL of CD45-Krome Orange.
3. Add 50 μL of PB to each tube according to the experimental scheme or patient's name (*see* **Note 4**).
4. Gently vortex.
5. Incubate samples at room temperature in the dark for 20 min (*see* **Note 5**).
6. Add 2 mL of 1× IO TEST 3 Lysing solution to each tube for red blood cell lysis.
7. Moderately vortex.
8. Incubate at room temperature in the dark for 10 min.
9. Centrifuge at 500 × *g* for 3 min. Discard supernatant and vortex to loosen cell pellets.
10. Resuspend samples in 500 μL of PBS or acquisition buffer (*see* **Note 6**).
11. Acquired stained samples within 2–4 h.

3.2 BM Neutrophil Subset Immunophenotyping

These two ten-color staining panels allow identification of principal BM leukocyte populations (Panel 1, T and B lymphocytes, NK and NKT cells, monocytes, neutrophils, HSPCs, and erythroid precursors), or investigation of neutrophil maturation stage and immunophenotypic aberrations (Panel 2).

1. For each experimental condition or patient, label two 5 mL polypropylene tubes as “Panel 1” and “Panel 2.”
2. For “Panel 1” tube, add the following antibodies at saturating concentrations to each tube (*see* **Note 3**):
   - 5 μL of HLA-DR-FITC;
   - 5 μL of CD117-PE;
   - 5 μL of CD56-ECD;
   - 5 μL of CD19-PC5.5;
   - 5 μL of CD7-PC7;
   - 5 μL of CD3-APC;
   - 5 μL of CD34-APC-AlexaFluor 700;
   - 5 μL of CD33-APC-AlexaFluor 750;
   - 5 μL of CD15-Pacific Blue;
   - 5 μL of CD45-Krome Orange.
3. For “Panel 2” tube, add the following antibodies at saturating concentrations to each tube (*see* **Note 3**):
   - 5 μL of CD36-FITC;
   - 5 μL of CD117-PE;
   - 5 μL of CD64-ECD;
   - 5 μL of CD13-PC5.5;
   - 5 μL of CD11b-PC7;
   - 5 μL of CD33-APC;
   - 5 μL of CD34-APC-AlexaFluor 700;
   - 5 μL of CD14-APC-AlexaFluor 750;
   - 5 μL of CD16-Pacific Blue;
   - 5 μL of CD45-Krome Orange.
4. Add 50 μL of BM blood to each tube according to the experimental scheme or patient’s name and proceed as described in Sect. 3.1, Steps 4–11.

3.3 Pleural and Ascitic Fluid Neutrophil Subset Immunophenotyping

This ten-color staining panel allows the identification of principal circulating leukocyte populations (T and B lymphocytes, NK and NKT cells, monocytes, neutrophils, and HSPCs), and neutrophil maturation stage (CD16 vs CD11b maturation curve).

1. Label 5 mL polypropylene tubes according to experimental schemes or patients’ names.

2. Add the following antibodies at saturating concentrations to each tube (*see* **Note 3**):
   - 5 μL of CD19-FITC;
   - 5 μL of CD14-PE;
   - 5 μL of CD56-ECD;
   - 5 μL of CD117-PC5.5;
   - 5 μL of CD11b-PC7;
   - 5 μL of CD33-APC;
   - 5 μL of CD34-APC-AlexaFluor 700;
   - 5 μL of CD5-APC-AlexaFluor 750;
   - 5 μL of CD16-Pacific Blue;
   - 5 μL of CD45-Krome Orange.
3. Add 100 μL of pleural or ascitic fluid specimen to each tube according to the experimental scheme or patient's name (*see* **Note 7**).
4. Proceed as described in Sect. 3.1, Steps 4–11.

3.4 In-Depth Neutrophil Subset Immunophenotyping

For characterization of LDGs and NDNs, an additional tube for each experimental condition or patient should be processed as follows.

1. Label 5 mL polypropylene tubes according to experimental schemes or patients' names.
2. Add the following antibodies at saturating concentrations to each tube (*see* **Note 3**):
   - 5 μL of HLA-DR-FITC;
   - 5 μL of CD15-PE;
   - 5 μL of CD33-PE-Dazzle;
   - 5 μL of CD117-PC5.5;
   - 5 μL of CD11b-PC7;
   - 5 μL of Lineage-APC;
   - 5 μL of CD34-APC-AlexaFluor 700;
   - 5 μL of CD14-APC-AlexaFluor 750;
   - 5 μL of CD16-Pacific Blue;
   - 5 μL of CD45-Krome Orange.
3. Add 50 μL of whole blood or 100 μL of pleural or ascitic fluid specimen to each tube according to the experimental scheme or patient's name (*see* **Note 7**).
4. Proceed as described in Sect. 3.1, Steps 4–11.

For immunophenotypic characterization of neutrophils in myelodysplastic syndromes and for maturation curve analysis, an additional tube for each experimental condition or patient should be processed as follows.

1. Label 5 mL polypropylene tubes according to experimental schemes or patients' names.
2. Add the following antibodies at saturating concentrations to each tube (*see* **Note 3**):
   - 5 μL of CD66b-FITC;
   - 5 μL of CD24-PE;
   - 5 μL of CD64-ECD;
   - 5 μL of CD13-PC5.5;
   - 5 μL of CD11b-PC7;
   - 5 μL of CD33-APC;
   - 5 μL of CD34-APC-AlexaFluor 700;
   - 5 μL of CD14-APC-AlexaFluor 750;
   - 5 μL of CD16-Pacific Blue;
   - 5 μL of CD45-Krome Orange.
3. Add 50 μL of whole blood or 100 μL of pleural or ascitic fluid specimen to each tube according to the experimental scheme or patient's name (*see* **Note 7**).
4. Proceed as described in Sect. 3.1, Steps 4–11.

For identification of glycosyl phosphatidylinositol (GPI) deficient neutrophils as a diagnostic marker of paroxysmal nocturnal hemoglobinuria, an additional tube for each experimental condition or patient should be processed as follows.

1. Label 5 mL polypropylene tubes according to experimental schemes or patients' names.
2. Add the following antibodies at saturating concentrations to each tube (*see* **Note 3**):
   - 5 μL of FLAER-AlexaFluor 488;
   - 5 μL of CD157-PE;
   - 5 μL of CD45-ECD;
   - 5 μL of CD15-PC5.5;
   - 5 μL of CD33-PC7.
3. Add 50 μL of whole blood or 100 μL of pleural or ascitic fluid specimen to each tube according to the experimental scheme or patient's name (*see* **Note 7**).
4. Proceed as described in Sect. 3.1, Steps 4–11.

3.5 BAL Neutrophil Subset Immunophenotyping

For BAL neutrophil subset immunophenotyping, cell viability and absolute cell count should be determined before further processing. Next, using a nine-color staining panel, principal pulmonary leukocyte populations can be identified, including resident monocytes/macrophages, and neutrophil maturation stage can also be investigated.

1. Approximately quantify the total specimen volume before starting the staining protocol by checking graduations on containers.
2. Filter the specimen using a 100 μm filter and a 50 mL Falcon tube for sample collection (*see* **Note 1**).
3. Label 5 mL polypropylene tubes according to experimental schemes or patients' names.
4. Add the following antibodies at saturating concentrations to each tube for cell viability analysis (*see* **Note 3**):
   - 5 μL of CD45-FITC;
   - 5 μL of 7-AAD.
5. Add 100 μL of filtered BAL specimen to each tube according to the experimental scheme or patient's name.
6. Gently vortex.
7. Incubate samples at room temperature in the dark for 20 min (*see* **Note 5**).
8. Add 2 mL of 1× IO TEST 3 Lysing solution to each tube for red blood cell lysis.
9. Moderately vortex.
10. Incubate at room temperature in the dark for 10 min.
11. Add 100 μL of beads (equal volume of starting sample) for absolute cell count.
12. Acquired samples just after staining.

If samples show good cellularity and viability, further immunophenotypic analysis could be performed as follows.

1. Centrifuge at 500 × *g* for 5 min. Discard the supernatant until the final volume is ten times less than the starting volume (e.g., from 30 mL of starting volume to 3 mL of final volume).
2. Resuspend cell pellets in residual volume.
3. Label 5 mL polypropylene tubes according to experimental schemes or patients' name.
4. Add the following antibodies at saturating concentrations to each tube (*see* **Note 3**):
   - 5 μL of CD11a-FITC;

- 5 μL of CD11c-PE
- 5 μL of CD64-ECD;
- 5 μL of CD13-PC5.5
- 5 μL of CD11b-PC7
- 5 μL of CD33-APC;
- 5 μL of CD14-APC-AlexaFluor 750;
- 5 μL of CD16-Pacific Blue;
- 5 μL of CD45-Krome Orange.

5. Add 100 μL of BAL specimen to each tube according to the experimental scheme or patient's name.
6. Proceed as described in Sect. 3.1, Steps 4–11.

3.6 Acquisition and Compensation

For sample acquisition, the cytometer should be equipped with appropriate lasers: violet, 405 nm; blue, 488 nm; and red laser, 628/640 nm. Filters should detect wavelengths for the following fluorochromes: FITC or AlexaFluor 488; PE; ECD or PE-Dazzle; PC5.5; PC7; APC; APC-AlexaFluor 700; APC-AlexaFLuor 750; Pacific Blue; and Krome Orange.

Pre-acquisition compensation can be calculated using:

a) Flow-Set and compensation kit (Beckman Coulter).
b) Single-color controls for FITC- or AlexaFluor 488-, PE-, ECD- or PE-Dazzle-, PC5.5-, PC7-, APC-, APC-AlexaFluor 700-, APC-AlexaFluor 750-, Pacific Blue-, and Krome Orange-conjugated antibodies.
c) Unstained sample as a negative control for setting PMT voltages.

A minimum of 100,000 single cells should be recorded for sample acquisition at low/medium speed. Samples can be saved as ".lmd" or ".fcs" format files.

3.7 Gating Strategy

Post-acquisition analysis can be carried out using FlowJo software (BD Biosciences) or Navios EX Software v2.0, or Kaluza C Analysis Flow Cytometry Software v1.1 (Beckman Coulter).

1. Check postacquisition compensation matrix. Check for any spill-over into other channels.
2. Apply postacquisition compensation matrix to samples.
3. First, cell populations are identified based on linear parameter [side scattered area (SSC-A)] and CD45 expression (Fig. 1A). Neutrophils are $CD45^+$ cells and have heterogeneous and high SSC (*see* **Note 8**).

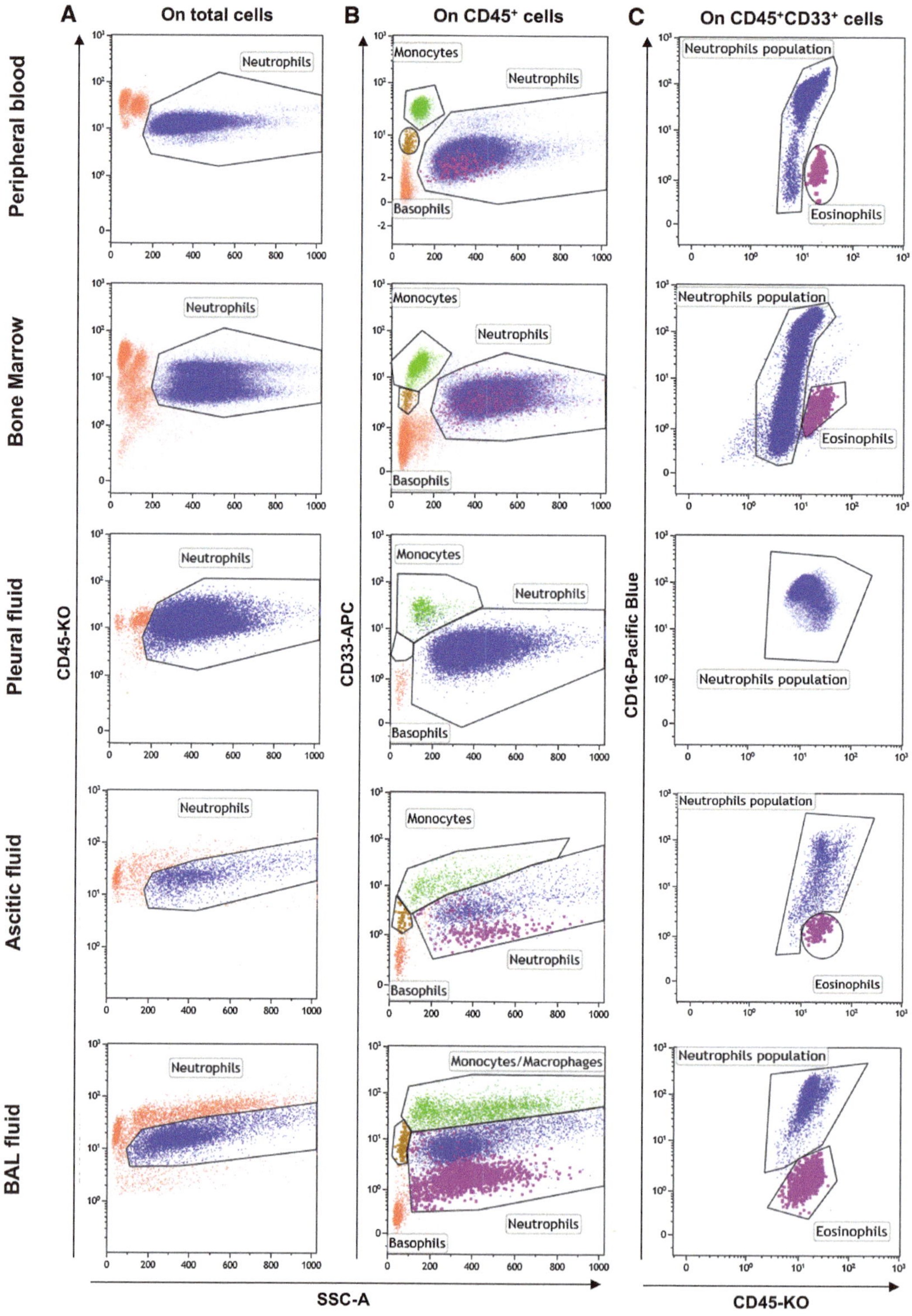
A
On total cells
B
On CD45⁺ cells
C
On CD45⁺CD33⁺ cells
Peripheral blood
Bone Marrow
Pleural fluid
Ascitic fluid
BAL fluid
CD45-KO
CD33-APC
CD16-Pacific Blue
Neutrophils
Monocytes
Basophils
Monocytes/Macrophages
Neutrophils population
Eosinophils
SSC-A
CD45-KO

4. On $CD45^+$ neutrophils, CD33 expression vs SSC-A is investigated to exclude the presence of residual monocytes and basophils (Fig. 1B) (*see* **Note 9**).
5. On $CD45^+CD33^+$ granulocytes, CD16 vs CD45 expression is studied to eliminate eosinophils, which are $CD16^-CD45^{++}$ cells with a partial CD33 expression and increased SSC (Fig. 1C) (*see* **Note 10**).
6. Next, for maturation curve analysis, CD16 vs CD11b expression should be investigated, and immature $CD16^-CD11b^-$, intermediate $CD16^-CD11b^{dim}$, and mature $CD16^+CD11b^+$ neutrophils can be identified (Fig. 2A) (*see* **Note 11**).
7. Additional maturation curves can be studied by CD11b vs CD64 (Fig. 2B) and CD13 vs CD16 expression (Fig. 2C).
8. Other immunophenotypic neutrophil markers are CD66b, CD24, CD15, CD10, CD11a, and CD11c (Fig. 3) (*see* **Note 12**).

For identification of GPI-deficient neutrophils:

1. First, cell populations are identified based on linear parameter (SSC-A) and CD45 expression (Fig. 4A). Neutrophils are CD45+ cells and have heterogeneous and high SSC.
2. On $CD45^+$ neutrophils, CD33 expression vs SSC-A is investigated to exclude the presence of residual monocytes and basophils (Fig. 4B).
3. On $CD45^+CD33^+$ granulocytes, CD15 vs CD45 expression is studied to separate eosinophils (Fig. 4C), which are $CD15^{dim}CD45^{++}$ cells, while neutrophils are $CD15^+CD45^+$.
4. On neutrophils, GPI^+ cells are identified based on FLAER positivity vs CD157 expression (Fig. 4C). The percentage of $FLAER^+$ cells should be similar to that of $CD157^+$ neutrophils (*see* **Note 13**).

For the identification and characterization of LDGs and NDNs, the following gating strategy could be employed (Fig. 5).

1. First, cell populations are identified based on linear parameter (SSC-A) and CD45 expression.

Fig. 1 Neutrophil identification in various human specimens. (**A**), First, neutrophils are identified based on CD45 expression and side scattered area (SSC-A) (dark blue population), and next (**B**) residual monocytes and basophils (dark orange) based on CD33 expression, and (**C**) residual eosinophils ($CD16^-CD45^{++}$ cells; purple population) are removed. Various cell distributions based on linear parameters and CD45/CD33 expression are presented for peripheral blood, bone marrow, pleural and ascitic fluids, and broncho-alveolar lavage (BAL) specimens

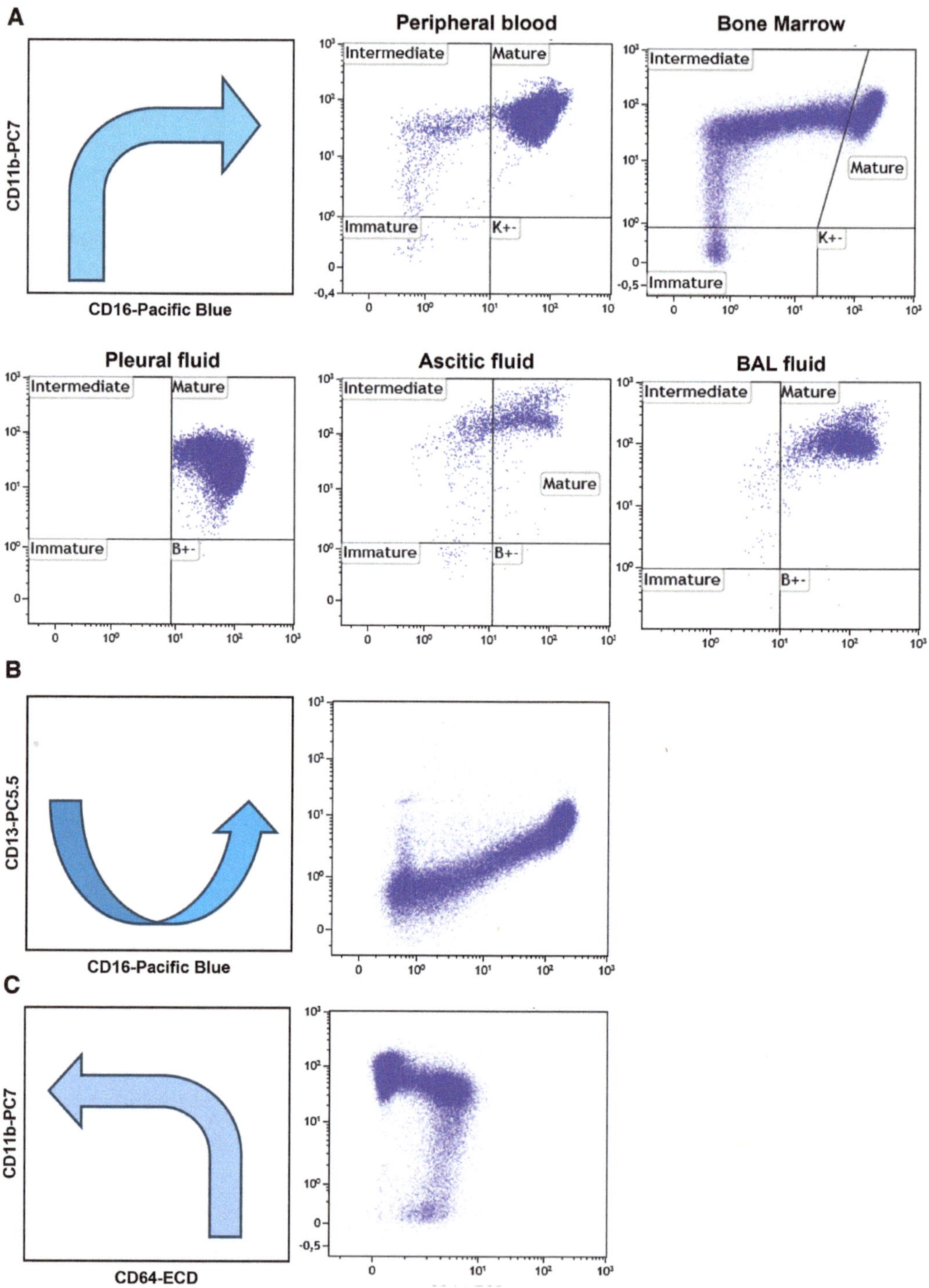

Fig. 2 Neutrophil maturation curve analysis. Neutrophil maturation status can be investigated by (**A**) CD11b vs CD16 expression, (**B**) CD13 vs CD16, or (**C**) CD11b vs CD64 expression. Various examples of maturation curves are reported for peripheral blood, bone marrow, pleural and ascitic fluids, and broncho-alveolar lavage (BAL) specimens

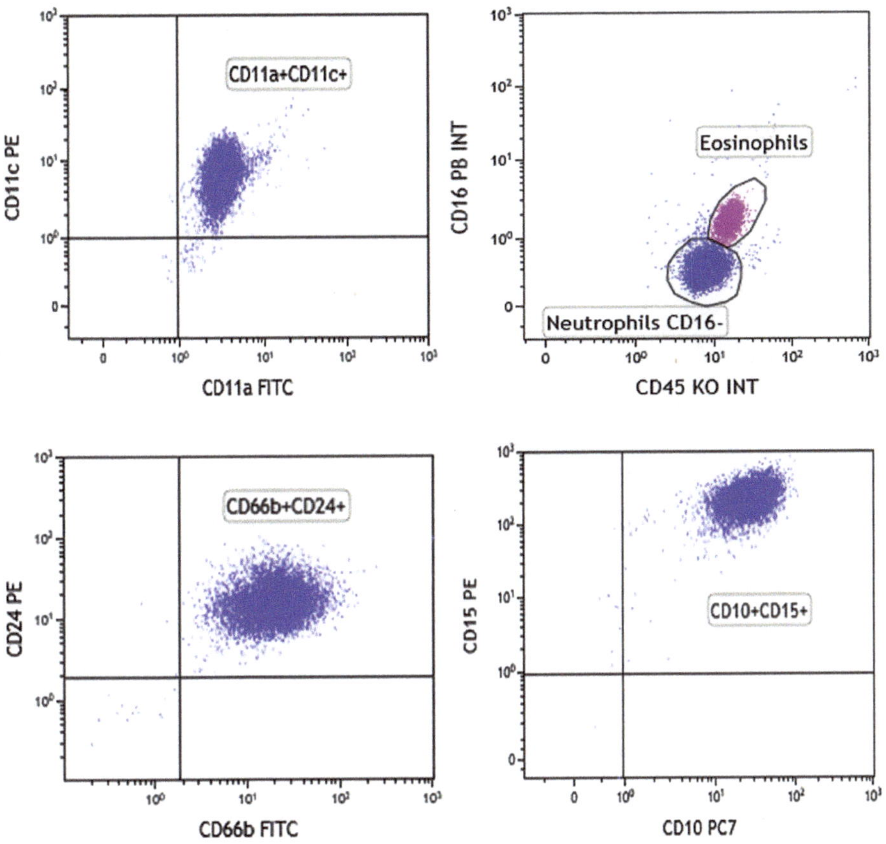

Fig. 3 Additional phenotypic markers. Neutrophils can be further studied for CD11c, CD11a, CD16, CD24, CD66b, CD10, and CD15 expression

2. Non-neutrophilic populations are excluded by sequential removal of Lineage^{+}CD34^{+}HLA-DR^{+} cells.
3. On Lineage^{-}CD34^{-}CD117^{-}HLA-DR^{-} cells, CD33 and CD15 expression is studied to identify CD33^{+}CD15^{dim} NDNs and CD33^{+}CD15^{+} neutrophils (*see* **Note 14**).
4. On NDNs, CD16 vs CD11b maturation curve can be studied for characterization of immature CD16^{-}CD11b^{-}, intermediate CD16^{-}CD11b^{dim}, and mature CD16^{+}CD11b^{+} neutrophils.
5. On CD33^{+}CD15^{+} neutrophils, CD14^{+} cells are excluded (CD15 vs CD14 expression).

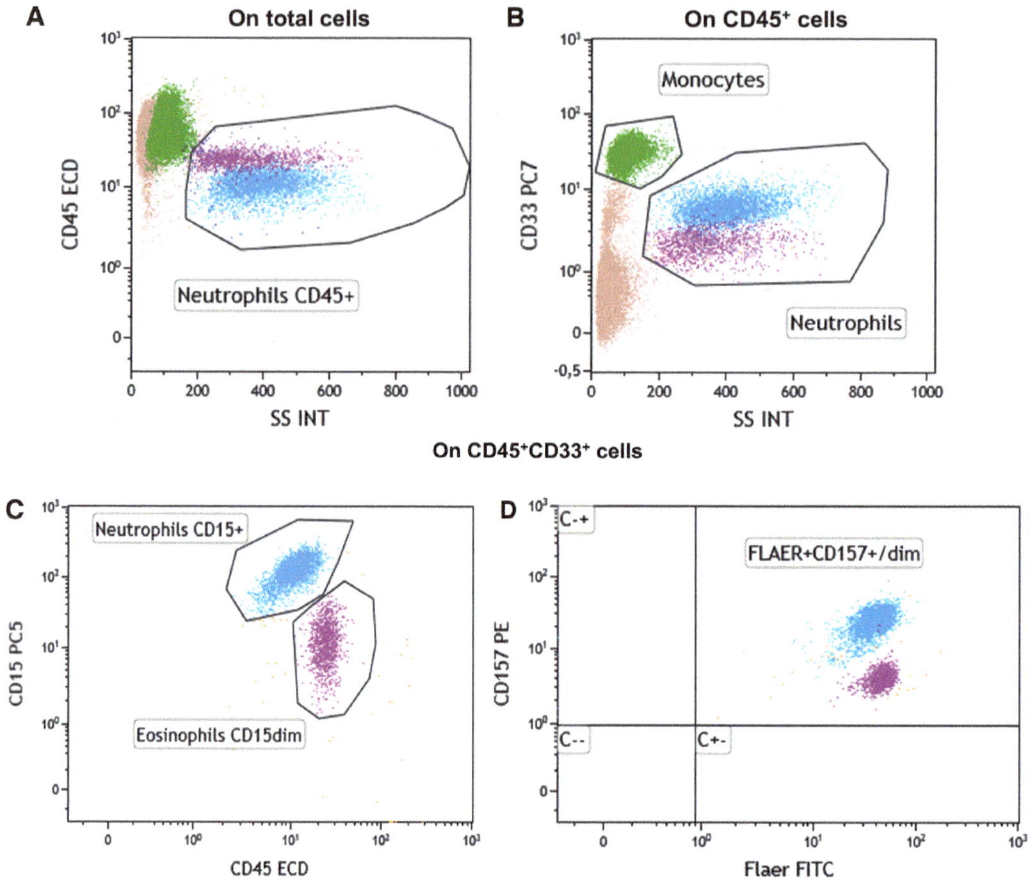

Fig. 4 Gating strategy for GPI-deficient neutrophils. (**A**) Cell populations are identified based on linear parameter (SSC-A) and CD45 expression, and next (**B**) CD33$^+$ cells are gated. (**C**) On CD45$^+$CD33$^+$ granulocytes, CD15 vs CD45 expression is studied to separate eosinophils (purple population) from neutrophils (light blue cells). (**D**) On these populations, CD157 and FLAER can be studied to identify GPI-positive cells

6. On CD33$^+$CD15$^+$CD14$^-$ neutrophils (LDGs), two maturation curves can be studied: CD16 vs CD11b curve for characterization of immature CD16$^-$CD11b$^-$, intermediate CD16$^-$CD11b^{dim}, and mature CD16$^+$CD11b$^+$ neutrophils; and CD15 vs CD16 curve for identification of CD15$^+$CD16$^-$ immature, CD15$^+$CD16^{dim} intermediated, and CD15$^+$CD16$^+$ mature LDGs (*see* **Note 15**).

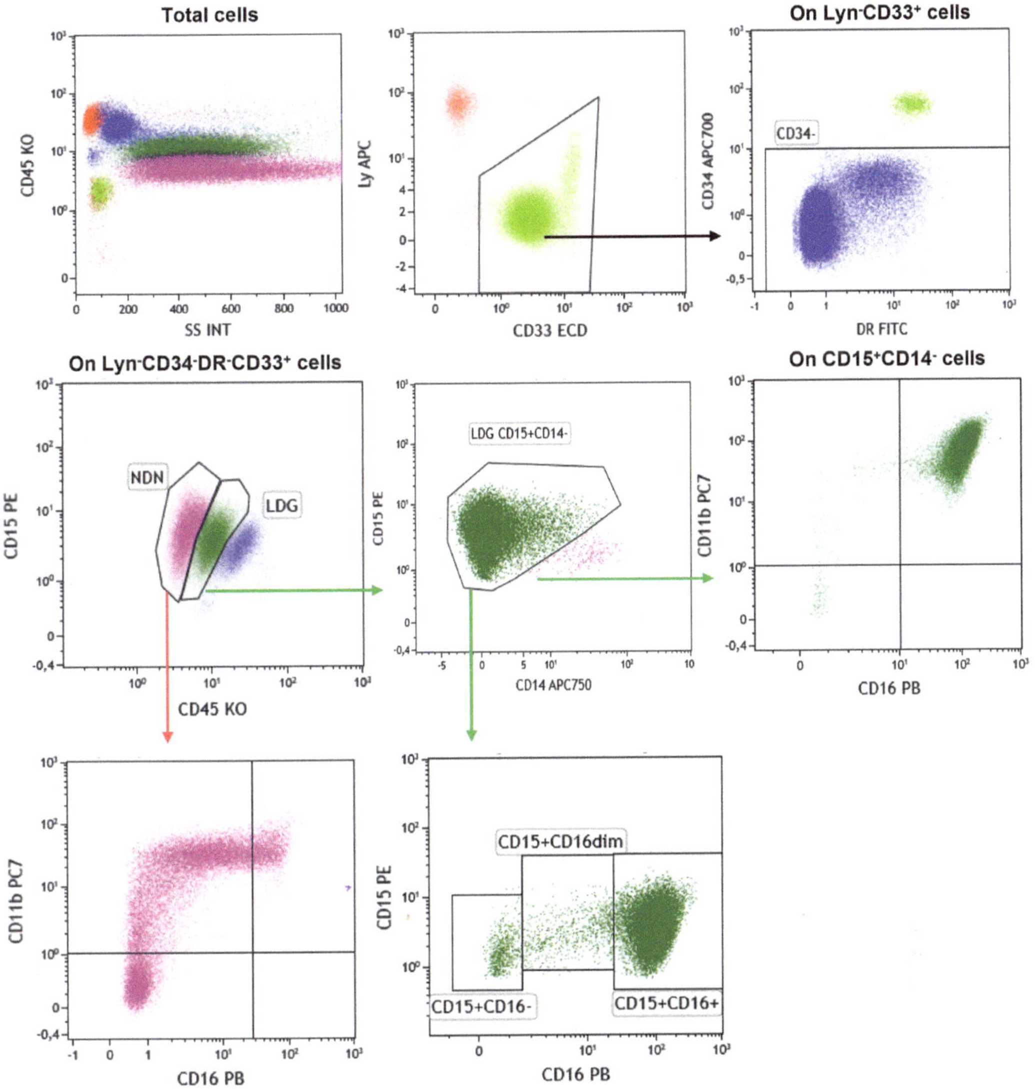

Fig. 5 Gating strategy for identification of low-density granulocytes (LDGs) and normal-density neutrophils (NDNs). Cell populations are identified based on linear parameters (red, lymphocytes; dark blue, monocytes; purple and dark green, neutrophils) and CD45 expression. Non-neutrophilic populations are excluded by sequential removal of Lineage$^+$ cells vs CD33 expression and on CD33$^+$ cells, by removing CD34$^+$ and HLA-DR$^+$ populations. On Lineage$^-$CD34$^-$CD117$^-$HLA-DR$^-$ cells, CD15 expression is studied to identify CD33$^+$CD15^{dim} NDNs and CD33$^+$CD15$^+$ neutrophils. On NDNs (purple population), CD16 vs CD11b maturation curve can be studied for characterization of immature CD16$^-$CD11b$^-$, intermediate CD16$^-$CD11b^{dim}, and mature CD16$^+$CD11b$^+$ neutrophils. On CD33$^+$CD15$^+$ neutrophils, CD14$^+$ cells are excluded (CD15 vs CD14 expression), and on LDGs (dark green population), two maturation curves can be studied: CD16 vs CD11b curve; and CD15 vs CD16 curve for identification of CD15$^+$CD16$^-$ immature, CD15$^+$CD16^{dim} intermediated, and CD15$^+$CD16$^+$ mature LDGs

4 Notes

1. If a membrane filter is not available, a sterile gauze can be used.
2. Additional channels can be employed to design an 11–13-color panel by combining Panel 1 and Panel 2, or by adding T or B lymphocyte markers, or other neutrophil maturation markers.
3. An antibody master mix can be prepared based on the number of samples. For each antibody, multiply the volume for a single test (e.g., 5 μL of CD19-FITC antibody) by the number of samples. For example, if ten samples must be stained, 5 μL of CD19-FITC x 10 → 50 μL of antibody. Add the appropriate volume of each antibody to a 15 mL Falcon tube. Next, calculate total antibody volume for each sample (e.g., 5 μL of each antibody in our panel for a total of ten antibodies → final antibody volume of 50 μL), and add that volume to each 5 mL polypropylene tube according to the experimental design or patient's name.
4. Store the remaining sample at +4°C, as additional staining could be required for more in-depth immunophenotyping.
5. Incubate at +4°C in case of room temperature exceeding + 20°C or for increasing antibody binding.
6. Resuspend in acquisition buffer with 2% paraformaldehyde if sample acquisition is not carried out within 1–2 h after staining.
7. For concentration, transfer samples to a 50 mL Falcon tube and centrifuge at 500 × *g* for 5–10 min. Discard supernatant and resuspend cell pellets in 1 mL of PBS. Add 100 μL of samples to each 5 mL polypropylene tube.
8. Neutrophils can show reduced SSC in certain pathological conditions, such as MDS, and could be erroneously gated together with monocytes.
9. Basophils have a similar CD33 expression compared to neutrophils, while SSC is lower. Moreover, basophils are $CD11b^{+}CD16^{-}$.
10. Eosinophils should be correctly removed, as they can inaccurately increase the percentage of immature and intermediate neutrophils.
11. In PB, in normal conditions, only mature $CD11b^{+}CD16^{+}$ neutrophils are present. Identification of circulating immature and intermediate forms could be associated with maturation block in the BM, inflammatory conditions, or autoimmune diseases.

12. CD16 absence on the neutrophil membrane could be associated with pathological conditions (e.g., MDS and AML) or autoimmune diseases. CD66b, CD24, CD10, CD15, CD11c, and CD11a are present on terminally differentiated and mature neutrophils.
13. Simultaneous investigation of FLAER and CD157 is important for the correct identification of GPI-deficient clones. Discordance between FLAER$^+$ and CD157$^+$ cell percentage should be further investigated by adding additional GPI-linked proteins to the immunophenotypic analysis, such as CD24 expression on neutrophils. On eosinophils, CD157 expression is decreased compared to neutrophils.
14. If backgated on CD45 vs SSC-A dot-plot, LDGs have a lower SSC compared to NDNs.
15. HLA-DR expression can be investigated on each neutrophil subset and reported as median fluorescence intensity.

References

1. Hong CW (2017) Current understanding in neutrophil differentiation and heterogeneity. Immune Netw 17(5):298–306. https://doi.org/10.4110/in.2017.17.5.298. Epub 2017 Oct 16. PMID: 29093651; PMCID: PMC5662779
2. Grieshaber-Bouyer R, Nigrovic PA (2019) Neutrophil heterogeneity as therapeutic opportunity in immune-mediated disease. Front Immunol 10:346. https://doi.org/10.3389/fimmu.2019.00346. PMID: 30886615; PMCID: PMC6409342
3. Silvestre-Roig C, Fridlender ZG, Glogauer M, Scapini P (2019) Neutrophil diversity in health and disease. Trends Immunol 40(7):565–583. https://doi.org/10.1016/j.it.2019.04.012. Epub 2019 May 31. PMID: 31160207; PMCID: PMC7185435
4. Chen T, Zhou Z, Liu Y, Xu J, Zhu C, Sun R, Hu H, Liu Y, Dai L, Holmdahl R, Herrmann M, Zhang L, Muñoz LE, Meng L, Zhao Y (2024) Neutrophils with low production of reactive oxygen species are activated during immune priming and promote development of arthritis. Redox Biol 78:103401. https://doi.org/10.1016/j.redox.2024.103401. Epub 2024 Oct 18. PMID: 39471640; PMCID: PMC11550370
5. Hsu BE, Tabariès S, Johnson RM, Andrzejewski S, Senecal J, Lehuédé C, Annis MG, Ma EH, Völs S, Ramsay L, Froment R, Monast A, Watson IR, Granot Z, Jones RG, St-Pierre J, Siegel PM (2019) Immature low-density neutrophils exhibit metabolic flexibility that facilitates breast cancer liver metastasis. Cell Rep 27(13):3902–3915.e6. https://doi.org/10.1016/j.celrep.2019.05.091. PMID: 31242422
6. Rahman S, Sagar D, Hanna RN, Lightfoot YL, Mistry P, Smith CK, Manna Z, Hasni S, Siegel RM, Sanjuan MA, Kolbeck R, Kaplan MJ, Casey KA (2019) Low-density granulocytes activate T cells and demonstrate a non-suppressive role in systemic lupus erythematosus. Ann Rheum Dis 78(7):957–966. https://doi.org/10.1136/annrheumdis-2018-214620. Epub 2019 Apr 30. PMID: 31040119; PMCID: PMC6585283
7. van den Hoogen LL, van der Linden M, Meyaard L, Fritsch-Stork RDE, van Roon JA, Radstake TR (2020) Neutrophil extracellular traps and low-density granulocytes are associated with the interferon signature in systemic lupus erythematosus, but not in antiphospholipid syndrome. Ann Rheum Dis 79(10):e135. https://doi.org/10.1136/annrheumdis-2019-215781. Epub 2019 Jun 8. PMID: 31177097
8. Ning X, Wang WM, Jin HZ (2022) Low-density granulocytes in immune-mediated inflammatory diseases. J Immunol Res 2022:1622160. https://doi.org/10.1155/2022/1622160. PMID: 35141336; PMCID: PMC8820945
9. Mauracher LM, Krall M, Roiß J, Hell L, Koder S, Hofbauer TM, Gebhart J, Hayden H, Brostjan C, Ay C, Pabinger I (2021) Neutrophil subpopulations and their activation potential in patients with antiphospholipid syndrome and healthy individuals. Rheumatology

60(4):1687–1699. https://doi.org/10.1093/rheumatology/keaa532. PMID: 33026085; PMCID: PMC8024003

10. Suárez A, Tobío-Parada U, Rodríguez-Carrio J, Martínez-Zapico A, Pérez-Álvarez ÁI, Suárez-Díaz S, Caminal-Montero L, López P (2024). Circulating levels of low-density granulocytes and cell-free DNA as predictors of cardiovascular disease and bone deterioration in SLE patients. Thromb Haemost 25(9):859–870. https://doi.org/10.1055/a-2467-6826. PMID: 39542026
11. López P, Rodríguez-Carrio J, Martínez-Zapico A, Pérez-Álvarez ÁI, Suárez-Díaz S, Mozo L, Benavente L, Caminal-Montero L, Suárez A (2020) Low-density granulocytes and monocytes as biomarkers of cardiovascular risk in systemic lupus erythematosus. Rheumatology **59**(7):1795. doi: 10.1093/rheumatology/keaa225. Erratum for: Rheumatology (Oxford). 2020;**59**(7):1752–1764. https://doi.org/10.1093/rheumatology/keaa016. PMID: 32447395
12. Bertolini A, Picone F, Ferrara I, Della Corte AM, Serio B, Gorrese M, Campana A, Simeon V, Luponio S, Marcucci R, Scala P, Langella M, Selleri C, Giudice V (2025) Immature forms of low density granulocytes are increased in acute myeloid leukemia and myelodysplastic syndromes. Sci Rep 15:8661. https://doi.org/10.1038/s41598-025-92513-8 PMID: 40082470; PMCID: PMC11906629.
13. Raskovalova T, Berger MG, Jacob MC, Park S, Campos L, Aanei CM, Kasprzak J, Pereira B, Labarère J, Cesbron JY, Veyrat-Masson R (2019) Flow cytometric analysis of neutrophil myeloperoxidase expression in peripheral blood for ruling out myelodysplastic syndromes: a diagnostic accuracy study. Haematologica 104(12):2382–2390. https://doi.org/10.3324/haematol.2018.202275. Epub 2019 Apr 19. PMID: 31004030; PMCID: PMC6959174
14. Monaghan SA, Surti U, Doty K, Craig FE (2012) Altered neutrophil maturation patterns that limit identification of myelodysplastic syndromes. Cytometry B Clin Cytom 82(4):217–28. https://doi.org/10.1002/cyto.b.21016. Epub 2012 Mar 19. PMID: 22431481
15. Bento LC, Correia RP, Pitangueiras Mangueira CL, De Souza Barroso R, Rocha FA, Bacal NS, Marti LC (2017) The use of flow cytometry in myelodysplastic syndromes: a review. Front Oncol 7:270. https://doi.org/10.3389/fonc.2017.00270. PMID: 29188193; PMCID: PMC5694750
16. Raskovalova T, Scheffen L, Jacob MC, Vettier C, Bulabois B, Szymanski G, Chevalier S, Gonnet N, Park S, Labarère J (2022) Comparative diagnostic accuracy between simplified and original flow cytometric gating strategies for peripheral blood neutrophil myeloperoxidase expression in ruling out myelodysplastic syndromes. PLoS One 17(11):e0276095. https://doi.org/10.1371/journal.pone.0276095. PMID: 36399441; PMCID: PMC9674135

Chapter 4

Natural Killer Cells and Innate Lymphoid Cells Immunoprofiling

Christelle Piperoglou and Frédéric Vély

Abstract

Phenotyping of natural killer (NK) cells by flow cytometry allows a detailed analysis of their functional diversity and activation status. This multiparametric approach assesses the expression of activating or inhibitory receptors and detects markers of maturation or exhaustion. It is crucial for understanding the role of NK cells in the immune response, particularly against infections and tumor cells. This phenotyping also allows us to monitor the effects of immunomodulatory treatments and to identify prognostic or predictive biomarkers in different clinical contexts.

Keywords NK cells, Flow cytometry, Immunoprofling

1 Introduction

NK cells are lymphocytes that are part of the innate immune system and play a crucial role in the body's defense against viral infections and tumor cells [1]. In contrast to the T and B lymphocytes of the adaptive system, NK cells do not require prior sensitization in order to recognize and eliminate abnormal cells. They exert their cytotoxic function by recognizing an imbalance between activating and inhibitory signals on the surface of the target cells, in particular by recognizing a low expression of HLA class I molecules, which is often altered in infected or cancerous cells. Once activated, NK cells release granules containing cytotoxic enzymes such as perforin and granzymes that induce cell death.

In addition to their direct cytotoxic function, NK cells secrete cytokines such as interferon-gamma (IFN-γ), which are involved in the activation of other components of immunity, including macrophages and dendritic cells, thus enhancing the overall immune response [2]. Their dual ability to kill and modulate the immune environment makes them key players in immune surveillance.

Angélique Biancotto and Pradeep K. Dagur (eds.), *Immunophenotyping*, Methods in Molecular Biology, vol. 3087, https://doi.org/10.1007/978-1-0716-5607-5_4,

A new NK cell nomenclature based on single-cell transcriptome data has just been proposed [3]. Pending validation of these signatures at the protein level, this chapter uses the classical nomenclature to define human NK cells as $CD3^{-}CD56^{+}$ lymphocytes with two populations distinguished by CD56 expression [4]. $CD56^{dim}$ NK cells predominate in the peripheral blood and have cytotoxic properties due to their high expression of CD16 and perforin, among others. $CD56^{bright}$ NK cells normally make up less than 10% of circulating NK cells and are described as immature forms.

In recent years, interest in NK cells in a therapeutic context has steadily increased. Their natural ability to fight cancer cells has led to the development of innovative immunotherapy strategies. These approaches include the ex vivo expansion of autologous or allogeneic NK cells, their genetic modification (in particular through chimeric CAR-NK receptors), or the use of antibodies to facilitate their recruitment and activation. These therapies are currently being investigated in numerous clinical trials, particularly for the treatment of leukemias, lymphomas, and solid tumors, and offer promising prospects for more targeted and less toxic treatments [5].

2 Materials

The following materials and reagents were used for immunophenotyping of T, B, and NK cells.

2.1 Biological Samples

- Fresh whole blood samples
- Peripheral blood mononuclear cells (PBMCs), either freshly isolated or cryopreserved

2.2 Reagents and Buffers

- BD Multitest™ six-Color TBNK reagent (BD Biosciences)
- BD Trucount™ tubes (BD Biosciences)
- BD FACS lysing solution (1×) (BD Biosciences)
- FACS Buffer: PBS 1× supplemented with 10% Fetal Bovine Serum (FBS) and 1 mM EDTA
- Complete medium: RPMI-1640 medium supplemented with 10% Fetal Bovine Serum (FBS), 1% Penicillin–Streptomycin, and 1% L-glutamine
- Live/Dead stain (Thermo Fisher Scientific)
- FcR Blocking Reagent (BD Biosciences)
- Standard density gradient medium, such as Ficoll-Paque™ PLUS
- Freezing medium: 90% FBS and 10% dimethyl sulfoxide (DMSO)

2.3 Equipment and Consumables

- Flow cytometer (BD LSR II Fortessa™)
- 12 × 75 mm polystyrene round-bottom FACS tubes
- 15 mL and 50 mL conical centrifuge tubes
- Refrigerated centrifuge
- Pipettes and sterile pipette tips
- Vortex mixer
- Automated cell counter
- Cryovials for cell storage
- Freezing container for controlled-rate freezing
- −80°C freezer and liquid nitrogen storage system

3 Methods

3.1 Immunophenotyping of TBNK Lymphocyte Subsets

Lymphocyte counts were performed on fresh whole blood samples.

3.1.1 Reagent Preparation and Sample Staining

1. Add 20 μL of BD Multitest six-color TBNK Reagent into a BD Trucount™ tube.
2. Add 50 μL of whole blood to the tube.
3. Vortex the tube thoroughly to ensure proper mixing.

3.1.2 Incubation and Red Blood Cell Lysis (Manual Method)

1. Incubate the tube for 15 min at room temperature, protected from light.
2. Add 450 μL of BD FACS Lysing Solution (1×).
3. Vortex the tube to mix the contents.
4. Incubate the tube for 10 min (8–15 min) at room temperature, protected from light.

3.1.3 Sample Acquisition

After incubation, the sample was ready for flow cytometric analysis. If the samples were not immediately analyzed, they were stored in the dark at room temperature until analysis. The TBNK immunophenotyping panel included CD45, CD3, CD4, CD8, CD56, CD16, and CD19 markers, which enabled the identification of T, B, and NK cells in the blood sample. A gating strategy was used to differentiate these cell populations based on their distinct surface marker profiles.

3.2 Extended NK Cell Immunophenotyping

3.2.1 Sample Preparation

The phenotyping of NK cells by flow cytometric analysis necessitates the preparation of single-cell suspensions containing viable cells.

Appropriate starting materials for NK cell phenotyping include whole blood (*see* **Note 1**), fresh PBMCs: isolated from whole blood using density gradient centrifugation (*see* **Note 2**), cryopreserved PBMCs: thawed and processed following standardized protocols (*see* **Note 3**) to maintain cell integrity and marker expression.

3.2.2 Panel Design

The NK cell phenotyping panel should be designed based on the study's objectives and the specifications of the available flow cytometer. The panel typically includes markers such as CD45, CD3, CD56, and CD16 to identify NK cells. Additional NK cell markers of interest may be included as needed (Table 1). A gating strategy specific to PBMCs is used for the identification of NK cells (Fig. 1). A graphical representation of the expression of different markers is shown in Fig. 2.

3.2.3 Cell Preparation

1. Adjust the cell suspension to a concentration of $1\text{–}2 \times 10^6$ cells/mL in RPMI supplemented with 10% FBS, 1% P–S, and 1% glutamine.
2. Transfer the required volume into FACS tubes to obtain $0.5\text{–}2 \times 10^6$ cells per tube.

3.2.4 Dead Cell Marker Staining

1. FACS buffer was added to adjust the final volume to 2 mL, followed by centrifugation at 645 × *g* for 2 min.
2. Discard the supernatant by inverting the tube and blotting on absorbent tissue.

Table 1

List of antibodies used in classical NK cell immunophenotyping. An example of an antibody panel is provided with all features: name of marker, fluorochrome, clone, provider, reference, and test sample

Markers	Fluorochrome	Ab clone	Source	Catalog no.	Volume
Live/Dead	APC-H7	n/a	Thermo	L347976	1000 μL[a]
CD45	V500	HI30	BD Biosciences	560777	2.5 μL
CD33	Percp-Cy5.5	WM53	Ozyme	303414	1 μL
CD3	BUV737	UCHT1	BD Biosciences	612750	5 μL
CD56	BUV395	B159	BD Biosciences	740299	5 μL
CD16	BUV496	3G8	BD Biosciences	564653	5 μL
CD57	FITC	HNK1	Ozyme	359604	5 μL
NKp46	BV605	9E2	Ozyme	331925	1.5 μL
NKG2D	PE-Cy7	1D11	Biolegend	320812	2.5 μL
NKp30	BV421		Biolegend	563385	5 μL
NKG2A	APC	REA110	Miltenyi	130-113-563	10 μL

[a]Live/ Dead was prediluted 1 μL of Live/ Dead in 1 mL of DPBS 1×

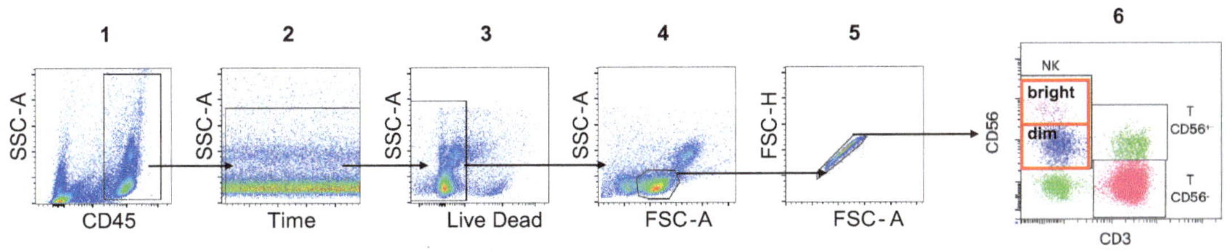

Fig. 1 Gating strategy for the identification of NK cells by flow cytometry. **1**. Gate on leukocyte population in CD45–SSC plot. **2**. Select the time region corresponding to stable acquisition in time–SSC plot. **3**. Gate on live cells in dead cell marker–FSC plot. **4**. Within the living cells, gate on the lymphocyte population in FSC–SSC plot. **5**. Exclude doublets by selecting a linear population in FSC-H/FSC-A plot. **6**. Within single-cell, use the CD3/CD56 dot plot to highlight NK cells as CD3⁻CD56⁺ lymphocytes. Both CD56^{dim} and CD56^{bright} NK cells can be visualized

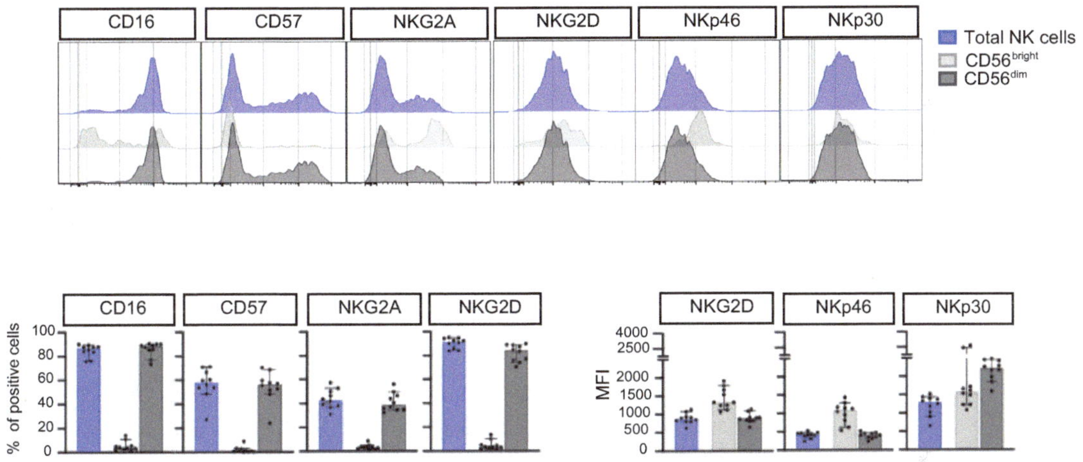

Fig. 2 Expression of CD16, CD57, NKG2A, NKG2D, NKp46, and NKp30 in resting NK cells from healthy volunteers. The upper panel shows a representative staining of these markers on all (blue), CD56^{dim} (dark gray), and CD56^{bright} (light gray) NK cells. The bottom panel shows the percentage and MFI of expression of these markers in each NK cell subset from ten healthy volunteers

3. Vortex each sample for 3 s to loosen the cell pellet.
4. Resuspend the cells in 1 mLof DPBS 1× and add 1 μL of Live/Dead marker according to the manufacturer's instructions.
5. Incubate the samples for 30 min at room temperature in the dark.
6. Add 1 mL of DPBS 1× and centrifuge at 645 ×*g* for 2 min.
7. Discard the supernatant by inverting the tube and blotting on absorbent tissue.
8. Add 2 mL of DPBS 1× and centrifuge at 645 ×*g* for 2 min.

3.2.5 Fc Block Saturation

1. Add 2.5 μg of Fc Block to the cell pellet of each sample for every 1×10^6 cells

2. Incubate the samples for 15 min at room temperature in the dark

3.2.6 Surface Marker Staining

1. Add a master mix containing surface-staining antibodies to the cells preincubated with Fc Block. To minimize nonspecific binding, each fluorochrome-conjugated antibody or reagent should be properly titrated before use.
2. Incubate the samples for 20 min at room temperature in the dark.
3. Repeat the washing steps to remove unbound antibodies.

3.2.7 Final Preparation for Flow Cytometric Acquisition

After washing, resuspend in 300 μL FACS buffer, and keep on ice until acquisition on the flow cytometer.

3.3 KIR Phenotyping Protocol on NK Cells

The activity of NK cells is influenced by genetic factors through the varying expression of polymorphic killer cell immunoglobulin-like receptors (KIRs), which interact with HLA class I molecules. Each individual inherits a set of KIR genes. The expression of these inherited genes is stochastic at the level of individual NK cells [6]. This leads to a high heterogeneity of KIR expression by NK cells. This loss of heterogeneity may be associated with clonal or oligoclonal expansion of NK cells [7].

3.3.1 Sample Preparation

KIR phenotyping is performed directly on 50 μL of whole blood.

3.3.2 Staining Protocol

Three tubes are prepared for analysis, each containing the following antibodies for NK cell characterization and KIR receptor detection (Table 2).
Common gating antibodies:
CD45 APC-H7: leukocyte marker.
CD56 V450: NK cell marker. Some T cells also express CD56.
CD3 AMCYAN: used to exclude T cells ($CD3^+$) and identify NK cells ($CD3^-CD56^+$).
For each tube, specific KIR antibodies are added as follows:
Tube 1:
CD158a, h PE (KIR2DL1/ KIR2DS1),
CD158a FITC (KIR2DL2),
CD158b1, b2, j PE-Cy7 (KIR2DL2, KIR2DL3, KIR2DS2).
Tube 2:
CD158e1, e2 PE (KIR2DL1/ KIR3DS1),
CD158e APC (KIR3DL1).
Tube 3:
CD158i PE (KIR2DS4).

3.3.3 Incubation

The antibodies are added to the whole blood in the respective tubes, followed by 20 min of incubation at room temperature, protected

Table 2
List of antibodies used in classical KIRotyping. An example of an antibody panel is provided with all features: name of marker, fluorochrome, clone, provider, reference, and test sample

Markers	Fluorochrome	Ab clone	Source	Catalog no.	Volume
CD45	APC-H7	2D1	BD Biosciences	641408	5 µL
CD3	Amcyan	SK7	BD Biosciences	339186	5 µL
CD56	V450	B159	BD Biosciences	560360	5 µL
CD158a,h	PE	EB6B	Beckman Coulter	A09778	10 µL
CD158a	FITC	HP-3E4	BD Biosciences	556062	10 µL
CD158e1, e2	PE	Z27.3.7	Beckman Coulter	IM3292	10 µL
CD158b1, b2, j	PE-Cy7	GL183	Beckman Coulter	A66901	10 µL
CD158e	APC	DX9	R&D	FAB1225A	10 µL
CD158i	PE	FES172	Beckman Coulter	IM3337	10 µL

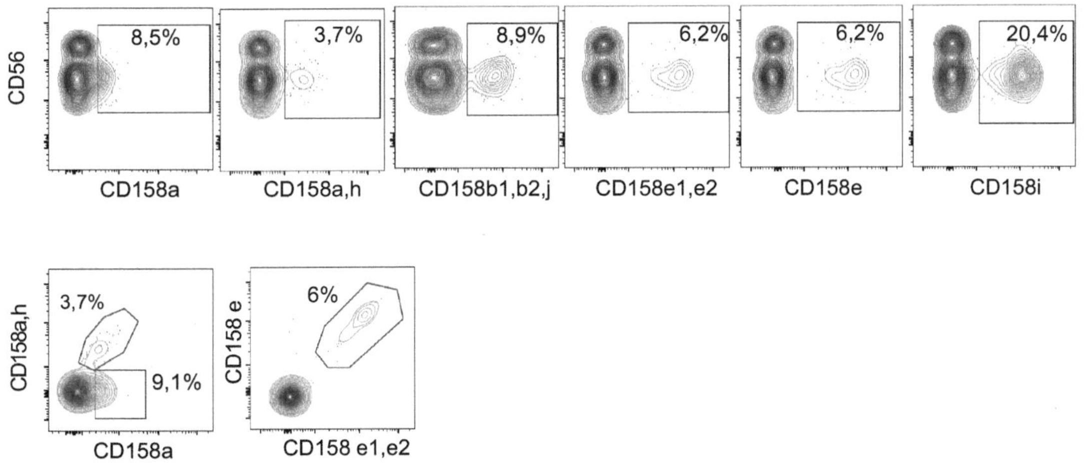

Fig. 3 KIRotyping of NK cells. Representative staining on NK cells using distinct specific anti-KIR antibodies. Normal distribution has been published previously [7]

from light. This step allows specific binding of antibodies to the KIR receptors on the surface of NK cells.

3.3.4 Lysis and Acquisition

After incubation, red blood cell lysis as previously described. Following lysis, data acquisition is carried out by flow cytometry, according to established parameters for detecting specific KIR markers.

The tubes are then analyzed on a flow cytometer to assess the expression of KIR receptors on NK cells (Fig. 3).

4 Notes

1. NK cell phenotyping can be performed using 100 μL of whole blood, corresponding to approximately 1×10^6 cells. The whole blood is first washed twice with 1× PBS and then incubated with a pre-diluted Live/Dead stain (1:1000 dilution). After a 15-min incubation in the dark at room temperature, the sample is washed with FACS buffer. The antibody mix is then directly added to the red blood cell pellet. After a 20-min incubation in the dark at room temperature, the cells are automatically lysed and immediately ready for acquisition on the flow cytometer.
2. Peripheral blood mononuclear cells (PBMCs) were isolated from whole blood using a standard density gradient medium centrifugation process. Briefly, blood was diluted 1:1 with DPBS and carefully layered over Ficoll in a 50 mL conical tube. Samples were centrifuged at $963 \times g$ for 30 min at room temperature without a brake. The PBMC layer was collected, washed twice with FACS Buffer by centrifugation at $300 \times g$ for 10 min, and resuspended in DPBS or culture medium. Cells were counted and either used immediately or cryopreserved for further analysis.
3. The cell suspension is centrifuged at $390 \times g$ for 5 min, and the pellet is resuspended in 1 mL of freezing medium, composed of 90% fetal bovine serum (FBS) and 10% dimethyl sulfoxide (DMSO), added dropwise on ice. After homogenization, the suspension is transferred to cryovials, placed in a freezing container at –80°C for 24 h, and then transferred to liquid nitrogen for long-term storage.

References

1. Spits H, Artis D, Colonna M et al (2013) Innate lymphoid cells—a proposal for uniform nomenclature. Nat Rev Immunol 13:145–149
2. Freud AG, Mundy-Bosse BL, Yu J et al (2017) The broad spectrum of human natural killer cell diversity. Immunity 47:820–833
3. Rebuffet L, Melsen JE, Escalière B et al (2024) High-dimensional single-cell analysis of human natural killer cell heterogeneity. Nat Immunol 25:1474–1488
4. Cooper MA, Fehniger TA, Caligiuri MA (2001) The biology of human natural killer-cell subsets. Trends Immunol 22:633–640
5. Vivier E, Rebuffet L, Narni-Mancinelli E et al (2024) Natural killer cell therapies. Nature 626:727–736
6. Parham P (2005) MHC class I molecules and KIRs in human history, health and survival. Nat Rev Immunol 5:201–214
7. Pascal V, Schleinitz N, Brunet C et al (2004) Comparative analysis of NK cell subset distribution in normal and lymphoproliferative disease of granular lymphocyte conditions. Eur J Immunol 34:2930–2940

Chapter 5

Immunophenotyping of Human Regulatory T-Cells from Peripheral Blood Mononuclear Cells

Maria Lopez-Ocasio, Wan-Chi Lin, Steven C. Hockman, Keyvan Keyvanfar and Pradeep K. Dagur

Abstract

Regulatory T cells, or Tregs, are a unique type of $CD4^+$ T cell that act as the immune system's brakes—they stop it from turning against the body's own tissues. For about twenty years now, researchers have been digging into what sets Tregs apart from other T cells, both in their composition and function (1). Markers like CD25 (IL-2Rα), Foxp3, and CTLA-4 usually show up when Tregs are active, but these markers aren't always reliable—they can shift depending on where the cells are and what they are doing. Foxp3 is a transcription factor that Tregs need to develop and function. Mutations on Foxp3 gene lead to serious autoimmune diseases like immune dysregulation, polyendocrinopathy, enteropathy, and X-linked (IPEX) syndrome. Lately, studies have uncovered even more variety within Tregs by finding subsets with markers like Helios, TIGIT, CD127 (which is usually low in Tregs), and neuropilin-1. Identifying and measuring Tregs accurately is important not only in research but also in the clinic. Identifying Tregs on patient samples provides a clear picture of what is going on with their immune system—whether it is an autoimmune disease, a transplant, or cancer treatment. Flow cytometry has become a go-to tool for identifying, typing, and subclassifying subsets. It helps showing how Tregs work, how stable they are, and how well they keep other immune cells in check. By profiling Tregs, doctors can group patients more accurately and find biomarkers that tell you if a treatment's working or if a disease is getting worse. Better protocols to test Treg mean more personalized immune therapies. Tregs are serious power players when it comes to immune suppression, and being able to profile them precisely—whether in animals or people—is important for both science and medical care.

Keywords CD4, Tregs, Immunophenotyping, CD25, Foxp3, CTLA-4, Flow cytometry, Suppression

1 Introduction

Immunophenotyping of human regulatory T cells (Tregs) is a necessary step to understand their role in immune regulation and diseases. Tregs have been described by their expression of CD4 and high CD25 expression, plus the FOXP3 transcription factor. These markers are essential for maintaining peripheral tolerance in check. However, this classical definition by markers does not capture the

Angélique Biancotto and Pradeep K. Dagur (eds.), *Immunophenotyping*, Methods in Molecular Biology, vol. 3087, https://doi.org/10.1007/978-1-0716-5607-5_5,

diversity and functional states of Tregs in human peripheral blood and tissues, as Tregs come in different flavors and functional states.

Recent advances in immunophenotyping have given us a much bigger molecular toolbox for sorting out Treg subsets. Today, we can accurately differentiate between Tregs that are suppressive and nonsuppressive.

For example, in autoimmune diseases, studies of type [1] diabetes, researchers have shown that Tregs from patients often retain FOXP3 expression but are impaired as they are unable to suppress effector T cells [2, 3]. These findings show limitations of relying only on FOXP3 as a marker of regulatory function, particularly during inflammation or disease settings.

Researchers now rely on multi-parametric flow cytometry and high-dimensional profiling as an approach for Treg studies. Markers like low CD27 [4], Helios [5, 6], CTLA-4 [7], CD39 and CD73 [8, 9], GARP and LAP [10], and LAG-3 [11] all provide important information regarding the stability, activity or suppressive functions of Tregs. The combination of these markers, rather than the expression of only one, is the best way to define functional Tregs in both peripheral blood and in tissues.

To maintain consistency across Treg studies, efforts to standardize immunophenotyping protocols have led to consensus marker panels and gating strategies [12, 13]. These protocols have aided researchers to select key Treg subsets, regardless of studies or patient's cohorts. OMIP-guided frameworks [14, 15] have sharpened things even more, allowing scientists to separate naïve, effector, and non-suppressive Treg populations with real precision.

There are more Treg subsets than the conventional $CD4^+CD25^+$ $FOXP3^+$ Tregs. Nonclassical groups like $CD4^+CD49b^+$ Tregs [16], $CD8^+$ regulatory populations [17], and tissue-derived $FoxP3^+CD56^+$ cells [18] are gaining attention now. These subsets do not only look different—they move, signal, and regulate in their own ways, which translates into tailoring marker panels to really understand what goes on in different tissues or diseases [19].

This chapter provides a practical guide to immunophenotyping human Tregs. It covers how to pick the right markers, how to tackle staining, and how to get the most out of your data—especially when using flow cytometry or similar platforms. By combining what you see on the cells with what they do, you get a much clearer picture of Treg biology in both health and disease.

2 Materials

2.1 Isolation of Lymphocytes

1. Fresh sample: 10 mL of human peripheral blood (collected in anticoagulant, e.g., EDTA or heparin) from healthy donors.

2. Refrigerated centrifuge, Allegra X14-R (Beckman Coulture with adaptors for 5 mL (5 mm round-bottom test tubes).
3. ACK Lysing solution (Quality Biologicals).
4. Wash buffer: 1× phosphate-buffered saline (PBS) (Quality Biologicals) + 1% BSA.
5. Falcon Round-bottom polystyrene 5 mL FACS test tubes (BD Falcon, Fischer Scientific).
6. Falcon Round-bottom polystyrene 5 mL FACS test tubes with cell strainer (BD Falcon, Fischer Scientific).

2.2 Staining PBMCs

1. Wash buffer: 1× phosphate-buffered saline (PBS) (Quality Biologicals) + 1% BSA.
2. Fluorochrome-conjugated human antibodies for Treg-cells (see Table 1).
3. Staining buffer: 1× PBS with 1% BSA and 10% FBS, filtered.
4. Disposable pipettes.
5. Refrigerated centrifuge, Allegra X14-R (Beckman Coulture with adaptors for 5 mL (5 mm round-bottom test tubes).
6. OneComp eBeads, Compensation Beads (ThermoFisher Scientific). Use as suggested in the manufacturer's instructions.
7. Fixation buffer: eBioscience™ Foxp3/Transcription Factor Staining Buffer Set (ThermoFisher Scientific) for Fixation/Permeabilization of intracellular molecules; or 1% paraformaldehyde in 1× PBS for after staining both surface and intracellular.
8. Single-color controls consisting of unstained PBMCs and single-color using beads are used for compensation in each experiment.
9. Samples analyzed on a Five Lasers Cytek Aurora Spectral analyzer (Cytek) and acquisition software (SpectroFlo); analysis software. Acquire at least 10,000 (or as many as possible) events in the CD4+ CD25+ gate.
10. After acquisition we export the files and perform the analyses later with the FlowJo 10.1 software. (Tree Star/BD Biosciences).
11. SpectroFlo beads (Cytek). For QC performance on the instrument.

Table 1
Antibody panel: antigen-specific mAb used for cellular antigens

Specificity	Fluorochrome	Ab clone	Vendor	Volume
CD3	BV570	UCHT1	Biolegend	10 μL
CD4	APC-Cy7	RPA-T4	BD Biosciences	5 μL
CD8	BUV737	SK1	BD Biosciences	5 μL
CD19	BUV496	SJ25C1	BD Biosciences	5 μL
CD25	BUV395	BC96	BD Biosciences	5 μL
Helios	PE-eF610	22F6	Invitrogen	5 μL
Foxp3	PE	259D/C7	BD Biosciences	10 μL
CD39	PE-Cy7	A1	Biolegend	5 μL
CD73	BV421	AD2	BD Biosciences	5 μL
CTLA-4	BV711	BNI3	Biolegend	5 μL
CD127	AF647	HIL-7R-M21	BD Biosciences	5 μL
CD38	PE-Cy5	HIT-2	BD Biosciences	5 μL
Lag3	RB780	T47-530	BD Biosciences	5 μL
CD45	AF700	HI30	Biolegend	5 μL
CD15	BV786	HI95	BD Biosciences	5 μL
HLA-DR	BV650	G46-6	BD Biosciences	5 μL
CD49b	BV510	AK-7	BD Biosciences	5 μL
CD24	BUV805	M1/69	BD Biosciences	5 μL
CD5	FITC	UCHT2	BD (Pharmingen) Biosciences	10 μL
CD56	RB613	B159	BD Biosciences	5 μL

Fluorochrome nomenclature: BV brilliant violet antibodies (BV421, BV510, BV570, BV650, BV711, BV786), APC-Cy7 allophycocyanin-cyanine H7 tandem, AF647 Alexa Fluor 647, AF700 Alexa Fluor 700, BUV brilliant ultraviolet (BUV396, BUV496, BUV737, BUV805), PE phycoerythrin, PE-Cy5 phycoerythrin-cyanine five tandem, PE-eF610 phycoerythrin-eFluor 610, PE-Cy7 phycoerythrin-cyanine seven tandem, RB real blue (RB613, RB780), FITC fluorescein isothiocyanate

3 Method

3.1 Sample Preparation

1. Blood collection and centrifugation. Peripheral blood was collected in either heparin or EDTA-coated tubes. Store the tubes at room temperature (RT) and place on a nutator to prevent clotting. Using a disposable pipette, transfer 10 mL of whole blood into a 50 mL Falcon tube. Centrifuge at 2000 ×*g* for 10 min at RT to separate plasma from cells of interest (e.g., lymphocytes).

2. Plasma removal. Carefully remove the plasma layer without disturbing the cell layer. Remove the cell layer and place in a clean 50 mL Falcon tube. Add 25 mL of RBC lysis buffer to the tube. Mix gently on a nutator mixer for 2 min. Centrifuge at 300 × *g* for 5 min at 4°C. Remove the supernatant and gently disrupt the cell pellet by slow vortexing or tapping. Repeat the RBC lysis step: Add another 25 mL of RBC lysis buffer. Mix on nutator for 1 min. Centrifuge at 300 × *g* for 5 min at 4°C. Remove lysis buffer and gently break the pellet again.
3. Washing and Preparation for Staining. Add 10 mL of staining buffer to wash the cell pellet. Centrifuge at 300 × *g* for 5 min. Aspirate the supernatant completely. Add fresh ice-cold staining buffer to adjust the cell concentration to 1–2 × 10^6 cells/mL.
    a. *Note: If large cell clumps are observed, pass the suspension through a 40–100 μm mesh strainer to obtain a single-cell suspension (see Note 1).*
4. Antibody Staining. Transfer the cell suspension into appropriate 5 mL Falcon tubes. Add the working amount of fluorochrome-conjugated human antibodies and Fc receptor blocker to reduce nonspecific binding. Incubate the cells at 4°C for 25 min in the dark.
5. Washing poststaining. Add 2 mL of staining buffer, then centrifuge at 300 × *g* for 5 min. Resuspend the cell pellet in 0.4–0.5 mL of staining buffer.
6. Intracellular staining buffers (Fix and Perm). Prepare a fresh Foxp3 Fixation/Permeabilization working solution by mixing one part of Foxp3 Fixation/Permeabilization Concentrate with three parts of Foxp3 Fixation/Permeabilization Diluent. One milliliter of the working solution is required for each sample if staining in tubes.
    a. Prepare a 1× working solution of Permeabilization (Perm) Buffer by mixing one part of 10× Permeabilization Buffer with nine parts of distilled water. 8.5 mL of the working solution is required for each sample if staining in tubes. (Original protocol from ThermoFisher).
7. Intracellular staining (nuclear staining). After fixation with 1× Fixation buffer for 45 min, wash the sample by adding 1 mL of perm buffer, then centrifuge at 300 × *g* for 5 min. Prepare a staining cocktail of intracellular antibodies in 1× Perm buffer. Incubate for 45 min. Add 1 mL of staining buffer, then centrifuge at 300 × *g* for 5 min. Resuspend the cell pellet in 0.4–0.5 mL of staining buffer.
8. Optional Fixation (for delayed acquisition). If sample acquisition will occur after 24 h, proceed with fixation: Add 0.5 mL

of fixation buffer (e.g., paraformaldehyde-based) (see Note 2). Incubate at RT for 15–20 min in the dark. Centrifuge at 500 × g for 5 min, then discard the fixation buffer supernatant.

Caution: Paraformaldehyde is toxic. Handle with care and dispose of according to safety regulations (see Note 3).

9. Final Resuspension and Storage. Resuspend the fixed cells in 0.3–0.5 mL of staining buffer (see Note 4). Keep the samples cold and protected from light until flow cytometric acquisition. It is recommended to acquire samples as soon as possible, preferably within 72 h of staining.

3.2 Data Collection on Spectral Flow Cytometer

1. Turn on cytometer, wait 30 min before running QC.
2. Open software by double-clicking the SpectroFlo icon (Cytek Biosciences). Log in.
3. Once logged in, on the landing page, select QC & Setup and perform a daily QC. Use Spectral Flo beads (300 μL 1× PBS + 1 drop of beads). Load tube on instrument. Start.
4. After QC is complete, go to the Acquisition module in the top right corner. Create a new experiment and assign a name. Define Controls (Reference Group). Select "Reference Group" to define your controls.
5. For each control:
6. Choose Control Type: *Cells* or *Beads*
7. Select the Unstained Control (must match sample type)
8. Assign each Single-Stained Control to the appropriate tag. Click Save.
9. Load and run each individual reference control (*see* **Note 5**).
10. Unmix your data (*see* **Note 6**). Visualize your reference (single-color) controls spectra with the unmixing wizard. Live unmix is used after recording all reference controls, followed by unmixing data by selecting it. Adjust gate placements if necessary.
11. Add Sample Groups and Tubes. Create groups for your experimental samples. Add and label sample tubes (e.g., "Sample_1", "Sample_2"). Add keywords to tubes if needed for tracking or gating. Click Next (*see* **Note** 7).
12. (Optional) Check "Autofluorescence Extraction" to reduce background. Click Apply (*see* **Note 8**).
13. SpectroFlo now applies live unmixing in real time to sample tubes.
14. Analyze your unmix data. Create and analysis worksheet and save it as a template to reuse and share.

15. Unmix data can be loaded into FlowJo (BD Biosciences) and further analyze by gating populations appropriately.

3.3 Data Analysis: Gating Strategy Using FlowJo Software

Launch FlowJo and create a new workspace.

1. **Add FCS Files**: Import compensation FCS files and sample FCS files. Organize them into separate group folders within the workspace for clarity and ease of analysis.
2. **Recalculate Compensation (if required)**: If compensation requires adjustment, use the compensation FCS files to recalculate it. Apply the newly calculated compensation matrix to the sample FCS files (*see* **Note 9**).
3. **Open FCS files** Begin data analysis by opening each FCS file and initiating gating steps.
4. **Define Gated Populations**:
   i. Refer to Fig. 1 for visual representation of the gating strategy. The following steps describe each gate in sequence.
5. **Lymphocyte Gate**:
   i. Create a polychromatic plot (dot plot, A.).
   ii. Set FSC-A (Forward Scatter Area) on the X-axis and SSC-A (Side Scatter Area) on the Y-axis.
   iii. Draw a polygonal gate around the lymphocyte population.
   iv. Ensure that the "Gate events inside" option is checked in the active gate checkbox to include only the events within the gate.
6. **Singlet Gate (Doublet Discrimination)**:
   i. Select the lymphocyte gate and right-click to "Drill Down" into that population.
   ii. Create another (bivariate) plot with FSC-A on the X-axis and FSC-H (Forward Scatter Height) on the Y-axis.
   iii. Gate the singlet population, which typically appears along a diagonal line (45° angle) when area scaling is properly set during acquisition.
   iv. Doublets, which deviate from the diagonal (toward FSC-A but not FSC-H), should be gated out.
   v. This step ensures that only single cells are included in further analysis.
7. **T-cell populations**
   i. From the single cell population, create a plot with CD19 on X-axis and CD3 on Y-axis and select a CD3+ CD19− population.

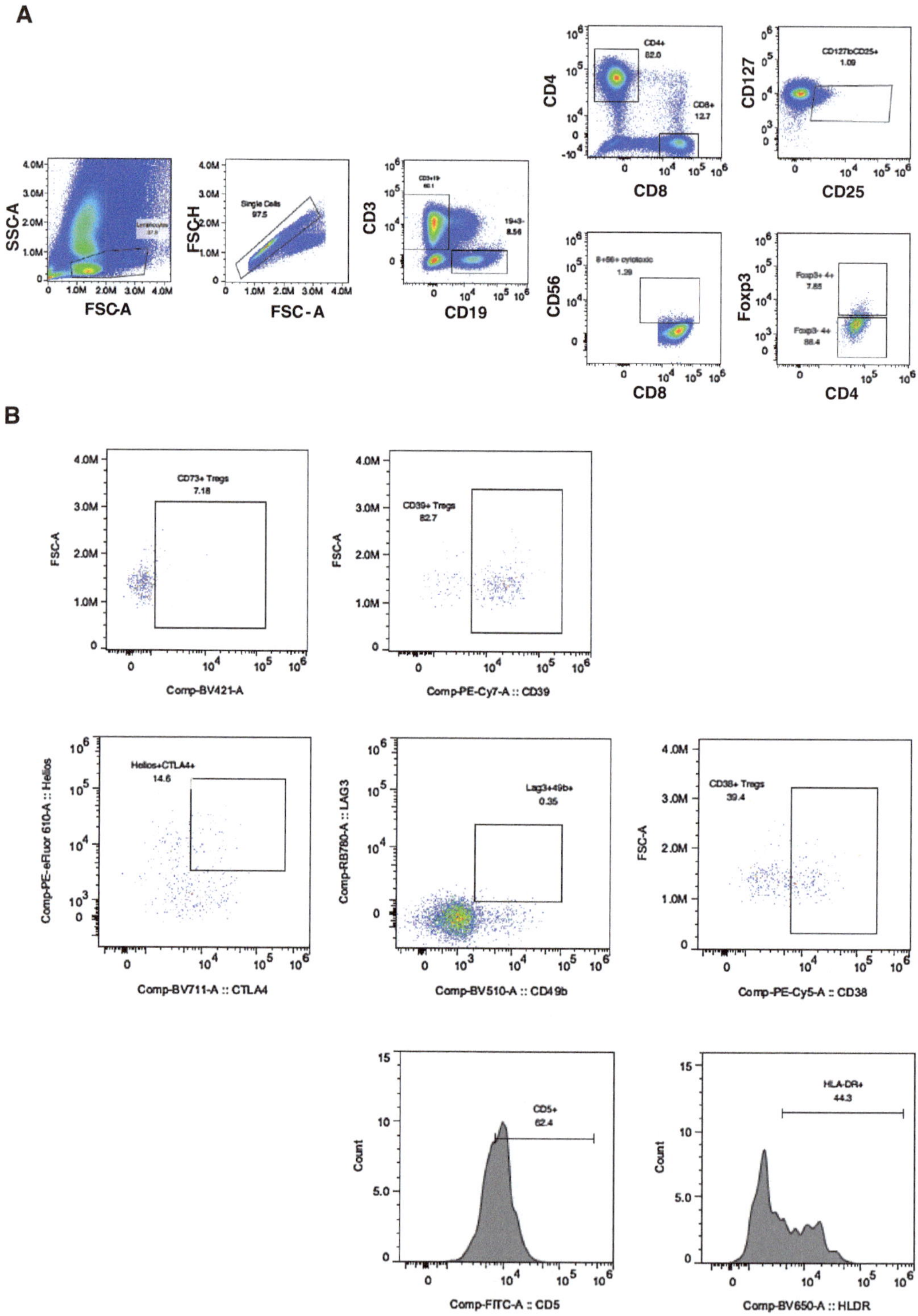

Fig. 1 Gating strategies and frequencies of Tregs cells in the peripheral blood from healthy subjects. The blood was processed, and cells were stained to evaluate the frequencies of Treg cells. Gating strategies and a representative dot plot from one individual are shown. (**A**) FSC/SSC dot plot used to isolate the lymphocytic population from all events, effectively excluding other leukocytic populations for subsequent analysis. (**B**) Gating of various Treg subsets

ii. From gated CD3+CD19− cells, draw a plot to display CD4 and CD8 populations. Draw two gates, on CD4+ CD8− and a CD8+CD4− populations.

iii. Select the CD4+CD8− cells and create a dot plot for Tregs with CD127 on Y axis and CD25 on X axis, analyze as follows: divide this population into two subsets based on the presence or absence of Foxp3: Foxp3+CD4+ and Foxp3-CD4+ T-cell subsets.

iv. From the CD8+CD4− T cells make a new plot on CD56 vs CD8 and gate on CD56+CD8+ cytotoxic cells.

v. From the Treg population, make gates to look at the expression of other critical marker of Treg cells. (B.) FSC-A vs CD73; FSC-A vs CD39; Helios vs CTLA-4; Lag-3 vs CD49b; FSC-A vs CD38, and the expression of CD5 and HLA-DR.

4 Notes

1. If large cell clumps are observed, pass the suspension through a 40–100 μm mesh strainer to obtain a single-cell suspension.
2. If samples cannot be acquired within 24 h and fixation is required, resuspend cells by vortexing before fixation.
3. Caution! Paraformaldehyde is toxic. Handle with care and dispose of according to safety regulations.
4. Resuspend cells in 0.3–0.5 mL of staining buffer depending on cell concentration.
5. Always use bright and clean single stains to ensure accurate unmixing.
6. Unmixed Data = each fluorochrome's contribution is separated using spectral reference controls.
7. Fixed samples can clump again; filter again if necessary.
8. Use Autofluorescence extraction if samples have high background (e.g., tissues, tumors).
9. Calculated compensation obtained from DIVA software can be modified in FlowJo by creating or modifying a compensation matrix.

References

1. Benoist C, Mathis D (2012) Treg cells, life history, and diversity. Cold Spring Harb Perspect Biol 4(9):a007021. https://doi.org/10.1101/cshperspect.a007021

2. Lawson JM, Tremble J, Dayan C, Beyan H, Leslie RD, Peakman M, Tree TI (2008) Increased resistance to CD4+CD25hi regulatory T cell-mediated suppression in patients with type 1 diabetes.

Clin Exp Immunol 154(3):353–9. https://doi.org/10.1111/j.1365-2249.2008.03810.x
3. Schneider A, Rieck M, Sanda S, Pihoker C, Greenbaum C, Buckner JH (2008) The effector T cells of diabetic subjects are resistant to regulation via CD4+ FOXP3+ regulatory T cells. J Immunol 181(10):7350–5. https://doi.org/10.4049/jimmunol.181.10.7350
4. Yu N, Li X, Song W, Li D, Yu D, Zeng X, Li M, Leng X, Li X (2012) CD4(+)CD25 (+)CD127 (low/-) T cells: a more specific Treg population in human peripheral blood. Inflammation 35(6):1773–80. https://doi.org/10.1007/s10753-012-9496-8
5. Elkord E, Abd Al Samid M, Chaudhary B (2015) Helios, and not FoxP3, is the marker of activated Tregs expressing GARP/LAP. Oncotarget 6(24):20026–36. https://doi.org/10.18632/oncotarget.4771
6. Sebastian M, Lopez-Ocasio M, Metidji A, Rieder SA, Shevach EM, Thornton AM (2016) Helios controls a limited subset of regulatory T cell functions. J Immunol 196(1):144–55. https://doi.org/10.4049/jimmunol.1501704
7. Ahn M, Dostal J, Hegde P, Lee DJ (2025) CTLA-4 expression on Tregs is needed for suppression of autoimmune uveitis. Sci Rep 15(1):17745. https://doi.org/10.1038/s41598-025-02816-z
8. Moncrieffe H, Nistala K, Kamhieh Y, Evans J, Eddaoudi A, Eaton S, Wedderburn LR (2010) High expression of the ectonucleotidase CD39 on T cells from the inflamed site identifies two distinct populations, one regulatory and one memory T cell population. J Immunol 185(1):134–43. https://doi.org/10.4049/jimmunol.0803474
9. Da M, Chen L, Enk A, Ring S, Mahnke K (2022) The multifaceted actions of CD73 during development and suppressive actions of regulatory T cells. Front Immunol 13:914799. https://doi.org/10.3389/fimmu.2022.914799
10. Abd Al Samid M, Chaudhary B, Khaled YS, Ammori BJ, Elkord E (2016) Combining FoxP3 and Helios with GARP/LAP markers can identify expanded Treg subsets in cancer patients. Oncotarget 7(12):14083–94. https://doi.org/10.18632/oncotarget.7334
11. Gertel S, Polachek A, Elkayam O, Furer V (2022) Lymphocyte activation gene-3 (LAG-3) regulatory T cells: an evolving biomarker for treatment response in autoimmune diseases. Autoimmun Rev 21(6):103085
12. Maecker HT, McCoy JP, Nussenblatt R (2012) Standardizing immunophenotyping for the human immunology project. Nat Rev Immunol 12(3):191–200. https://doi.org/10.1038/nri3158
13. Santegoets SJ, Dijkgraaf EM, Battaglia A, Beckhove P, Britten CM, Gallimore A, Godkin A, Gouttefangeas C, de Gruijl TD, Koenen HJ, Scheffold A, Shevach EM, Staats J, Taskén K, Whiteside TL, Kroep JR, Welters MJ, van der Burg SH (2015) Monitoring regulatory T cells in clinical samples: consensus on an essential marker set and gating strategy for regulatory T cell analysis by flow cytometry. Cancer Immunol Immunother 64(10):1271–86. https://doi.org/10.1007/s00262-015-1729-x
14. Biancotto A, Dagur PK, Fuchs JC, Langweiler M, McCoy Jr JP (2012) OMIP-004: in-depth characterization of human T regulatory cells. Cytometry A 81(1):15–6. https://doi.org/10.1002/cyto.a.21158
15. Nowatzky J, Stagnar C, Manches O (2019) OMIP-053: identification, classification, and isolation of major FoxP3 expressing human $CD4^+$ Treg subsets. Cytometry A. 95(3):264–267. https://doi.org/10.1002/cyto.a.23704
16. Vicente R, Quentin J, Mausset-Bonnefont AL, Chuchana P, Martire D, Cren M, Jorgensen C, Louis-Plence P (2016) Nonclassical CD4+ CD49b+ regulatory T cells as a better alternative to conventional CD4+CD25+ T cells to dampen arthritis severity. J Immunol 196(1):298–309. https://doi.org/10.4049/jimmunol.1501069
17. Hu D, Weiner HL, Ritz J (2013) Identification of cytolytic CD161- CD56+ regulatory CD8 T cells in human peripheral blood. PLoS One 8(3):e59545. https://doi.org/10.1371/journal.pone.0059545
18. Mitchell J, Kelly J, Kvedaraite E, von Bahr Greenwood T, Henter JI, Pellicci DG, Berzins SP, Kannourakis G (2020) Foxp3+ Tregs from Langerhans cell histiocytosis lesions co-express CD56 and have a definitively regulatory capacity. Clin Immunol 215:108418
19. Gootjes C, Zwaginga JJ, Roep BO, Nikolic T (2024) Defining human regulatory T cells beyond FOXP3: the need to combine phenotype with function. Cells 13(11):941. https://doi.org/10.3390/cells13110941
20. Feng X, Zhang L, Acharya C, An G, Wen K, Qiu L, Munshi NC, Tai YT, Anderson KC (2017) Targeting CD38 suppresses induction and function of T regulatory cells to mitigate immunosuppression in multiple myeloma. Clin Cancer Res 23(15):4290–4300. https://doi.org/10.1158/1078-0432.CCR-16-3192

Check for updates

Chapter 6

Immunophenotyping of Human B Lymphocytes from Peripheral Blood Mononuclear Cells

Maria Lopez-Ocasio, Wan-Chi Lin, Steven C. Hockman, Keyvan Keyvanfar and Pradeep K. Dagur

Abstract

Flow cytometry has made it possible to identify human B cell subsets based on surface and intracellular antigens, a process known as immunophenotyping. CD19 and CD20 are commonly expressed in all B cells; the presence or absence of CD27 separates memory B cells from naïve cells, while IgD and IgM are used to distinguish naïve from transitional B cells. Markers such as CD38, CD27, and CD24 are helpful to identify plasmablasts and immature B cells. CD21 points to mature B cells, CD10 is on immature B cells, CD138 marks plasma cells, and CD5 is high in regulatory B cells. Memory B cells have less CD23 than naïve B cells. B cells are multifaceted. They present antigens (APCs), secrete antibodies, and keep the immune system in check. Identifying B-cell subsets in clinical settings—especially for diagnosing immune disorders, autoimmune diseases, and blood cancers like leukemias and lymphomas. Knowing which B cells are present in a certain disease setting helps clinicians understand how far a disease has progressed and how active it is. With this information, a treatment that fits the patient's needs can be selected, ones that target B cells directly, like anti-CD20 antibodies. With flow cytometry as a tool, we can spot B-cell malignancies like chronic lymphocytic leukemia (CLL), follicular lymphoma, mantle cell lymphoma, and multiple myeloma. Flow cytometry is a great tool for both clinicians and researchers. It is fast, reliable, and sensitive. In our protocol, we used a 19-fluorochrome panel to successfully sort out different B cell subsets from human blood. Advances like these not only help classify the disease better, but they also make it possible to personalize treatment decisions in ways that matter for the patients.

Keywords Multi-color flow cytometry, CD19, CD20, Memory B cells, Plasmablasts, B-cell malignancies

1 Introduction

B cells are a critical part of the adaptive immune system, required at various stages such as antigen presentation, immunomodulation, and antibody secretion. Identifying subsets of these cells throughout their developmental stages and/or diseases provides a useful tool toward an understanding of both, normal immune function and pathological conditions, including autoimmune and immunodeficiency disorders

Angélique Biancotto and Pradeep K. Dagur (eds.), *Immunophenotyping*, Methods in Molecular Biology, vol. 3087, https://doi.org/10.1007/978-1-0716-5607-5_6,

and hematologic malignancies. Flow cytometry is a necessary tool for B-cell immunophenotyping, offering high-throughput and multiparametric analysis with minimal sample requirements.

Identifying human B-cell subsets depends on the expression of distinct surface and intracellular antigens that reflect their developmental as well as their functional stages. Pan-B-cell antigens such as CD19 and CD20 enable the identification of the B-cell compartment. Further identification of functional subsets is done using markers like CD27 to differentiate memory from naïve B cells; IgD and IgM to describe naïve and transitional populations; and CD138, CD38, CD24, or CD10 to identify plasmablasts and immature B cells [1]. Other markers, including CD21, CD5, and CD23, further separate the B-cell population into subsets of maturation, regulation, or activation state [2–4].

Studies have shown the importance of these B-cell markers in health and disease. As an example, B regs or regulatory B cells with the $CD24^{hi}$ $CD38^{hi}$ phenotype have been identified to play a role in the autoimmune pathogenesis and therapy response in rheumatoid arthritis patients [5]. On the other hand, marginal zone-like B cells with a high expression of CD24 help protect against atherosclerosis formation [6]. On the opposite side, abnormal expression of CD5 or CD23 has been linked to change B cell receptor signaling in B-cell malignancies such as chronic lymphocytic leukemia and mantle cell lymphoma [4, 7].

New technologies such as spectral cytometry has made it possible to design flow cytometry panels with 20 or more markers to dissect B-cell diversity [8, 9]. These panels have become key tools in both clinical and research settings. They can be use them for immunomonitoring, tracking biomarkers, and shaping precision therapies. With breakthrough B-cell-targeted treatments, like antiCD20 monoclonal antibodies (e.g., rituximab), now more than ever, it is important to have accurate immunophenotyping methods, ones that help measuring how well therapies are working.

In this chapter, we present a 19-color flow cytometry panel optimized for comprehensive profiling of human B-cell subsets from peripheral blood (Fig. 1). This method allows the simultaneous evaluation of critical phenotypic markers, facilitating robust identification of both conventional and functionally distinct B-cell populations.

2 Materials

2.1 Isolation of Lymphocytes

1. Fresh sample: 10 mL of human peripheral blood (collected in anti-coagulant, e.g., EDTA or heparin) from healthy donors.
2. Refrigerated centrifuge, Allegra X14-R (Beckman Coulture, with adaptors for 5 mL (5 mm round-bottom test tubes).

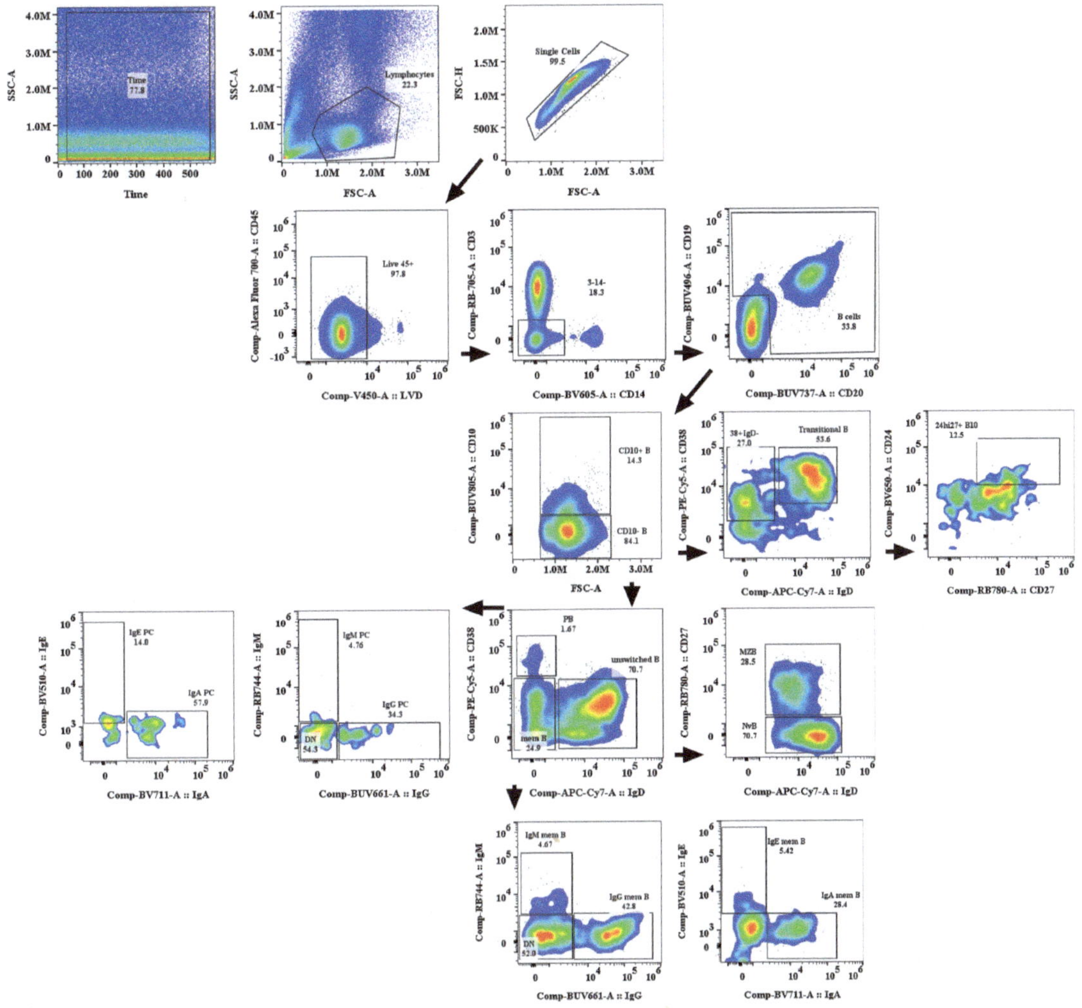

Fig. 1 Gating strategies and frequencies of B cells in the peripheral blood from healthy subjects. The blood was processed, and cells were stained to evaluate the frequencies of B cells. Gating strategies and a representative dot plot from one individual are shown. A time plot was used to remove noise, follow by a FSC/SSC dot plot used to isolate the lymphocytic population from all events, effectively excluding other leukocytic populations for subsequent analysis

3. ACK Lysing solution (Quality Biologicals).
4. Falcon Round-bottom polystyrene 5 mL FACS test tubes (BD Falcon, Fischer Scientific).
5. Falcon Round-bottom polystyrene 5 mL FACS test tubes with cell strainer (BD Falcon, Fischer Scientific).

2.2 Staining of PBMCs

1. Wash buffer: 1× phosphate buffered saline (PBS) (Quality Biologicals) + 1% BSA.

2. Fluorochrome-conjugated human antibodies for B cells (see Table 1).
3. Staining buffer: 1× PBS with 1% BSA and 10% FBS, filtered.
4. Disposable pipettes.
5. Refrigerated centrifuge, Allegra X14-R (Beckman Coulture, with adaptors for 5 mL (5 mm round-bottom test tubes).
6. OneComp eBeads, Compensation Beads (ThermoFisher Scientific). Use as suggested in the manufacturer's instructions.
7. Fixation buffer: 1% paraformaldehyde in 1× PBS.

Table 1
Antibody panel: antigen-specific mAb used for cellular antigens

Specificity	Fluorochrome	Ab clone	Vendor	Volume
CD3	RB705	UCHT1	BD Biosciences	5 μl
CD20	BUV 737	2H7	BD Biosciences	5 μL
CD19	BUV496	SJ25C1	BD Biosciences	5μL
Fixable Viability Stain 450	V450		BD Biosciences	100 μL[a] Working Conc
CD10	BUV805	Hl10a	BD Biosciences	5 μL
CD14	BV605	M5E2	BD Biosciences	5 μL
CD24	BV650	ML5	BD Biosciences	10 μL
CD23	PE	EBVC5-5	Biolegend	5 μL
CD27	BV421	AD2	BD Biosciences	5 μL
CD45	AF700	HI30	Biolegend	5 μL
CD138	AF647	HIL-7R-M21	BD Biosciences	5 μL
CD38	PE-Cy5	HIT-2	BD Biosciences	5 μL
IgM	RB780	T47-530	BD Biosciences	5 μL
IgD	APC-Cy7	IA6-2	BD Biosciences	5 μL
IgG	BV510	AK-7	BD Biosciences	5 μL
HLA-DR	BV650	G46-6	BD Biosciences	5 μL
IgE	BV510	AK-7	BD Biosciences	5 μL
IgA	BUV805	M1/69	BD Biosciences	5 μL
CD5	FITC	UCHT2	BD Biosciences	10 μL

Fluorochrome nomenclature: BV brilliant violet antibodies (BV510, BV605, BV650, BV711, BV786), *APC-Cy7* allophycocyanin-cyanine H7 tandem, *AF700* Alexa Fluor 700, *BUV* brilliant ultraviolet (BUV496, BUV661, BUV737, BUV805), *PE* phycoerythrin, *PE-Cy5* phycoerythrin-cyanine 5 tandem, *RB* real blue (RB613, RB705, RB744, RB780), *FITC* fluorescein isothiocyanate [a]FVS450 Fixable Viability Stain 450 was prediluted 1 μL of stock in 1 mL of FACS buffer

8. Single-color controls consisting of unstained PBMCs and single-color using beads are used for compensation in each experiment.
9. Samples analyzed on a Five Lasers Cytek Aurora Spectral analyzer (Cytek) and acquisition software (SpectroFlo). Acquire at least 10,000 (or as many as possible) events in the CD4+ CD25+ gate.
10. After acquisition, we export the files and perform the analyses later with the FlowJo 10.1 software. (Tree Star/BD Biosciences).
11. SpectroFlo beads (Cytek). For QC performance on the instrument.

3 Methods

3.1 Blood Collection and Sample Preparation

1. Peripheral blood was collected in either heparin or EDTA-coated tubes. Store the tubes at room temperature (RT) and place on a nutator to prevent clotting. Using a disposable pipette, transfer 10 mL of whole blood into a 50 mL Falcon tube. Centrifuge at 2000 × *g* for 10 min at RT to separate plasma from cells of interest (e.g., lymphocytes).
2. Carefully remove the plasma layer without disturbing the cell layer. Remove the cell layer and place it in a clean 50 mL Falcon tube. Add 25 mL of RBC lysis buffer to the tube. Mix gently on a nutator mixer for 2 min. Centrifuge at 300 × *g* for 5 min at 4°C. Remove the supernatant and gently disrupt the cell pellet by slow vortexing or tapping. Repeat the RBC lysis step: Add another 25 mL of RBC lysis buffer. Mix on nutator for 1 min. Centrifuge at 300 × *g* for 5 min at 4°C. Remove lysis buffer and gently break the pellet again.
3. Add 10 mL of staining buffer to wash the cell pellet. Centrifuge at 300 × *g* for 5 min. Aspirate the supernatant completely. Add fresh ice-cold staining buffer to adjust the cell concentration to $1–2 \times 10^6$ cells/mL (*see* **Note 1**).
4. Transfer the cell suspension into appropriate 5 mL Falcon tubes. Add the working amount of fluorochrome-conjugated human antibodies, live/dead staining and Fc receptor blocker to reduce nonspecific binding. Incubate the cells at 4°C for 25 min in the dark.
5. Add 2 mL of staining buffer, then centrifuge at 300 × *g* for 5 min. Resuspend the cell pellet in 0.4–0.5 mL of staining buffer.

6. Optional Fixation (for delayed acquisition): If sample acquisition will occur after 24 h, proceed with fixation: Add 0.5 mL of fixation buffer (e.g., paraformaldehyde-based) (see Note 2). Incubate at RT for 15–20 min in the dark. Centrifuge at 500 × *g* for 5 min, then discard the fixation buffer supernatant (*see* **Note 3**).
7. Resuspend the fixed cells in 0.3–0.5 mL of staining buffer (*see* **Note 4**). Keep the samples cold and protected from light until flow cytometric acquisition. It is recommended to acquire samples as soon as possible, preferably within 72 h of staining.

3.2 Data Collection on Spectral Flow Cytometer

1. Turn on cytometer and wait 30 min before running QC.
2. Open software by double-clicking the SpectroFlo icon (Cytek Biosciences). Log in.
3. Once logged in, on the landing page, select QC & Setup and perform a daily QC. Use Spectral Flo beads (300 μL 1× PBS + 1 drop of beads). Load tube on instrument. Start.
4. After QC is complete, go to the Acquisition module in the top right corner. Create a new experiment and assign a name. Define Controls (Reference Group). Select "Reference Group" to define your controls.
5. For each control:

    i. Choose Control Type: *Cells* or *Beads*
    ii. Select the Unstained Control (must match sample type)
    iii. Assign each Single-Stained Control to the appropriate tag. Click Save.

6. Load and run each individual reference control (*see* **Note 5**).
7. Unmix your data (*see* **Note 6**). Visualize your reference (single-color) controls spectra with the unmixing wizard. Live unmix is used after recording all reference controls, followed by unmixing data by selecting it. Adjust gate placements if necessary.
8. Add Sample Groups and Tubes. Create groups for your experimental samples. Add and label sample tubes (e.g., "Sample_1", "Sample_2"). Add keywords to tubes if needed for tracking or gating. Click Next (*see* **Note 7**).
9. (Optional) Check "Autofluorescence Extraction" to reduce background. Click Apply (*see* **Note 8**).
10. SpectroFlo now applies live unmixing in real time to sample tubes.
11. Analyze your unmix data. Create an analysis worksheet and save it as a template to reuse and share.

12. Unmix data can be loaded into FlowJo (BD Biosciences) and further analyzed by gating populations appropriately.

3.3 Data Analysis: Gating Strategy

1. Launch FlowJo and create a new workspace.
2. **Add FCS Files**: Import compensation FCS files and sample FCS files. Organize them into separate group folders within the workspace for clarity and ease of analysis.
3. **Recalculate Compensation (if required)**: If compensation requires adjustment, use the compensation FCS files to recalculate it. Apply the newly calculated compensation matrix to the sample FCS files (*see* **Note 9**).
4. **Open FCS files** Begin data analysis by opening each FCS file and initiating gating steps.
5. **Define Gated Populations**:
   i. Refer to Fig. 1 for a visual representation of the gating strategy. The following steps describe each gate in sequence.
6. **Lymphocyte Gate**:
   i. Create a polychromatic plot (dot plot).
   ii. Set FSC-A (Forward Scatter Area) on the X-axis and SSC-A (Side Scatter Area) on the Y-axis.
   iii. Draw a polygonal gate around the lymphocyte population.
   iv. Ensure that the "Gate events inside" option is checked in the active gate checkbox to include only the events within the gate.
7. **Singlet Gate (Doublet Discrimination)**:
   i. Select the lymphocyte gate and right-click to "Drill Down" into that population.
   ii. Create another (bivariate) plot with FSC-A on the X-axis and FSC-H (Forward Scatter Height) on the Y-axis.
   iii. Gate the singlet population, which typically appears along a diagonal line (45° angle) when area scaling is properly set during acquisition.
   iv. Doublets, which deviate from the diagonal (toward FSC-A but not FSC-H), should be gated out.
   v. This step ensures that only single cells are included in further analysis.
8. **B-cell populations**
   i. Draw a gate on Live CD45+ cells, excluding "dead" or compromised non-CD45 cells.

ii. From the Live CD45+ population (pop), create another plot with CD14 on X-axis and CD3 on Y-axis, selecting a CD3− CD14− negative population.

iii. On gated CD3−CD14− cells, draw another plot to display CD19 and CD20 populations. Select all CD19+ CD20+, and exclude double negatives on the plot.

iv. Select CD10 on the Y axis and FSC-A on X axis for double-positive B cells and analyze as follows: divide this population into two subsets based on the presence or absence of CD10 (CD10+ or CD10−) B-cell subsets.

v. From the CD10+ B cells make a new plot on IgD and CD38 and make a Transitional (T) B gate on CD38+ and IgD+ population and a gate on CD38+ IgD- population.

vi. Make another plot from the CD38+ IgD− pop displaying CD27 on X-axis and CD24 on Y-axis; make a gate of B10 cells by selecting the CD24hi and CD27+ pop.

vii. From the CD10− B-cell populations, make a gate with CD38 on the Y-axis and IgD on the X-axis. Three pops will be gated: CD38+ IgD− (plasma cells or PCs); CD38− IgD+ (unswitched B cells); and CD38− IgD− (memory B cells). From PCs and memory B cell gates, IgM vs IgG, and, IgA vs IgE populations are display, respectively.

4 Notes

1. If large cell clumps are observed, pass the suspension through a 40–100 μm mesh strainer to obtain a single-cell suspension.
2. If samples cannot be acquired within 24 h and fixation is required, resuspend cells by vortexing before fixating.
3. Caution! Paraformaldehyde is toxic. Handle with care and dispose of according to safety regulations.
4. Resuspend cells in 0.3–0.5 mL of staining buffer depending on cell concentration.
5. Always use bright and clean single stains to ensure accurate unmixing.
6. Unmixed Data = each fluorochrome's contribution is separated using spectral reference controls.
7. Fixed samples can clump again; filter again if necessary.
8. Use autofluorescence extraction if samples have high background (e.g., tissues, tumors).
9. Calculated compensation obtained from DIVA software/ SpectroFlo can be modified in FlowJo by creating or modifying a compensation matrix.

References

1. Veneri D, Ortolani R, Franchini M, Tridente G, Pizzolo G, Vella A (2009) Expression of CD27 and CD23 on peripheral blood B lymphocytes in humans of different ages. Blood Transfus 7(1):29–34. https://doi.org/10.2450/2008.0007-08
2. Carsetti R, Terreri S, Conti MG, Fernandez Salinas A, Corrente F, Capponi C, Albano C, Piano Mortari E (2022) Comprehensive phenotyping of human peripheral blood B lymphocytes in healthy conditions. Cytometry A 101(2):131–139. https://doi.org/10.1002/cyto.a.24507
3. Kumánovics A, Sadighi Akha AA (2022) Flow cytometry for B-cell subset analysis in immunodeficiencies. J Immunol Methods 509:113327. https://doi.org/10.1016/j.jim.2022.113327
4. Taher TE, Bystrom J, Mignen O, Pers JO, Renaudineau Y, Mageed RA (2020) CD5 and B lymphocyte responses: multifaceted effects through multitudes of pathways and channels. Cell Mol Immunol 17(11):1201–1203. https://doi.org/10.1038/s41423-020-0490-z
5. Chen L, Xu J, Sun B et al (2025) The vital role of $CD19^{+}CD24^{hi}CD38^{hi}$ regulatory B cells in rheumatoid arthritis: correlations with disease activity and treatment response. Adv Rheumatol 65:43. https://doi.org/10.1186/s42358-025-00474-3
6. Pattarabanjird T, Nguyen AT, McSkimming C et al (2023) Human circulating $CD24^{hi}$ marginal zone B cells produce IgM targeting atherogenic antigens and confer protection from vascular disease. Nat Cardiovasc Res 2:1003–1014. https://doi.org/10.1038/s44161-023-00356-1
7. Liu C, Richard K, Wiggins M et al (2016) CD23 can negatively regulate B-cell receptor signaling. Sci Rep 6: 25629. https://doi.org/10.1038/srep25629
8. Liechti T, Günthard HF, Trkola A (2018) OMIP-047: high-dimensional phenotypic characterization of B cells. Cytometry A 93(6): 592–596. https://doi.org/10.1002/cyto.a.23488
9. Liechti T, Roederer M (2019) OMIP-051 – 28-color flow cytometry panel to characterize B cells and myeloid cells. Cytometry 95:150–155. https://doi.org/10.1002/cyto.a.23689

Check for updates

Chapter 7

Immunophenotyping Studies in Nonhuman Primate Models

Taha B. Hayal and Chuanfeng Wu

Abstract

Nonhuman primate (NHP) models are important in biomedical research due to their close evolutionary relationship with humans, which provides a high degree of anatomical, genetic, and physiological similarity. This close resemblance is particularly valuable for studying complex biological processes and diseases that are difficult to model in other animals. NHPs have hematopoietic and immune systems that closely mirror those of humans, making them crucial for researching hematopoietic stem cell biology, immunology, infectious diseases, and vaccine development, offering insights that are more likely to translate to human health. With the rapid development of immunotherapy, the NHP model serves as an invaluable tool for preclinical studies of immunotherapy. Immunophenotyping, the process of characterizing and quantifying hematopoietic cell populations and their functional states, is critical in studies related to the hematopoietic and immune system. While the NHP model is the most relevant animal model for human biology, the immunophenotyping methods used may not always align with those employed for human samples. This chapter provides a comprehensive overview of key considerations for conducting immunophenotyping studies using NHP models, particularly in the rhesus macaque model. It covers methodological aspects, including sample preparation, the selection and validation of appropriate antibody clones, the application of flow cytometry techniques, and the data analysis to ensure high-quality and reproducible results. By providing a comprehensive overview of these considerations, this chapter aims to equip researchers with the knowledge needed to effectively utilize rhesus macaque NHP models in immunophenotyping studies, ultimately enhancing the translational impact of their findings on human health.

Keywords Nonhuman primate, Immunophenotyping, Monoclonal antibodies, Cell surface markers, Intracellular markers, Peripheral blood mononuclear cells (PBMCs), Hematopoietic lineage cells, Cross-reactivity

1 Introduction

Nonhuman primates (NHPs) serve as essential models in immunological research due to their close physiological and immunological resemblance to humans [1]. The NHP model enables the exploration of complex biological processes, immune dynamics, disease mechanisms, and the evaluation of therapeutic interventions in a system that closely mirrors human biology [2–5]. As NHP models bridge the gap between preclinical studies and clinical applications,

Angélique Biancotto and Pradeep K. Dagur (eds.), *Immunophenotyping*, Methods in Molecular Biology, vol. 3087,
https://doi.org/10.1007/978-1-0716-5607-5_7,

the need for effective methodologies to assess immune cell populations becomes very important. Immunophenotyping is an essential technique identifying immune cell populations and characterizing their functional states [6]. This process utilizes specific cell markers to identify and quantify various cell types, providing valuable insights into their functions and interactions within the immune system. While immunophenotyping is well-established in human and rodent models, its application in NHPs offers unique advantages and challenges [7]. To maximize the utility of NHP models, researchers must navigate several species-specific considerations, including the selection and validation of appropriate antibodies and markers, the optimization of flow cytometry protocols, and the interpretation of data within the context of NHP immunology compared to human immunology. This chapter aims to provide a comprehensive overview of the key considerations and best practices for immunophenotyping studies in NHP models, specifically in the rhesus macaque model. We will explore methodological aspects such as sample preparation, marker selection and panel design, antibody clone selection, cell staining protocol, data collection and data analysis, while addressing the unique challenges associated with working with NHPs. By addressing these factors, this chapter seeks to guide researchers in leveraging NHP models to advance our understanding of immune system function and to develop innovative therapeutic approaches.

2 Materials

For a comprehensive immunophenotyping study in NHP models, the preparation and use of various buffers, media, solutions, supplies, and the use of some special equipment are crucial. Below is a detailed list of these components:

2.1 Materials Needed for Isolation of Granulocytes and Mononuclear Cells (MNCs) from Peripheral Blood (PB) or Bone Marrow (BM)

1. Blood collection tubes that prevent blood clotting with acid citrate dextrose (ACD), ethylenediaminetetraacetic acid (EDTA), or Heparin (*see* **Note 1**).
2. 15 and 50 ml conical polypropylene centrifuge tubes.
3. 2, 5, 10, 25, and 50 ml sterile serological pipettes.
4. Electric pipette aids (Drummond Scientific, USA).
5. Ficoll density gradient media.
6. Microcentrifuge tubes.
7. Phosphate-buffered saline (PBS).
8. Red blood cells (RBC) Lysing buffer (ACK Lysing Buffer).
9. Cell strainer with a pore size of 100 μm, 70 μm, and 40 μm.
10. Cryopreservation media: Either CryoStor® CS10 Cell Freezing Medium or FBS supplemented with 10% DMSO.

2.2 Materials Needed for Cell Staining

1. FACS buffer: PBS supplemented with 0.5%–2% of fetal bovine serum (FBS) or bovine serum albumin (BSA) to reduce non-specific antibody binding.
2. FACS Buffer plus EDTA: 2 mM EDTA is added to FACS buffer to prevent cell clumping when thawing frozen cells.
3. Fixation buffer: 1% paraformaldehyde (PFA) in PBS to preserve cell morphology and stained markers for a prolonged time.
4. Permeabilization buffers that contain saponin, methanol, or Triton X-100 for enabling the antibodies to stain the intracellular markers (*see* **Note 2**).
5. Fluorochrome-Conjugated Antibodies: Specific antibodies to various surface markers and intracellular antigens (e.g., CD4, CD8, CD19, etc.) (*see* **Note 3**).
6. Live/Dead Staining Dyes: Such as LIVE/Dead™ Fixable Aqua dead cell stain kit, or propidium iodide (PI) or 7-AAD to distinguish viable cells from dead cells.

2.3 Equipment and Supplies Needed for Immunophenotyping

1. Flow Cytometer.
2. Fluorescence-Activated Cell Sorter (FACS): Essential for analyzing and sorting cells based on fluorescence markers.
3. Centrifuge that should have an adjustable deceleration setting is essential for spinning down cells and separating cell fractions.
4. Hemocytometer or automated cell counter to count the cells while evaluating the viability.
5. Pipettes and pipette tips.
6. Freezer for cryopreservation of the cells, such as −80°C or liquid nitrogen.

This list encompasses the core components needed for effective immunophenotyping in NHP models. Specific requirements may vary depending on the experimental protocols and the particular NHP species used.

3 Methods

In this section, we provide detailed, step-by-step protocols along with key considerations for immunophenotyping in the NHP model.

3.1 Isolation of Granulocytes and MNCs from PB or BM

3.1.1 Blood Collection Tube Selection and Handling Consideration

Choosing the appropriate blood collection tube and proper handling protocols are essential for obtaining high-quality blood or bone marrow samples and ensuring reliable, reproducible data. Na-Heparin tubes are known to better preserve the expression of cell surface markers and maintain population distribution over extended periods. EDTA tubes excel in preserving absolute leukocyte counts [8]. ACD tubes prevent platelet activation, thereby maintaining sample integrity during storage and transport. They provide more consistent results when conditions such as transportation time, temperature, or handling procedures are variable, making them preferable for clinical and biobanking applications [9]. Notably, overnight storage of blood in ACD tubes at 4°C may increase the NK-cell fraction in Ficoll–Hypaque-separated PBMCs [10]. Selecting the appropriate blood collection tube based on experimental needs is a critical consideration during experimental planning. In nonhuman primate (NHP) models, these three types of blood collection tubes are commonly used. While freshly collected samples are preferred, storing them overnight at 4°C is acceptable for shipping or temporary storage prior to cell purification from PB or BM samples.

3.1.2 Protocol for Purification of Granulocytes and MNCs from PB or BM

Ficoll separation is a widely used technique for isolating and purifying NHP cells. This method relies on density gradient centrifugation and is particularly effective for separating lymphocytes and granulocytes from whole blood and BM aspirates. Below is the detailed protocol, with the entire procedure illustrated in Fig. 1.

1. Sample Preparation Transfer approximately 10 ml of blood or BM samples into 50 ml centrifuge tubes and add 20 ml of PBS to dilute them.
2. Ficoll Addition: Using the lowest speed on the pipette aid, carefully add 15 ml of Ficoll to the bottom of each tube to ensure the formation of a Ficoll layer.
3. Centrifugation: Centrifuge at 1600 rpm (600 × *g*) for 30 min at room temperature (RT) with high acceleration [9] and low deceleration (0–2) settings.
4. Layer Separation: After centrifugation, the following compartments, including the plasma layer, MNCs layer (buffy coat), Ficoll layer, and granulocyte and erythrocyte layer, should be visible (as illustrated in Fig. 1).
5. MNCs Collection: Collect the MNCs layer into a new 50 ml tube, add PBS to bring the volume to 50 ml, and centrifuge at 1600 rpm at RT for 10 min. Carefully aspirate the supernatant, avoiding disturbance of the MNCs pellet.
6. RBC Lysis: To address red cell contamination, resuspend the MNCs pellet in 1 ml of the remaining supernatant and fill the tube to 50 ml with RBC Lysing buffer ACK lysing buffer.
7. Incubation: Incubate at RT for 15 min with gentle shaking.
8. Centrifugation: Centrifuge at 1600 rpm at RT for 10 min.
9. MNCs Resuspension: Aspirate the supernatant and resuspend the MNC pellet in 20 ml PBS.

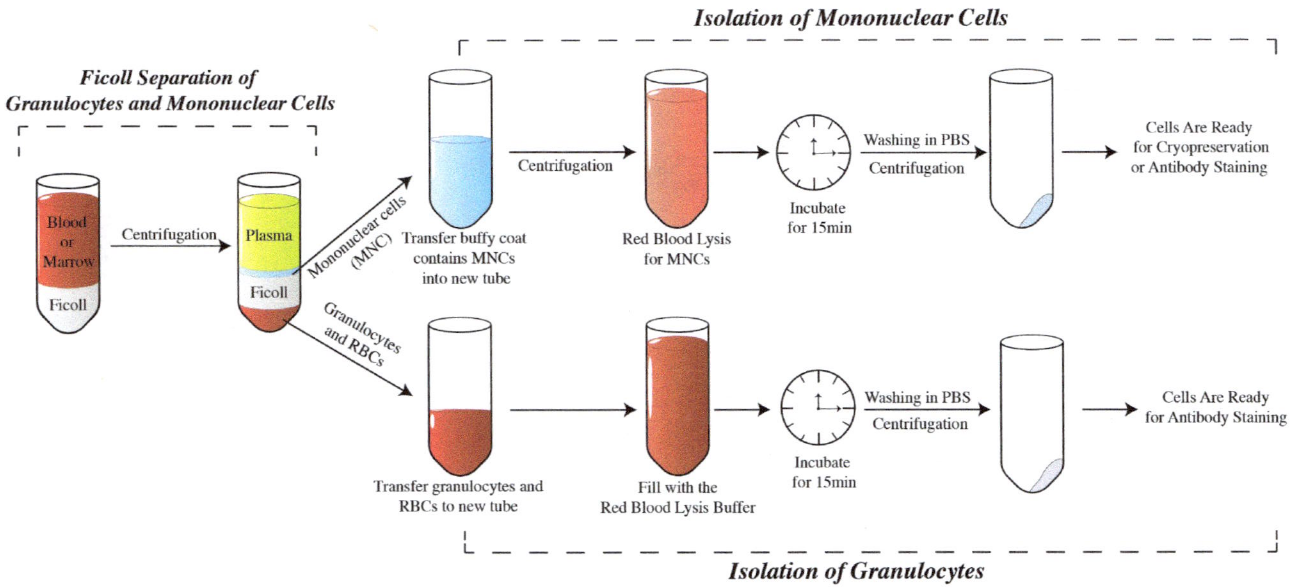

Fig. 1 Diagram illustrating the procedures for the purification of granulocytes and mononuclear cells (MNCs) from peripheral blood (PB) or bone marrow (BM)

10. Cell Filtration: Filter the cells through a cell strainer with a pore size of 70–100 μm.
11. Cell Counting: Count the cells.
12. Final MNCs Centrifugation: Centrifuge at 1600 rpm for 10 min at RT.
13. MNCs Resuspension for Analysis: Resuspend the MNCs in appropriate media for either cryopreservation or flow cytometry analysis.
14. Granulocyte Isolation: To isolate granulocytes, remove all layers above the granulocyte and erythrocyte layer. Add RBC ACK lysing buffer to this compartment to lyse the RBCs (*see* **Note 4**).
15. Incubation: Incubate at RT for 15 min with gentle shaking.
16. Granulocyte Centrifugation: Centrifuge at 1600 rpm at RT for 10 min.
17. Granulocyte Resuspension: Aspirate the supernatant and resuspend the granulocyte pellet in 20 ml PBS.
18. Granulocyte Filtration: Filter the cells through a cell strainer with a pore size of 70–100 μm.
19. Granulocyte Counting: Count the cells.
20. Final Granulocyte Centrifugation: Centrifuge at 1600 rpm for 10 min at RT.
21. Granulocyte Resuspension for Analysis: Resuspend the granulocytes in FACS buffer for flow cytometry analysis (*see* **Note 5**).

3.2 Panel Design Consideration

The panel design for immunophenotyping of hematopoietic lineage cells in PB, BM, or tissue in NHP should be tailored to the study's focus and the available flow cytometer. The most important considerations are to select meaningful markers and specific assays.

3.2.1 Essential Surface Markers for Identifying and Characterizing Major Hematopoietic Lineage Cells in Rhesus Macaque

To effectively identify and characterize the major hematopoietic lineage cells in rhesus macaques from PB, BM, or tissue cells, the following key markers are recommended and widely accepted:

For T cells, the general markers CD3, CD4, and CD8 are essential [11], with subsets including naïve T cells identified as CD95−CD28+, central memory (CM) T cells marked as CD95+CD28+, and effector memory (EM) T cells identified as CD95+CD28− [1, 2]. The chemokine receptor CCR7 and CCR5 can help refine EM and CM subpopulations [12]. T cell receptor (TCR) classification can further identify αβ and γδ T cells as needed [13]. To identify regulatory T cells (Tregs), markers such as CD127, CD25, CD27, and forkhead box protein 3 (FoxP3) are utilized [12, 14]. B cells are marked by CD20 and can be subdivided into naïve B cells (IgD+

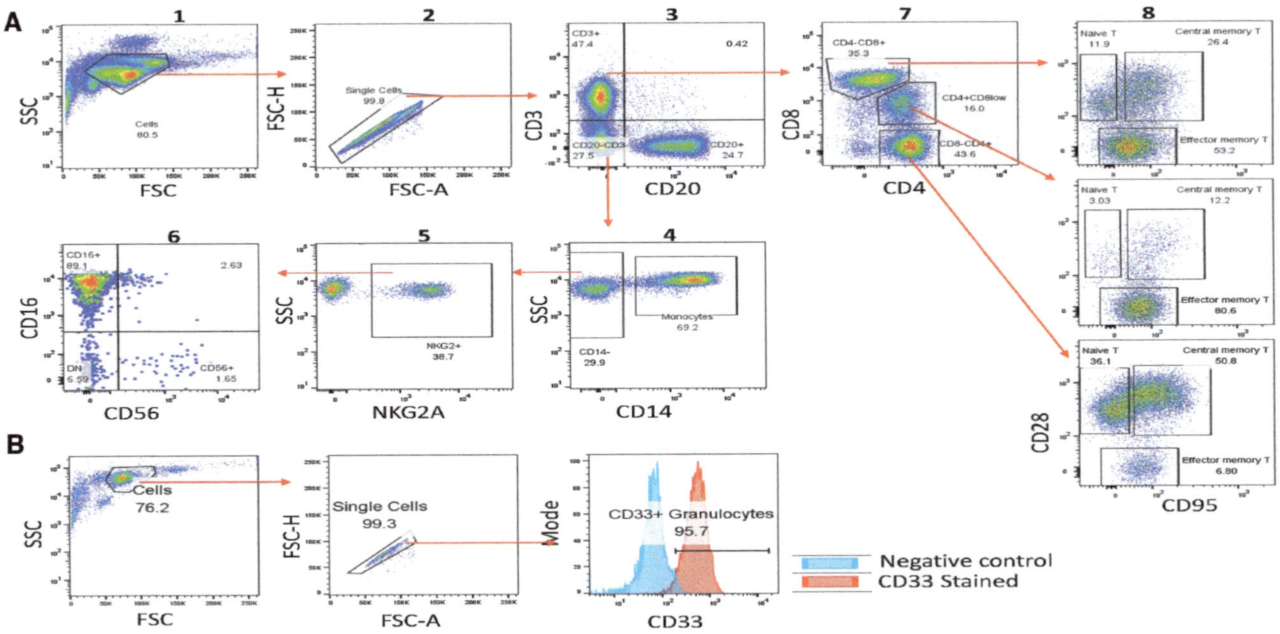

Fig. 2 (**A**) Identification of lineage cells in rhesus macaque PBMCs. Rhesus macaque PBMCs are stained and analyzed by flow cytometry using the following gating strategy: (1) Gate on the cell population in the FSC-SSC plot. (2) Gate on the linear population in the FSC-A vs. FSC-H plot to include the single cells. (3) Identify T cells and B cells as CD3+ T and CD20+ B cells, respectively. (4) Gate on CD14+ monocytes from the CD3-CD20− population. (5) Gate on NKG2+ bulk NK cells from the CD3−CD20−CD14− population. (6) Identify NK subsets based on CD16 and CD56 expression within the bulk NKG2+ NK population. (7) Identify T cell subsets, including CD4+CD8- T, CD4-CD8+ T, and CD4+CD8low T cells, based on CD4 and CD8 expression. (8) Identify naïve T, central memory T, and effector memory T cells based on CD95 and CD28 expression. (**B**) Identification of rhesus granulocytes based on CD33 expression

CD27−), memory B cells (IgD−CD27+), and marginal-zone-like B cells (IgD+CD27+) [13]. Monocytes and macrophages in rhesus macaques play crucial roles in tissue repair, as well as in innate and adaptive immune responses, contributing significantly to tissue regeneration [15, 16], with CD14 serving as a general marker for monocytes and HLA-DR, CD11b, and CD163 for defining macrophages. Additional markers such as CD11c, CD68, CD169, CD204, and CD206 can be used to further characterize tissue macrophage subsets [16]. CD33 is considered a primary marker for identifying granulocytes in rhesus [7, 17]. Rhesus natural killer (NK) cells can be identified using the bulk marker NKG2A (with clone Z199 recognizing NKG2A and NKG2C isoforms) [18–20], and the majority of the NK cells also express NKP80 and CD8α [18, 21]. Major NK subpopulations can be distinguished with CD56 and CD16 [20, 22, 23]. Rhesus dendritic cells (DCs), though a minor population in PBMCs, play a critical role in the innate immune system, with the key markers HLA-DR, CD123, and CD11c used to identify two major DC subsets: HLA-DR+CD123−CD11c+ Myeloid DCs (mDCs) and HLA-DR+CD123+CD11c− Plasmacytoid DCs (pDCs) [13, 24–27]. The detailed antibody staining protocol can be found in Sects. 3.4 and 3.5. The gating strategy to identify major hematopoietic lineage cells in rhesus macaque is shown in Fig. 2. And the antibody information is provided in Table 1.

3.2.2 Functional Immunophenotyping

Functional immunophenotyping by flow cytometry enables researchers to assess cell activation levels, proliferation, survival status, and evaluates functional responses such as degranulation and cytokine production in immune cells. In the rhesus macaque NHP model, in addition to the previously discussed surface markers, we utilize the Ki-67 marker to assess overall cell proliferation [22]. Assays using Annexin V and propidium iodide (PI) can determine cell viability and apoptosis. A key function of immune cells is cytokine production following immunological stimulation. After stimulating cells with antigens, cytokines, or mitogens, intracellular cytokines such as IFN-γ and TNF-α can be detected using specific permeabilization and fixation protocols. We investigated the cytokine production function of rhesus NK cells following overnight treatment with IL-12 (10 ng/ml), IL-15 (10 ng/ml), and IL-18 (100 ng/ml), as well as stimulation with K562 cells; detailed protocols for this can be found in a previous report [22]. Additionally, CD107a levels are commonly assessed to evaluate degranulation in T and NK cells in the rhesus macaque NHP model.

3.3 Antibody Clone Selection Considerations

There are significant challenges in utilizing NHP models when the project requires immunophenotyping, as very few antibodies have been developed against nonhuman primate antigens, and databases detailing cross-reactivity between species are incomplete. As a result, researchers often rely on human antibodies for immunophenotyping

Table 1
Summary of antibodies and clones used for immunophenotyping major hematopoietic lineages in rhesus macaques

Antigen	Cell subset	Vendor	Catalog number	Clone
CD3	T cells	BD Biosciences	563918	SP34-2
CD14	Monocytes	BD Biosciences	558121	M5E2
CD20	B cells	BD Biosciences	335794	L27
CD16	NK cells	Biolegend	302012	3G8
CD56	NK cells	Beckman Coulter	A79388	N901
NKG2A	NK cells	Beckman Coulter	B10246	Z199
CD33	Granulocytes	Miltenyi Biotec	130-113-349	AC104.3E3
CD8	Cytotoxic T cells (CD8+ T cells) and NK cells	BD Biosciences	563795	RPA-T8
CD4	Helper T cells (CD4+ T cells)	BD Biosciences	563875	SK3
CD95	Memory T cells (Fas receptor)	BD Biosciences	556640	DX2
CD28	T cells	BD Biosciences	556622	CD28.2
CCR7	Naïve and central memory T cells	BD Biosciences	561143	150503
CCR5	Memory T cells (effector)	BD Biosciences	550632	3A9
CD127	Memory T cells (IL-7 receptor)	Invitrogen	25-1278-42	eBioRDR5
CD25	Regulatory T cells (Tregs) (IL-2 receptor)	Invitrogen	48-0259-42	BC96
CD27	Memory B cells	Biolegend	356413	M-T271
FoxP3	Regulatory T cells (Tregs)	Invitrogen	35-4776-42	PCH101
IgD	Naïve B cells	Southern Biotech	2030-02	Polyclonal
HLA-DR	Antigen-presenting cells (APCs), Monocytes, B cells, Dendritic cells	Biolegend	307618	L243
CD11b	Monocytes, Macrophages, Granulocytes	Biolegend	101208	M1/70
CD163	Macrophages	Biolegend	333605	GHI/61
CD11c	Dendritic cells, Monocytes	BD Biosciences	570574	3.9
CD68	Macrophages	BD Biosciences	565595	Y1/82A
CD169	Macrophages	Biolegend	346009	7-239
CD206	Macrophages	BD Biosciences	551135	19.2
NKP80	NK cells	Miltenyi Biotec	130-125-849	4A4-D10
CD8	Cytotoxic T cells and NK cells	Biolegend	344734	SK1
CD45	All leukocytes	BD Biosciences	564098	D058-1283
Ki67	Proliferating cells	BD Biosciences	563756	B56

in NHP studies. However, not all the human antibodies are cross-reactive with NHP cells. Researchers have to select appropriate antibody clones that will be able to cross-react with NHP cells. When selecting antibody clones, several factors must be considered to ensure the accuracy and relevance of findings. Differences in antibody staining intensities may result from variations in cross-reactivity with different epitopes on cell populations across different species. Epitope variations between species can arise from differences in gene sequences, protein structures, or post-translational modifications. And even within a species, individual donors can exhibit distinct epitope patterns. Therefore, careful titration of antibody is essential to identify the appropriate antibodies for effective immunoprofiling in NHP research.

As shown in Fig. 3, first, leveraging resources such as the NIH-supported cross-reactivity database (www.nhpreagents.org) is a great source. This database includes information on antibodies and their cross-reactivity with various NHP species, aiding in identifying clones that are likely to perform well in specific immunophenotyping tasks. However, this database does not include all the antibodies you might need to use, or there might be multiple antibody clones suggested without detailed information about their performance. If an appropriate antibody clone is not found in the NHP reagents database or if any details are overlooked, a thorough literature search is necessary. For example, studies like the comprehensive cross-species immune atlas highlight unique expression patterns of certain markers in NHP models compared to humans and mice, such as differences in CD33 and CD16 expression profiles [7]. Third, recommendations

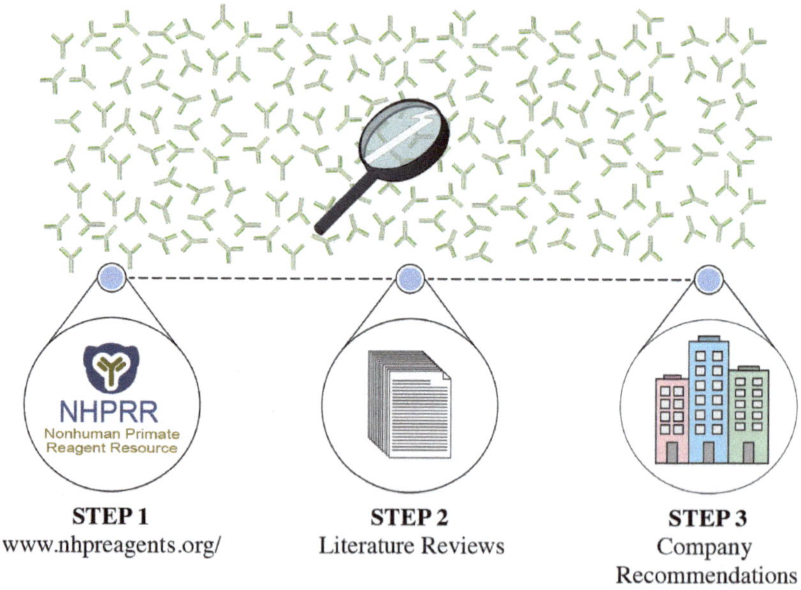

Fig. 3 Integration of comprehensive resources for antibody clone selection

from companies that produce reagents tailored for NHP research, including those validated in rhesus or cynomolgus macaques, can help ensure the selected clones meet the necessary sensitivity and specificity requirements for accurate profiling. To make informed decisions on antibody clone selection, integration of these three aspects—NIH nonhuman primate resources, literature insights, and company recommendations—is important to get reliable data for NHP immunological studies.

3.4 Protocol for Surface Markers Staining

1. Resuspend the desired amount of the cells in 100 μl of FACS buffer (*see* **Note 6**).
2. Add an appropriate amount of antibody targeting specific antigens. For analyzing MNCs and granulocytes, the panel commonly used by our laboratory for rhesus macaques is listed in Table 2 (*see* **Note 7**).
3. Incubate the antibody-cell suspension for 25 min at RT with gentle agitation.
4. Wash the unbound antibodies by adding 3 ml of FACS buffer to the tubes.
5. Centrifuge at 1600 rpm for 5 min at RT.
6. Resuspend the cell pellet in 400 μl of FACS buffer and pass the suspension through the strainer cap of the flow tube.
7. The samples are ready for flow cytometry analysis or can proceed with intracellular staining as needed (*see* **Note 8**).

Table 2
A Typical Lineage Staining Panel (CD33 staining is performed separately on the granulocytes)

Antigen	Conjugation	Cell subset	Clone	Volume per tube
CD3	BV786	T cells	SP34-2	3 μl
CD4	BV650	T cells	SK3	3 μl
CD8	BV711	T cells	SK1	3 μl
CD28	PE	T cells	CD28.2	3 μl
CD95	APC	T cells	DX2	3 μl
CD14	Pacific Blue	Monocytes	M5E2	2 μl
CD20	APC-Cy7	B cells	L27	2 μl
CD16	BV605	NK cells	3G8	2 μl
CD56	PE-Cy5.5	NK cells	N901	2 μl
NKG2A	PE-Cy7	NK cells	Z199	2 μl
CD33	PE	Granulocytes	AC104.3E3	2 μl

3.5 Protocol for Intracellular Markers Staining

To stain intracellular markers, cells must be fixed and permeabilized to allow antibodies to enter. It is important to note that permeabilization can disrupt the epitopes of surface markers; therefore, surface markers should be stained before the permeabilization step. There is a risk of staining unpresented surface markers after permeabilization, so avoid staining any surface markers post-permeabilization to ensure reliable results. Although various approaches exist for the fixation and permeabilization of cells, this section outlines the method for the Foxp3/Transcription Factor Staining Buffer Set (Thermo Fisher), which is optimized for staining nuclear and many secreted proteins. For staining secreted cytokines and chemokines, it is recommended to use the Intracellular Fixation & Permeabilization Buffer Set (Thermo Fisher).

1. After the final wash from surface marker staining, resuspend the cells in 1 ml of Foxp3 Fixation/Permeabilization working solution, prepared by mixing one part Fixation/Permeabilization Concentrate with three parts Fixation/Permeabilization Diluent. Vortex to mix thoroughly.
2. Incubate the samples for 30 min at room temperature while protecting them from light.
3. Add 2 ml of 1× Permeabilization Buffer (dilute the 10× concentrate in distilled water) to each tube.
4. Centrifuge at 1600 rpm for 5 min at room temperature.
5. Discard the supernatant and add another 2 ml of 1× Permeabilization Buffer. Centrifuge again at 1600 rpm for 5 min at room temperature.
6. Resuspend the pellet in 100 μl of 1× Permeabilization Buffer and add 2 μl of normal mouse/rat serum to block the samples.
7. Incubate the samples for 15 min at room temperature (*see* **Note 9**).
8. Directly add the recommended amount of conjugated intracellular marker antibody to the cells without washing.
9. Incubate the samples for 30 min at room temperature in the dark.
10. Add 2 ml of 1× Permeabilization Buffer to each tube and centrifuge at 1600 rpm for 5 min at room temperature.
11. Discard the supernatant and repeat the washing step by adding 2 ml of 1× Permeabilization Buffer to the pellet and centrifuging at 1600 rpm for 5 min.
12. Resuspend the pellet in 400 μl of FACS buffer and pass through the strainer cap of the flow tube.
13. The samples are now ready for flow cytometric analysis (*see* **Note 10**).

3.6 Data Collection on Flow Cytometer

3.6.1 Instrument Setup

Turn on the flow cytometer and allow it to warm up according to the manufacturer's instructions.

Perform daily quality control using calibration beads or control samples to ensure instrument performance.

3.6.2 Configure Settings

Set the appropriate parameters (e.g., voltage, compensation) for each channel based on the fluorophores used.

3.6.3 Single-Color Controls and Compensation Setup

To collect data on flow cytometer, proper single-color controls are needed to setup the protocol and calculate the compensation for panel on the machine.

3.6.4 Run Samples

Load the stained cell samples into the flow cytometer.

Collect data by triggering events based on the forward scatter (FSC) and side scatter (SSC) to distinguish live cells and debris.

3.6.5 Data Collection

Collect at least 10,000–100,000 events for statistical significance, depending on the cell type and experimental design. Monitor the acquisition in real-time, checking for any abnormalities in the data (*see* **Note 11**).

3.6.6 Data Export

Save the acquired data in FCS format for further analysis

3.6.7 Cleaning

Ensure all samples are disposed of properly following biosafety regulations.

3.6.8 Data Analysis

Import the FCS files into data analysis software (e.g., FlowJo). Apply appropriate gating strategies to identify and quantify specific cell populations based on their fluorescent markers. Analyze results and generate figures or reports as needed.

4 Notes

1. The type of tube should be chosen depending on further analysis. For details, please see Sect. 3.1.1.
2. In Sect. 3.5., Foxp3/Transcription Factor Staining Buffer Set (ThermoFisher) was detailed.
3. Antibody clone selection and panel design based on the project are discussed in Sects. 3.2 and 3.3.
4. For granulocyte collection, transfer the granulocyte and RBC compartment into a new 50 ml centrifuge tube for RBC lysis. Do not use the original Ficoll tube, as this could lead to contamination of the granulocytes with MNCs.

5. Cryopreservation of granulocytes is not recommended, as they are difficult to thaw after freezing.
6. $0.5–1 \times 10^6$ cells are usually enough for flow cytometric analysis. However, for rare populations, it is better to start with a larger number of cells.
7. Titration of the antibodies prior to use is the best way to determine the volume needed per sample.
8. Alternatively, the cells can be fixed in 1% paraformaldehyde (PFA) for 30 min at room temperature and stored at 4°C for up to 1 week before analysis.
9. The blocking step is optional but recommended to reduce background. Omit this step if desired.
10. A fluorescence minus one (FMO) control is often needed for intracellular markers.
11. More cell events need to be collected if the target cells belong to a rare cell population.

Acknowledgments

All work was supported by the intramural research programs of the National Heart, Lung, and Blood Institute (NHLBI). The authors would like to thank the NHLBI Nonhuman Primate Program, the NIH Nonhuman Primate Reagent Resource, and the NHLBI Flow Cytometry Core for their invaluable support.

References

1. Yost OC, Downey A, Ramos KS (eds) (2023) Nonhuman primate models in biomedical research: state of the science and future needs. The National Academies Collection: reports funded by National Institutes of Health. Washington (DC)
2. Sibal LR, Samson KJ (2001) Nonhuman primates: a critical role in current disease research. ILAR J 42(2):74–84
3. Lemaitre J, Naninck T, Delache B, Creppy J, Huber P, Holzapfel M et al (2021) Nonhuman primate models of human respiratory infections. Mol Immunol 135:147–164
4. Prescott MJ, Clark C, Dowling WE, Shurtleff AC (2021) Opportunities for refinement of non-human primate vaccine studies. Vaccines (Basel) 9(3):284
5. Estes JD, Wong SW, Brenchley JM (2018) Nonhuman primate models of human viral infections. Nat Rev Immunol 18(6):390–404
6. Preglej T, Brinkmann M, Steiner G, Aletaha D, Goschl L, Bonelli M (2023) Advanced immunophenotyping: a powerful tool for immune profiling, drug screening, and a personalized treatment approach. Front Immunol 14:1096096
7. Bjornson-Hooper ZB, Fragiadakis GK, Spitzer MH, Chen H, Madhireddy D, Hu K et al (2022) A comprehensive atlas of immunological differences between humans, mice, and non-human primates. Front Immunol 13:867015
8. Diks AM, Bonroy C, Teodosio C, Groenland RJ, de Mooij B, de Maertelaere E et al (2019) Impact of blood storage and sample handling on quality of high dimensional flow cytometric data in multicenter clinical research. J Immunol Methods 475:112616
9. Gyorgy B, Paloczi K, Kovacs A, Barabas E, Beko G, Varnai K et al (2014) Improved circulating microparticle analysis in acid-citrate

dextrose (ACD) anticoagulant tube. Thromb Res 133(2):285–292
10. Kim DW, Jang YY, Shin MG, Shin JH, Suh SP, Ryang DW et al (2013) Overnight storage of blood in ACD tubes at 4degreesC increases NK cell fraction in peripheral blood mononuclear cells. Ann Clin Lab Sci 43(3):267–273
11. Macchia I, Gauduin MC, Kaur A, Johnson RP (2006) Expression of CD8alpha identifies a distinct subset of effector memory CD4+ T lymphocytes. Immunology 119(2):232–242
12. Vaccari M, Franchini G (2010) Memory T cells in Rhesus macaques. Adv Exp Med Biol 684:126–144
13. Messaoudi I, Estep R, Robinson B, Wong SW (2011) Nonhuman primate models of human immunology. Antioxid Redox Signal 14(2):261–273
14. Anderson A, Martens CL, Hendrix R, Stempora LL, Miller WP, Hamby K et al (2008) Expanded nonhuman primate tregs exhibit a unique gene expression signature and potently downregulate alloimmune responses. Am J Transplant 8(11):2252–2264
15. Sugimoto C, Merino KM, Hasegawa A, Wang X, Alvarez XA, Wakao H et al (2017) Critical role for monocytes/macrophages in rapid progression to AIDS in pediatric simian immunodeficiency virus-infected rhesus macaques. J Virol 91(17):e00379-17
16. Rahmberg AR, Wu C, Shin T, Hong SG, Pei L, Markowitz TE et al (2024) Ongoing production of tissue-resident macrophages from hematopoietic stem cells in healthy adult macaques. Blood Adv 8(3):523–537
17. Kim MY, Yu KR, Kenderian SS, Ruella M, Chen S, Shin TH et al (2018) Genetic inactivation of CD33 in hematopoietic stem cells to enable CAR T cell immunotherapy for acute myeloid leukemia. Cell 173(6):1439–53 e19
18. Mavilio D, Benjamin J, Kim D, Lombardo G, Daucher M, Kinter A et al (2005) Identification of NKG2A and NKp80 as specific natural killer cell markers in rhesus and pigtailed monkeys. Blood 106(5):1718–1725
19. LaBonte ML, Choi EI, Letvin NL (2004) Molecular determinants regulating the pairing of NKG2 molecules with CD94 for cell surface heterodimer expression. J Immunol 172(11):6902–6912
20. Wu C, Espinoza DA, Koelle SJ, Yang D, Truitt L, Schlums H et al (2018) Clonal expansion and compartmentalized maintenance of rhesus macaque NK cell subsets. Sci Immunol 3(29):eaat9781.
21. Webster RL, Johnson RP (2005) Delineation of multiple subpopulations of natural killer cells in rhesus macaques. Immunology 115(2):206–214
22. Wu C, Liang JA, Brenchley JM, Shin T, Fan X, Mortlock RD et al (2022) Barcode clonal tracking of tissue-resident immune cells in rhesus macaque highlights distinct clonal distribution pattern of tissue NK cells. Front Immunol 13:994498
23. Mortlock RD, Wu C, Potter EL, Abraham DM, Allan DSJ, Hong SG et al (2021) Tissue trafficking kinetics of rhesus macaque natural killer cells measured by serial intravascular staining. Front Immunol 12:772332
24. Chung E, Amrute SB, Abel K, Gupta G, Wang Y, Miller CJ et al (2005) Characterization of virus-responsive plasmacytoid dendritic cells in the rhesus macaque. Clin Diagn Lab Immunol 12(3):426–435
25. Coates PT, Barratt-Boyes SM, Zhang L, Donnenberg VS, O'Connell PJ, Logar AJ et al (2003) Dendritic cell subsets in blood and lymphoid tissue of rhesus monkeys and their mobilization with Flt3 ligand. Blood 102(7):2513–2521
26. Teleshova N, Jones J, Kenney J, Purcell J, Bohm R, Gettie A et al (2004) Short-term Flt3L treatment effectively mobilizes functional macaque dendritic cells. J Leukoc Biol 75(6):1102–1110
27. Brown KN, Barratt-Boyes SM (2009) Surface phenotype and rapid quantification of blood dendritic cell subsets in the rhesus macaque. J Med Primatol 38(4):272–278

Chapter 8

Immunophenotypic Assessment of Natural Killer Cell Degranulation and Monocyte Function from Cryopreserved Peripheral Blood Mononuclear Cells

Abhinav Saurabh, Laurel G. Mendelsohn, Tiffany M. Powell-Wiley and Yvonne Baumer

Abstract

Natural killer (NK) cells and monocytes are central components of the innate immune system, playing critical roles in host defense, immune surveillance, and inflammation. Functional alterations in these cell populations are implicated in cancer, infectious diseases, cardiovascular disease, and responses to environmental and psychosocial stressors. Accurate measurement of NK cell and monocyte distribution and function is therefore essential for both basic and translational immunology. In large-scale or longitudinal studies, cryopreserved peripheral blood mononuclear cells (PBMCs) are commonly used to enable standardized analyses across time and sites; however, cryopreservation and thawing can adversely affect cell viability and function. This chapter presents a reproducible protocol for the cryopreservation, thawing, and functional assessment of NK cells and monocytes from human PBMCs. PBMCs are cryopreserved using a 9:1 ratio of fetal bovine serum (FBS) and dimethyl sulfoxide (DMSO), allowing long-term storage at ultra-low temperatures while preserving cellular integrity. The thawing process utilizes a nuclease to minimize cell clumping, enhancing cell recovery and viability. Following thawing, PBMCs are prepared for downstream functional assays. NK cell function is assessed using a standard degranulation assay following coculture with K562 target cells, providing a robust measure of cytotoxic capacity. Together, this protocol offers a practical guide for the reliable measurement of NK cell and monocyte distribution and function from cryopreserved PBMCs, supporting high-quality immunological analyses in clinical and population-based research.

Keywords Immunophenotyping, NK cells, Degranulation, Flow cytometry, Cryopreserved PBMCs

1 Introduction

The immune system is a complex network within the human body, comprising a diverse population of immune cells that continuously monitor the body for both external and internal threats. However, this delicate balance of the immune system can be disrupted by various factors, including lifestyle, environmental, medical, and biological factors. This protocol focuses on two specialized immune cells that are part of innate immunity—natural killer (NK cells) and monocytes.

Angélique Biancotto and Pradeep K. Dagur (eds.), *Immunophenotyping*, Methods in Molecular Biology, vol. 3087,
https://doi.org/10.1007/978-1-0716-5607-5_8,

Natural killer (NK) cells serve as the body's first line of defense, with the primary role of identifying and neutralizing external threats, such as viruses, and internal threats, including tumor cells [1]. NK cells play a crucial role in cancer [2] and various metabolic disorders [3], including obesity or diabetes [4]. Interestingly, the literature provides conflicting evidence for increased [5] as well as decreased [6] NK cell numbers or proportions in cardiovascular disease (CVD), whereas NK cell function is reported to be decreased in human CVD [7, 8]. It has become increasingly clear that NK cells are present within human atherosclerotic lesions [9, 10], indicating their key yet understudied role. Furthermore, NK cells can kill tumor cells through death ligands and the release of cytotoxic granules [11]. However, exposure to tumor cells and their microenvironment has been shown to impair NK cell effector function [12]. Therefore, NK cells have in recent years gained attention as a potential immunotherapy for cancer treatment [13].

Mature NK cells express CD56 (neural cell adhesion molecule) and CD16 (Fc gamma receptor III), allowing characterization of NK cell subsets with cytolytic and proliferative capacities. The cytolytic NK cells ($CD56^+CD16^+$), which constitute 90%–95% of total circulating NK cells, perform receptor-mediated cellular cytotoxicity by forming an immunological synapse with the target cells, with subsequent release of perforin and granzymes. The $CD56bright^+CD16^-$ NK cell subset, constituting 5%–10% of total circulating NK cells, on the other hand, secretes cytokines such as IFN-γ and TNF-α [1, 14, 15]. Additionally, NK cells secrete other chemokines, including GM-CSF (granulocyte-macrophage colony-stimulating factor), indicating their involvement in recruitment, migration, and the support of other immune cells [16]. Researchers often use a degranulation assay to examine NK cell function by culturing NK cells with a cancer cell line and evaluating degranulation markers (CD107a, perforin, and granzyme) by flow cytometry [4] or a chromium release assay determining NK cell killing capacity [17].

In clinical translational studies with larger patient populations, isolating NK cells and consistently and reliably determining NK cell function from freshly isolated cells pose challenges (e.g., daily assay variations). Therefore, cryopreservation of peripheral blood mononuclear cells (PBMCs) is advantageous for evaluating NK cell function at a later time point, as it minimizes variation in daily laboratory work, decreases blood volume requirements, and is cost-effective when managing a large number of samples [18]. However, compared to T cells, NK cells are more sensitive to freezing and thawing; notably, their viability is the first parameter affected by cryopreservation [19]. Previous studies have demonstrated that the protocol, duration of cryopreservation, and/or type of donor can reduce viability and cytotoxicity [20, 21]. Therefore, it is essential to develop cryopreservation and thawing protocols to assess the fundamental properties of immune cells, enabling researchers to generate high-throughput, reproducible data.

Like NK cells, the present protocol also provides an opportunity to examine monocytes and their function. Essentially, peripheral monocytes do not constitute a homogeneous cell population; instead, three distinct subsets have been characterized based on CD14 and CD16 expression. The three subsets are represented as classical ($CD14^{bright}CD16^{-}$), nonclassical ($CD14^{dim}CD16^{+}$), and intermediate monocytes ($CD14^{bright}CD16^{+}$) [22, 23]. These subsets play varied roles, differing under healthy and pathological states. Classical monocytes have a proinflammatory phenotype, secreting cytokines that facilitate the differentiation of monocyte-derived dendritic cells (DCs) and macrophages, thereby bridging innate and adaptive immunity. They are also capable of phagocytosis and migration through the expression of chemokine receptors. Nonclassical monocytes display a surveillance and anti-viral nature, often patrolling the vasculature. Intermediate monocytes, on the other hand, have the capacity to process and present antigens and are also associated with inflammation, activation, and differentiation [24–27]. Monocytes and macrophages play a crucial role in health and disease, including obesity, cardiovascular diseases, and type 2 diabetes [28–30]. Furthermore, unlike NK cells, cryopreservation of monocytes has been reported to induce apoptosis, leading to low cell recovery after thawing and a decrease in cell numbers over time [31, 32], without affecting their functionality [33]. This has compelled researchers to standardize cryopreservation and postthaw processes while retaining the capacity for differentiation to accelerate workflows [34].

Cryopreservation of cells is an essential technical tool for researchers, providing a means for long- or short-term storage of biospecimens, including cells. Our work contributes to this endeavor by proposing and establishing a protocol to assess the distribution and function of NK cells and monocytes. The proposed protocol not only retains cell viability and function but also provides a 13-color flow cytometry panel to address NK cells and monocytes from cryopreserved PBMCs (Fig. 1).

2 Materials and Methods

Participants were enrolled under IRB-approved clinical trial (NCT01143454). All study participants provided written informed consent and were deidentified before samples were received. Study approval was obtained from the IRB at the National Heart, Lung, and Blood Institute and NIH in accordance with the principles of the Declaration of Helsinki. The guidelines for Good Clinical Practice and the Belmont Report (National Commission for the Protection of Human Subjects of Biomedical and Behavioral Research) were

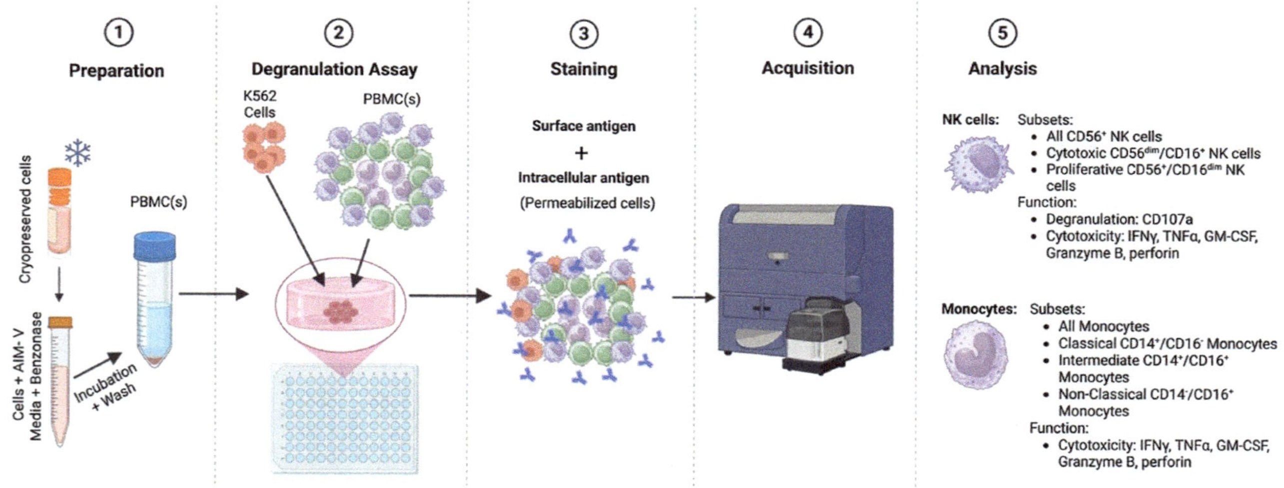

Fig. 1 Graphical summary of the protocol summarized in this book chapter

Table 1
List of reagents utilized in this book chapter, listed with catalogue number and supplier

Reagent	Catalogue number	Company
A. Thawing of cryopreserved cells and isolation of PBMCs		
Heparin CPT blood collection tubes	362753	BD Biosciences
Cryovials	607101	NEST
Fetal bovine serum (FBS)	97068-085	Avantor Seradigm
Dimethyl sulfoxide (DMSO)	470300-92	Ward's Science
AIM-V™ Media	12055-091	Gibco, USA
Benzonase	71205-3	Merck Millipore, USA
15 mL conical tubes	430766	Corning
BD Vacutainer® CPT™ Mononuclear cell preparation tubes	362753	BD Biosciences
Corning cell strainer, 70 μm Nylon	431751	Corning Incorporated
B. Degranulation assay		
MACS buffer	130-091-221	Miltenyi Biotec
Cell activation cocktail	54761 ML	Fisher Scientific
BD GolgiStop™ protein transport inhibitor	554724	BD Biosciences
96-well cell culture plate	3799	Corning Incorporated
BD Cytofix/Cytoperm	554722	BD Biosciences
BD Perm/Wash	554723	BD Biosciences
5 ml polystyrene round-bottom tube	352054	Falcon

followed. Peripheral blood was collected in Heparin CPT tubes (Cat: 362753, BD). All materials, reagents, and antibodies used in this protocol are summarized in Tables 1 and 2.

2.1 Isolation and cryopreservation of PBMCs from Peripheral Blood

- Centrifuge blood-filled CPT tubes at 1200×*g* for 20 min at RT.
- Isolate the PBMC layer and transfer it into a 50 mL conical tube. Wash with 20 mL PBS + 1% FBS under sterile conditions in a laminar airflow hood.
- Count PBMCs using a cell counter or manually by using a trypan blue dye exclusion protocol [35].
- Spin for 5 min at 340 ×*g*, then aspirate the supernatant.
- For cryopreservation, suspend PBMCs in a cryopreservation medium (Notes: FBS and DMSO at a ratio of 9:1, freshly prepared) at a concentration of 1–2 × 10^6/mL.

Table 2
List of fluorochrome-conjugated antibodies utilized in this protocol, listed with clone, catalogue number, and supplier

Fluorochrome	Clone	Catalogue number	Supplier
BUV805 Mouse Anti-Human CD45	HI30	612891	BD Biosciences
Brilliant Violet 605™ anti-human CD3	UCHT1	300460	BioLegend
Hu CD19 BUV395	SJ25C1	563549	BD Biosciences
Brilliant Violet 711™ anti-human CD14	M5E2	301838	BioLegend
Brilliant Violet 650™ anti-human CD16	3G8	302042	BioLegend
Hu CD56 BV786	NCAM16.2	564058	BD Biosciences
PE anti-human CD107a (LAMP-1)	H4A3	328608	BioLegend
FITCanti-human TNF-α	MAb11	502906	BioLegend
APC anti-human IFN-γ	4S.B3	502512	BioLegend
APC/Cyanine7 anti-human/mouse Granzyme B Recombinant	QA16A02	372228	BioLegend
PE/Cyanine7 anti-human Perforin	dG9	308126	BioLegend
PerCP/Cyanine5.5 anti-human GM-CSF	BV D2-21C11	502312	BioLegend
LIVE/DEAD™ Fixable Yellow Dead Cell Stain Kit, for 405 nm excitation	n. a.	L34968	Thermo Fisher Scientific

- Label the cryovial according to your institution's regulations.
- Transfer 1 ml of the suspended PBMCs into a cryovial. Subsequently, place the PBMC-filled cryovial in a Mr Frosty (or similar device).
- Immediately transfer Mr Frosty to a −80°C freezer for 24 h. Afterwards, place the cryopreserved PBMCs into an appropriate liquid nitrogen container for long-term storage.

2.2 Thawing of Cryopreserved PBMCs

- Prepare thawing media: Add benzonase (25 units/mL) to 10 mL of AIM-V media and prewarm at 37°C/5%CO_2 for 15–20 min (*see* **Note 1**) in a 15 mL conical tube.
- Transfer a vial of cryopreserved PBMCs on dry ice to a water bath at 37°C to allow thawing for 30 s.
- Place the cryovial in the laminar air hood and spray with 70% ethanol to ensure sterile working conditions. Before opening the vial, let the ethanol dry completely.
- Add 1 mL of thawing media slowly and dropwise into the cryovial, and mix the cells gently by pipetting up and down.

- Transfer the cell suspension into the 15 mL conical tube with 9 mL of thawing media, centrifuge at 340 x g, aspirate supernatant, and resuspend the cell pellet in fresh thawing media.
- Place the cell suspension in the incubator at 37°C and 5% CO_2 for a 1-h resting period.
- After 1 h of resting, filter the cell suspension through a 70 μm filter into a new conical tube and count the PBMCs using a cell counter or manually using a trypan blue dye exclusion protocol [35].
- Centrifuge the cells at 340 × *g* for 5 min.
- Remove supernatant and resuspend cells in thawing media.

2.3 Degranulation Assay

- To evaluate NK cell function in cryopreserved PBMCs, the degranulation assay can be standardized by varying the effector (PBMC) to target (K562 cells) ratio (*see* **Note 2**).
- Seed PBMCs with K562 cells in a 96-well round-bottom plate at a ratio of 10:1 (450,000:50,000), 5:1 (400,000:100,000), and 1:1 (250,000:250,000) [effector:target (E:T)].
- For every patient sample, include a well with a cell activator cocktail, in addition to the PBMC and K562 coculture (Fig. 2). This will help to determine each donor's maximum response capacity.
- Spin plate at 340 × g at RT, discard supernatant, and resuspend each well in 200 μL of thawing media.

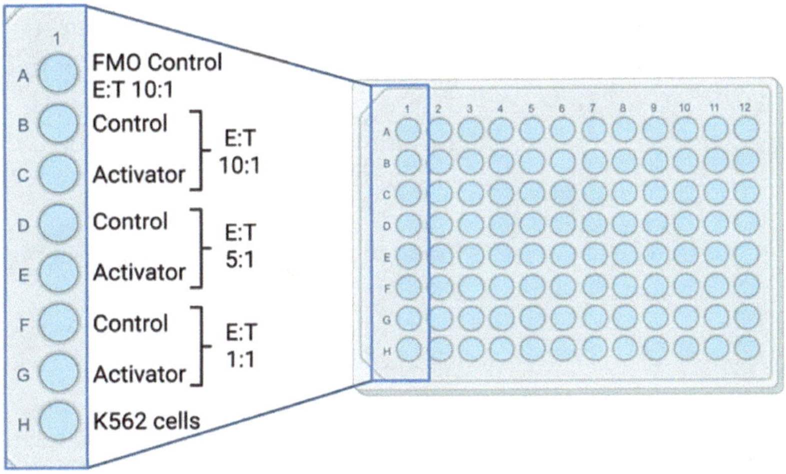

Fig. 2 Graphical display of the plate setup. For each patient sample, 7 wells in a 96-well round-bottom plate are required, plus an additional well containing only K562 cells. The Fluorescence minus one (FMO) well is utilized to determine the gating scheme for CD107a and all intracellular markers in the analysis portion of the experiment. With this plate layout, three different effector:target (E:T) cell ratios are tested under control and activating conditions

- Add the cell activator at a 1:500 dilution per the manufacturer's recommendation to the respective wells.
- Incubate the cell culture for a total of 5 h at 37°C/5%CO_2.
- After 1 h of incubation, add the Golgi transport inhibitor at a 1:1,500 dilution per the manufacturer's recommendation (Table 1) to all wells.
- Incubate the coculture for another 4 h at 37°C/5%CO_2.
- After 5 h of incubation, proceed with supernatant harvest and cell staining.

2.4 Supernatant Harvest

- Centrifuge the plate with the cocultures for 5 min at 340 ×*g*.
- Collect supernatants without disturbing the pellet by carefully removing 150 μL to a sterile 96-well cell culture plate utilizing a multichannel pipette. Store the supernatants at −80°C for further use. These supernatants can be used to evaluate secreted soluble markers like IFNγ or TNFα by ELISA (*see* **Note 3**).
- To the staining plate, add 200 μL of staining buffer (Table 1) and spin again for 5 min at 340 ×*g*. Proceed for staining.

2.5 Staining for Subsequent Flow Cytometry Analysis

- After centrifugation, flick the plates to discard supernatants.
- The staining can be carried out in the 96-well plate and is a two-step process of the initial labeling of extracellular markers, followed by labeling of the intracellular markers.
- To all the wells, except the wells marked as "unstained," add 50 μL of a cocktail of optimal doses of fluorochrome-conjugated antibodies (live–dead, CD3, CD45, CD56, CD16, CD14, and CD107a) (Table 2) suspended in staining buffer (Table 1) (*see* **Notes 4 and 5**).
- Resuspend each well and incubate the cells suspended in the antibody cocktail for 20 min at 4°C protected from light (*see* **Notes 6 and 7**).
- After incubation, add 150 μL staining buffer to all wells as a washing step.
- Centrifuge the cells for 5 min at 340 × g, then discard the supernatant. Repeat the past two steps twice for additional washing.
- After the final wash, resuspend the cell pellet in 50 μL per well of Cytofix/Cytoperm solution (fixation and permeabilization buffer) (Table 1) and incubate for 10 min at room temperature in the dark.
- After fixation, add 150 μL permeabilization wash buffer (included in the Cytofix/Cytoperm kit) to all wells for washing, then centrifuge the plate for 5 min at 340 ×*g*. Gently flick the plates to discard supernatant.

- To all the wells, except the well-marked as "unstained," add 50 μL intracellular staining cocktail of optimal doses of fluorochrome-conjugated antibodies (IFN-γ, TNF-α, Perforin, GM-CSF, and Granzyme B) (Table 2) suspended in permeabilization wash buffer.
- Incubate the cells with the intracellular staining antibody cocktail for 20 min at room temperature in the dark.
- Subsequently, wash the cells with 200 μL staining buffer per well.
- Centrifuge the cells for 5 min at 340 ×*g*. Gently flick the plates to discard supernatant and repeat this washing step one more time.
- Finally, suspend the cells in 1% paraformaldehyde buffer (staining buffer with 1% paraformaldehyde) by adding 200 μL to each well. These stained cells should be acquired immediately or within 72 h of staining (stored at 4°C in the dark).

3 Data Acquisition on a Flow Cytometer

- This 13-color panel was designed for acquisition in the BD Symphony flow cytometer. However, any commercially available flow cytometer can be used for designing and acquiring a multicolor flow panel.
- Acquisition is established by using appropriate forward-scatter and side-scatter to weed out cellular aggregates and debris.
- Before acquiring experimental data-setup and compensation: Single-color tubes (positive control) for each fluorochrome are prepared using compensation beads (OneComp, Invitrogen beads), in addition to a negative control using unstained cells. (*see* **Notes 8 and 9**). These control tubes are then used to make a compensation matrix for a 13-color flow cytometry panel. The BD Symphony flow cytometer utilizes the BD Diva automated software to generate a compensation matrix. However, before using the software, ensure that each fluorochrome is brighter in its respective PMTs (photo multiplier tubes).
- All control tubes are acquired using BD FACSDIVA v9.3 acquisition and analysis software.
- Connect the BD Symphony flow cytometer by opening BD FACSDIVA software. A proper connection is ensured when a green light appears at the bottom right corner of the DIVA software window.
- Click on the browser window to open a new experiment and add the number of tubes to be acquired.

- After the workstation is connected to the cytometer and an experiment is open, a gray pointer icon appears next to tubes in the browser. Click the icon next to the tube to set it as the current tube pointer and to activate acquisition control. As the icon turns green, the tube becomes the active tube in the acquisition dashboard.
- Click on the tab view to select: Browser, Inspector, Cytometer, Acquisition dashboard, and Worksheet panels.
- On the "Cytometer" window, click on the "Parameters" tab, and select all the parameters that you need (for the 13-color-flow panel). "Add" or "delete" button can be used to modify the panel, if needed.
- The experiment layout can be used to create labels and acquisition criteria for each tube.
- Next, click on experiment → compensation setup → create compensation controls. This can also be done by right-clicking on the experiment (*see* **Note 10**).
- It is recommended to "include a separate unstained control tube/well." Check this option if using an unstained sample (negative control).
- Start recording at least 5000 events for each compensation control, ensuring each positive control (single color) produces a bright signal in its respective PMT.
- At the end of the acquisition, click on Experiment → Compensation setup → Calculate compensation.
- Name the compensation matrix, then click Link and Save, followed by "OK." Using "Apply Only," will apply the compensation values to the cytometer settings of the experiment.
- Click on "Global Worksheet," and then using the plot/menu button, allows for selection of either dot, contour, and/or histogram plot to create respective plots on the sheet to create a hierarchical gating plan.
- Please ensure that a live-dead exclusion and a doublet-discrimination (FSC-A vs FSC-H, in this study) gate are included in the hierarchy.
- Each "Specimen" should be renamed as per the experimental setup.
- Click on "Tube 1" and "acquire" the sample tubes. Adjust the FSC & SSC gate and flow rate to 3000–5000 events per second. Record the data.
- At the end of acquisition, right-click on experiments and select Export fcs files to an appropriate folder to save the files, ensuring both compensation and sample files are transferred.

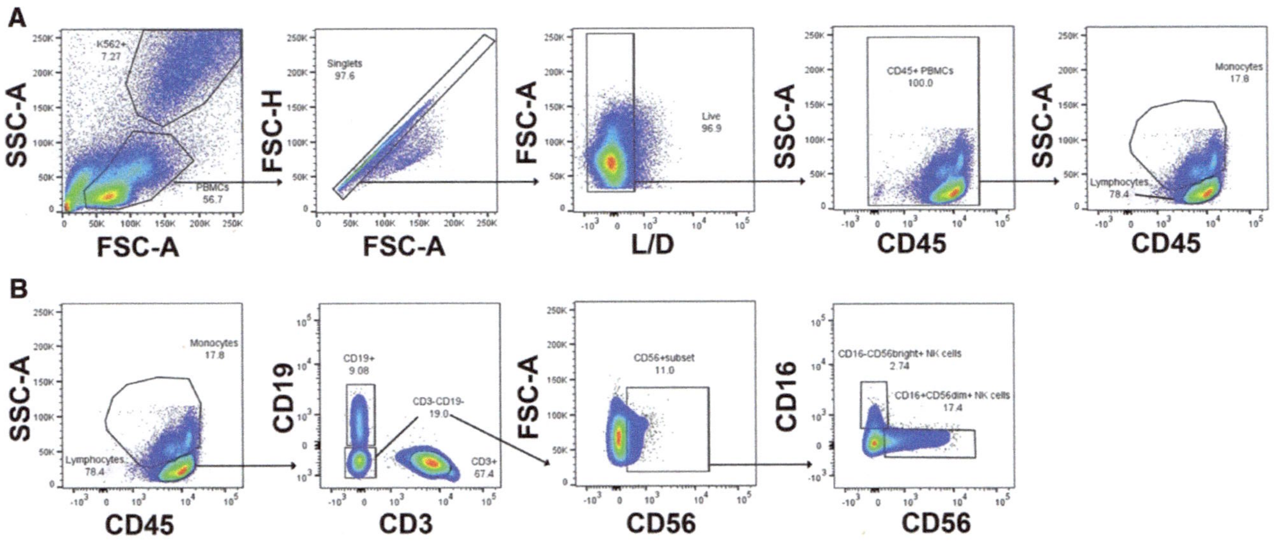

Fig. 3 Gating scheme of the initial steps of the analysis. (**A**) Displays the gating scheme from the initial coculture by focusing on the PBMC population, followed by exclusion of duplicates and subsequent selection of live cells only. Afterwards, all leukocytes will be selected by focusing on $CD45^+$ cells. Within the $CD45^+$ leukocytes, lymphocytes are found in the $CD45^+$ and SSC-A low population, while monocytes are characterized by CD45+ expression and higher SSC-A. As these samples are derived from cryopreserved PBMCs, no granulocyte population is found. (**B**) The gating scheme to determine the $CD3^-/CD19^-/CD56^+$ NK cell population is shown. By focusing on CD45+ and SSC-A low lymphocytes and further exclusion of $CD19^+$ B cells and $CD3^+$ T cells, further analysis of the $CD19^-/CD3^-$ population will generate a $CD56^{dim\text{-}high}$ cell population (NK cells). In a last step, for the general NK cell gating scheme, NK cell subsets can be identified by CD56/CD16 expression

4 Data Analysis in FlowJo Software

- Open the FlowJo software and drag the FCS files of both samples and compensation controls into two separate folders of a new workspace. The compensation controls can be used to create a new compensation matrix on the FlowJo software for further analysis of the samples, if needed [36].
- Double-click on the FMO control FCS file to open a pseudocolor flow cytometry bivariate plot (FSC-A vs SSC-A), create a polygonal gate, as shown in Fig. 3A, to demarcate the population containing lymphocytes and monocytes, and name it.
- Double-click on this population to open another pseudocolor flow cytometry bivariate plot with FSC-A (X-axis) vs FSC-H (Y-axis) to demarcate single cells (singlets). Doublets are usually located toward FSC-A. Create a rectangular gate to include singlets, as shown in Fig. 3A.
- Double-click on this "Single Cell" population to open Live-Dead (LD)-A vs. FSC-A plot. Create a rectangular gate to include only live cells in the analysis.
- Double-click on the CD45 subset-A vs. SSC-A population to open CD45-A vs SSC-A plot and create polygonal plots to demarcate the lymphocyte and monocyte populations as shown in Fig. 3A.
- For analysis of either NK cells or monocytes, double-click on the specific subsets.
- First, double-click on the lymphocyte population to open the CD3-A vs CD19-A bivariate plot. Using polygonal gates, demarcate $CD19^+$ (B cells), $CD3^+$ (T cells), and $CD3^-CD19^-$ populations.
- Within the $CD3^-CD19^-$ subset, double-click to open the CD56-subset-A vs FSC-A bivariate plot. Now, create a rectangular gate to demarcate the CD56-subset as shown in Fig. 3B.
- Double-click on the CD56-subset and open CD107a-A vs FSC-A on an FMO control. Use this gate to copy on the $CD107a^+$ stained sample (control and cell-activator-stimulated cells). This will provide the % frequency of $CD107a^+$ expressing CD56 subset.
- Repeat the above steps on the FMO-controls to create gates for other degranulation markers (Granzyme-B & Perforin) in addition to cytokine-producing CD56 subsets (IFN-γ, TNF-α, and GM-CSF) (Fig. 4A and B).
- Similarly, monocytes and their subsets can be gated out. Double-click on the monocyte population of the CD45-A vs SSC-A plot to gate out the $CD14^+$ vs $CD16^+$ pseudocolor plot (Fig. 5).

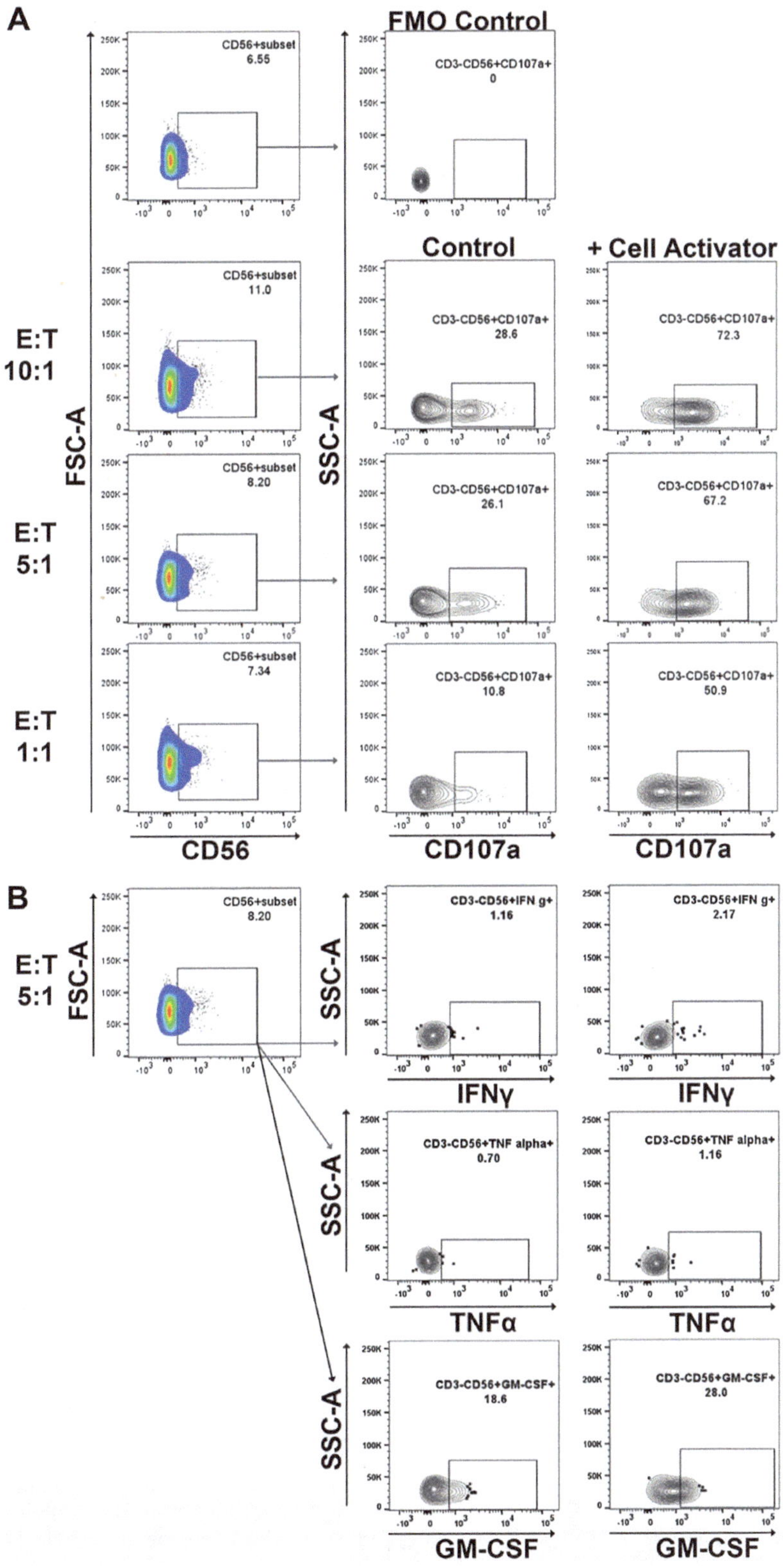
A
FMO Control
Control
+ Cell Activator
E:T 10:1
E:T 5:1
E:T 1:1
FSC-A
SSC-A
CD56+subset 6.55
CD56+subset 11.0
CD56+subset 8.20
CD56+subset 7.34
CD3-CD56+CD107a+ 0
CD3-CD56+CD107a+ 28.6
CD3-CD56+CD107a+ 72.3
CD3-CD56+CD107a+ 26.1
CD3-CD56+CD107a+ 67.2
CD3-CD56+CD107a+ 10.8
CD3-CD56+CD107a+ 50.9
CD56
CD107a
B
E:T 5:1
CD56+subset 8.20
CD3-CD56+IFN g+ 1.16
CD3-CD56+IFN g+ 2.17
IFNγ
CD3-CD56+TNF alpha+ 0.70
CD3-CD56+TNF alpha+ 1.16
TNFα
CD3-CD56+GM-CSF+ 18.6
CD3-CD56+GM-CSF+ 28.0
GM-CSF

- Double-click on the $CD14^+$ vs $CD16^+$ pseudocolor plot to open up a clean $CD14^+$ vs $CD16^+$ plot. On this population, use polygonal gates to demarcate nonclassical ($CD14^-CD16^+$), intermediate ($CD14^+CD16^+$), and classical ($CD14^+CD16^-$) monocytes.
- Double-click on this $CD14^-CD16^+$ and open TNF-α A vs SSC-A. Use this gate to copy on the TNF-α^+ stained sample (control and cell-activator stimulated cells). This will provide the % frequency of TNF-α^+ producing nonclassical monocytes.
- Repeat the above steps on the FMO-controls to create gates for other markers like IFN-γ.

5 Discussion

Cryopreserving NK cells in freezing media with 10%–20% DMSO has been shown to decrease their viability and function (11). The freeze–thaw process appears to have a lesser impact when NK cells are expanded in IL-2-containing medium. Additionally, about 50% of the initial population has been shown to survive when grown in media containing IL-2. However, concerns remain regarding the loss of cytotoxicity and degranulation markers during the long-term cryopreservation of NK cells [37, 38]. Additionally, we performed a degranulation assay immediately after thawing the cryopreserved cells and found that exposure to K562 cells increased CD107a expression (degranulation) in all samples, and this effect was further exacerbated by the addition of the activator cocktail. This indicates that our method of cryopreservation, thawing, and coculture is highly robust and consistently applicable for evaluating immune function in cryopreserved cells (*see* **Note 11**).

In our protocol, we have observed viability rates of 62%–91% for long-term cryopreserved cells (>4 years). Similar viability results were reported by Lapteva et al., showing that NK cell viability post-thaw was 91% after short-term storage but 51%–95% after long-term storage (12 months) [20]. The added advantage of our technique is that we could employ a 13-color flow cytometry panel in our assay, containing markers for monocytes (CD14, CD16), T cells (CD3), and NK cells (CD56). In addition to assessing NK cell function using

Fig. 4 Gating scheme of NK cell analysis. (**A**) Displays an example of the gating scheme when testing various Effector (E): Target (T) ratios. With decreasing effector cells, fewer $CD56^+$ NK cells can be observed, which are displaying at a 1:1 ratio very low degranulation potential (CD107a+ cells) under baseline (control) conditions. The addition of cell activator during treatment accelerated degranulation potential for all E:T ratios. (**B**) The gating scheme to determine the cytokine and chemokine expression of $CD3^-$/$CD19^-$/$CD56^+$ NK cell population at a 5:1 E:T ratio

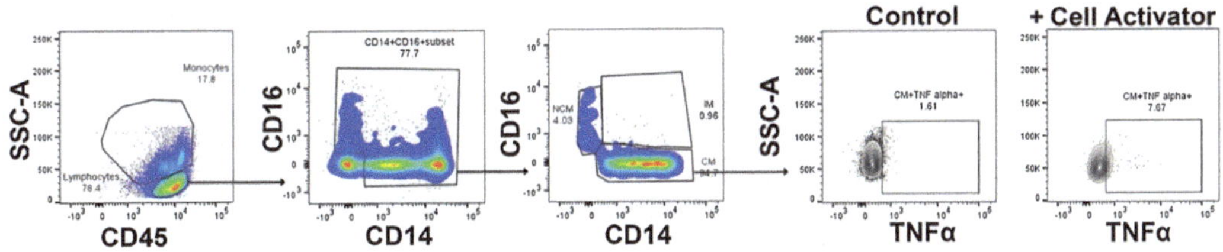

Fig. 5 Gating scheme for monocyte analysis. An example is shown for downstream monocyte analysis. After gating the CD45+ monocyte population, monocytes can be identified by their expression of CD14 and CD16. Cells that are CD14 and/or CD16 positive are gated as the "all monocyte" subset. A further subdivision into three monocyte subsets can be made within the "all monocyte" gate. CD14+/CD16− monocytes are classical monocytes (CM), while CD14+/CD16+ monocytes are intermediate monocytes (IM) and CD14-/CD16+ monocytes are nonclassical monocytes (NCM). Subsequently, each subset can be analyzed for intracellular cytokine expression, as shown here for TNFα in control or cell-activator-treated PBMCs

the degranulation assay, we could also evaluate monocyte responses to cell activator stimulation. This panel could be expanded to include T cell evaluation as well. In the present protocol, we successfully retained viable and functional cells after thawing and isolated viable cells from cryopreserved PBMCs obtained from 8 mL of peripheral blood from study participants. Therefore, this protocol allows evaluation of immune cell function after PBMC cryopreservation from a small volume of study participant blood.

6 Notes

1. It is recommended to use a nuclease, such as benzonase, to degrade DNA from dead cells, preventing cell clumping.
2. K562 is a target cancer line that can be used to check the degranulation capacity of NK cells. It is essential to keep sufficient K562 cells in culture for the proposed target number in this assay.
3. The supernatants can be utilized to detect secretion of cytokines like IFN-γ or TNF-α by ELISA.
4. It is essential to titrate fluorochrome-conjugated antibodies for optimal dose before staining to avoid false positive data.
5. We recommend setting up a fluorescence minus one (FMO) control for better data analysis of CD107a and all intracellular markers.
6. We recommend making a total volume of 50 μL staining cocktail that includes an optimal dose of fluorochrome-conjugated antibodies, supplemented by staining buffer.

7. To avoid photobleaching and antibody capping, it is recommended to stain at 4°C in the dark.
8. If using a live–dead marker in the flow panel, then a single stain control needs to be added to the compensation panel utilizing cells as compensation beads will not be labeled by live–dead fixable dyes (Table 2).
9. Avoid washing after adding compensation beads to prevent loss of beads.
10. Compensation controls can only be acquired on a normal worksheet, not on a global worksheet.
11. Contrary to previous findings, we have observed better viability and function when cryopreserving cells in FBS with 10% DMSO.

References

1. Abel AM et al (2018) Natural killer cells: development, maturation, and clinical utilization. Front Immunol 9:1869
2. Wolf NK, Kissiov DU, Raulet DH (2023) Roles of natural killer cells in immunity to cancer, and applications to immunotherapy. Nat Rev Immunol 23(2):90–105
3. Li Y et al (2021) Natural killer cells: friend or foe in metabolic diseases?. Front Immunol 12:614429
4. Baumer Y et al (2024) Obesity modulates NK cell activity via LDL and DUSP1 signaling for populations with adverse social determinants. JCI Insight 10(2). https://pubmed.ncbi.nlm.nih.gov/39718832/
5. Clerc G, Rouz PM (1997) Lymphocyte subsets in severe atherosclerosis before revascularization. Ann Intern Med 126(12):1004–5
6. Hak L et al (2007) NK cell compartment in patients with coronary heart disease. Immun Ageing 4:3
7. Kumric M et al (2020) The role of natural killer (NK) cells in acute coronary syndrome: a comprehensive review. Biomolecules 10(11). https://pmc.ncbi.nlm.nih.gov/articles/PMC7694449/
8. Jonasson L, Backteman K, Ernerudh J (2005) Loss of natural killer cell activity in patients with coronary artery disease. Atherosclerosis 183(2):316–21
9. Barcia Duran JG et al (2024) Immune checkpoint landscape of human atherosclerosis and influence of cardiometabolic factors. Nat Cardiovasc Res 3(12):1482–1502
10. Horstmann H et al (2024) Cross-species single-cell RNA sequencing reveals divergent phenotypes and activation states of adaptive immunity in human carotid and experimental murine atherosclerosis. Cardiovasc Res 120(14):1713–1726
11. Ramirez-Labrada A et al (2022) All about (NK cell-mediated) death in two acts and an unexpected encore: initiation, execution and activation of adaptive immunity. Front Immunol 13:896228
12. Portale F, Di Mitri D (2023) NK cells in cancer: mechanisms of dysfunction and therapeutic potential. Int J Mol Sci 24(11). https://pubmed.ncbi.nlm.nih.gov/37298470/
13. Page A et al (2024) Development of NK cell-based cancer immunotherapies through receptor engineering. Cell Mol Immunol 21(4):315–331
14. Huntington ND, Vosshenrich CA, Di Santo JP (2007) Developmental pathways that generate natural-killer-cell diversity in mice and humans. Nat Rev Immunol 7(9):703–14
15. Shi FD et al (2011) Organ-specific features of natural killer cells. Nat Rev Immunol 11(10):658–71
16. Paul S, Lal G (2017) The molecular mechanism of natural killer cells function and its importance in cancer immunotherapy. Front Immunol 8:1124
17. Brunner KT et al (1968) Quantitative assay of the lytic action of immune lymphoid cells on 51-Cr-labelled allogeneic target cells in vitro; inhibition by isoantibody and by drugs. Immunology 14(2):181–96
18. Market M et al (2020) A method of assessment of human natural killer cell phenotype and function in whole blood. Front Immunol 11:963

19. Saultz JN, Otegbeye F (2023) Optimizing the cryopreservation and post-thaw recovery of natural killer cells is critical for the success of off-the-shelf platforms. Front Immunol 14:1304689
20. Lapteva N et al (2012) Large-scale ex vivo expansion and characterization of natural killer cells for clinical applications. Cytotherapy 14(9):1131–43
21. Damodharan SN et al (2020) Analysis of ex vivo expanded and activated clinical-grade human NK cells after cryopreservation. Cytotherapy 22(8):450–457
22. Ziegler-Heitbrock L et al (2010) Nomenclature of monocytes and dendritic cells in blood. Blood 116(16):e74–80
23. Patel AA et al (2017) The fate and lifespan of human monocyte subsets in steady state and systemic inflammation. J Exp Med 214(7):1913–1923
24. Zawada AM et al (2011) SuperSAGE evidence for CD14++CD16+ monocytes as a third monocyte subset. Blood 118(12):e50–61
25. Wong KL et al (2012) The three human monocyte subsets: implications for health and disease. Immunol Res 53(1-3):41–57
26. Gren ST et al (2015) A single-cell gene-expression profile reveals inter-cellular heterogeneity within human monocyte subsets. PLoS One 10(12):e0144351
27. Williams H et al (2023) Monocyte differentiation and heterogeneity: inter-subset and interindividual differences. Int J Mol Sci 24(10). https://pubmed.ncbi.nlm.nih.gov/37240103/
28. Williams H et al (2021) Nature versus number: monocytes in cardiovascular disease. Int J Mol Sci 22(17). https://pubmed.ncbi.nlm.nih.gov/34502027/
29. Devevre EF et al (2015) Profiling of the three circulating monocyte subpopulations in human obesity. J Immunol 194(8):3917–23
30. Hotamisligil GS (2017) Inflammation, metaflammation and immunometabolic disorders. Nature 542(7640):177–185
31. Lund PK et al (2001) Flow cytometric evaluation of apoptosis, necrosis and recovery when culturing monocytes. J Immunol Methods 252(1-2):45–55
32. Pardali E et al (2016) Cryopreservation of primary human monocytes does not negatively affect their functionality or their ability to be labelled with radionuclides: basis for molecular imaging and cell therapy. EJNMMI Res 6(1):77
33. Seager Danciger J et al (2004) Method for large scale isolation, culture and cryopreservation of human monocytes suitable for chemotaxis, cellular adhesion assays, macrophage and dendritic cell differentiation. J Immunol Methods 288(1-2):123–34
34. Liu T et al (2023) Optimization of differentiation and transcriptomic profile of THP-1 cells into macrophage by PMA. PLoS One 18(7):e0286056
35. Baumer Y et al (2020) Immune cell phenotyping in low blood volumes for assessment of cardiovascular disease risk, development, and progression: a pilot study. J Transl Med 18(1):29
36. Saxena A, Dagur PK, Biancotto A (2019) Multiparametric flow cytometry analysis of naive, memory, and effector T cells. Methods Mol Biol 2032:129–140
37. Miller JS et al (2014) Expansion and homing of adoptively transferred human natural killer cells in immunodeficient mice varies with product preparation and in vivo cytokine administration: implications for clinical therapy. Biol Blood Marrow Transplant 20(8):1252–7
38. Domogala A, Madrigal JA, Saudemont A (2016) Cryopreservation has no effect on function of natural killer cells differentiated in vitro from umbilical cord blood CD34(+) cells. Cytotherapy 18(6):754–9

Part II

Clinical

Check for updates

Chapter 9

Immunophenotyping of Acute Myeloid Leukemia

Pallavi Kanwar Galera and Raul Braylan

Abstract

Immunophenotyping by multiparameter flow cytometry is a rapid and efficient technique to simultaneously assess and correlate multiple individual cell properties like size and internal complexity along with antigen expression in a population of cells. This method is utilized for rapid characterization of the blasts and classification of acute myeloid leukemia (AML) in both the peripheral blood (PB) and bone marrow (BM). This technique is not only useful in the initial diagnosis but also in monitoring and determining the prognosis of the disease through minimal residual disease (MRD) testing. This chapter provides an overview of procedures for specimen processing, staining, and immunophenotyping of AML and describes the principles of data analysis for AML classification and MRD testing.

Keywords Immunophenotype, Acute myeloid leukemia, Myelodysplastic syndrome, Minimal residual disease, Flow cytometry

1 Introduction

AML is a clonal myeloid precursor neoplasm derived from neoplastic hematopoietic stem cells characterized by an increased number of blasts or blast equivalents in the BM and PB. Myeloid blasts include myeloblasts, monoblasts, erythroblasts, megakaryoblasts, and blast equivalents include promyelocytes and promonocytes. The WHO Classification of Haematolymphoid tumors fifth edition [1], subdivides AML into AML with defining genetic abnormalities, AML defined by differentiation, and myeloid sarcoma. A separate category of secondary myeloid neoplasms including myeloid neoplasms and proliferations associated with antecedent or predisposing conditions including postcytotoxic therapy, germline predisposition and Down syndrome is now recognized. These categories are based on morphological, immunophenotypic, genetic, and clinical data with immunophenotype of the leukemic blasts being an important component. Historically, AML was defined by the presence of greater than 20% blasts or blast equivalents in a BM or PB specimen; however, this prerequisite is not required anymore for AMLs with defining

Angélique Biancotto and Pradeep K. Dagur (eds.), *Immunophenotyping*, Methods in Molecular Biology, vol. 3087,
https://doi.org/10.1007/978-1-0716-5607-5_9,

genetic abnormalities with the exception of AML with BCR::ABL1 fusion and AML with CEBPA mutation.

Flow cytometric immunophenotyping can be utilized for rapid identification and characterization of the blasts and blasts equivalents, in both the PB and BM [2, 3] and, at the same time, it allows us to assess the maturation patterns of the various BM elements, further refining the classification of AML [4–6]. With adequate population selection (gating) strategies and appropriate antibody panels, flow cytometry can also be very useful in the enumeration of blasts, especially in PB. However, while flow cytometry is an excellent means for detection and characterization of blasts, it may not be as accurate in quantitating blasts in BM specimens primarily due to uncontrolled hemodilution [7], leading to underestimation of blast numbers and making morphology/immunohistochemistry still the gold standard for blast counting in BM samples. Other confounding factors, like loss of nucleated red blood cells (RBC) during lysis and processing, may artificially increase the blast percentage [8, 9]. These issues do not apply to blast counts in PB.

Flow cytometry is also of paramount importance in monitoring disease and prognosticating it through minimal or measurable residual disease (MRD) testing [6, 10–12]. MRD is the presence of leukemic blasts at levels lower than the limit of detection by conventional morphologic evaluation. Multiple studies have demonstrated that detection of leukemic blasts following induction and consolidation therapy identifies patients who are at a higher risk of relapse and thus a worse outcome [13–16]. Due to its affordability, widespread availability, and quick turnaround time, flow cytometry is extensively used for MRD testing [17–19].

This chapter outlines the protocol used at the Department of Laboratory Medicine, Clinical Center, National Institutes of Health (NIH), providing an approach for diagnosing, and classifying AML using a 10-parameter (eight-color) measuring flow cytometer (FACS-Canto II) with FACSDiva data acquisition software (both from Becton Dickinson [BD]). It also describes the various strategies utilized for data analysis using FCS Express software (De Novo).

It should be emphasized that there are plethora of approaches and guidelines that can be utilized for these purposes [20, 21], and other laboratories may use different procedures to obtain equivalent results.

2 Materials

2.1 Reagents

1. Phosphate-buffered saline (PBS) (KD Medical)/bovine serum albumin (BSA) (EMD Millipore Corp.) buffer: 500 mL 1× PBS and 0.5 g bovine serum albumin, fraction V (fatty acid free). Mix well and filter through a 45 μm Millipore filter. Store at 4°C for up to 1 month.

2. FACSLyse Buffer (BD): Prepare a 1:10 dilution of the FACSLyse buffer using deionized water. Store at room temperature in a glass bottle for up to 1 month.
3. Stabilizing Fixative (BD): Prepare a 1:3 dilution of the Stabilizing Fixative using deionized water. The solution can be kept for 1 month when stored in a polystyrene container at room temperature.
4. Erythrocyte Lysis Buffer (Qiagen).
5. Fix Perm Reagent A and B (Caltag).

2.1.1 Precautions

All PB and BM products should be considered biohazardous and treated with proper precautions as if they were capable of transmitting infection. The Stabilizing Fixative contains 3% paraformaldehyde. Appropriate disposal methods should be used.

2.2 Monoclonal

2.2.1 Antibody Panels

Table 1 describes the antibodies utilized in the NIH laboratory for detection, characterization, and classification of AML with appropriate corresponding fluorochromes (*see* **Note 1**).

2.3 Specimens

1. PB (anticoagulated in EDTA or Heparin).
2. BM Aspirate (anticoagulated in EDTA or Heparin).

Table 1
Antibodies and fluorochromes utilized in the NIH laboratory[a]

FITC	PE	PerCP-Cy5.5	PE-Cy7	APC	APC-Alexa 750	APC-H7	Pacific Blue	V500
CD16	CD10	CD5	CD19	CD10	CD34	CD14	HLA-DR	CD45
CD42b + CD61	CD13	CD11c	CD56	CD56	CD38		CD2	
CD36	CD19	CD117	CD34	CD11b	CD71		CD4	
cMPO[b]	CD64	CD34	CD15	CD7	CD3		CD19	
	CD5	CD33		CD123			cCD3[b]	
	cCD79a[b]	CD3		cTDT[b]				

APC Allophycocyanin, *APC-Alexa* 750 Allophycocyanin Alexa Fluor 750, *APC-H7* Allophycocyanin H7, *FITC* Fluorescein isothiocyanate, *MPO* Myeloperoxidase, *PB* Pacific blue, *PE* Phycoerythrin, *PE-Cy7* (PC7) Phycoerythrin cyanine-7, *PerCP-Cy5.5* Peridinin chlorophyll protein cyanine 5.5, *TDT* Terminal deoxynucleotidyl transferase, V500 BD Horizon dye V500 [a]Other laboratories may utilize different antibody and fluorochrome combinations [b]c: Intracytoplasmic antibodies

3 Methods

3.1 Labeling with Monoclonal Antibodies Against Surface Antigens

It is also critical to appropriately titrate the amount of antibody to get an optimal antibody dilution for maximum signal-to-noise ratio [22, 23].

1. Label 12×75 mm disposable polystyrene tubes and add the antibodies.
2. Wash the PB or BM sample at least twice as follows: To wash the sample, in a 15 mL disposable conical tube, add at least 2 mL of PBS/BSA to the sample and spin at 311×*g* for 8 min. Aspirate the supernatant and repeat again by adding at least 2 mL of PBS/BSA, mixing, and spinning at 311×*g* for 8 min. Aspirate the supernatant. Resuspend by adding PBS/BSA to the sample to make a final white cell concentration of 5000–10,000 cells/μL.
3. Add 100 μL of cell suspension (white cell concentration 5000–10,000 cells/μL) to each 12×75 mm disposable polystyrene tube with antibodies.
4. Incubate the tubes that contain the antibodies and the sample for 15 min to 1 h at room temperature in the dark (*see* **Note 2**).
5. Add 2 mL 1× FACSLyse Buffer to each tube to lyse the red blood cells. Vortex.
6. Incubate for 10 min at room temperature in the dark.
7. Spin at 216×*g* for 5 min.
8. Decant supernatant. Mix.
9. Add 2 mL PBS/BSA to each tube.
10. Repeat Steps 7–9 (*see* **Note 3**).
11. Add 400 μL 1_ Stabilizing Fixative to each tube (*see* **Note 4**).
12. The cells are now ready for analysis on the flow cytometer.

3.2 Labeling with Monoclonal Antibodies Against Combined Surface and Intracellular Antigens

This method allows the detection of cytoplasmic antigens, along with the assessment of expression of surface antigens. It utilizes a cross-linking fixative to stabilize cellular contents prior to labeling with antibodies against cytoplasmic antigens. The procedure outlined here utilizes Erythrocyte Lysis Buffer (Qiagen) and Fix Perm Reagent A and B (Caltag), but other products may be used. It is also critical to appropriately titrate the amount of antibody to get an optimal antibody dilution for maximum signal-to-noise ratio [22, 23].

1. Lyse RBCs from sample using Erythrocyte Lysis Buffer: To lyse the RBCs add 2 mL of the Erythrocyte Lysis Buffer to 200 μL of cell suspension in a 12×75 mm disposable

polystyrene tube. Allow to lyse for approximately 10 min. Spin at 216 × *g* for 5 min. Decant supernatant.

2. Wash the sample: To wash the sample, add 2 mL of PBS/BSA and then spin at 216 _ g for 5 min. Decant supernatant.
3. Add appropriate antibodies to the 12 _ 75 mm disposable polystyrene tube containing the cells for cell surface staining.
4. Incubate for 15 min.
5. Wash the sample: To wash the sample, add 2 mL of PBS/BSA and then spin at 216 _ g for 5 min. Decant supernatant.
6. Add 100 μL of Fix Perm Reagent A.
7. Incubate for 10 min.
8. Wash the sample: To wash the sample, add 2 mL of PBS/BSA and then spin at 216 _ g for 5 min. Decant supernatant.
9. Add 100 μL of Fix Perm Reagent B.
10. Add intracellular antibodies.
11. Incubate for 15 min.
12. Wash the sample: To wash the sample, add 2 mL of PBS/BSA and then spin at 216 _ g for 5 min. Decant supernatant.
13. Resuspend in 400 μL of Stabilizing Fixative.

3.3 Instrument Settings

A variety of instruments are available, and different laboratories may use different instruments to perform flow cytometric analysis. Settings and calibrations are instrument-dependent, and manufacturer's guidelines should be employed. The number of cells acquired for analysis per tube may vary depending on the level of detection sensitivity desired. In our laboratory we acquire at least 100,000 cells/tube for routine analysis; however, higher numbers are necessary for MRD testing.

3.4 Data Analysis

3.4.1 Data Quality and Viability

The initial evaluation of results consists of an assessment of the quality of data and viability of the specimen. One strategy to evaluate the quality and consistency of the acquired data is to observe data over time to identify air aspiration artifacts due to sample exhaustion and fluidic instabilities (*see* **Note 5**) (Fig. 1). The second step in ensuring the quality of data collected is to make certain that the data analyzed is an actual measure of a single cell at a time. Frequently, high cell concentration, high flow rates, or nonspecific antibody bindings can induce cell aggregates and clump formations. To analyze exclusively single cells, a combination of forward light scatter area (FSC-A) and forward light scatter width (FSC-W) and/or height (FSC-H) is utilized to exclude cell doublets and aggregates from the data (Fig. 1).

Finally, to evaluate the viability of the specimen and to exclude debris and degenerated cells, viability dyes (propidium iodide, 7AAD)

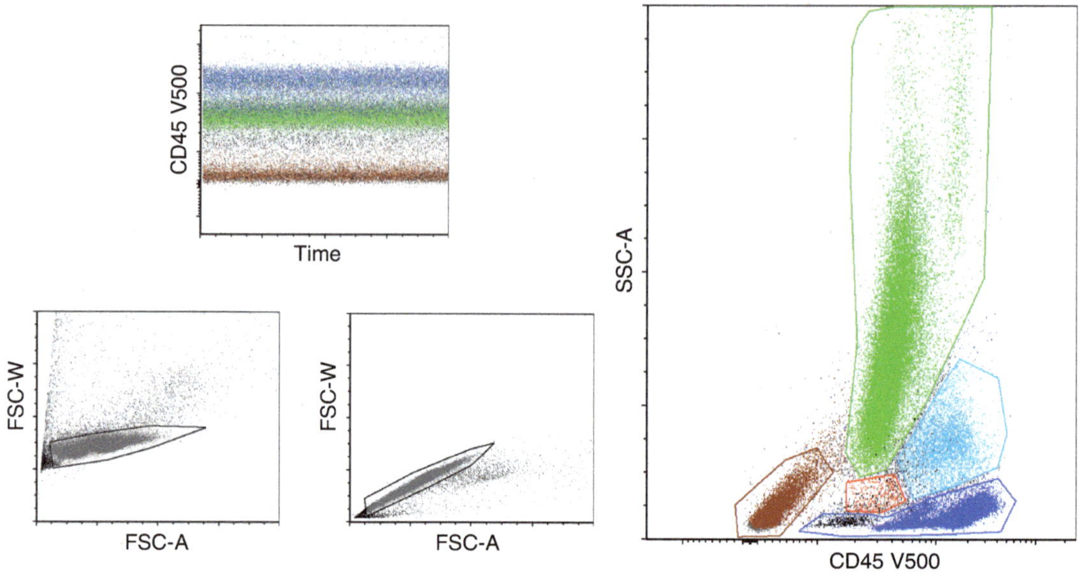

Fig. 1 Normal bone marrow. Data quality: Data are observed over time to identify and exclude fluidic instabilities such as air aspiration artifacts (upper left graph). A combination of forward light scatter area (FSC-A) and forward light scatter width (FSC-W) (lower left graph), and/or forward light scatter height (FSC-H) (lower middle graph) is utilized to exclude doublets and to ensure that the data analyzed are a measure of a single cell at a time. Initial gating (right graph): Based on the side light scatter (SSC-A) and CD45 expression the marrow populations can be broadly selected as red blood cells (brown), granulocytes (green), lymphocytes (blue), monocytes (aqua), and cells in the blast gate (red)

can be utilized. Alternatively, for practical purposes, nonviable cells may be excluded utilizing light scatter characteristics [20, 24, 25].

3.4.2 Initial Cell Population Gating

The initial gating, after excluding all elements that are not single viable cells, is performed to define broad categories of cell populations. This is done by examining side light scatter (SSC-A) and CD45 signals (*see* **Note 6**) [26]. Using this combination, the following populations can be broadly identified: erythroid (low SSC-A and negative CD45), granulocytes (moderate to high SSC-A and moderate CD45), lymphocytes (low SSC-A and bright CD45), monocytes (intermediate SSC-A between lymphocytes and granulocytes and bright CD45), and blasts (low SSC-A and dim CD45) (Fig. 1) [27]. Though an effective initial gate, the "blast gate" based on SSC-A and CD45 may be inaccurate [28] since there may be other populations that fall in this gate, namely basophils, hypogranular myeloid cells, plasmacytoid dendritic cells, and immature monocytes. Also, there would be instances where the blasts may have a brighter or dimmer CD45 and a varied SSC-A. Other populations may also, to a certain extent, be relatively ill-defined using this initial gating approach. Therefore, a panel of additional markers is required to further delineate the blasts as well as other distinct cell populations.

3.4.3 Strategies for Blast Identification

To assess the presence of myeloblasts, commonly used antigens include CD34 (Fig. 2) and CD117 since blasts usually express these markers. However, some myeloblasts may be negative for CD34 and/or CD117 and not even fall within the typical location in the CD45 and SSC-A graphs, making their recognition difficult. In these instances, other reagents and strategies are necessary for their detection. For example, blasts in acute promyelocytic leukemia are CD34 dim or negative and may mimic normal granulocytes, but they usually have distinct phenotypic features [29]. Also, monoblasts usually express no or dim CD34, requiring the use of reagents for monocyte-associated markers for their recognition.

3.4.4 Strategies for Lineage Assignment

Myeloid lineage markers commonly used include CD13, CD15, CD33, MPO, and CD16 along with markers expressed during monocytic differentiation such as CD11b, CD64, CD14, and CD4 (Fig. 2). Erythroid precursors express CD71, CD105, CD117, Glycophorin A (CD 235a), and CD36, and megakaryocytic precursors express platelet-associated glycoproteins such as CD61 and CD42b. Assessment of lymphoid markers (CD3, CD2, CD4, CD5, CD7, CD19, CD56, CD22, and CD79a) should be included at the time of initial diagnosis, to assign lineage or to look for aberrant antigen expression. Nonlineage-specific markers (CD45RA, CD90, CD371, CD38, CD56, CD123, and HLA-DR) are also helpful to further delineate blasts and aberrancies.

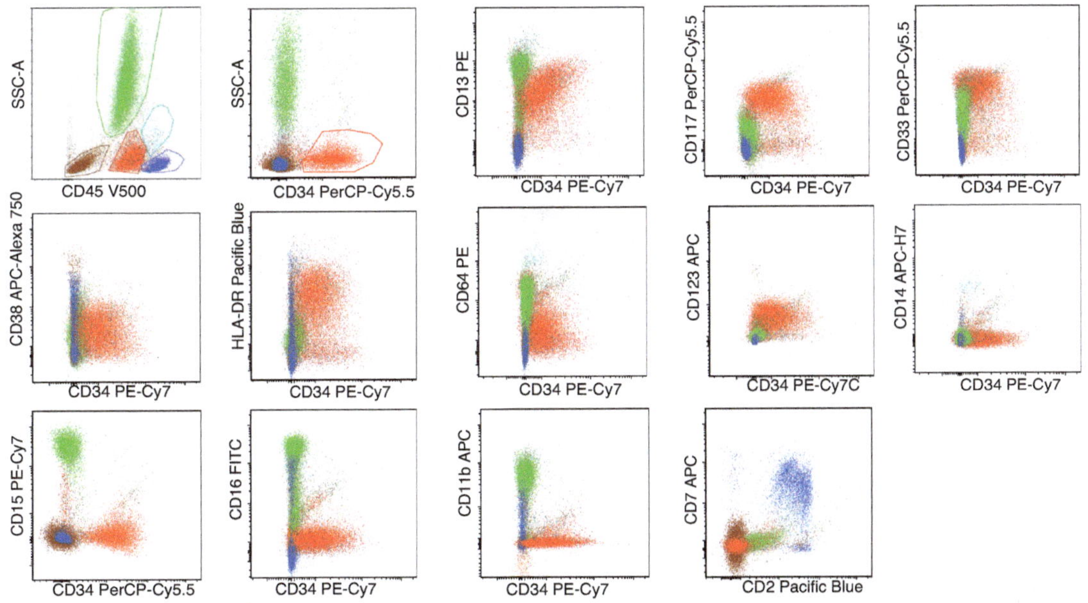

Fig. 2 Bone marrow with increased, phenotypically abnormal myeloblasts, comprising 23% of total cells. The blasts (red dots) express CD45, CD34, CD13, CD117, CD33 (bright), CD38 (dim/negative), HLA-DR, CD64 (dim/negative), and CD123 (dim). Blasts are negative for CD14, CD15, CD16, CD11b, CD7, and CD2

Immunophenotypic aberrancies include over or under expression of antigens, or absence of antigen expression (e.g., dim or absent CD38, CD13, or CD33 expression on myeloblasts), asynchronous antigen expression (expression of markers of mature elements in early precursors such as CD11b, CD64, or CD15 in myeloblasts), abnormally homogenous expression of certain markers (e.g., of CD38 or CD33), and expression of nonmyeloid antigens on myeloblasts (CD2, CD7, CD56, and CD19) (Figs. 3, 4, and 5) [8, 30].

3.4.5 Background Myelopoiesis

In acute myeloid leukemias, it may be useful to evaluate the background myelopoiesis to determine the presence of possible dysplastic changes. For this purpose, knowledge of normal maturation patterns utilizing various combinations of antigens is essential. Patterns generated by CD45 and SSC-A (SSC-A reflecting granularity), CD13 and CD16, CD11b and CD64, and CD11b and CD16 have been well studied for assessment of myeloid maturation and are reproducible [4, 5] (Fig. 5). Alterations of these patterns and other aberrancies like loss of CD10 on mature granulocytes, or expression of CD56 on myeloid and monocytic cells are helpful clues into probable underlying myelodysplasia [8, 20, 31].

3.4.6 AML with Defining Genetic Abnormalities, Special Considerations

Immunophenotyping can be helpful as an adjunct in classifying AML with defining genetic abnormalities; however, cytogenetic and molecular studies remain the mainstay in the diagnostic algorithm. Although the association between the immunophenotype and the genetic abnormality is not absolute, a few of the AML with defining genetic abnormalities show a strong correlation with the immunophenotype including AML with RUNX1::RUNX1T1, acute promyelocytic leukemia with PML::RARA, and AML with KMT2A

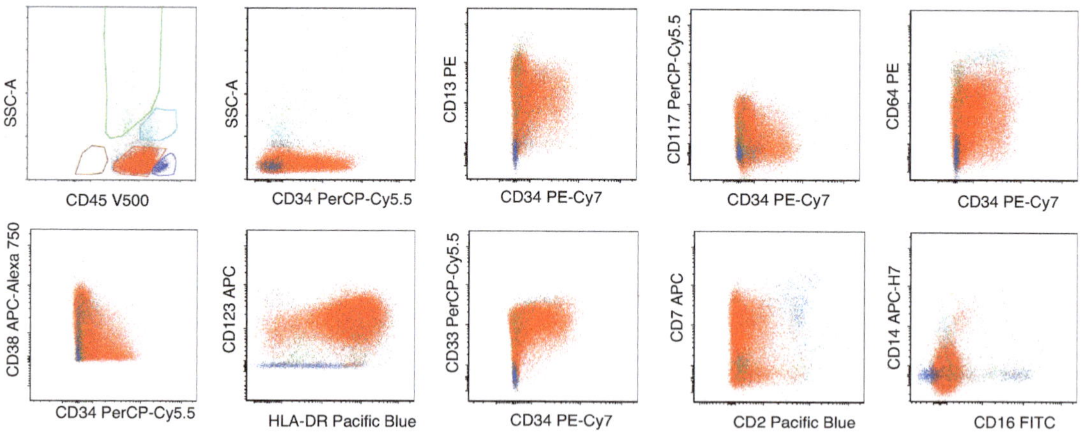

Fig. 3 Bone marrow involved by AML with *FLT3* mutation and phenotypically abnormal myeloblasts comprising 95% of total cells. The blasts (red dots) express CD45, CD34 (dim to negative), CD13, CD117 (variable), CD64 (variable), CD38 (partial), HLA-DR, CD123, CD33, CD7 (partial aberrant). Blasts are negative for CD16, CD14, and other T-cell and B-cell markers (not shown)

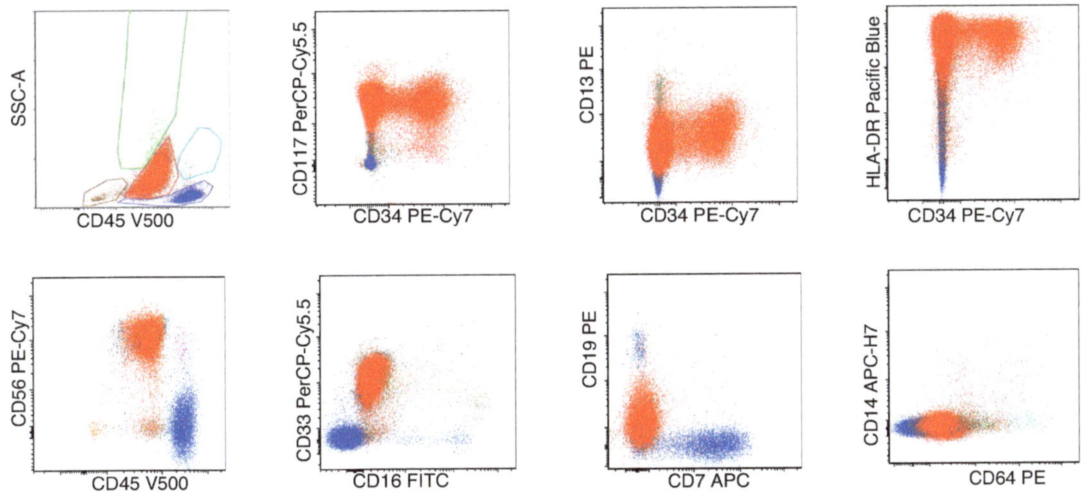

Fig. 4 A patient with a history of autoimmune lymphoproliferative syndrome with *FAS* mutation. Bone marrow aspirate shows two populations of myeloblasts (red dots) based on CD34 expression, with the minority of the blasts (approximately 25%) expressing CD34 and the majority of the blasts are negative for CD34. The blasts also express CD45, CD117, CD13, HLA-DR (bright), CD56 (aberrant), and CD33. A small subset with dim expression of CD19 is noted. The blasts are negative for CD7, CD14, and CD64. They were also negative for CD10, CD20, CD3, and CD5 (not shown)

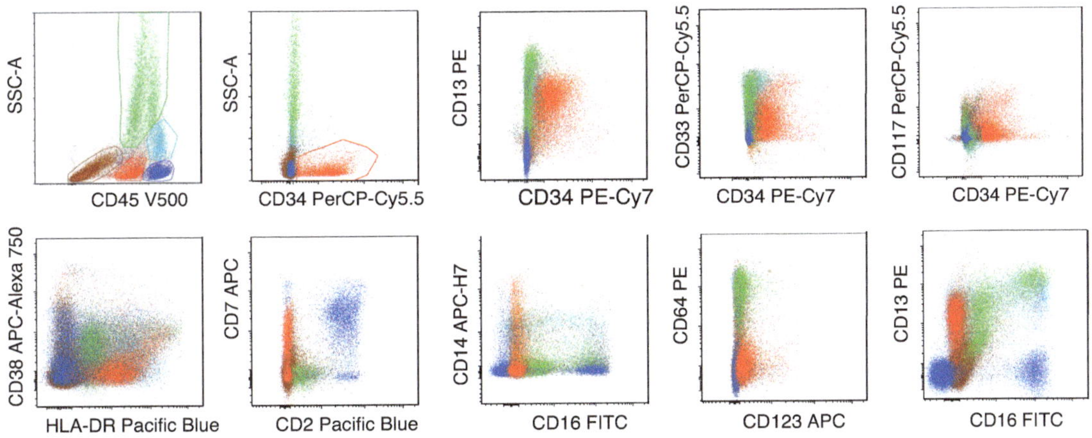

Fig. 5 A patient with history of BRCA1 and BRCA2 mutations and therapy for breast cancer 7 years prior to presentation. Bone marrow aspirate shows abnormal myeloblasts (red dots), comprising 13% of total cells. The blasts express CD45, CD34, CD13, CD33, CD117 (dim to negative), HLA-DR, CD7 (partial and aberrant) and are negative for CD38 (aberrant), CD2, CD16, CD14, CD64, and CD123. Granulocytes (green dots) reveal a disrupted maturation pattern based on the expression of CD13 and CD16

rearrangement. Some salient immunophenotypic features of AML with defining genetic abnormalities are described below:

1. Acute promyelocytic leukemia (APL) with PML::RARA; t(15;17)(q22;q11-12): Abnormal promyelocytes (blast

equivalents) demonstrate relative high SSC-A, homogeneous bright expression of CD33, heterogeneous expression of CD13, CD117 and MPO, and low or absent expression of HLA-DR, CD34, CD11a, and CD11b. CD64 expression is common. In the microgranular variant there is frequent expression of CD34 and CD2 by at least some cells [32]. CD2 expression in APL has been associated with FLT3-ITD [33]. CD56 expression is seen in a subset of cases and has been associated with a worse outcome [34].

2. AML with RUNX1::RUNX1T1; t(8;21)(q22;q22.1): Blasts usually show high expression of CD34, HLA-DR, MPO, and CD13, and weak expression of CD33. Neutrophilic differentiation and maturation as assessed by expression of CD15 and/or CD65, maturation asynchrony (coexpression of CD34 and CD15) and frequent coexpression of B-cell markers (CD19, and cytoplasmic CD79a) [35, 36] are noted (*see* **Note 7**). CD56 expression is seen in a subset of cases, and this may portend an adverse prognosis [37, 38]. The adverse prognosis could be secondary to the fact that cases with KIT mutations have higher CD56 expression [39].
3. AML with CBFB::MYH11; inv(16)(p13.1q22); or t(16;16)(p13.1;q22): Blasts express CD34 and CD117 intensely. Blast populations differentiating toward granulocytic lineage (positive for CD13, CD33, CD15, CD65, and MPO) and the monocytic lineage (positive for CD14, CD4, CD11b, CD11c, CD64, and CD36) are seen. Maturation asynchrony (coexpression of CD34 and CD15) and coexpression of CD2 with myeloid markers are frequently noted [1, 40].
4. AML with DEK::NUP214; t(6;9)(p23;q34.1): The blasts have a nonspecific myeloid immunophenotype, with expression of MPO, CD9, CD13, CD33, CD38, CD123, and HLA-DR. Most of the cases show an expression of CD34, CD117, and CD15. A subset of cases is TdT positive. Associated basophilia is often present and can be detected with basophils uniformly expressing CD123, CD33, and CD38 but no HLA-DR [41–44].
5. AML with RBM15::MRTFA (MKL1); t(1;22)(p13.3;q13.1): This is characterized by megakaryocytic differentiation and the blasts express one or more of the platelet glycoproteins: CD41 (glycoprotein IIb/IIIa), CD61 (glycoprotein IIIa), and CD42b (glycoprotein Ib) (*see* **Note 8**) along with variable expression of myeloid markers CD13 and CD33. They also express CD36. CD34, CD45, and HLA-DR are often negative [45–47].
6. AML with BCR::ABL1 fusion; t(9;22)(q34;q11.2): The blasts express CD34, HLA-DR, and myeloid antigens (CD13,

CD33, and CD117). Expression of CD7, CD19, and TdT is also common [48–50].

7. AML with KMT2A rearrangement: Most adult AML cases reveal monocytic differentiation (CD14, CD4, CD11b, CD11c, CD64, and CD36,) with a variable expression of markers of immaturity (CD34, CD117), and CD56 [51] (Fig. 6). In children, there is a high expression of CD33, CD65, CD4, and HLADR, with a lower expression of CD13, CD34, and CD14 [52].
8. AML with MECOM rearrangement; inv(3)(q21.3q26.2) or t(3;3)(q21.3;q26.2): The blasts express CD34, CD33, CD13, CD117, HLA-DR, and CD38 (commonly), and a frequent aberrant CD7. A subset of cases may express megakaryocytic markers such as CD41 and CD61 [53, 54].
9. AML with NUP98 rearrangement: The blasts usually express stem cell and myeloid markers, including CD34, CD117, CD13, CD33, HLA-DR, and MPO. Other markers that may be expressed include CD123, CD11b, CD11c, CD14, CD36, CD7, CD15, and CD64. Blasts in approximately 1/3rd of the patients <3 years of age show a megakaryocytic differentiation with expression of CD41, CD42b, and CD61

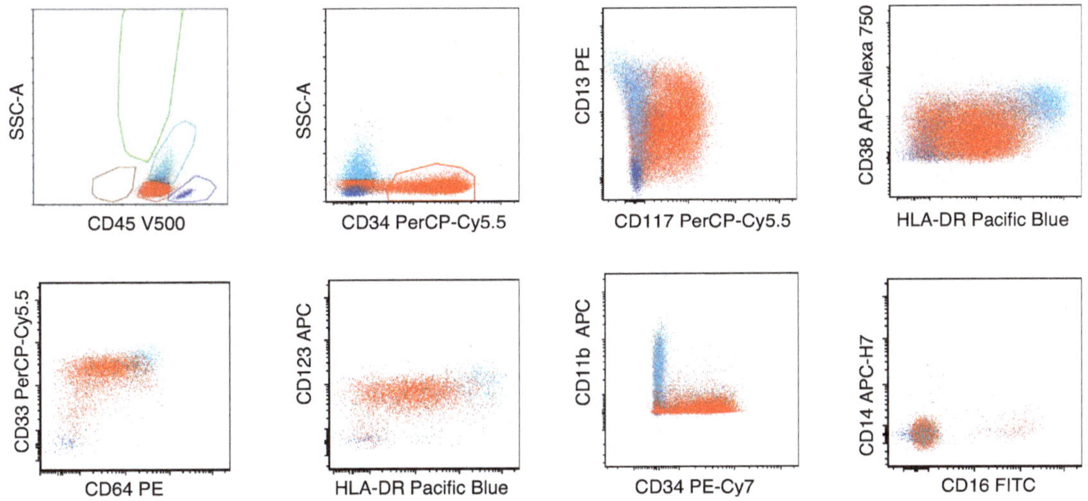

Fig. 6 AML with t(9;11)(p21.3;q23.3); *KMT2A-MLLT3*. Bone marrow aspirate reveals over 90% blasts, with two different immunophenotypically abnormal populations: The first population is of immature myeloblasts (red dots), representing approximately 81% of total cells and expressing CD45(dim), CD34 (partial, 72% of total cells), CD117, CD13, HLA-DR, CD33 (bright), CD123, and CD64 (dim to negative). These blasts are negative for CD14, CD16, and CD11b. They are also negative for CD15, CD61/CD42b, T-cell, and B-cell markers (not shown). The second population is of monoblasts (aqua dots), representing approximately 10% of total cells with intermediate SSC-A and expressing CD45 (brighter than first population), CD13, HLA-DR (bright), CD38, CD33 (bright), CD64 (bright), CD11b (variable), and CD36 (not shown). These are negative for CD34, CD117, CD16, and CD14

[55, 56]. Erythroid differentiation is also common in children [57].

10. AML with NPM1 mutation: The blasts show high expression of CD33 and variable (often low) CD13 expression along with CD117 and CD123 [58]. CD34 and HLA-DR are often negative (*see* **Note 9**). The blasts can either have an immature myeloid or a monocytic immunophenotype with expression of CD36, CD64, and CD14 [59]. A blast population with expression of CD34, CD25, CD123, and CD99 is associated with FLT3-ITD mutations [60].
11. AML with CEBPA mutation: The blasts usually express HLA-DR and CD34. Expression of CD7 is also commonly seen. The cases with biallelic mutation are reported to have a higher frequency of HLA-DR, CD7, and CD15 expression and a lower frequency of CD56 expression in comparison to cases with a single mutation [61, 62]. Monocytic markers such as CD14 and CD64 are usually absent.
12. AML with other defined genetic alterations:

    CBFA2T3::GLIS2: The blasts often show the RAM immunophenotype with bright CD56 expression (at minimum 2 log10 units greater than normal myeloid progenitors), dim to negative CD45 and CD38 and lack of HLA-DR [63–65].
    KAT6A::CREBBP: AML often shows monocytic/myelomonocytic differentiation with expression of CD4, CD14, CD13, CD33, CD56, and HLA-DR, and without CD34 and CD117 [66, 67].

3.4.7 Minimal Residual Disease

The detection limit of leukemic blasts by flow cytometry ranges from 0.1% to 0.01% of leukocytes. The detection limit is dependent on various factors, including the number of cells collected, antibody panel, flow cytometer, and distinct immunophenotypic properties of the leukemic blasts. Utilizing a single tube with a higher number of antibodies and broader panels allows for efficient and reliable MRD assessment while reducing the reagent cost and technical labor [6]. For a detection sensitivity of 0.01%, at least 1,000,000 cells should be acquired.

Two strategies are commonly employed in MRD testing for AML by flow cytometry. The first strategy is identifying a combination of aberrant markers and their expression levels on leukemic blasts that are usually absent in normal hematopoietic precursors. This strategy is widely known as "leukemia-associated immunophenotype" or LAIP. At diagnosis a broad, extensive antigen panel is utilized to define the LAIP. Subsequently, after treatment, the selected antibody panel is used to identify blasts with pretreatment LAIP. The advantages of this approach are that the data analysis is relatively simple, and sensitivity for each LAIP (defined based on the background expression in nonleukemic specimens) is known. The disadvantages include the requirement to know the complete phenotype of the pretreatment leukemic sample, the need for customized antibody panels for subsequent evaluations after the evaluation of the initial diagnostic specimen and, finally, the immunophenotype of the leukemic blasts might change after treatment [3, 6, 10, 12, 21, 24].

The second strategy known as "difference from normal" identifies leukemic blasts based on immunophenotypic differences between the leukemic blasts and normal hematopoietic precursors. There is no requirement of defining or having prior knowledge of LAIPs at diagnosis (which is a hindrance in many tertiary or referral centers that do not have this information available); the same standardized panel can be used for all evaluations and since this method is based on differences between blasts and normal myeloid precursors, immunophenotypic shifts and change of LAIPs after treatment do not interfere with the assessment. However, extensive knowledge of antigen expression patterns of normal hematopoietic precursors and their maturation patterns is essential for an accurate evaluation [3, 6, 10, 12, 21, 24]. In practice, a combination of the two strategies is often utilized.

4 Notes

1. A variety of antibodies with a variety of fluorochrome conjugates are available by various manufacturers. The antibody and fluorochrome selections and combinations may vary depending upon the instrument, number of detectors, and the approach of the particular lab. Additional antibodies that are not frequently used such as CD99, CD105, CD45RA, Clec12a, and CD41 can also be added to aid in the diagnosis of AML [6 ,45–47, 53, 54, 60].
2. Prolonged fixation and exposure to fluorescent light may degrade fluorochromes. To minimize fluorochrome degradation, labeling of the specimen with antibodies prior to red cell lysis along with limited fixation and decreased light exposure is recommended.
3. Addition of the wash step reduces background fluorescence due to unbound antibody.
4. The use of a fixative helps in better preservation of the light scatter properties of the cells and provides a better separation between viable and nonviable cells.
5. In case of the air bubble artifact secondary to aspiration of air due to sample exhaustion, the error can be corrected by simply excluding the data during those time periods.
6. The advantage of using SSC-A and CD45 for initial gating is that by just adding a single common antibody across all tubes,

the various populations can be compared with ease across all tubes.

7. AML with RUNX1::RUNX1T1; t(8;21)(q22;q22.1) can be misclassified as mixed phenotype acute leukemia due to frequent expression of B-cell markers (CD19, and cytoplasmic CD79a) on the blasts. Presence of CD56- and CD19-positive myeloid blasts should raise a possibility of AML with RUNX1::RUNX1T1. Cases of acute monoblastic leukemia may show a higher expression of CD14 [68].
8. In AML, with megakaryocytic differentiation, cytoplasmic expression of CD41 or CD61 is more specific and sensitive than surface staining.
9. Cases of AML with NPM1 mutations are often negative for CD34 and HLA-DR, similar to blasts equivalents seen in APL; however, they have a lower SSC-A and fall in the blast gate. CD34-positive cases do occur and have been associated with an adverse prognosis [69, 70].

References

1. Khoury JD, Solary E, Abla O et al. (2022) The 5th edition of the World Health Organization Classification of haematolymphoid tumours: myeloid and histiocytic/dendritic neoplasms. Leukemia 36(7):1703–1719. https://doi.org/10.1038/s41375-022-01613-1
2. Craig FE, Foon KA (2008) Flow cytometric immunophenotyping for hematologic neoplasms. Blood 2008:111. https://doi.org/10.1182/blood-2007-11-120535
3. Chen X, Cherian S (2017) Acute myeloid leukemia immunophenotyping by flow cytometric analysis. Clin Lab Med 37(4):753–769
4. Van Lochem E, Van der Velden V, Wind H, Te Marvelde J, Westerdaal N, Van Dongen J (2004) Immunophenotypic differentiation patterns of normal hematopoiesis in human bone marrow: reference patterns for age-related changes and disease-induced shifts. Cytometry B Clin Cytom 60(1):1–13
5. Gorczyca W, Sun Z-Y, Cronin W, Li X, Mau S, Tugulea S (2011) Immunophenotypic pattern of myeloid populations by flow cytometry analysis. Meth Cell Biol 103:221–266
6. Roshal M, Gao Q (2025) Flow cytometry evaluation of acute myeloid leukemia minimal residual disease based on an understanding of the normal maturation patterns in the blast compartments. Am J Clin Pathol 163(5):775–793. https://doi.org/10.1093/ajcp/aqae187
7. Loken MR, Chu SC, Fritschle W (2009) Normalization of bone marrow aspirates for hemodilution in flow cytometric analyses. Cytometry B Clin Cytom 2009:76. https://doi.org/10.1002/cyto.b.20429
8. Wood BL (2007) Myeloid malignancies: myelodysplastic syndromes, myeloproliferative disorders, and acute myeloid leukemia. Clin Lab Med 27(3):551–575
9. Wood BL, Arroz M, Barnett D (2007) 2006 Bethesda international consensus recommendations on the immunophenotypic analysis of hematolymphoid neoplasia by flow cytometry: optimal reagents and reporting for the flow cytometric diagnosis of hematopoietic neoplasia. Cytometry B Clin Cytom 2007:72. https://doi.org/10.1002/cyto.b.20363
10. Wood BL (2016) Principles of minimal residual disease detection for hematopoietic neoplasms by flow cytometry. Cytometry B Clin Cytom 90(1):47–53
11. Xu J, Jorgensen JL, Wang SA (2017) How do we use multicolor flow cytometry to detect minimal residual disease in acute myeloid leukemia. Clin Lab Med 37(4):787–802
12. Zhou Y, Wood BL (2017) Methods of detection of measurable residual disease in AML. Curr Hematol Malign Rep 12(6):557–567
13. Dohner H, Estey E, Grimwade D (2017) Diagnosis and management of AML in adults: 2017 ELN recommendations from an international expert panel. Blood 2017:129. https://doi.org/10.1182/blood-2016-08-733196
14. Terwijn M, Putten WL, Kelder A (2013) High prognostic impact of flow cytometric minimal residual disease detection in acute myeloid

leukemia: data from the HOVON/SAKK AML 42A study. J Clin Oncol 2013:31. https://doi.org/10.1200/JCO.2012.45.9628
15. Venditti A, Buccisano F, Del Poeta G et al. (2000) Level of minimal residual disease after consolidation therapy predicts outcome in acute myeloid leukemia: Presented in part at the 41st annual meeting of the American Society of Hematology, December 3–7, 1999, New Orleans, LA. Blood 96(12):3948–3952
16. Zhu H-H, Zhang X-H, Qin Y-Z et al. (2013) MRD-directed risk stratification treatment may improve outcomes of t (8; 21) AML in the first complete remission: results from the AML05 multicenter trial. Blood 121(20):4056–4062
17. Freeman SD, Virgo P, Couzens S (2013) Prognostic relevance of treatment response measured by flow cytometric residual disease detection in older patients with acute myeloid leukemia. J Clin Oncol 2013:31. https://doi.org/10.1200/JCO.2013.49.1753
18. Walter RB, Gooley TA, Wood BL (2011) Impact of pretransplantation minimal residual disease, as detected by multiparametric flow cytometry, on outcome of myeloablative hematopoietic cell transplantation for acute myeloid leukemia. J Clin Oncol 2011:29. https://doi.org/10.1200/JCO.2010.31.8121
19. Getta BM, Devlin SM, Levine RL (2017) Multicolor flow cytometry and multigene next-generation sequencing are complementary and highly predictive for relapse in acute myeloid leukemia after allogeneic transplantation. Biol Blood Marrow Transplant 2017:23. https://doi.org/10.1016/j.bbmt.2017.03.017
20. Van Dongen J, Lhermitte L, Böttcher S et al. (2012) EuroFlow antibody panels for standardized n-dimensional flow cytometric immunophenotyping of normal, reactive and malignant leukocytes. Leukemia 26(9):1908–1975
21. Schuurhuis GJ, Heuser M, Freeman S (2018) Minimal/measurable residual disease in AML: a consensus document from the European LeukemiaNet MRD working party. Blood 2018:131. https://doi.org/10.1182/blood-2017-09-801498
22. Reilly JT (1996) Use and evaluation of leukocyte monoclonal antibodies in the diagnostic laboratory: a review. Clin Lab Haematol 1996:18. https://doi.org/10.1111/j.1365-2257.1996.tb00728.x
23. Owens MA, Loken MR (1995) Quality control of flow cytometer and reagents. In: Owens MA, Loken MR (Eds.)Flow cytometry: principles for clinical laboratory practice quality assurance for quantitative immunophenotyping. Wiley
24. Wood BL (2013) Flow cytometric monitoring of residual disease in acute leukemia. Methods Mol Biol 2013:999. https://doi.org/10.1007/978-1-62703-357-2_8
25. Chen W, Luu HS (2017) Immunophenotyping by multiparameter flow cytometry. Methods Mol Biol 2017:1633. https://doi.org/10.1007/978-1-4939-7142-8_4
26. Stelzer GT, Shults KE, Loken MR (1993) CD45 gating for routine flow cytometric analysis of human bone marrow specimens. Ann N Y Acad Sci 1993:677. https://doi.org/10.1111/j.1749-6632.1993.tb38783.x
27. Borowitz MJ, Guenther KL, Shults KE (1993) Immunophenotyping of acute leukemia by flow cytometric analysis. Use of CD45 and right-angle light scatter to gate on leukemic blasts in three-color analysis. Am J Clin Pathol 1993:100. https://doi.org/10.1093/ajcp/100.5.534
28. Harrington AM, Olteanu H, Kroft SH (2012) A dissection of the CD45/side scatter "blast gate". Am J Clin Pathol 2012:137. https://doi.org/10.1309/AJCPN4G1IZPABRLH
29. Orfao A, Chillón MC, Bortoluci AM (1999) The flow cytometric pattern of CD34, CD15 and CD13 expression in acute myeloblastic leukemia is highly characteristic of the presence of PML-RARalpha gene rearrangements. Haematologica 1999:84
30. Kussick SJ, Wood BL (2003) Using 4-color flow cytometry to identify abnormal myeloid populations. Arch Pathol Lab Med 2003:127
31. van de Loosdrecht AA, Alhan C, Bene MC et al. (2009) Standardization of flow cytometry in myelodysplastic syndromes: report from the first European LeukemiaNet working conference on flow cytometry in myelodysplastic syndromes. Haematologica 94(8):1124
32. Exner M, Thalhammer R, Kapiotis S (2000) The "typical" immunophenotype of acute promyelocytic leukemia (APL-M3): does it prove true for the M3-variant. Cytometry 2000:42. doi:3.0.CO;2-S
33. Takenokuchi M, Kawano S, Nakamachi Y (2012) FLT3/ITD associated with an immature immunophenotype in PML-RARα leukemia. Hematol Rep 2012:4. https://doi.org/10.4081/4380
34. Montesinos P, Rayón C, Vellenga E (2011) Clinical significance of CD56 expression in patients with acute promyelocytic leukemia treated with all-trans retinoic acid and anthracycline-based regimens. Blood 2011:117. https://doi.org/10.1182/blood-2010-04-277434
35. Hurwitz CA, Raimondi SC, Head D (1992) Distinctive immunophenotypic features of t(8;21)(q22;q22) acute myeloblastic leukemia in children. Blood 1992:80
36. Kita K, Nakase K, Miwa H (1992) Phenotypical characteristics of acute myelocytic leukemia associated with the t(8;21)(q22;q22) chromosomal abnormality: frequent expression of immature B-cell antigen CD19 together with stem cell antigen CD34. Blood 1992:80

37. Baer MR, Stewart CC, Lawrence D (1997) Expression of the neural cell adhesion molecule CD56 is associated with short remission duration and survival in acute myeloid leukemia with t(8;21)(q22;q22). Blood 1997:90
38. Iriyama N, Hatta Y, Takeuchi J (2013) CD56 expression is an independent prognostic factor for relapse in acute myeloid leukemia with t(8;21). Leukemia Res 2013:37. https://doi.org/10.1016/j.leukres.2013.05.002
39. De J, Zanjani R, Hibbard M (2007) Immunophenotypic profile predictive of KIT activating mutations in AML1-ETO leukemia. Am J Clin Pathol 2007:128. https://doi.org/10.1309/JVALJNL4ELQMD536
40. Ouyang J, Goswami M, Peng J (2016) Comparison of multiparameter flow cytometry immunophenotypic analysis and quantitative RT-PCR for the detection of minimal residual disease of core binding factor acute myeloid leukemia. Am J Clin Pathol 2016:145. https://doi.org/10.1093/ajcp/aqw038
41. Alsabeh R, Brynes RK, Slovak ML (1997) Acute myeloid leukemia with t(6;9) (p23;q34): association with myelodysplasia, basophilia, and initial CD34 negative immunophenotype. Am J Clin Pathol 1997:107. https://doi.org/10.1093/ajcp/107.4.430
42. Lai YY, Li Y, Shi Y (2012) Characteristics of 11 patients with acute myeloid leukemia accompanied with karyotype aberration t(6;9). Zhongguo Shi Yan Xue Ye Xue Za Zhi 2012:20
43. Oyarzo MP, Lin P, Glassman A (2004) Acute myeloid leukemia with t(6;9) (p23;q34) is associated with dysplasia and a high frequency of flt3 gene mutations. Am J Clin Pathol 2004:122. https://doi.org/10.1309/5DGB59KQA527PD47
44. Slovak ML, Gundacker H, Bloomfield CD (2006) A retrospective study of 69 patients with t(6;9)(p23;q34) AML emphasizes the need for a prospective, multicenter initiative for rare "poor prognosis" myeloid malignancies. Leukemia 2006:20. https://doi.org/10.1038/sj.leu.2404233
45. Bernstein J, Dastugue N, Haas OA (2000) Nineteen cases of the t(1;22)(p13;q13) acute megakaryblastic leukaemia of infants/ children and a review of 39 cases: report from a t(1;22) study group. Leukemia 2000:14. https://doi.org/10.1038/sj.leu.2401639
46. Carroll A, Civin C, Schneider N (1991) The t(1;22) (p13;q13) is nonrandom and restricted to infants with acute megakaryoblastic leukemia: a pediatric oncology group study. Blood 1991:78
47. Inaba H, Zhou Y, Abla O (2015) Heterogeneous cytogenetic subgroups and outcomes in childhood acute megakaryoblastic leukemia: a retrospective international study. Blood 2015:126. https://doi.org/10.1182/blood-2015-02-629204
48. Cuneo A, Ferrant A, Michaux JL et al. (1996) Philadelphia chromosome-positive acute myeloid leukemia: cytoimmunologic and cytogenetic features. Haematologica 81(5):423–7
49. Paietta E, Racevskis J, Bennett JM, Neuberg D, Cassileth PA, Rowe JM, Wiernik PH (1998) Biologic heterogeneity in Philadelphia chromosome-positive acute leukemia with myeloid morphology: the Eastern Cooperative Oncology Group experience. Leukemia 12(12):1881–5. https://doi.org/10.1038/sj.leu.2401229
50. Soupir CP, Vergilio JA, Dal Cin P, Muzikansky A, Kantarjian H, Jones D, Hasserjian RP (2007) Philadelphia chromosome-positive acute myeloid leukemia: a rare aggressive leukemia with clinicopathologic features distinct from chronic myeloid leukemia in myeloid blast crisis. Am J Clin Pathol 127(4):642–50. https://doi.org/10.1309/b4nver1ajj84ctuu
51. Muñoz L, Nomdedéu JF, Villamor N (2003) Acute myeloid leukemia with MLL rearrangements: clinicobiological features, prognostic impact and value of flow cytometry in the detection of residual leukemic cells. Leukemia 2003:17. https://doi.org/10.1038/sj.leu.2402708
52. Creutzig U, Harbott J, Sperling C (1995) Clinical significance of surface antigen expression in children with acute myeloid leukemia: results of study AML-BFM-87. Blood 1995:86
53. Medeiros BC, Kohrt HE, Arber DA (2010) Immunophenotypic features of acute myeloid leukemia with inv(3)(q21q26.2)/t(3;3)(q21;q26.2). Leukemia Res 2010:34. https://doi.org/10.1016/j.leukres.2009.08.029
54. Raya JM, Martín-Santos T, Luño E (2015) Acute myeloid leukemia with inv(3)(q21q26.2) or t(3;3)(q21;q26.2): clinical and biological features and comparison with other acute myeloid leukemias with cytogenetic aberrations involving long arm of chromosome 3. Hematology 2015:13
55. Hara Y, Shiba N, Yamato G et al. (2020) Patients aged less than 3 years with acute myeloid leukaemia characterize a molecularly and clinically distinct subgroup. Brit J Haematol 188(4):528–539. https://doi.org/10.1111/bjh.16203
56. Noort S, Wander P, Alonzo TA et al. (2021) The clinical and biological characteristics of NUP98-KDM5A in pediatric acute myeloid leukemia. Haematologica 106(2):630–634. https://doi.org/10.3324/haematol.2019.236745
57. Chisholm KM, Heerema-McKenney AE, Choi JK et al. (2020) Acute erythroid leukemia is enriched in NUP98 fusions: a report from

the Children's Oncology Group. Blood Adv 4(23):6000–6008. https://doi.org/10.1182/bloodadvances.2020002712
58. Nomdedéu J, Bussaglia E, Villamor N (2011) Immunophenotype of acute myeloid leukemia with NPM mutations: prognostic impact of the leukemic compartment size. Leukemia Res 2011:35. https://doi.org/10.1016/j.leukres.2010.05.015
59. Liu YR, Zhu HH, Ruan GR (2013) NPM1-mutated acute myeloid leukemia of monocytic or myeloid origin exhibit distinct immunophenotypes. Leukemia Res 2013:37. https://doi.org/10.1016/j.leukres.2013.03.009
60. Angelini DF, Ottone T, Guerrera G (2015) A leukemia-associated CD34/CD123/CD25/CD99+ immunophenotype identifies FLT3-mutated clones in acute myeloid leukemia. Clin Cancer Res 2015:21. https://doi.org/10.1158/1078-0432.CCR-14-3186
61. Lin LI, Chen CY, Lin DT (2005) Characterization of CEBPA mutations in acute myeloid leukemia: most patients with CEBPA mutations have biallelic mutations and show a distinct immunophenotype of the leukemic cells. Clin Cancer Res 2005:11. https://doi.org/10.1158/1078-0432.CCR-04-1816
62. Hou HA, Lin LI, Chen CY (2009) Reply to "heterogeneity within AML with CEBPA mutations; only CEBPA double mutations, but not single CEBPA mutations are associated with favorable prognosis". Brit J Cancer 2009:101. https://doi.org/10.1038/sj.bjc.6605207
63. Zangrando A, Cavagnero F, Scarparo P et al. (2021) CD56, HLA-DR, and CD45 recognize a subtype of childhood AML harboring CBFA2T3-GLIS2 fusion transcript. Cytometry A J Int Soc Analyt Cytol 99(8):844–850. https://doi.org/10.1002/cyto.a.24339
64. Eidenschink Brodersen L, Alonzo TA, Menssen AJ et al. (2016) A recurrent immunophenotype at diagnosis independently identifies high-risk pediatric acute myeloid leukemia: a report from Children's Oncology Group. Leukemia 30(10):2077–2080. https://doi.org/10.1038/leu.2016.119
65. Masetti R, Bertuccio SN, Pession A, Locatelli F (2019) CBFA2T3-GLIS2-positive acute myeloid leukaemia. A peculiar paediatric entity. Brit J Haematol 184(3):337–347. https://doi.org/10.1111/bjh.15725
66. Haferlach T, Kohlmann A, Klein HU et al. (2009) AML with translocation t(8;16)(p11;p13) demonstrates unique cytomorphological, cytogenetic, molecular and prognostic features. Leukemia 23(5):934–43. https://doi.org/10.1038/leu.2008.388
67. Xie W, Hu S, Xu J, Chen Z, Medeiros LJ, Tang G (2019) Acute myeloid leukemia with t(8;16)(p11.2;p13.3)/KAT6A-CREBBP in adults. Ann Hematol 98(5):1149–1157. https://doi.org/10.1007/s00277-019-03637-7
68. Haferlach T, Kohlmann A, Klein HU (2009) AML with translocation t(8;16)(p11;p13) demonstrates unique cytomorphological, cytogenetic, molecular and prognostic features. Leukemia 2009:23. https://doi.org/10.1038/leu.2008.388
69. Dang H, Chen Y, Kamel-Reid S (2013) CD34 expression predicts an adverse outcome in patients with NPM1-positive acute myeloid leukemia. Human Pathol 2013:44. https://doi.org/10.1016/j.humpath.2013.03.007
70. Chen CY, Chou WC, Tsay W (2013) Hierarchical cluster analysis of immunophenotyped classify AML patients with NPM1 gene mutation into two groups with distinct prognosis. BMC Cancer 2013: 13. https://doi.org/10.1186/1471-2407-13-107

Chapter 10

Flow Cytometric Immunophenotyping of Acute Lymphoblastic Leukemia

Dana Delgado, Jolene L. Cardinali and Joseph A. DiGiuseppe

Abstract

Immunophenotyping by flow cytometry is an important component in the diagnostic evaluation of patients with acute lymphoblastic leukemia. This technique further permits the detection of minimal residual disease after therapy, a robust prognostic factor that may guide individualized treatment. We describe here laboratory methods for both the initial characterization of lymphoblasts at diagnosis and the detection of rare leukemic lymphoblasts after treatment. In addition to antibody combinations suitable for diagnosis and detection of minimal residual disease, we describe procedures for peripheral blood and bone marrow sample preparation, procedures for labeling of cell-surface and intracellular proteins with fluorochrome-conjugated antibodies, and approaches to analysis of immunophenotypic data, including those obtained in patients following CD19-targeted therapies.

Keywords Immunophenotyping, Flow cytometry, Acute lymphoblastic leukemia, Minimal residual disease, Antibodies, Targeted therapy

1 Introduction

Acute lymphoblastic leukemia (ALL) comprises a group of malignant neoplasms of committed B-lineage or T-lineage precursor lymphoid cells termed lymphoblasts [1]. Diagnosis and classification of ALL require an integrated approach incorporating morphology, immunophenotype, cytogenetics, and molecular genetics [2, 3]. Immunophenotyping by flow cytometry is an important component in the diagnostic evaluation of patients with ALL [4, 5], enabling its rapid distinction from morphologic mimics such as acute myeloid leukemia (AML) and acute leukemia of ambiguous lineage, including mixed phenotype acute leukemia (MPAL) [2, 3]. In addition, immunophenotyping by flow cytometry permits the detection and quantification of rare leukemic blasts that may persist after treatment of ALL (reviewed in [5–8]). The presence of such minimal residual disease (MRD) following therapy is among the strongest prognostic

Angélique Biancotto and Pradeep K. Dagur (eds.), *Immunophenotyping*, Methods in Molecular Biology, vol. 3087,
https://doi.org/10.1007/978-1-0716-5607-5_10,

factors in ALL [9] and may potentially inform subsequent individualized therapies. Here we describe methods for immunophenotyping of ALL by flow cytometry. We describe antibody combinations suitable for diagnosis and detection of MRD, procedures for peripheral blood and bone marrow sample preparation, procedures for labeling of cell-surface and intracellular proteins with fluorochrome-conjugated antibodies, and approaches to analysis of immunophenotypic data, including those obtained in patients following CD19-targeted therapies.

2 Materials

2.1 Reagents

1. *PBS/azide (Phosphate-buffered saline with 0.1% sodium azide)*: Add 120.0 g NaCl, 3.0 g KCl, 17.25 g anhydrous Na_2HPO_4, 3.0 g KH_2PO_4, and 15.0 g of sodium azide to a 15 L container, and fill to 15 L with distilled water. Mix thoroughly, allow to sit overnight, and verify pH between 7.2 and 7.4. Store at room temperature for up to 3 months.
2. *0.5% PBS/BSA/azide (Phosphate-buffered saline with 0.5% Bovine serum albumin (w/v) and 0.1% sodium azide)*: Add 7.5 g bovine serum albumin (BSA) to 1500 mL PBS/azide (Reagent 1) in a 1500-mL bottle. Mix thoroughly until the BSA is completely dissolved. Store at 2–8°C for up to 3 months.
3. *2.5% PBS/BSA/azide (Phosphate-buffered saline with 2.5% Bovine serum albumin (w/v) and 0.1% sodium azide)*: Add 5.0 g BSA to 200 mL PBS/azide (Reagent 1) in a 250-mL Erlenmeyer flask. Mix thoroughly, and store at 2–8°C for up to 3 months.
4. *5% PBS/BSA/azide (Phosphate-buffered saline with 5% Bovine serum albumin (w/v) and 0.1% sodium azide)*: Add 10.0 g BSA to 200 ml PBS/azide (Reagent 1) in a 250-mL Erlenmeyer flask. Mix thoroughly, and store at 2–8°C for up to 3 months.
5. *RPMI/FCS (Roswell Park Memorial Institute 1640 medium with 10% fetal calf serum (v/v))*: Incubate 50 mL of fetal calf serum at 37°C for 30 min, and add to 450 mL of RPMI. Store at 2–8°C for up to 1 month.
6. *2% Formaldehyde solution*: Add 200 mL of 10% methanol-free formaldehyde to a 1000-mL volumetric flask. Bring to volume with 0.5% PBS/azide, and adjust pH to 7.2–7.4. Store at 2–8°C for up to 6 months.
7. *FACS™ Lysing Solution (BD Biosciences)*: Add 20 mL of 10× FACS™ Lysing Solution to a 200-mL volumetric flask. Bring to volume with distilled water. Store in a light-resistant bottle at 2–8°C for up to 1 month.

2.2 Antibody Combinations

Antibody combinations suitable for the diagnosis and detection of minimal residual disease (MRD) in acute lymphoblastic leukemia (ALL) are presented in Table 1. These combinations were developed for use on flow cytometers equipped with three lasers (excitations at 405 nm, 488 nm, and 633 nm), and capable of detecting at least eight different fluorochromes. Not all the antibody combinations in Table 1 are required in each case. Tubes 1, 2, and 7–9 would suffice to establish a diagnosis of B lymphoblastic leukemia, and tubes 5–9 would be sufficient for a diagnosis of T lymphoblastic leukemia; in both instances, the assessment of intracellular antigens (tubes 7–9) permits exclusion of MPAL. For MRD assessment in B-lineage ALL (tubes 2–4) and T-lineage ALL (tubes 5 and 6), more focused panels are appropriate.

3 Methods

Cell suspensions can be prepared from peripheral blood or bone marrow samples anticoagulated with either EDTA or heparin (*see* **Note 1**). Universal precautions should be observed in handling specimens.

3.1 Preparation of Cell Suspension from Peripheral Blood

1. Pipet an anticoagulated peripheral blood specimen into a 10-mL specimen tube labeled with two patient identifiers, and denote the meniscus with a permanent marker.
2. Dilute the peripheral blood with 0.5% PBS/BSA/azide to 10 mL.
3. Centrifuge at 800 RCF for 5 min.
4. Remove the supernatant, and reconstitute to 10 mL with 0.5% PBS/BSA/azide.
5. Centrifuge at 800 RCF for 5 min.
6. Repeat Steps 4 and 5.
7. Resuspend in 5.0% PBS/BSA/azide to the designated initial volume.
8. Adjust to a WBC concentration of 10,000/μL or less (*see* **Note 2**). For experiments in which cell-surface immunoglobulin light chains will not be labeled (see tubes 2–9, Table 1), this cell suspension is used in Procedures 3.3 and 3.4.
9. If cell-surface immunoglobulin light chains will be labeled (see tube 1, Table 1), aliquot 1 mL of the cell suspension prepared above into an empty 10-mL specimen tube labeled with two patient identifiers, denote the meniscus with a permanent marker, dilute to 10 mL with FCS/RPMI, and incubate at 37°C for 30 min (*see* **Note 3**). Centrifuge, wash twice, and resuspend the incubated specimen as in Steps 3–8, prior to use in Procedures 3.3 and 3.4.

Table 1
Antibody combinations suitable for diagnosis and detection of minimal residual disease in acute lymphoblastic leukemia

Tube	FITC	PE	PerCP-Cy5.5	PE-Cy7	APC	APC-AF750	PB, V450 or BV421	V500
1	Kappa	Lambda	CD20	CD19	CD10	CD38	CD5	CD45
2	CD20	CD22	CD34	CD19	CD13+CD33	CD38	CD10	CD45
3	CD20	CD49f	CD34	CD19	CD58	CD38	CD10	CD45
4	CD24	CD304	CD34	CD19	CD86	CD38	CD10	CD45
5	CD16	CD56	CD5	CD3	CD7	CD8	CD4	CD45
6	CD7	CD1a	CD3	CD2	CD5	CD8	CD4	CD45
7	cyMPO	cyCD3	CD34	-	CD7	-	HLA-DR	CD45
8	cyMPO	cyCD22	cyCD79a	CD19	CD34	-	HLA-DR	CD45
9	nTdT	cyCD3	cyCD79a	CD19	CD34	-	HLA-DR	CD45

FITC fluorescein isothiocyanate, *PE* phycoerythrin, *PerCP-Cy5.5* peridinin-chlorophyll-Cy5.5, *APC* allophycocyanin, *APC-AF750* APC-Alexa Fluor 750, *PB* Pacific Blue, *V450* BD Horizon™ V450, *BV421* Brilliant Violet™ 421, *V500* BD Horizon™ V500

3.2 Preparation of Cell Suspension from Bone Marrow

1. Filter anticoagulated bone marrow specimen through a 53-μM nylon mesh into a labeled specimen cup.
2. Gently dislodge cells from marrow particles collected on the mesh using the bottom of a polystyrene tube or bulb of a transfer pipette, and wash any freed cells through the mesh with 2.5% PBS/BSA/azide.
3. Transfer the filtered specimen to an appropriately labeled 10-mL specimen tube. Cap and mix well. Denote the meniscus with a permanent marker. Obtain a cell count of the suspension.
4. Centrifuge at 800 RCF for 10 min.
5. Remove the supernatant, resuspend in 0.5% PBS/BSA/azide, centrifuge at 800 RCF for 5 min, and decant the supernatant. Wash once more in this manner.
6. Resuspend in 5.0% PBS/BSA/azide to achieve a final WBC concentration of 10,000/μL. For experiments in which cell-surface immunoglobulin light chains will not be labeled (see tubes 2–9, Table 1), this cell suspension is used in Procedures 3.3 and 3.4.
7. If cell-surface immunoglobulin light chains will be labeled (see tube 1, Table 1), aliquot 1 mL of the washed cell suspension prepared above into an empty 10-mL specimen tube labeled with two patient identifiers, denote the meniscus with a permanent marker, add 10 mL of RPMI/FCS, and incubate at 37°C for 30 min (*see* **Note 3**).
8. Centrifuge, wash twice, and resuspend the incubated specimen as in Steps 3–8 of Procedure 3.1, prior to use in Procedures 3.3 and 3.4.

3.3 Labeling Procedure for Cell-Surface Antigens

1. Label 12 × 75-mm polystyrene tubes with two patient identifiers and the antibody combination to be added.
2. Deliver 100 μL of the appropriate antibody combination (cocktail) to each tube (*see* **Note 4**).
3. Deliver 100 μL of the cell suspension (or preincubated cell suspension in the case of immunoglobulin light chain studies) as prepared in Procedures 3.1 and 3.2 to each tube, and vortex (*see* **Note 5**).
4. Incubate the tubes for 15 min at room temperature in the dark.
5. Pipette 2 mL of FACS™ Lysing Solution into each tube, and vortex.
6. Incubate for 10 min at room temperature in the dark.
7. Centrifuge for 5 min at 800 RCF, and remove the supernatant.
8. Dilute to a volume of 2 mL with 0.5% PBS/BSA/azide, and vortex.

9. Centrifuge for 5 min at 800 RCF, and remove the supernatant.
10. Fix with 500 μL of 0.5% PBS/BSA/azide and 500 μL of 2% formaldehyde.
11. Acquire the labeled cells on the flow cytometer (*see* **Note 6**). Acquisition of 100,000 viable cells is sufficient for initial diagnosis; for MRD detection, acquisition of 1,000,000 cells is attempted (*see* **Note 7**). To mitigate carryover during acquisition for MRD detection, tubes containing distilled water should be positioned between sample tubes.

3.4 Combined Labeling Procedure for Cell-Surface and Intracellular Antigens

Abnormal patterns of cell-surface antigen expression (including expression of antigens that are normally absent, loss of antigens that are normally expressed, and expression of antigens or combinations of antigens at inappropriate developmental stages) are common among the acute leukemias, including B-lineage and T-lineage acute lymphoblastic leukemia. By comparison, cytoplasmic expression of certain antigens confers improved lineage specificity; incorporation of intracellular studies into the diagnostic panel is therefore useful. The following procedure, which permits simultaneous detection of cell-surface and intracellular antigens, should be used for antibody combinations 7–9 in Table 1 (*see* **Note 8**). We use the Fix & Perm cell fixation and cell permeabilization kit (Thermo Fisher), but similar products are manufactured by others.

1. Label 3 12×75-mm polystyrene tubes with two patient identifiers, and add 100 μL of cocktail containing antibodies directed against the cell-surface antigens designated in tubes 7–9 (Table 1).
2. Add 100 μL of cell suspension to each tube, and vortex.
3. Incubate for 15 min at room temperature in the dark.
4. Add 100 μL of Reagent A to each tube, and vortex.
5. Incubate for 15 min at room temperature in the dark.
6. Add 2 mL of 0.5% PBS/BSA/azide and centrifuge for 5 min at 800 RCF.
7. Decant the supernatant, add 100 μL of Reagent B to each tube, and vortex until the cell pellet is thoroughly resuspended.
8. Add the appropriate volumes of antibodies against the intracellular epitopes designated in tubes 7–9 (Table 1), and vortex.
9. Incubate for 15 min at room temperature in the dark.
10. Add 2 mL of 0.5% PBS/BSA/azide and centrifuge for 5 min at 800 RCF.
11. Decant the supernatant, and add 500 μL of 0.5% PBS/BSA/azide and 500 μL of 2% formaldehyde.

12. Acquire the labeled cells on the flow cytometer (*see* **Note 6**). Acquisition of 100,000 viable cells is sufficient for initial diagnosis; for MRD detection, acquisition of 1,000,000 cells is attempted (*see* **Note 7**). To mitigate carryover during acquisition for MRD detection, tubes containing distilled water should be positioned between sample tubes.

3.5 Data Analysis

At both diagnosis and assessment of MRD after therapy, the goal of data analysis is the identification of normal and abnormal cell populations on the basis of their immunophenotypic and light-scatter properties. Data analysis is an iterative process, which requires a thorough understanding of normal and abnormal maturation patterns [7, 10], and familiarity with the technical artifacts that may produce spurious immunophenotypic patterns [11]. The magnitude of this challenge is underscored by a study of MRD detection in B lymphoblastic leukemia, in which analytic precision among a cohort of experienced practitioners using identical data sets was disappointing [12]. Although the development of algorithms capable of analyzing complex multiparametric data sets holds the promise of improving analytic precision [13, 14], such approaches have yet to be widely adopted in routine practice.

3.5.1 Exclusion of Artifacts

Assessment of immunophenotypic patterns is facilitated by exclusion of artifacts due to fluidic perturbations during acquisition and focusing the analysis on viable singlet cells. A display of all events acquired over time (Fig. 1) may reveal sample exhaustion or fluidic perturbations; if present, such compromised data may be excluded from further analysis. Coincident events may yield unexpected combinations of antigen expression and thereby simulate abnormal populations. A display of peak area vs. peak height for a single parameter

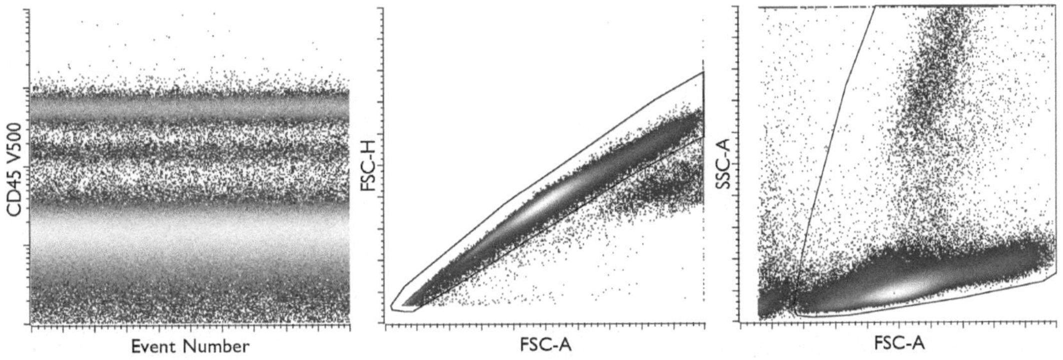

Fig. 1 Gating strategies to exclude overt data artifacts. (*Left*) Assessment of acquisition using event number to detect potential fluidic perturbations. (*Center*) Exclusion of coincident events (cellular doublets or higher-ordered aggregates) by gating on singlets defined by FSC-A and FSC-H. (*Right*) Exclusion of debris and nonviable cells by gating on cells with the FSC and SSC properties of viable cells

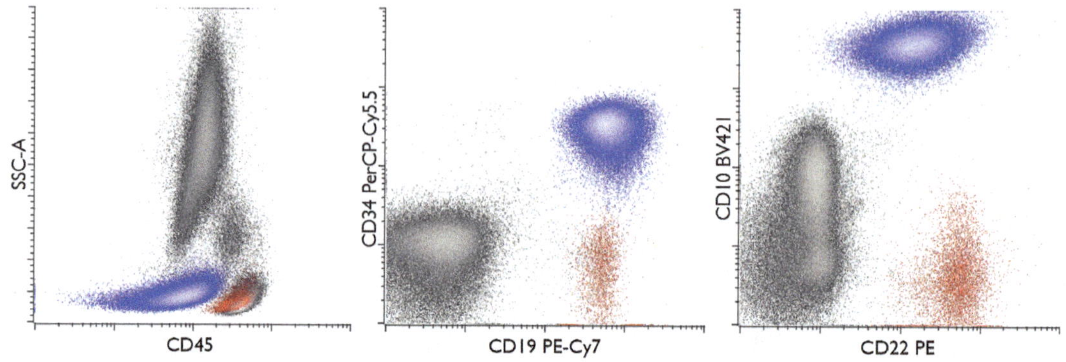

Fig. 2 Identification of blasts at diagnosis in B lymphoblastic leukemia (peripheral blood). The B-lymphoblasts (blue) have dim CD45 relative to mature B cells (red) and low SSC. In addition to B-lineage antigens (CD19, CD22), the blasts express CD10 and CD34. (All viable singlets are shown.)

enables gating on singlet events and exclusion of coincident events from subsequent analysis (Fig. 1). Finally, since degenerating or nonviable cells may be bound non-specifically by fluorochrome-conjugated antibodies, the analysis may be facilitated by gating on cells with the light-scatter properties of viable cells (Fig. 1; *see* **Note 9**).

3.5.2 Identification of the Abnormal Population: Diagnosis

The next step in data analysis is identification of the abnormal population. A display of CD45 vs. SSC is very useful in this regard, as lymphoblasts commonly occupy a region defined by dim CD45 and low SSC (Figs. 2 and 3). In the case of B lymphoblastic leukemia, CD45 expression may be virtually undetectable (Fig. 1), while in the case of T lymphoblastic leukemia, overlap of CD45 expression between the blasts and mature T cells is more common (Fig. 2). Bivariate plots of antigen vs. light scatter, or antigen vs. antigen, are then used to refine phenotypic characterization of the blast population (Figs. 2 and 3). Identification of the abnormal population is usually straightforward at the time of diagnosis, since blasts typically comprise a substantial fraction of all cells.

3.5.3 Identification of the Abnormal Population: MRD

By comparison, identification of a rare persistent blast population after treatment may be challenging. In many cases, knowledge of the patient's diagnostic immunophenotype may facilitate subsequent MRD detection. However, the patient's diagnostic studies may not be available at the time of MRD assessment. Moreover, alterations in phenotype during therapy are well described in both B-lineage [15] and T-lineage ALL [16]. In contrast, identification of populations whose composite immunophenotypes differ from those of normal cells is not subject to these limitations. An important caveat in the identification of MRD using either approach is the recognition of normal precursor lymphoid cells. In the case of MRD detection in B lymphoblastic leukemia, in particular, the presence of normal B-cell precursors [17] in bone marrow may complicate the analysis [12].

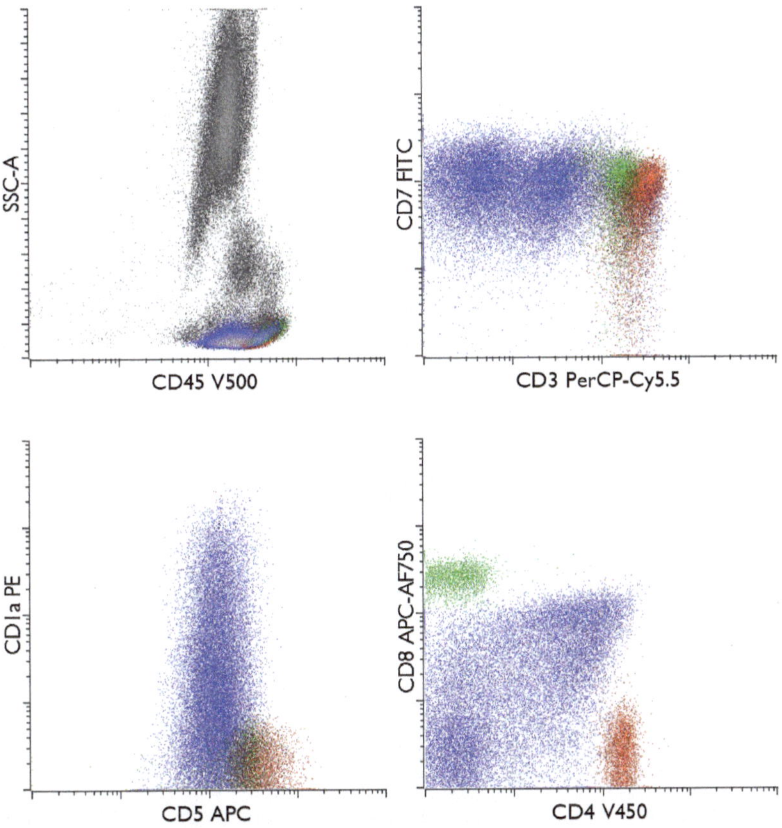

Fig. 3 Identification of blasts at diagnosis in T lymphoblastic leukemia (peripheral blood). The T-lymphoblasts (blue) express CD45 at levels slightly dimmer than (but partially overlapping with) those of the normal CD4+ (red) and CD8+ (green) T cells. The blasts are phenotypically heterogeneous, with absent or dim surface CD3 and partial, variable CD1a. Subsets of the blasts are CD4−/CD8−, CD4−/CD8(dim)+, CD4(dim)+/CD8−, or CD4+/CD8+. (All viable singlets are shown in the upper left panel; only T cells are shown in the remaining panels.)

In analyzing the data for potential abnormal populations after therapy, it is helpful to draw an initial gate using CD19 vs. SSC for B lymphoblastic leukemia (Fig. 4; see also Sect. 3.5.4) or CD7 vs. SSC for T lymphoblastic leukemia, as these antigens are expressed in virtually all cases. This initial gate is then refined in an iterative fashion using various combinations of antigen expression to isolate a cluster of events whose multiparametric phenotypic properties differ from those of normal precursor cells (Figs. 4 and 5). In the illustrated example of MRD detected in a patient treated for B lymphoblastic leukemia (Fig. 4), the initial CD19+ gate includes a preponderance of normal B-cell precursors (blue), with dim CD45, little or no CD20, and substantial CD34, all of which are immunophenotypic properties shared by the patient's MRD (red, highlighted). However, the MRD in this case differs from the normal B-cell precursors in its expression of one or more myeloid antigens (CD13+CD33), lack of CD10 and

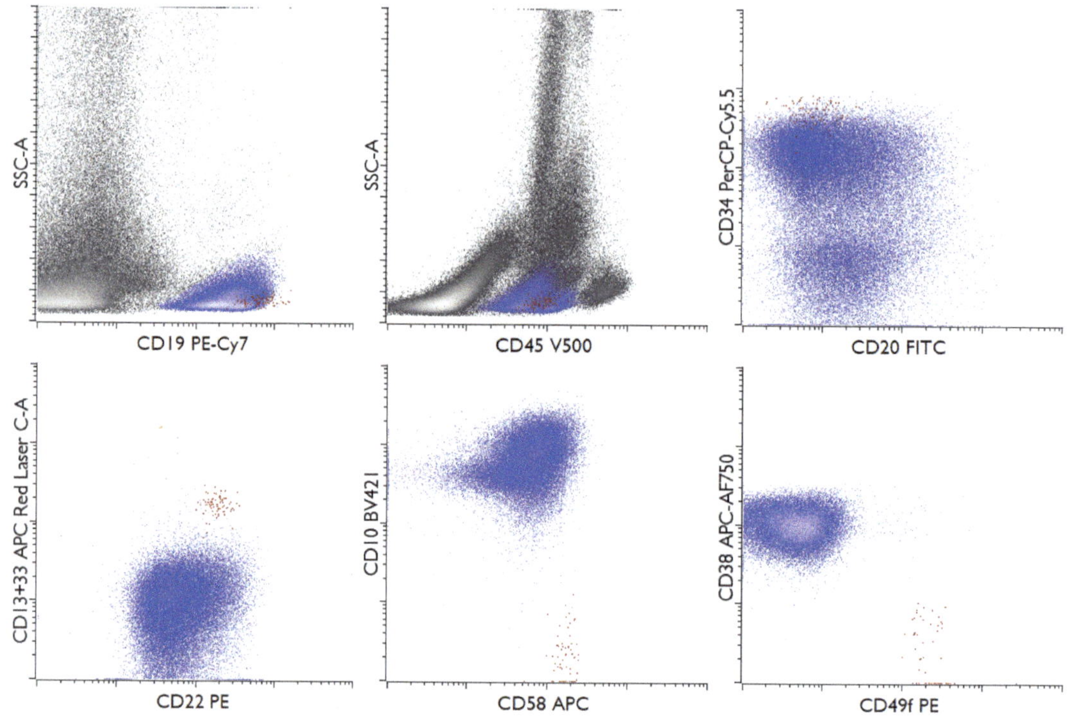

Fig. 4 Identification of MRD in B lymphoblastic leukemia (bone marrow). The initial CD19+ gate includes a preponderance of normal B-cell precursors (blue), with dim CD45, little or no CD20, and substantial CD34; in contrast, the MRD (red, highlighted) differs from the normal B-cell precursors in its expression of one or more myeloid antigens (CD13+CD33), lack of CD10 and CD38, and overexpression of CD49f and CD58. (All viable singlets are shown in the upper left and center panels; only B cells are shown in the remaining panels.)

CD38 expression, and overexpression of CD49f (*see* **Note 10**) and CD58 (Fig. 4). In the example of MRD detected in a patient treated for T lymphoblastic leukemia (Fig. 5), iterative gating discloses an abnormal T-cell population (blue, highlighted) with dim CD5 and CD45, absent surface CD3, and coexpression (albeit weak) of CD4 and CD8. These phenotypic features differ from those of the normal CD4+ and CD8+ T cells (red and green, respectively), but resemble those of a subset of blasts seen in the patient's diagnostic specimen (Fig. 5, lower right panel).

3.5.4 MRD Assessment in B-ALL Following CD19-Targeted Therapies

Therapeutic agents targeted against CD19 (e.g., blinatumomab [18, 19], anti-CD19 chimeric antigen receptor (CAR)-T cells [20]) represent a significant advance in the treatment of patients with B-ALL. Since treatment with these agents may result in loss of the epitope [21, 22], antibody combinations that do not rely on detection of CD19 should be used for MRD assessment in these patients. A robust method for the flow cytometric detection of MRD in B-ALL patients following CD19-targeted therapies using the alternative B-cell markers, CD22 and CD24, has been reported [23]. Importantly, the

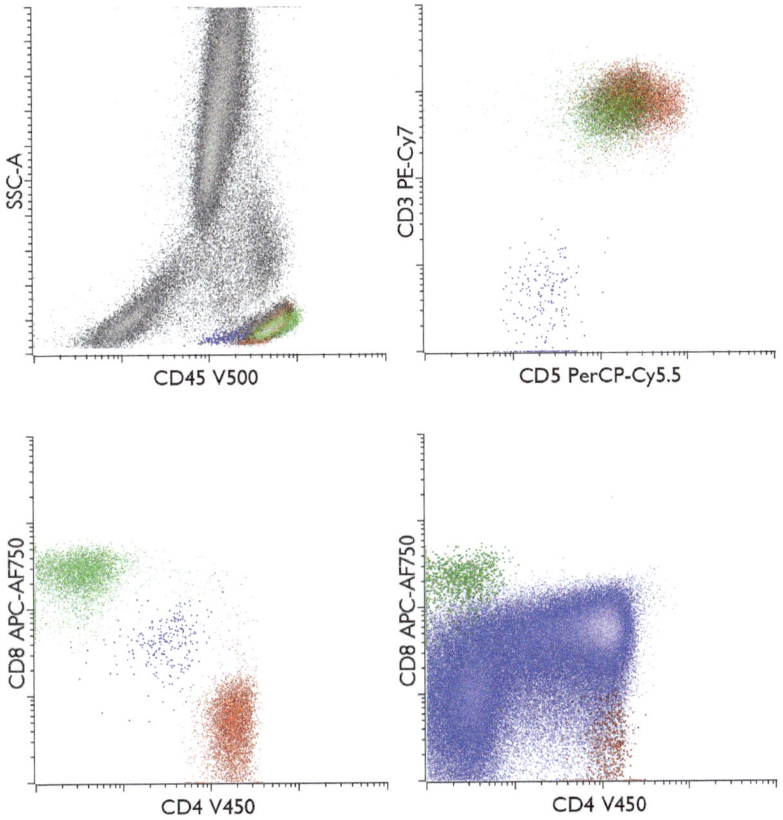

Fig. 5 Identification of MRD in T lymphoblastic leukemia (bone marrow). An abnormal T-cell population (blue, highlighted) is identified with dim CD5 and CD45, absent surface CD3, and coexpression (albeit weak) of CD4 and CD8. These phenotypic features differ from those of the normal CD4+ and CD8+ T cells (red and green, respectively) but resemble those of a subset of blasts seen in the patient's diagnostic specimen (*Lower Right*). (All viable singlets are shown in the upper left panel; only T cells are shown in the remaining panels.)

inclusion of CD24 in addition to CD22 permits detection of disease when CD22 is targeted therapeutically [24], either alone or in combination with CD19. The panel used in our laboratory for patients treated with CD19-targeted therapies is presented in Table 2. The analysis strategy for this panel employs a composite gate including viable singlets that are positive for CD22 and/or CD24 without CD66b, which permits exclusion of neutrophils [25] (Fig. 6). This composite CD22+ and/or CD24+ gate includes immature B cells at different stages of maturation, as well as mature B cells, and serves as the starting point for minimal residual disease (MRD) assessment.

In contrast with CD19 gating, this alternative gating strategy does not include plasma cells, which are predominantly negative for CD22 and CD24 [26]. However, the composite CD22+ and/or CD24+ gate is not entirely specific for B cells, since basophils and plasmacytoid dendritic cells are variably CD22+ with a number of

Table 2
Antibody combination suitable for detection of minimal residual disease in B-ALL patients treated with CD19-directed agents

FITC	PE	PerCP-Cy 5.5	PE-Cy7	APC	APC-H7	BV421	BV510
CD66b	CD22	CD34	CD19	CD24	CD45	CD10	CD38

FITC fluorescein isothiocyanate, *PE* phycoerythrin, *PerCP-Cy5.5* peridinin-chlorophyll-Cy5.5, *PE-Cy7* phycoerythrin-cyanine 7, *APC* allophycocyanin, *APC-H7*, *BV421* Brilliant Violet™ 421, *BV510* Brilliant Violet™ 510

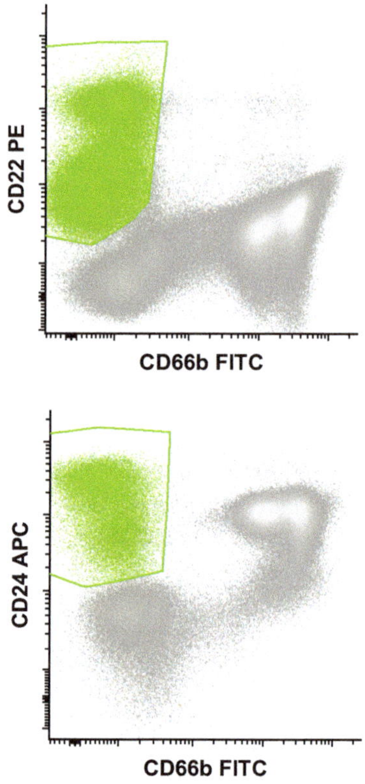

Fig. 6 Modified gating strategy for residual disease evaluation in B-lymphoblastic leukemia, posttargeted therapy. A combination gate was performed utilizing Boolean strategies that include cells positive for CD22 but not CD66b (green gate, upper plot) and/or cells positive for CD24 but not CD66b (green gate, lower plot). (All viable singlets are shown in both plots.)

clones [27]. Also important to recognize within the CD22+ and/or CD24+ gate are three normal CD19− negative populations that are seen at relatively low levels in most normal marrow specimens, but which may be increased following CD19-directed therapies, regardless of the presence of MRD [22, 28]. The first of these is an early progenitor cell of uncertain lineage (Fig. 7, red), which is brightly

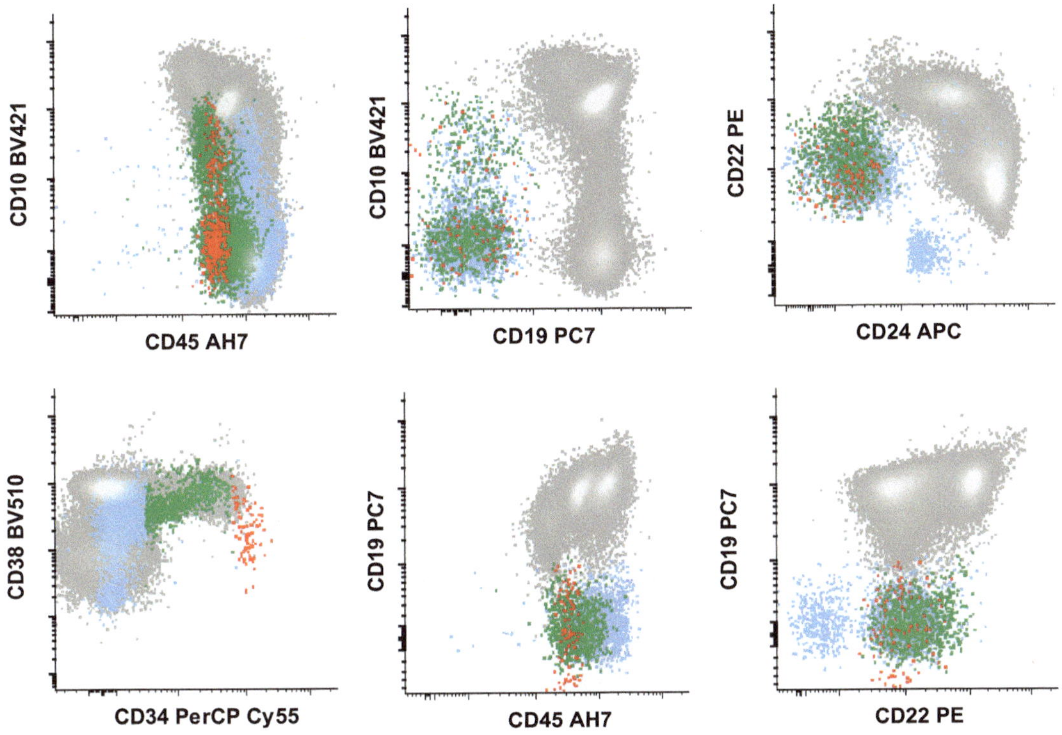

Fig. 7 Normal cell populations in modified gating strategy for evaluation of MRD in B-lymphoblastic leukemia, post therapy (bone marrow). Hematogones and mature B cells are shown in gray. Three CD19-negative populations are highlighted: (i) an early progenitor population of uncertain lineage (red), (ii) early (CD19−) B-cell precursors (green), and a heterogeneous nonB-cell population (light blue). The early progenitor cells of uncertain lineage (red) are brightly positive for CD34, with dim expression of CD22 and CD38, and negative for CD10 and CD24. The second population (green) is an early (CD19−) B-cell precursor, which expresses CD34 and CD38 at levels similar to those of normal hematogones and is further positive for CD10 (variable), CD20 (dim, variable), CD22, and CD79a. The third heterogeneous population (light blue) includes basophils, plasmacytoid dendritic cells, and cells of undetermined lineage. These are negative for CD10 and CD34 and show bimodal, reciprocal expression of CD22 and CD24. (Only cells within the CD22+ and/or CD24+ gate described in Fig. 6 are shown)

positive for CD34, with dim expression of CD22 and CD38, and predominantly negative for CD10 and CD24. The second population (Fig. 7, green) is an early B-cell precursor, prior to acquisition of CD19, which expresses CD34 and CD38 at the levels of normal hematogones, as well as CD22 at a level comparable with that of mature B cells, and is further positive for CD10 (partial), CD20 (dim, variable), and CD79a [21, 28]. The third gate (Fig. 7, light blue) includes a heterogeneous population of "non-B-cells" that includes basophils, plasmacytoid dendritic cells, and cells of undetermined lineage [28, 29]. This third (light blue) population is negative for CD10 and CD34 and is predominantly positive for CD22 without CD24; a smaller subset shows CD24 expression without CD22. Although the

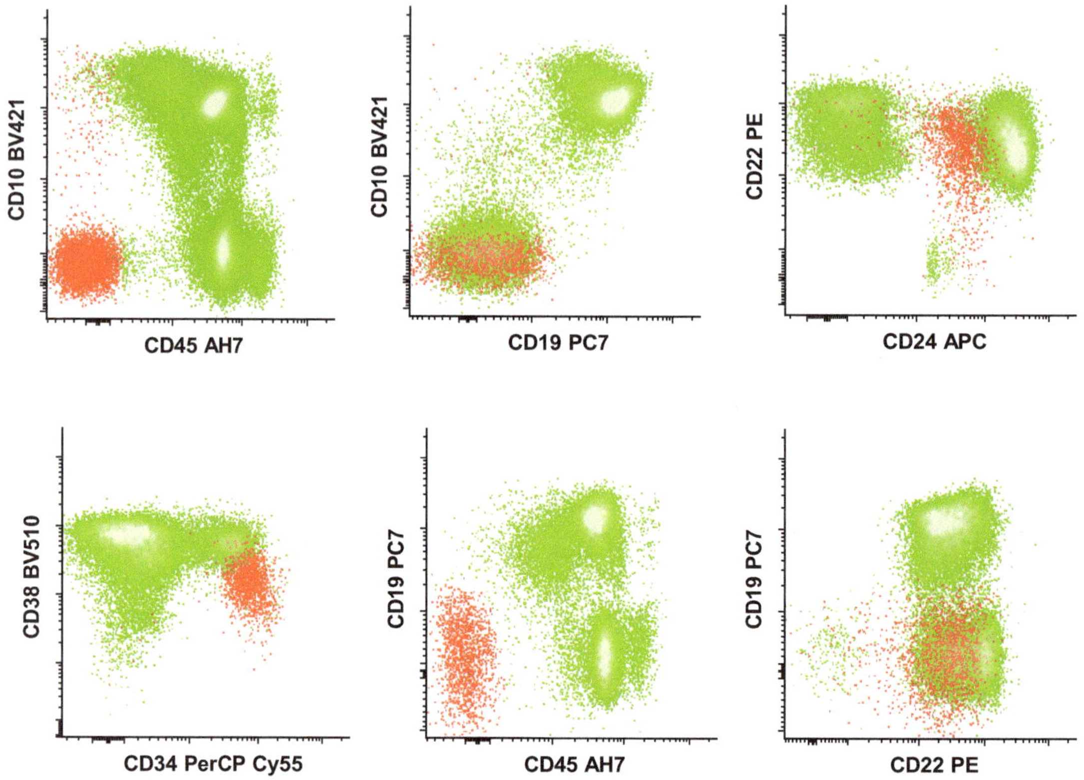

Fig. 8 Identification of MRD in B-lymphoblastic leukemia (bone marrow), posttargeted therapy. The abnormal B-lymphoblast population (red) is negative for CD19 and shows bright CD34 expression with slightly dimmer CD38, similar to the early progenitor population previously described in Fig. 7 (red). The key distinguishing features of the abnormal B-lymphoblasts are their dim expression of CD24, near-total lack of CD10, and complete absence of CD45. (Only cells within the CD22+ and/or CD24+ gate described in Fig. 6 are shown.)

identity of some of these cells is uncertain, it is essential to distinguish them from lymphoblasts when evaluating for CD19-negative disease.

The subsequent steps in identifying CD19-negative MRD within the CD22+ and/or CD24+ gate are similar to those discussed above. Normal B-cell maturation patterns and an example of MRD detected with this panel are illustrated in Fig. 8 (*see* **Note 11**).

4 Notes

1. In general, samples should be processed as soon after collection as is feasible to minimize loss of cell viability. However, if a delay in preparation of the cell suspension cannot be avoided, the specimen may be diluted 1:2 with RPMI and held overnight at 2–8°C.

2. Ideally, the final WBC concentration of the cell suspension will be 10,000/μL, enabling the delivery of 1,000,000 cells for labeling upon pipetting a volume of 100 μL. Should the final concentration of the cell suspension be 5000/μL or less, 200 μL of cell suspension may be added to 200 μL of cocktail.
3. Preincubation at 37°C allows for Fc-receptor-mediated endocytosis of bound immunoglobulin molecules, thereby reducing background staining in studies with fluorochrome-conjugated antibodies directed against immunoglobulin light chains.
4. The individual fluorochrome-conjugated antibodies that comprise previously characterized antibody combinations (see Table 1) may be pooled to produce "cocktails" with reagents sufficient for use in multiple (e.g., 50) assays. These are prepared such that the appropriate final concentration of each antibody will be achieved upon addition of 100 μL of cell suspension to 100 μL of cocktail. However, the laboratory should verify that the performance of the pooled reagents does not differ significantly from that observed with individually added reagents and determine the stability of each cocktail. We aliquot cocktails into light-resistant glass vials and have found these aliquots to be stable for up to 4 weeks at 2–8°C. To verify that the correct reagents have been added during its preparation, each new cocktail is tested with an appropriate normal specimen. Fluorescence intensities and patterns of antigen expression observed using the new cocktail should be visually indistinguishable from those seen with the previous cocktail (i.e., lot-to-lot comparison).
5. For MRD analysis, in which the goal is to collect up to 1,000,000 labeled cells, 200 μL of washed cell suspension (containing a total of 2,000,000 cells) should be added to 200 μL of cocktail.
6. Immediate acquisition of fixed samples is desirable, since breakdown of tandem fluorochromes is enhanced by exposure to heat, light, and fixatives. In our experience with the antibody combinations listed in Table 1, storage of fixed samples overnight at 2–8°C in the dark prior to acquisition on the flow cytometer does not result in substantial tandem degradation. However, tandem dyes and their conjugates vary widely in their propensity to degrade, and each laboratory should verify that storage of fixed, labeled samples prior to acquisition does not yield significant tandem degradation.
7. Visual recognition of a rare abnormal population comprising fewer than 20 cellular events is difficult, and the threshold for positivity in most studies demonstrating the clinical significance of MRD in ALL has been 0.01%, or 1 in 10,000

cells [9]. At a minimum, then, 200,000 viable cells would be required to recognize a cluster of 20 abnormal cells at a frequency of 0.01%. The ability to recognize a potentially abnormal cluster as a distinct cellular population is also influenced by the magnitude of antigenic deviation from normal, the extent to which antigen expression is homogeneous, and the number of admixed normal events; in certain instances, populations comprising greater than 20 cells may not be apparent. We therefore attempt to acquire 1,000,000 events for each antibody combination in MRD assays.

8. By enabling the simultaneous detection of a relatively myeloid-specific antigen (myeloperoxidase) and T-lineage (cytoplasmic CD3-ε; Tube 7) or B-lineage (cytoplasmic CD22 and CD79a; Tube 8) antigens, Tubes 7 and 8 facilitate the identification of mixed phenotype acute leukemia (MPAL), T/myeloid and B/myeloid, both of which should be distinguished from ALL. Tube 9 includes the nuclear antigen, TdT, in conjunction with T-lineage and B-lineage antigens; TdT expression (though not specific for ALL) distinguishes precursor lymphoid neoplasms from their mature counterparts.
9. By eliminating much of the background/nonspecific staining associated with degenerating or nonviable cells, selective gating on events with the light scatter properties of viable cells may facilitate recognition of normal and abnormal cellular populations during data analysis. However, care must be taken to ensure that cells of interest are not inadvertently excluded from subsequent analysis.
10. Overexpression of CD49f is among the most common recurrent phenotypic aberrations in B lymphoblastic leukemia [14, 30, 29], though we have observed an apparent diminution in CD49f expression in samples held for extended periods (e.g., greater than a day) prior to processing. Extended delays in sample processing should therefore be avoided.
11. Detection of MRD in patients with *KMT2A*-rearranged B-ALL treated with CD19-directed therapies may be particularly challenging, as these leukemias are frequently negative for CD10 [31] and CD24 [32] and may express very dim CD22 [33]. In these cases, MRD detection using NGS-based methods may be preferable.

Acknowledgments

The authors gratefully acknowledge Dr Brent Wood, Children's Hospital of Los Angeles, for his generous gift of WoodList software.

References

1. Jaffe ES, Arber DA, Campo E, Quintanilla-Fend L, Orazi A, Rimsza LM et al (2025) Hematopathology, 3rd edn. Elsevier
2. WHO (2024) Haematolymphoid tumours. IARC Publications
3. Arber DA, Borowitz MJ, Cook JR, de Leval L, Goodlad JR, Hasserjian RP et al (2025) The international consensus classification of myeloid and lymphoid neoplasms. Wolters Kluwer
4. Miller KP, Kovach AE (2024) Immunophenotyping by flow cytometry of acute lymphoblastic leukemia/lymphoma. In: Schmitz JL, Detrick B, O'Gorman MRG (eds) Manual of molecular and clinical laboratory immunology, 9th edn, ASM Press
5. DiGiuseppe JA, Wood BL (2019) Applications of flow cytometric immunophenotyping in the diagnosis and posttreatment monitoring of B and T lymphoblastic leukemia/lymphoma. Cytom B Clin Cytom 96(4):256–265
6. Chen X, Gao Q, Roshal M, Cherian S (2023) Flow cytometric assessment for minimal/measurable residual disease in B lymphoblastic leukemia/lymphoma in the era of immunotherapy. Cytom B Clin Cytom 104(3):205–223
7. Wood BL (2016) Principles of minimal residual disease detection for hematopoietic neoplasms by flow cytometry. Cytom B Clin Cytom 90:47–53
8. Gaipa G, Basso G, Biondi A, Campana D (2013) Detection of minimal residual disease in pediatric acute lymphoblastic leukemia. Cytom B Clin Cytom 84(6):359–369
9. Borowitz MJ, Devidas M, Hunger SP, Bowman WP, Carroll AJ, Carroll WL et al (2008) Clinical significance of minimal residual disease in childhood acute lymphoblastic leukemia and its relationship to other prognostic factors: a Children's Oncology Group study. Blood 111(12):5477–5485
10. Wood B (2004) Multicolor immunophenotyping: human immune system hematopoiesis. In: Darzynkiewicz Z, Roederer M, Tanke HJ (eds) Cytometry, 4th edn. New developments. Methods in cell biology, vol 75. Elsevier, San Diego, CA, pp 559–576
11. Wood B (2006) 9-color and 10-color flow cytometry in the clinical laboratory. Arch Pathol Lab Med 130(5):680–690
12. Keeney M, Wood BL, Hedley BD, DiGiuseppe JA, Stetler-Stevenson M, Paietta E et al (2017) A QA program for MRD testing demonstrates that systematic education can reduce discordance among experienced interpreters. Cytom B Clin Cytom 94:239–249
13. DiGiuseppe J, Tadmor MD, Pe'er D (2015) Detection of minimal residual disease in B lymphoblastic leukemia using viSNE. Cytom B Clin Cytom 88B:294–304
14. Lebecque B, Besombes J, Dannus LT, De Antonio M, Cacheux V, Greze V et al (2024) Faster clinical decisions in B-cell acute lymphoblastic leukaemia: a single flow cytometric 12-colour tube improves diagnosis and minimal residual disease follow-up. Br J Haematol 204(5):1872–1881
15. Gaipa G, Basso G, Aliprandi S, Migliavacca M, Vallinoto C, Maglia O et al (2008) Prednisone induces immunophenotypic modulation of CD10 and CD34 in nonapoptotic B-cell precursor acute lymphoblastic leukemia cells. Cytom B Clin Cytom 74(3):150–155
16. Roshal M, Fromm JR, Winter S, Dunsmore K, Wood BL (2010) Immaturity associated antigens are lost during induction for T cell lymphoblastic leukemia: implications for minimal residual disease detection. Cytom B Clin Cytom 78:139–146
17. McKenna RW, Washington LT, Aquino DB, Picker LJ, Kroft SH (2001) Immunophenotypic analysis of hematogones (B-lymphocyte precursors) in 662 consecutive bone marrow specimens by 4-color flow cytometry. Blood 98(8):2498–2507
18. Litzow MR, Sun Z, Mattison RJ, Paietta EM, Roberts KG, Zhang Y et al (2024) Blinatumomab for MRD-negative acute lymphoblastic leukemia in adults. N Engl J Med 391(4):320–333
19. Gupta S, Rau RE, Kairalla JA, Rabin KR, Wang C, Angiolillo AL et al (2024) Blinatumomab in standard-risk B-cell acute lymphoblastic leukemia in children. N Engl J Med
20. Park JH, Riviere I, Gonen M, Wang X, Senechal B, Curran KJ et al (2018) Long-term follow-up of CD19 CAR therapy in acute lymphoblastic leukemia. N Engl J Med 378:449–459
21. Cherian S, Stetler-Stevenson M (2018) Flow cytometric monitoring for residual disease in B lymphoblastic leukemia post T cell engaging targeted therapies. Curr Protoc Cytom 86(1):e44
22. Maude SL, Teachey DT, Porter DL, Grupp SA (2015) CD19-targeted chimeric antigen receptor T-cell therapy for acute lymphoblastic leukemia. Blood 125(26):4017–4023
23. Cherian S, Miller V, McCullouch V, Dougherty K, Fromm JR, Wood BL (2018) A novel flow cytometric assay for detection of residual disease in patients with B-lymphoblastic leukemia/lymphoma post anti-CD19 therapy. Cytom B Clin Cytom 94(1):112–120
24. Kantarjian HM, DeAngelo DJ, Stelljes M, Martinelli G, Liedtke M, Stock W et al (2016)

Inotuzumab ozogamicin versus standard therapy for acute lymphoblastic leukemia. N Engl J Med 375(8):740–753

25. Elghetany MT, Patel J (2002) Assessment of CD24 expression on bone marrow neutrophilic granulocytes: CD24 is a marker for the myelocytic stage of development. Am J Hematol 71(4):348–349
26. Mejstrikova E, Hrusak O, Borowitz MJ, Whitlock JA, Brethon B, Trippett TM et al (2017) CD19-negative relapse of pediatric B-cell precursor acute lymphoblastic leukemia following blinatumomab treatment. Blood Cancer J 7(12):659
27. Reineks EZ, Osei ES, Rosenberg A, Auletta J, Meyerson HJ (2009) CD22 expression on blastic plasmacytoid dendritic cell neoplasms and reactivity of anti-CD22 antibodies to peripheral blood dendritic cells. Cytom B Clin Cytom 76(4):237–248
28. Mikhailova E, Semchenkova A, Illarionova O, Kashpor S, Brilliantova V, Zakharova E et al (2021) Relative expansion of CD19-negative very-early normal B-cell precursors in children with acute lymphoblastic leukaemia after CD19 targeting by blinatumomab and CAR-T cell therapy: implications for flow cytometric detection of minimal residual disease. Br J Haematol 193(3):602–612
29. Shinoda-Matsuoka H, DiGiuseppe JA (2009) Overexpression of CD49f in precursor B-cell acute lymphoblastic leukemia: efficiency of minimal residual disease (MRD) detection by flow cytometric immunophenotyping compared with commonly used markers. Mod Pathol 89:388A
30. DiGiuseppe JA, Fuller SG, Borowitz MJ (2009) Overexpression of CD49f in precursor B-cell acute lymphoblastic leukemia: potential usefulness in minimal residual disease detection. Cytom B Clin Cytom 76(2):150–155
31. Pui CH, Frankel LS, Carroll AJ, Raimondi SC, Shuster JJ, Head DR et al (1991) Clinical characteristics and treatment outcome of childhood acute lymphoblastic leukemia with the t(4;11) (q21;q23): a collaborative study of 40 cases. Blood 77(3):440–447
32. Schwartz S, Rieder H, Schlager B, Burmeister T, Fischer L, Thiel E (2003) Expression of the human homologue of rat NG2 in adult acute lymphoblastic leukemia: close association with MLL rearrangement and a CD10(-)/CD24(-)/CD65s(+)/CD15(+) B-cell phenotype. Leukemia 17(8):1589–1595
33. Shah NN, Stevenson MS, Yuan CM, Richards K, Delbrook C, Kreitman RJ et al (2015) Characterization of CD22 expression in acute lymphoblastic leukemia. Pediatr Blood Cancer 62(6):964–969

Check for updates

Chapter 11

Clinical Flow Cytometric Testing in Chronic Lymphocytic Leukemia

Nadia Nasir, Maryalice Stetler-Stevenson and Hao-Wei Wang

Abstract

Flow cytometry is the cornerstone for establishing the diagnosis of chronic lymphocytic leukemia (CLL), owing to its characteristic and well-defined immunophenotype that enables accurate distinction from other leukemias and lymphomas. Beyond diagnosis, flow cytometry provides essential prognostic information and allows sensitive detection of minimal residual disease (MRD), a strong predictor of clinical outcome. CLL MRD assessment is increasingly used to guide risk stratification, therapeutic decision-making, and treatment duration in the era of targeted therapies and immunotherapies. This chapter reviews best practices for specimen collection, processing, staining, and data analysis and summarizes the principles of flow cytometric MRD assessment in CLL.

Keywords Flow cytometry, Chronic lymphocytic leukemia, Minimal residual disease

1 Introduction

Chronic lymphocytic leukemia/small lymphocytic lymphoma (CLL/SLL) is a B-cell neoplasm consisting of small mature B cells with distinct morphologic and immunophenotypic features [1]. CLL/SLL involves the peripheral blood, bone marrow, and lymphoid tissues such as lymph nodes, spleen, and tonsils. The diagnosis is established by clinical evaluation, morphological assessment, and immunophenotyping by flow cytometry (FC) [1]. The diagnostic criteria for CLL/SLL are the same in the current editions of the International Consensus Classification (ICC) and the World Health Organization (WHO) Classification. [2–4]. A diagnosis of CLL requires $\geq 5 \times 10^9/l$ of monoclonal B cells in the peripheral blood that exhibit the characteristic morphology and immunophenotype, while a tissue-based diagnosis of SLL requires organ involvement by neoplastic B cells. Although CLL and SLL represent the same disease, the latter term is used for cases with predominantly nodal, splenic, or extramedullary involvement and $<5 \times 10^9/l$ circulating

Angélique Biancotto and Pradeep K. Dagur (eds.), *Immunophenotyping*, Methods in Molecular Biology, vol. 3087,
https://doi.org/10.1007/978-1-0716-5607-5_11,

CLL cells [1]. For those cases with <5×10^9/l circulating monoclonal B cells and no evidence of lymphadenopathy, organomegaly, and other extramedullary disease, a diagnosis of monoclonal B-cell lymphocytosis (MBL) is rendered. The CLL-type MBL is considered a precursor condition to CLL, and it has an immunophenotype identical to CLL [5]. MBL is further subdivided into low-count and high-count MBL, based on a cut-off value of 0.5×10^9/l circulating monoclonal B cells. High-count MBL is associated with a risk of progression to CLL at a rate of 1%–2% per year, while low-count MBL carries minimal risk of progression [6–9].

The demonstration of the typical CLL immunophenotype by FC is vital for diagnosis. CLL characteristically expresses CD5, CD19, dim CD20, dim CD22, CD23, bright CD43, dim CD45, dim to negative CD79b, dim CD81, CD200, ROR1, and dim monoclonal surface immunoglobulin [Ig] (see Fig. 1 and **Notes 1 and 2**) and is typically negative for CD10, CD103, CD123, and other T- cell and myeloid antigens [10, 11]. The differential diagnosis of CLL/SLL primarily includes mantle cell lymphoma and other low-grade B-cell neoplasms that occasionally co-express CD5. The expression of CD200 and CD23 in combination with dim expression of CD20, CD22, CD45, CD79b, and CD81, as well as bright CD43, differentiates CLL from MCL [12]. Infrequently, CLL cases may have an atypical immunophenotype, such as a lack of CD5, CD23, and CD43; normal intensity of CD20, CD22, CD79b, and surface Ig;

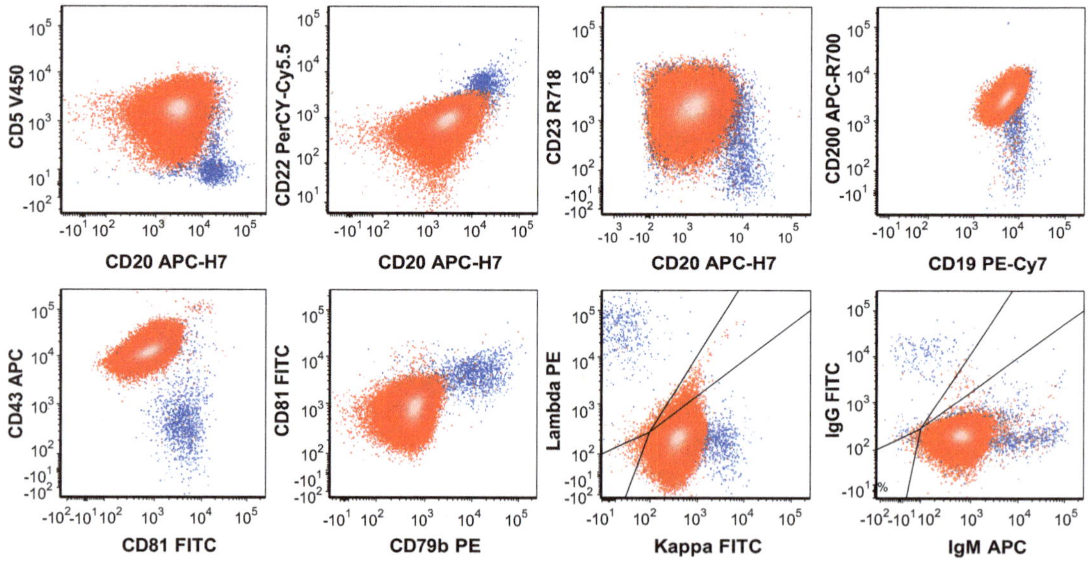

Fig. 1 Typical immunophenotype of CLL/SLL. Abnormal CLL cells are highlighted in red and normal B cells in blue. CLL cells are positive for CD5, CD19, CD23, CD43, and CD200, and have dim expression of CD20, CD22, CD79b, and CD81. The CLL cells are monoclonal, expressing dim IgM/Kappa surface immunoglobulin in this case

and aberrant expression of T-cell, myeloid, or other B-cell antigens (see **Note 3**) [12–14].

In addition to its diagnostic utility, FC also provides important prognostic information. CD38 expression in CLL has been found to be associated with an unmutated immunoglobulin heavy chain variable gene (IGHV) and an aggressive clinical course [15, 16]. Therefore, CD38 expression (using 30% as the cutoff) has been commonly used as a prognostic indicator in CLL, along with other biomarkers, including the mutational status of the IGHV gene and ZAP-70 positivity [17, 18]. However, the expression of CD38 may vary during the disease courses, and there have also been other studies that showed no correlation with IGHV mutational status or prognosis [18–20].

CD49d has emerged as a superior prognostic indicator in CLL that outperforms other FC-based markers, including CD38 and ZAP-70 [21, 22]. CD49d expression is stable over time, and determination of a positive or negative status is generally straightforward (see Fig. 2) [23, 24]. A cutoff value of 30% has been used to determine the positivity of CD49d expression, and a positive expression strongly correlates with an inferior response to both chemoimmunotherapy and BTK inhibitors. Additionally, approximately 20% of the CLL cases demonstrate a bimodal pattern of CD49d expression, with concomitant presence of a distinct CD49d-positive subpopulation and a CD49d-negative subpopulation. Irrespective of whether the CD49d-positive subpopulation exceeded the 30% cutoff or not, cases with bimodal CD49d had worse clinical outcomes similar to cases with homogeneous CD49d [25, 26]. Thus, the prognostic value of CD49d is improved by including the bimodal CD49d expression pattern.

FC detection of minimal residual disease (MRD) levels during and after therapy has also been shown to be an independent predictor of progression-free and overall survival in CLL [24, 27]. The outcome of CLL has significantly improved due to the development of more effective treatment options for CLL. These include targeted therapies such as inhibitors of Bruton tyrosine kinase (BTK) and BCL2, and immunotherapies including anti-CD20 antibodies and chimeric antigen receptor (CAR) T-cell therapy. Because of this, more patients can achieve complete remission (CR) without morphological evidence of diseases. However, a subset of patients who achieve CR still relapse even after a rigorous treatment regimen, especially those with detectable disease at the MRD level [27]. As a result, MRD tests have been increasingly used to predict outcomes in CLL, and many clinical trials have shown that therapies able to achieve MRD-negative CR usually result in improved clinical outcomes [28–30]. Additionally, the MRD status has also been used to guide subsequent treatment based on the probability of relapse. For instance, a personalized treatment scheme with duration of therapy

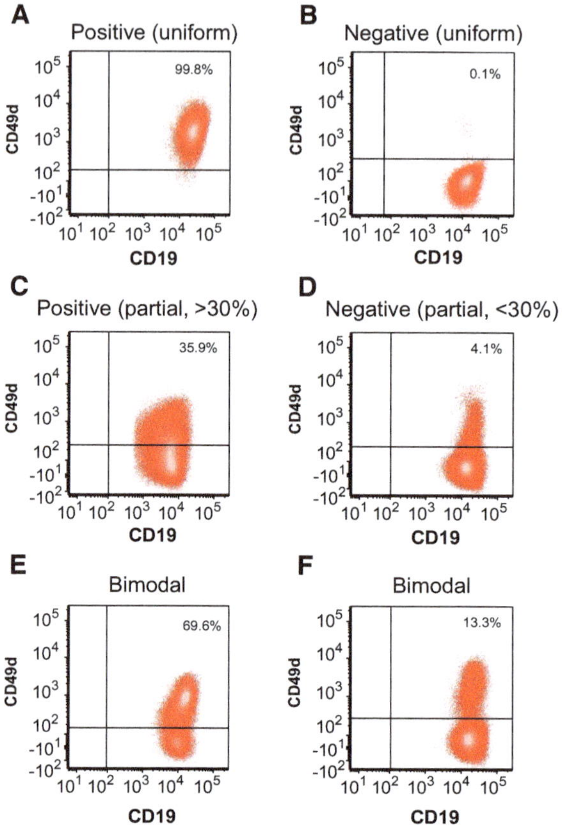

Fig. 2 Common expression patterns of CD49d in CLL/SLL. (**A**), Homogeneously positive expression of CD49d in CLL cells. (**B**), Homogeneously negative expression of CD49d in CLL cells. (**C**), Partial expression of CD49d in greater than 30% of the CLL cells, defined as CD49-positive. (**D**), Partial expression of CD49d in less than 30% of the CLL cells, defined as CD49d-negative. (**E and F**), Bimodal CD49d pattern with concomitant presence of a CD49d-positive subpopulation separate from a CD49d-negative subpopulation, regardless of the percentages of the positive subpopulation

determined according to the time taken to achieve MRD-negative status has demonstrated improved outcome [31, 32]. Thus, not only is it useful for predicting outcomes, but MRD evaluation in CLL can also be harnessed to tailor treatment.

2 Materials

2.1 Lysing Solution

1. To make 1 l of lysing solution, weigh 8.29 g ammonium chloride (NH_4Cl) and 1 g potassium bicarbonate ($KHCO_3$) and place both into a 1000 ml glass or clear plastic bottle.
2. Add 4 ml 0.5 M EDTA solution and fill with distilled water (H_2O) to a volume of 1000 ml. Cap and mix well.

3. Store at room temperature for up to 6 months.

2.2 Wash Solution: Phosphate Buffered Saline (PBS) with 10% Bovine Serum Albumin (BSA) and Sodium Azide

1. Weigh 4 g BSA and add to 4 L of 1× PBS with 0.1% sodium azide. Cap and mix well by inverting several times. Allow the wash solution to sit at room temperature for 30 min, mix again, and check to ensure that BSA has dissolved.
2. Perform pH testing after BSA is in solution and while the solution is at room temperature. Adjust pH to 7.3–7.5 by adding small amounts of concentrated HCL (if the pH is >7.5) or concentrated NaOH (if the pH is <7.3) (*see* **Notes 4 and 5**).
3. Store at 4°C for 2 months from preparation.

2.3 Fixative: 0.5% Formalin Fixative Solution

1. Mix 50 ml of 10% buffered formaldehyde with 950 ml of PBS in a bottle, cap, and mix well (*see* **Note 6**).
2. Wrap the bottle with foil and tape to protect from light. Check pH and adjust to 7.3–3.5 (see Sect. 2.2).
3. Store at room temperature for up to 1 year.

3 Methods

3.1 Cocktails and Panels

The panel in Table 1 is an example of an eight-color/ten-color panel that can be used to diagnose CLL and monitor disease post-therapy. Each cocktail is made by the addition of the appropriate volumes of different antibodies (Abs) as recommended by the vendor and/or titration. The amount of the cocktail to be added during staining tube (Sect. 3.6, Step 1 below) is based on the sum of the different Ab volumes in this cocktail. The cocktail should be stored at 4°C and light exposure minimized. The stability of these cocktails should be determined by each clinical laboratory.

3.2 Specimen Collection

PB and BM are the primary samples to be examined, although lymph node (LN) biopsy or fine needle aspiration (FNA) may be indicated. PB and BM specimens collected in sodium heparin are stable at room temperature for up to 72 and 24 h, respectively, while EDTA PB and BM specimens are stable for 24 and 12 h, respectively. Improper or lengthy specimen storage can negatively impact results. Overnight shipment of specimens with the possibility of exposing the specimen to temperature extremes has been shown to alter antigen expression [33]. An LN biopsy must be made into a cell suspension using mechanical tissue disaggregation (using commercial devices or manual tools). FNA samples usually do not require additional disaggregation. FNA or cell suspension can be stored in RPMI with 10% fetal bovine serum for up to 18 h at 4°C.

Table 1

Antibody panel for diagnosis and detection of minimal residual disease in chronic lymphocytic leukemia

Tube	FITC	PE	PerCP-Cy5.5 or PerCP	PE-Cy7	APC	APC-R700 or R718	APC-H7	V450	V500	BV605
1	Kappa	CD25	CD123	CD19	103	–	CD20	CD11c	CD45	Lambda
2	CD81	CD79b	CD22	CD19	CD43	–	CD20	CD5	CD3	–
3	Kappa	Lambda	CD20	CD19	CD5	CD3	CD14	CD38	CD45	CD10
4	Lambda	CD22	CD5	CD19	Kappa	CD200	CD20	CD11c	CD45	CD103
5	IgG	CD49d	CD22	CD19	IgM	–	CD20	CD5	CD3	–

FITC fluorescein isothiocyanate, *PE* phycoerythrin, *PerCP–Cy5.5* peridinin-chlorophyll–Cy5.5, *APC* allophycocyanin, *APC-R700* BD Horizon™ APC-R700, *R718* BD Horizon™ Red 718, *APC-H7* BD™ APC-H7, *V450* BD Horizon™ V450, *V500* BD Horizon™ V500, *BV605* BD Brilliant Violet™ 605

3.3 Prelysis, Cell Count, and Viability Assessment *[33,34]*

Minimizing sample treatment before staining and timely sample processing greatly reduces the risk of losing cells of interest (especially in the case of CLL MRD) and maintains cell viability and integrity.

1. Transfer ≤5 ml of either PB or BM into labeled 50 ml conical tube(s).
2. Add ammonium chloride lysing solution (see Sect. 2.1) to fill each conical tube. Cap and invert several times to mix. Incubate 10 min at room temperature, inverting to mix occasionally, and at the end.
3. Centrifuge tubes at 500 × *g* for 5 min, check for pellet, and aspirate or decant supernatant into waste container (*see* **Note** 7).
4. Cap and vortex tube(s) gently for less than 5 s to dislodge pellet(s). Fill the tube with PBS, cap, and invert to mix. Repeat Step 3 above (*see* **Note 8**).
5. Cap and vortex tube gently to dislodge the entire cell pellet. Add PBS to the cell pellet according to the pellet size (0.2–2 ml for small pellets and 3–5 ml for larger pellets), and flick or vortex gently to mix. Record the final cell suspension volume to be used later to adjust the cell count.
6. Assess viability and perform a cell count by using a viability dye that is excluded from viable cells and automated cell counters or a hemocytometer [35]. Samples with less than 75% viability should be rejected unless the sample is irreplaceable [33].

3.4 Cell Concentration Adjustment

1. Divide the total WBCs count (obtained by multiplying the cell count in the "count" tube by the dilution factor multiplied by the preparation volume in Sect. 3.3, Step 5) by the target WBCs count to get the amount of the required final volume. Example: if the total WBCs count = 15.0×10^6 and the target count is 20×10^6/ml, then the final volume should be 0.75 ml.
2. Adjust the final volume of the cell suspension to reach the target cell concentration of 20×10^6 cells/ml.

3.5 Staining Cells with Antibodies *[33,34]*

1. Deliver the appropriate volume of the premade cocktail (Sect. 3.1) to the bottom of the labeled 12 × 75 mm tubes using an automatic pipette.
2. Add 100 μl patient cell preparation to each tube and vortex briefly to mix. Incubate 30 min at room temperature, protected from the light.
3. Wash stained tubes by adding 3–4 ml wash solution to each tube; vortex to mix, then centrifuge at 500 × *g* for 5 min. Check for pellet and aspirate or decant tube into waste container, and flick or vortex tube to dislodge the entire cell pellet.
4. Repeat the wash Step 3 above.

5. Add 200–300 µl formalin fixative and vortex to mix and store stained and fixed cells at 4°C until acquisition.

3.6 Acquisition

Data is acquired on a cytometer. Care should be taken not to exclude low FSC events. Acquisition of ungated data is stored as list mode data. 1,000,000–2,000,000 events (depending on the lower limit of MRD detection required) are acquired when cellular content is adequate. In specimens with low cell count (e.g., FNA), acquire as many events as possible. Acquiring a high number of events is crucial for MRD detection. At the minimum, all fluorescence channels, time, FCS-A, FSC-H, and SSC-A should be acquired.

3.7 Analysis

Sequential gating of the list mode data is performed to distinguish abnormal cells of interest from normal cells based on light scatter and antigen profile. An example of a typical sequence of hierarchical analysis gating follows:

1. The Time vs. FSC-A (or SSC-A) plot of ungated cells is evaluated for discontinuity, indicating fluidic disturbances (e.g., bubbles, clogs, or in the case of low cell number, the tube being acquired dry). A time gate is drawn to exclude such events (Fig. 3).
2. A singlet gate is drawn on the viable gated FSC-A vs. FSC-H plots to exclude doublets (Fig. 3).
3. A viable gate is drawn on the time-gated FSC-A vs. SSC-A plot to exclude debris and inviable cells with very low scatter properties (Fig. 3).
4. The viable cells are displayed on an FSC-A vs. SSC-A plot to create a lymphocyte gate that includes all lymphocytes (Fig. 3).
5. A clean B-cell gate is drawn to identify CD19-expressing B-cells, excluding CD3-positive T cells and/or doublets not cleared by doublet exclusion (see Fig. 3 and **Note 9**).
6. The antigen profile of the lymphoid cells is evaluated under the Lymphocyte and B-cell gates to identify the CLL immunophenotype (see Fig. 1 and **Notes 10 and 11**).
7. For MRD assessment, low-level CLL involvement must be detected among a predominately polyclonal B-cell background. In addition to the general analysis approaches to identify the CLL immunophenotype (Fig. 1), a Boolean gating strategy proposed by the European Research Initiative in CLL (ERIC) can also be applied [11, 27, 36]. The ERIC CLL MRD assay has undergone several updates. In 2016, a standardized single-tube assay was validated, containing six essential markers, including CD19, CD20, CD5, CD43, CD79b, and CD81, which can reliably detect CLL MRD to the level of 1×10^{-5} [10]. Since most modern flow cytometry instruments offer

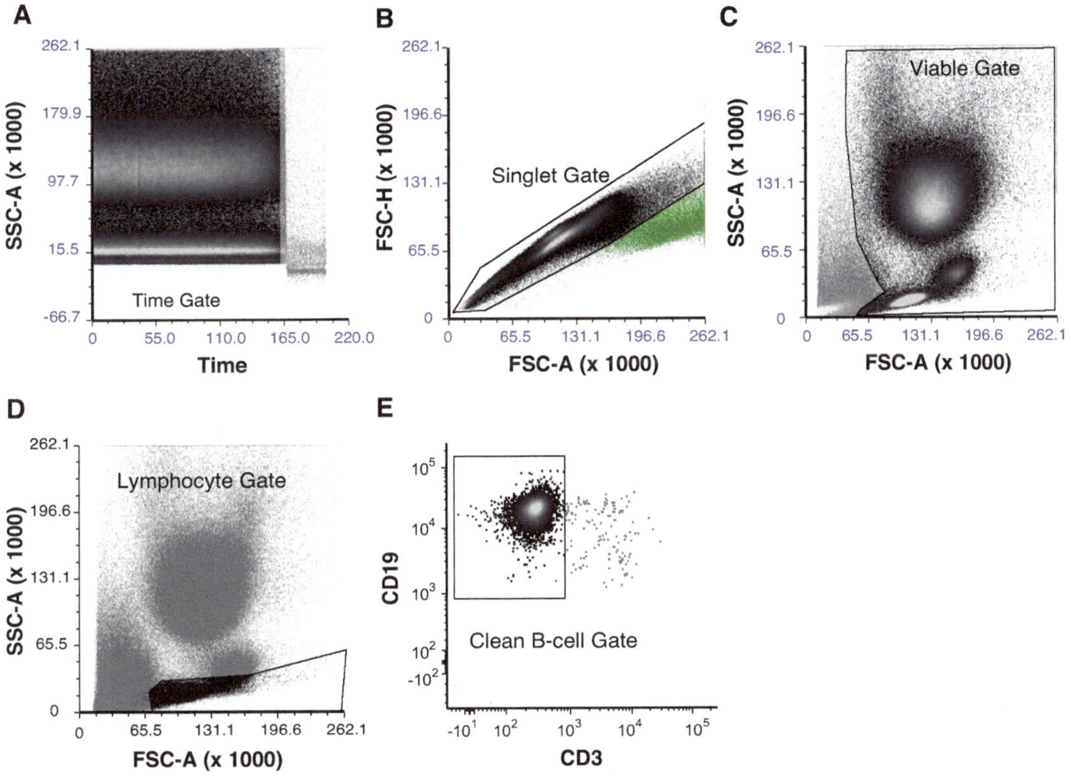

Fig. 3 General strategies of B-cell gating. (**A**), A time gate is set to exclude discontinuity caused by bubbles, clogs, or other fluidic disturbances. (**B**), Singlets gated to exclude doublets (shown in green). (**C**), Viable gate to exclude debris and inviable cells. (**D**), The viable cells are gated on FSC-A vs. SCC-A plot, which include all lymphocytes in the lymphocyte gate. (**E**), A clean B-cell gate is drawn to identify CD19-expressing B-cells and exclude CD3-positive T cells and doublets not cleared by doublet exclusion

eight or more colors, the six-marker core panel can be combined with additional markers as required. The ERIC method utilizes a Boolean gating strategy in which CD19-positive B-cells are evaluated and a series of analysis gates are created in areas where normal B-cell populations are not observed (Fig. 4). A B-cell gate that includes CD19-positive and CD3-negative cells (Fig. 3) is analyzed with pairs of antigens including CD5/CD20, CD5/CD22, CD5/CD79b, CD43/CD81, CD20/CD200, etc. (Fig. 4). The CLL MRD gate is defined by the combination of these antigen gates. Evaluation of surface light chain is not required for MRD; however, it may be useful in confirming the B-cell clonality, especially when rare disease is present.

8. The CD49d expression in CLL is evaluated. A positive expression is defined by an expression in >30% of CLL cells (Fig. 2). A bimodal pattern in which there is concomitant presence

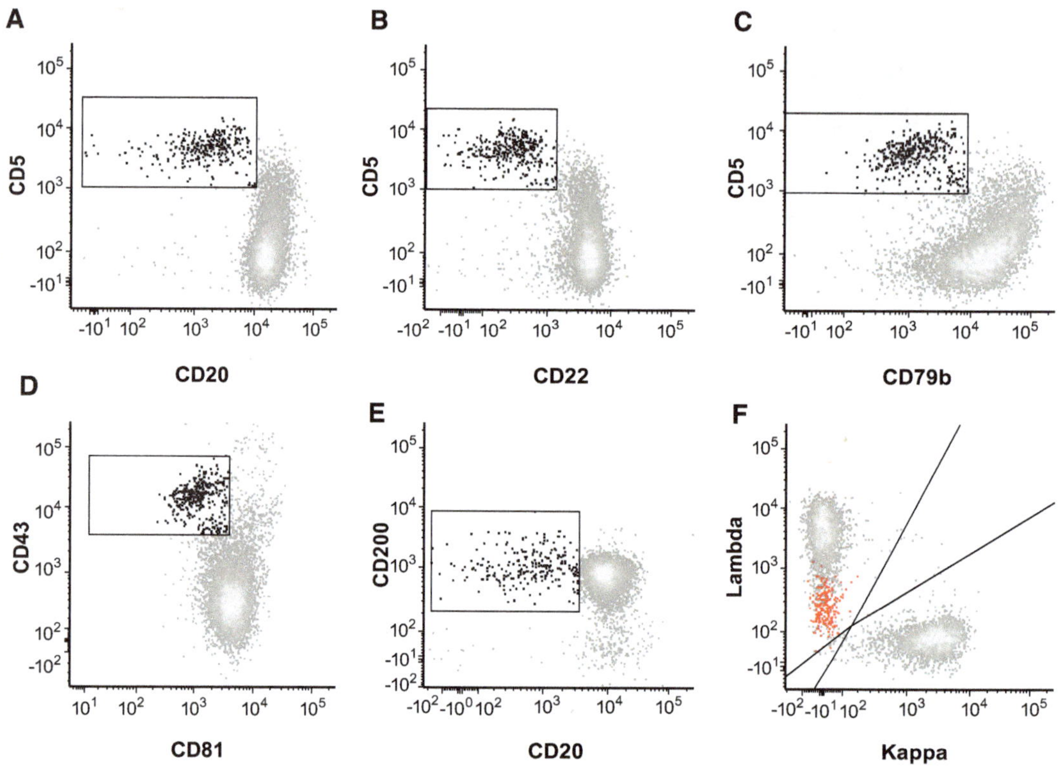

Fig. 4 Gating strategies for CLL MRD analysis (ERIC method). The clean B cell gate is analyzed with pairs of antigens. CLL MRD is detected by Boolean combinations of antigen gates that include **(A)** CD5-positive/CD20-dim, **(B)** CD5-positive/CD22-dim, **(C)** CD5-positive/CD79b-dim, **(D)** CD43-positive/CD81-dim and, **(E)** CD200-positive/CD20-dim events (black). A restricted light chain expression pattern is useful in confirming clonality of the CLL cells (red)

of a CD49d-positive subpopulation separate from a CD49d-negative subpopulation should be documented.

4 Notes

1. CD23 is labile, and its surface express can decrease with prolonged specimen storage prior to processing or during the sample processing [33, 37].
2. The great majority of CLL cases express dim IgM and IgD, while rare cases may show a predominant isotype-switched phenotype and express IgG. The IgG-switched CLL exhibits a distinct immunogenetic signature from IgM/IgD-expressing CLL [38].

3. CLL may show an atypical immunophenotype. For example, CLL with trisomy 12 often demonstrates higher expression of CD19, CD20, CD22, CD79b, and surface Ig and lower expression of CD43 [39].
4. The wash solution should be checked with a fresh control PB sample before use in patient samples (see Sect. 3). The supernatant from the lysed control blood should be clear. If there is evidence of hemolysis, discard the solution and prepare a new one.
5. Sodium azide is toxic through inhalation, skin contact, and ingestion. The vapor from sodium azide-containing solution should not be breathed, and it should not be poured down the sink where explosive conditions may develop.
6. Formaldehyde is a known carcinogen and acutely hazardous; do not inhale, and upon skin contact, flush with water.
7. Post removal of supernatant during red cell lysis, gauze or a cotton swab can be used to carefully wipe debris from the inside of the tube(s); do not disturb the pellet before wiping inside the tube.
8. Cytophilic immunoglobulin (bound immunoglobulin from the serum) may be bound to B cells and other cells with high levels of Fc receptors, which may lead to a high background or spurious pattern of surface light chain staining. Washing cells with PBS prior to staining removes the cytophilic immunoglobulins in most cases. An additional 30-min to 1-h incubation at 37° and washing again with PBS or culture medium is more effective and may be required in some cases. However, this procedure may also remove other labile antigens such as CD23 [33].
9. Anti-CD19 immunotherapy has been increasingly used in the treatment of B-cell neoplasms, including CLL, which may lead to loss of CD19 when the disease recurs. An alternative gating strategy is necessary to detect CLL MRD in this context. This can be done by additional B-cell gating using CD20, CD22, or other antigens such as CD40 [40]. Alternative approaches that begin with non-antigen-based gates, such as lymphocyte gates defined by scatter properties, should also be employed routinely in supplement to the B-cell antigen-based gating to avoid missing CLL MRD with post-therapy antigen loss.
10. The transitional B cells are B cells at the intermediate stage that leave the bone marrow and develop into naive B cells in the peripheral lymphoid organs and are usually found at low levels in the PB and BM. Since the transitional B cells also express CD5 and CD23, they may be mistaken as CLL. However, they can be distinguished from CLL with careful phenotypic examination. The transitional B cells typically

co-express CD10, CD38, and bright CD81 and are negative for CD43 [41, 42].

11. The immunophenotype of CLL may change after therapies. For example, CD20 loss is commonly seen post-anti-CD20 therapy. Additionally, a reduction in the expression of CD5, CD23, and CD200 is frequently seen after targeted therapies such as BTK inhibitors [43].

References

1. Campo E et al (2017) Chronic lymphocytic leukemia/small lymphocytic lymphoma. In: Swerdlow SH et al (eds) WHO classification of tumours of hematopoietic and lymphoid tissues. IARC, Lyon, pp 216–221
2. Falini B, Martino G, Lazzi S (2023) A comparison of the International Consensus and 5th World Health Organization classifications of mature B-cell lymphomas. Leukemia 37(1):18–34
3. Alaggio R et al (2022) The 5th edition of the World Health Organization classification of haematolymphoid tumours: lymphoid neoplasms. Leukemia 36(7):1720–1748
4. Campo E et al (2022) The international consensus classification of mature lymphoid neoplasms: a report from the Clinical Advisory Committee. Blood 140(11):1229–1253
5. Tang C et al (2022) Monoclonal B-cell lymphocytosis—a review of diagnostic criteria, biology, natural history, and clinical management. Leuk Lymphoma 63(12):2795–2806
6. Rawstron AC et al (2008) Monoclonal B-cell lymphocytosis and chronic lymphocytic leukemia. N Engl J Med 359(6):575–583
7. Vardi A et al (2013) Immunogenetics shows that not all MBL are equal: the larger the clone, the more similar to CLL. Blood 121(22):4521–4528
8. Fazi C et al (2011) General population low-count CLL-like MBL persists over time without clinical progression, although carrying the same cytogenetic abnormalities of CLL. Blood 118(25):6618–6625
9. Criado I et al (2018) Low-count monoclonal B-cell lymphocytosis persists after seven years of follow up and is associated with a poorer outcome. Haematologica 103(7):1198–1208
10. Rawstron AC et al (2018) Reproducible diagnosis of chronic lymphocytic leukemia by flow cytometry: an European Research Initiative on CLL (ERIC) & European Society for Clinical Cell Analysis (ESCCA) Harmonisation project. Cytometry B Clin Cytom 94(1):121–128
11. Rawstron AC et al (2007) International standardized approach for flow cytometric residual disease monitoring in chronic lymphocytic leukaemia. Leukemia 21(5):956–964
12. Ahmad E et al (2008) Immunophenotypic features distinguishing familial chronic lymphocytic leukemia from sporadic chronic lymphocytic leukemia. Cytometry B Clin Cytom 74(4):221–226
13. Kampalath B, Barcos MP, Stewart C (2003) Phenotypic heterogeneity of B cells in patients with chronic lymphocytic leukemia/small lymphocytic lymphoma. Am J Clin Pathol 119(6):824–832
14. Kingma DW et al (2002) CD2 is expressed by a subpopulation of normal B cells and is frequently present in mature B-cell neoplasms. Cytometry 50(5):243–248
15. Damle RN et al (1999) Ig V gene mutation status and CD38 expression as novel prognostic indicators in chronic lymphocytic leukemia. Blood 94(6):1840–1847
16. Balakrishna J et al (2021) Intensity of antigen expression reflects IGHV mutational status and Dohner-defined prognostic categories in chronic lymphocytic leukemia, monoclonal B-cell lymphocytosis, and small lymphocytic lymphoma. Leuk Lymphoma 62(8):1828–1839
17. Durig J et al (2002) CD38 expression is an important prognostic marker in chronic lymphocytic leukaemia. Leukemia 16(1):30–35
18. Hamblin TJ et al (2002) CD38 expression and immunoglobulin variable region mutations are independent prognostic variables in chronic lymphocytic leukemia, but CD38 expression may vary during the course of the disease. Blood 99(3):1023–1029
19. Hamblin TJ et al (2000) Immunoglobulin V genes and CD38 expression in CLL. Blood 95(7):2455–2457
20. Thunberg U et al (2001) CD38 expression is a poor predictor for VH gene mutational status and prognosis in chronic lymphocytic leukemia. Blood 97(6):1892–1894
21. Dal Bo M et al (2016) CD49d prevails over the novel recurrent mutations as independent prognosticator of overall survival in chronic lymphocytic leukemia. Leukemia 30(10):2011–2018

22. Bulian P et al (2014) CD49d is the strongest flow cytometry-based predictor of overall survival in chronic lymphocytic leukemia. J Clin Oncol 32(9):897–904
23. Gattei V et al (2008) Relevance of CD49d protein expression as overall survival and progressive disease prognosticator in chronic lymphocytic leukemia. Blood 111(2):865–873
24. Gooden CE et al (2018) CD49d shows superior performance characteristics for flow cytometric prognostic testing in chronic lymphocytic leukemia/small lymphocytic lymphoma. Cytomet B Clin Cytom 94(1):129–135
25. Tissino E et al (2020) CD49d promotes disease progression in chronic lymphocytic leukemia: new insights from CD49d bimodal expression. Blood 135(15):1244–1254
26. Alsadhan A et al (2023) CD49d expression identifies a biologically distinct subtype of chronic lymphocytic leukemia with inferior progression-free survival on BTK inhibitor therapy. Clin Cancer Res 29(18):3612–3621
27. Rawstron AC et al (2001) Quantitation of minimal disease levels in chronic lymphocytic leukemia using a sensitive flow cytometric assay improves the prediction of outcome and can be used to optimize therapy. Blood 98(1):29–35
28. Bottcher S et al (2012) Minimal residual disease quantification is an independent predictor of progression-free and overall survival in chronic lymphocytic leukemia: a multivariate analysis from the randomized GCLLSG CLL8 trial. J Clin Oncol 30(9):980–988
29. Kovacs G et al (2016) Minimal residual disease assessment improves prediction of outcome in patients with chronic lymphocytic leukemia (CLL) who achieve partial response: comprehensive analysis of two phase III studies of the German CLL Study Group. J Clin Oncol 34(31):3758–3765
30. Moreton P et al (2005) Eradication of minimal residual disease in B-cell chronic lymphocytic leukemia after alemtuzumab therapy is associated with prolonged survival. J Clin Oncol 23(13):2971–2979
31. Munir T et al (2024) Chronic lymphocytic leukemia therapy guided by measurable residual disease. N Engl J Med 390(4):326–337
32. Romero D (2024) Benefit with MRD-guided treatment in CLL. Nat Rev Clin Oncol 21(2):84
33. Stetler-Stevenson M et al (2007) Clinical flow cytometric analysis of neoplastic hematolymphoid cells; approved guideline (CLSI H43). Clinical and Laboratory Standards Institute, Wayne, Pennsylvania
34. Stelzer GT et al (1997) U.S.-Canadian Consensus recommendations on the immunophenotypic analysis of hematologic neoplasia by flow cytometry: standardization and validation of laboratory procedures. Cytometry 30(5):214–230
35. Cadena-Herrera D et al (2015) Validation of three viable-cell counting methods: manual, semi-automated, and automated. Biotechnol Rep (Amst) 7:9–16
36. Rawstron AC et al (2013) Improving efficiency and sensitivity: European Research Initiative in CLL (ERIC) update on the international harmonised approach for flow cytometric residual disease monitoring in CLL. Leukemia 27(1):142–149
37. Gu B, Bendall LJ, Wiley JS (1998) Adenosine triphosphate-induced shedding of CD23 and L-selectin (CD62L) from lymphocytes is mediated by the same receptor but different metalloproteases. Blood 92(3):946–951
38. Vardi A et al (2014) IgG-switched CLL has a distinct immunogenetic signature from the common MD variant: ontogenetic implications. Clin Cancer Res 20(2):323–330
39. Autore F et al (2018) Morphological, immunophenotypic, and genetic features of chronic lymphocytic leukemia with trisomy 12: a comprehensive review. Haematologica 103(6):931–938
40. Goshaw JM et al (2021) 14-Color single tube for flow cytometric characterization of CD5+ B-LPDs and high sensitivity automated minimal residual disease quantitation of CLL/SLL. Cytometry B Clin Cytom 100(4):509–518
41. Carsetti R et al (2022) Comprehensive phenotyping of human peripheral blood B lymphocytes in healthy conditions. Cytometry A 101(2):131–139
42. Agrawal S et al (2013) Transitional B cell subsets in human bone marrow. Clin Exp Immunol 174(1):53–59
43. Gao Q, Roshal M (2022) Minimal/measurable disease analysis in chronic lymphocytic leukemia/small lymphocytic lymphoma by flow cytometry. Curr Protoc 2(8):e503

Check for updates

Chapter 12

Measurable Residual Disease

Prashant R. Tembhare, Nilesh Deshpande and Sitaram Ghogale

Abstract

Measurable residual disease (MRD) serves as a critical biomarker of prognosis, treatment efficacy, and clinical outcome. It captures the presence of residual tumor cells below the detection threshold of conventional microscopy. Multiparametric flow cytometry (MFC) offers a rapid, cost-efficient, and widely applicable platform for MRD detection through leukemia-associated immunophenotypes (LAIPs) and deviation-from-normal (DfN) antigen maturation patterns. This technique underpins a wide range of clinical applications, including risk stratification, therapeutic decision-making, and post-transplant surveillance. This chapter outlines essential technical parameters for achieving high-sensitivity MRD detection across hematologic malignancies such as B-ALL, T-ALL, AML, MM, and CLL. Key topics covered include material required, reagent preparation, antibody panel design, sample processing and acquisition, gating strategies, and illustrative examples. Special emphasis is given to adaptations necessary for MRD monitoring following CD19-targeted therapies in B-ALL. By integrating rigorous methodology with evolving innovations, MFC continues to advance as a precise and expedient tool for MRD assessment, contributing significantly to personalized treatment strategies.

Keywords Flow cytometry, Measurable residual disease, Acute leukemia, Multiple myeloma, Chronic lymphocytic leukemia

1 Introduction

Measurable residual disease (MRD), formerly termed minimal residual disease, refers to a minute population of leukemic cells persisting during or after therapy at levels undetectable by morphology [1–4]. Conventional remission criterion (<5% blasts in bone marrow aspirates) lacks the sensitivity required for accurate prognostication, as it fails to detect submicroscopic disease burden [3–12]. Sensitive techniques such as multiparametric flow cytometry (MFC) and molecular assays, including quantitative polymerase chain reactions (qPCR), digital droplet PCR (ddPCR), or next-generation sequencing (NGS), detect clinically relevant MRD even in patients achieving morphologic remission [13]. MRD negativity is a strong, independent predictor of better prognosis and prolonged disease-free survival across leukemia

Angélique Biancotto and Pradeep K. Dagur (eds.), *Immunophenotyping*, Methods in Molecular Biology, vol. 3087,
https://doi.org/10.1007/978-1-0716-5607-5_12,

subtypes, risk categories, and treatment protocols, including hematopoietic stem cell transplantation (HSCT) [3–6, 12, 14–22]. It also allows timely interventions such as therapy intensification, targeted therapy, or HSCT [1, 6, 23–26]. MRD levels guide dynamic risk-adapted therapy and are now widely used as surrogate endpoints in clinical trials, reshaping modern trial design and treatment algorithms [6, 17, 27–39].

The concept of MRD emerged in the 1980s, when preclinical studies first demonstrated a link between leukemic burden and survival [40–43]. Initial methodologies used two to three color immunofluorescence microscopy, followed by allele-specific PCR (ASO-PCR) and MFC [40–43]. Technological advances have since transformed the field, enabling transition from the 3–4 to >30 color flow cytometry panels and from conventional (0.01%) to highly sensitive (0.0002%) or next-generation flow (NGF) assays, improving the accuracy and sensitivity of MRD assays [44–47]. MFC is among the most sensitive and broadly applicable MRD tools, detecting as few as two leukemic cells in 100,000 hematopoietic cells through aberrant immunophenotypes. MFC offers several practical advantages, including broad applicability (>90% of cases), cost-effectiveness, rapid turnaround time, simple infrastructure, widespread availability, the ability to assess bone marrow hemodilution, and knowledge of therapeutically targetable protein expression, etc. [3, 44–50]. MFC-MRD is applicable across most acute leukemia subtypes, except in scenarios where RQ-PCR can leverage highly specific genetic markers (e.g., *NPM1* mutations in AML or *PML::RARA* in APL) [35, 51]. MFC also enables detection of immunophenotypically divergent subclones—a critical advantage over targeted molecular assays [1, 35, 52, 53]. MFC limitations include operator dependence and interlaboratory variability, mitigated by standardization [1, 3, 35, 45–47]. It may miss MRD in mature cells harboring genetic lesions such as BCR::ABL1 or NPM1 (in AML) and PML::RARA (in APL). Combined MFC and molecular methods improve prognostic accuracy and relapse prediction further [1, 45, 46, 54–56]. Thus, MFC remains a cornerstone of MRD evaluation within a multimodal diagnostic framework.

1.1 Approach to MRD Analysis

The success of flow cytometric MRD (FCM-MRD) relies on achieving robust multidimensional resolution to distinguish leukemic blasts from immunophenotypically similar normal bone marrow populations (primarily normal progenitors and rare hematopoietic subsets). Optimal separation of residual leukemic cells from other marrow elements is achieved through two complementary strategies [35, 57–62].

1. *DfN Different-from-Normal (DfN) approach:* Identifies tumor cells by deviations from the normal antigen expression or maturation pattern of lineage-specific reference cells. This

method demands in-depth knowledge of physiological maturation stages and the immunophenotypic shifts induced by chemotherapy.

2. *Leukemia-Associated Immunophenotype (LAIP) approach*: Detects aberrant antigen expression or downregulation in tumor cells compared to normal lineage-matched reference cells.

1.2 Steps Involved in Establishing MRD Assays by MFC

1. Designing an antibody panel for distinguishing MRD from other BM cells,
2. Standardizing sample processing protocols to achieve the required sensitivity,
3. Designing software-based data analysis approaches:
   - Fixing the gating strategy for identifying target cells and
   - Mapping the normal maturation sequences of precursor cells to mature cells
   - Evaluation of abnormalities or "deviation-from-normal" in each sample
   - Establishment of clonality
   - Calculation and reporting of MRD.

Here, we outline optimized antibody combinations (panels) used for MRD detection, standardized protocols for sample processing, methodologies for labeling cell-surface and intracellular proteins with fluorochrome-conjugated antibodies, and analytical strategies for interpreting immunophenotypic data.

2 Materials

The following is a list of common reagents required for flow cytometric MRD monitoring using bulk-lyse and staining protocol in acute leukemia (B-ALL, T-ALL, and AML), chronic lymphocytic leukemia (CLL), and multiple myeloma (MM). Among these diseases, MRD monitoring in T-ALL and MM involves processing for intracellular staining steps.

1. Stock 5× PBS (Phosphate Buffered Saline): with 0.2% Bovine serum albumin and 0.1% sodium azide: Add 40.0 g NaCl, 10 g KCL, 10.5 g anhydrous Na_2HPO_4, albumin 10 g, 1 g KH_2PO_4, 1 g NaN_3 to a 1 L container, and make up the volume to 1000 mL with distilled water. Mix thoroughly until it is completely dissolved. Verify that pH is between 7.0 and 7.5. Filter stock solution with Whatman filter paper no. 1 and store at 2–8°C for up to 1 month.

2. Working 1× PBS: Add 100 mL of 5× PBS in 500 mL container and make up the volume up to 500 mL with distilled water. Mix thoroughly. Filter working PBS with Whatman filter paper no. 1. Verify that pH is between 7.0 and 7.5. Validate on a normal sample before use. Prepare a fresh working PBS every time.
3. Stock 5× Erythrocyte (RBC) lysing Reagent: Add 41.3 g NH_4Cl, 5 g $KHCO_3$, 0.185 g Tetra Sodium EDTA to a 1 L container and make the volume 1000 mL with distilled water. Mix thoroughly until it is completely dissolved. Verify that pH is between 7.2 and 7.4. Filter stock solution with Whatman filter paper no. 1 and store at 2–8°C for up to 1 month.
4. Working 1× Erythrocyte (RBC) lysing Reagent: Add 40 mL of 5× Erythrocyte lysing reagent to a 200 mL measuring cylinder and make it volume to 200 mL with distilled water. Filter working RBC lysing reagent with Whatman filter paper no. 1. Verify that pH is between 7.2 and 7.4. Validate on a normal sample before use. Prepare the working RBC lysing reagent freshly and store it at room temperature before use.
5. Stock 4% [8×] Paraformaldehyde: Add 4 g paraformaldehyde to a 100 mL aluminum foil-covered flask. Add 100 mL 1× working PBS reagent. Mix thoroughly by heating at 65–70°C (on a hot plate with a magnetic stirrer). Make sure that the flask is covered tightly with aluminum foil while heating. Allow to cool and store at 2–8°C for up to 2 weeks.
6. Working 0.5% Paraformaldehyde: Add 25 mL of 4% methanol-free paraformaldehyde to a 200 mL measuring cylinder and make the volume to 200 mL with 1× working PBS. Validate on a normal sample before use and store at 4°C.
7. Syto13 Reagent (Molecular probes S7575 Invitrogen): Thaw Syto13 (stored at −20°C) vial at room temperature. Spin the vial in a microcentrifuge for 5 min at 15,000 rpm to bring DMSO to the bottom of the tube. Make a dilution of stock solution of 1 in 200 (1 μl Syto13 + 199 μl 1× working PBS), being careful not to aspirate DMSO. Label this dilution as Stock Syto13 and vortex this dilution thoroughly. Make aliquots of 40 μL each from Stock Syto13 and store at −20°C to resist multiple freeze/thaw cycles, and it is stable for up to 1 year.
8. Working Syto13 dye: Add 160 μL of 1× PBS to aliquot. Store at 4°C for daily use. Use 2 μL/1 million cells. A new lot of working Syto13 needs to be validated for the volume of dye.
9. Intracellular staining reagent: fixative and permeabilization solution from Invitrogen, ready to use.

3 Methods

We describe a common bulk-lyse and staining protocol (described by Euroflow) used for high-sensitivity MFC-MRD monitoring for acute leukemia (B-ALL, T-ALL, and AML), chronic lymphocytic leukemia (CLL), and multiple myeloma (MM). Protocols for MFC-MRD monitoring in T-ALL and MM involve additional processing for intracellular staining. In T-ALL MRD, cytoplasmic CD3 staining and in MM MRD, cytoplasmic light chains (kappa and lambda) staining is performed. The steps involved in the cytoplasmic staining have been described after the surface staining protocol.

3.1 Sample Preparation and Cell Viability Assessment

3.1.1 Preparation of Cell Suspensions

Blood or bone marrow samples, anticoagulated with EDTA or heparin, can be used for MRD analysis. Cell suspensions can be prepared using the Bulk Lyse–Stain–Wash method as described by Euroflow Consortium [63].

3.1.2 Cell Count

It is highly recommended to take the cell count in the sample to determine the optimal lysing volume required to obtain the desired number of acquisition events. White blood cell (WBC) counts are obtained using an automated hematology analyzer.

- Dilute the sample with phosphate-buffered saline (PBS) or an appropriate diluent for a sample to diluent ratio of 1:5 (e.g., 40 μ L sample + 160 μL diluent).
- Take a WBC count and multiply it by a factor of five to get the corrected WBC count per μL. For calculating the required sample volume for MRD processing, one should take up to 70% cell loss during sample processing into account. For example, to acquire five million cells, the volume of the sample should be taken to get approximately sixteen million cells.
- *Note:* It should be ensured that the bone marrow sample taken for cell counts is free from clots and clumps. If necessary, filter the sample with 40–60 μ filter before counting or remove the clots using a pipette.

3.1.3 Viability

Immunophenotyping is ideally performed as early as possible after sample collection, preferably within 48 h of collection. For delayed processing—either from inpatient samples or those received from external centers—viability testing is performed before analysis. Low viability does not automatically warrant specimen rejection; however, if viability is <80%, it should be documented in the report with an appropriate rider. Sample viability can be assessed using

7-aminoactinomycin D (7-AAD; BD Biosciences, Franklin Lakes, NJ, USA) staining.

3.2 Preparation of Cell Suspension for MRD

1. Transfer the sample, as per the calculated lysing volume (based on cell count/μL), to a 15 mL conical-bottom Falcon tube.
2. Add working 1× Erythrocyte (RBC) Lysing Reagent: Use a maximum of 1 mL of sample per 15 mL of lysing solution. Note: If >1 mL of sample volume needs processing (e.g., in cases of low cell concentration), use additional 15 mL tubes to lyse the extra volume.
3. Mix well and incubate for 10 min at room temperature (RT).
4. Centrifuge for 5 min at 540 × *g*. Discard the supernatant using a laboratory aspirator or pipette without disturbing the pellet. Note: If RBCs are resistant to lysis, gently break the pellet, add 15 mL of lysing solution, and repeat Steps 2 and 3.
5. Gently break the pellet and add approximately 2 mL of PBS, mixing thoroughly by pipette or vortex.
6. Adjust the volume to 15 mL with PBS.
7. Centrifuge for 5 min at 540 × *g*.
8. Discard the supernatant and gently break the pellet. Add approximately 2 mL of PBS, mixing well.
9. Transfer the suspension to a 5 mL Falcon tube and adjust the volume to 4 mL with PBS.
10. Centrifuge for 2 min at 540 × *g*.
11. Discard the supernatant without disturbing the pellet.
12. Gently tap to loosen and mix the pellet.
13. Perform a cell count and add PBS based on the number of tubes × 50 μL.
14. Adjust the cell suspension to yield ten to twenty million cells per tube.

3.3 Antibody Staining Procedure for Cell Surface Antigens

1. Label Falcon tubes with the appropriate marker panel identification.
2. Add the appropriate volume of cell surface antibody cocktails or individual antibodies (as determined by titration) to each tube.
3. Add 50 μL of the prepared cell suspension to each tube.
4. Mix gently and incubate for 20 min at RT in the dark.
5. Add 2 mL of PBS, mix gently, and centrifuge for 2 min at 540 × *g*.
6. Discard the supernatant without disturbing the pellet.
7. Resuspend the pellet in 0.5–1 mL of 0.5% paraformaldehyde solution (depending on cell concentration) and store at 4°C.

8. Samples can be stored for up to 3 days before acquisition.
9. Acquire the sample on a flow cytometer with a predefined acquisition template and instrument settings for the MRD protocol.

3.4 Combined Labelling Procedure for Cell-Surface and Intracellular Antigens

1. Perform the surface staining procedure as described above up to Step 6.
2. Add 100 µL of reagent A (cell fixation, Invitrogen Cat # GAS001S100).
3. Incubate for 15 min at RT in the dark.
4. Add 2 mL of PBS and mix well.
5. Centrifuge for 2 min at 540 × *g*.
6. Discard the supernatant without disturbing the pellet.
7. Gently tap the pellet.
8. Add 100 µL of reagent B (permeabilization solution, Invitrogen Cat # GAS002S100).
9. Add the titrated volume of intracellular antibodies (such as cyCD3 or cyKappa/cyLambda) to the tube and mix gently.
10. Incubate for 15–20 min at RT in the dark.
11. Add 2 mL of PBS to the tube, centrifuge for 2 min at 540 × *g*, and discard the supernatant without disturbing the pellet.
12. Tap the pellet gently and resuspend it in 0.5–1 mL of 0.5% paraformaldehyde solution (depending on cell concentration).
13. Store at 4°C. Samples can be stored for up to 3 days before acquisition.
14. Acquire the sample on a flow cytometer with a predefined acquisition template and instrument settings for the MRD protocol.

4 B-ALL MRD

4.1 Antibody Panel for B-ALL MRD

We use a 12-color MFC-BALL MRD panel that includes commonly used antibodies shown to be effective in MRD detection in more than 95% of B-ALL patients (Table 1). This panel is designed for the detection of B-ALL MRD post-chemotherapy and includes B-cell lineage markers CD19 and CD22 for gating B-cells, including normal and abnormal precursors. Antibodies against CD10, CD20, CD34, CD38, and CD45 are included to map B-cell precursors maturation and additional leukemia-specific markers, i.e., CD73, CD81, CD86, CD123, and CD304, that help in establishing immunophenotypic aberrancies—LAIPs. Panel includes an additional combination of

Table 1

B-ALL MRD antibody panel

Tube-1: sytodye tube													
Fluiro chrome	**BV786**	**BV650**	**BV605**	**BV510**	**BV421**	**FITC/ Vio-bright B515**	**PE**	**ECD/ PECF594**	**PC5.5**	**PC7**	**APC**	**APC AF_700**	**APC AF_750**
Antibody	X	X	X	X	X	Syto13	X	X	X	X	CD19	CD45	
Clone						Green-fluorescent					HIB19	J.33	
Cat No.						S7575					555415	A79390	
Company						Invitrogen					BD	BC	
Tube-2: MRD main tube													
Antibody	**CD81**	**CD200**	**CD86**	**CD20**	**CD123**	**CD304 (BDCA4)**	**CD22**	**CD73**	**CD34**	**CD10**	**CD19**	**CD45**	**CD38**
Clone	JS-81	MRC OX-104	BU63	2H7	9F5	AD517F6	REA340	AD2	8G12	ALB1	HIB19	J.33	LS198-4-3
Cat No.	568522	743085	747529	563067	562 517	130-113-519	130-123-245	562817	347203	A46527	555415	A79390	B49200
Company	BD	BD	BD	BD	BD	MB	MB	BD	BD	BC	BD	BC	BC

BC Beckman Coulter Ltd., *BD* Becton Dickinson Biosciences, *Cat No* catalog number, *MB* Miltenyi Biotec

Syto13 (a nuclear dye), CD19, and CD45 with the purpose of calculating MRD levels out of all nucleated cells (described below).

Methods: Reagents, sample collection, cell counting, red cell lysis, and surface staining are performed using the same protocol described above.

4.2 Analysis Approach—B-ALL MRD

MFC-MRD analysis primarily requires the predefined template with the predefined gating strategy on a combination of markers in dotplots using the knowledge of expression patterns of these antigens on normal B-cell progenitors (BCPs) during the maturation process. Leukemic blasts universally deviate from the antigen expression patterns of normal BCPs.

4.2.1 Gating Strategy

Post-chemotherapy, primary B-cell gating employs pan-B-cell markers (CD19/CD22). This approach (Fig. 1) involves:

Step-1: Initial gating:

- Exclude doublets using FSC-H vs. FSC-A dotplot.
- Exclude granulocytes and isolate mononuclear cells (MNC) using CD38/SSC dotplot (Note: One can use CD45/SSC dotplot too, but CD38 vs. SSC provides better separation of granulocytes from MNCs).
- Check the expression of common sensitive markers such as CD10, CD34, and CD38 using dotplots of CD10 vs. CD38, CD38 vs. CD34, and CD10 vs. CD34. It provides expression of these markers in internal positive control population, e.g., CD10 expression on mesenchymal stem cells/endothelial cells and BCPS, CD34 expression in CD38 expression in plasma cells (10/38 PC and 38PC gates), BCPs (10/38 Hg gate and 10/34 levels) and myeloid blasts (CD34 gate—olive dots), etc. This provides the quality of sample processing and instrument setting.

Step-2: B-cell enrichment:

- It is done using a combination of CD19 and/or CD22 with CD45. So, B cells are isolated from the rest of the MNCs using CD19 vs. CD22 or CD19 vs. CD45.

Step-3: Mimics and artifacts exclusion:

- B-cell gate usually includes some CD19 and or CD22 positive non-B cells and plasma cells, which can interfere with analysis. CD38bright+/CD10− plasma cells are excluded using the CD10/CD38 dotplot. CD45bright non-B lymphocytes (e.g., NK cells with CD19 trogocytosis) are excluded using the CD20/CD45 dotplot. Exclude the debris or non-viable B-cells using "Clean-B" gate. Such an approach results in a yield of clean B-cells for further analysis

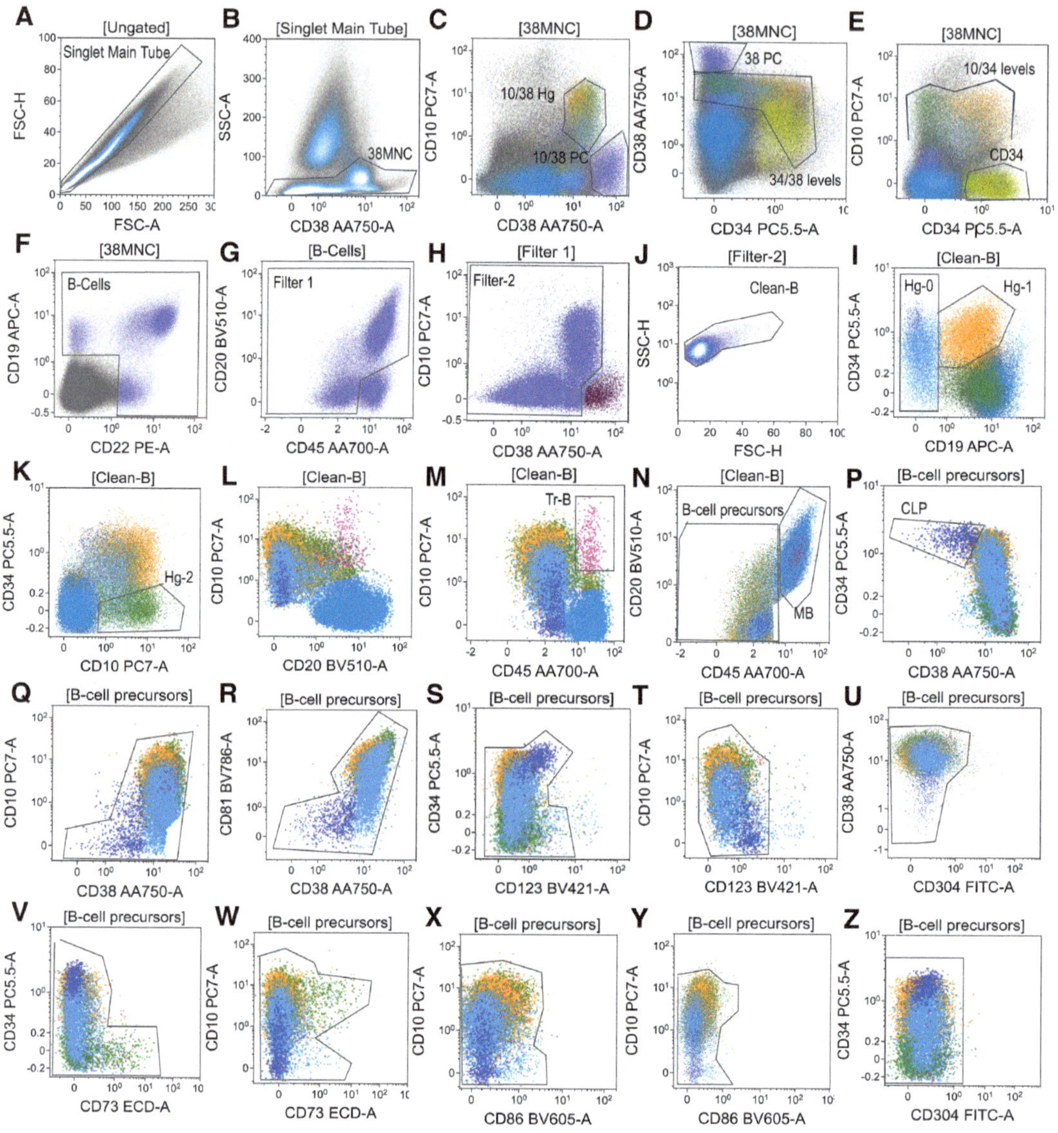

- Note: Exclusion gates must be context-specific. For example, CD10−B-ALL MRD (e.g., in *KMT2A/ZNF384*-rearranged B-ALL) should not be inadvertently gated out with such exclusion gates.
- Hence, it is advised to include all cells initially, and after evaluating the possible risk of the presence of MRD events based on the knowledge of diagnostic immunophenotype and expression patterns of other markers, one can exclude these cell populations.

Step-4: Mapping BCP maturation patterns:

Isolated B cells are analyzed for normal maturation patterns. Normal B-cell maturation includes progressive development of BCPs through common lymphoid progenitors (CLP) or early B-cell precursors type-I, followed by hematogones stage "0" (Hg-0) or early B-cell precursors type-II, then hematogones stage-1, hematogones stage-2, transitional B-cells, and mature B-cells. The key immunophenotypic features of these BCPs are as follows:

- Common lymphoid precursors or Early B-cell precursors (CLP, dark blue dots): bright CD34, dim CD38, dim CD81, dim to moderate CD22, moderate CD45, but negative for CD10, CD19, CD20, CD24.
- Hematogones stage "0" or BCP-0 (Hg-0, light blue dots): bright CD34, moderate CD38, CD81 and CD22, negative to dim CD10 and CD20, and dim CD45 but negative for CD19 and CD24.
- Hematogones stage-1 or BCP-1 (Hg-1, orange dots): bright CD34, moderate CD10, CD38 and CD81, dim CD22, dim CD19 negative to dim CD20, dim CD45 and negative to moderate CD24.
- Hematogones stage-2 or BCP-2 (Hg-2, green dots): moderate CD10, CD38 CD81, CD22, CD19 and CD45, negative to dim CD20, moderate CD24 but negative CD34.
- Transitional B-cells (pink dots): dim CD10, dim CD38, dim CD81, moderate CD22, moderate CD19, moderate CD20, moderate CD24 and bright CD45 but negative CD34.
- Mature B-cells (cobalt blue dots): moderate CD22, CD19, CD20, and CD24 and bright CD45 but negative CD10, CD34, CD38, CD81.
- Using these key features, these stages of BCPs are given different colors to track them in various dotplots. Gates are drawn

Fig. 1 Gating strategy for measurable residual disease (MRD) detection in B-acute lymphoblastic leukemia (B-ALL) postconventional chemotherapy MRD strategy. This figure illustrates the stepwise flow cytometric approach for identifying residual leukemic cells in B-ALL using immunophenotypic deviation from normal B-cell maturation. Doptplot-**A** excluded doublets, followed by granulocytes in dotplot-**B**. Mononuclear cells (MNCs) are identified using CD38 vs. SSC and applied to subsequent dotplots **C–F**. Consistent expression levels for CD10, CD34, and CD34 were evaluated in dotplots **C–E** as an internal quality check. The initial broad B-lineage gate is created using CD19 vs. CD22 (dotplot **F**). Subsequent gating steps are applied to exclude normal hematopoietic components, such as non-B lymphocytes (CD45−bright+ in dotplot-**G**) and plasma cells (CD38bright+/CD10− in dotplot-**H**). Non-viable B-cells and debris were removed using a "clean B cells" gate, which subsequently are subjected to template-based analysis. The different stages of BCP maturation are color-coded using a combination of CD10, CD19, CD20, CD34, and CD45. Here, early B-lymphoid progenitors (common lymphoid progenitors; CLP) are indicated with dark blue dots, next stage BCPs, i.e., stage "0" precursors (Hg-0) with light-blue, stage-1 BCPs (Hg-1) with orange, stage-2 BCPs (Hg-2) with green, transitional-B cells with pink, and mature B-cells with cobalt blue color. BCPs were isolated using a gate "B-cell precursors" which is subsequently applied to dotplots **P–Z** to map the maturation sequences (patterns) using various combinations of markers, which were outlined with well-defined gates. Residual leukemic populations can be identified (shown in Fig. 2) with deviation from normal B-cell maturation. Calculation of MRD is demonstrated in Fig. 2 using an MRD-positive example

around the boundaries of these BCPs with various combinations of these markers highlighting the different stages of BCPs using control BM samples (staging controls or regenerative non-B-ALL BM samples).

- Normal BCPs fall within the boundaries of gates, and any population that deviates from normal maturation sequences falls outside these gates with any combination of markers highlighting its immunophenotypic abnormalities and deviation from normal maturation.
- Additional gates are drawn to define boundaries for the expression of other markers that are usually absent or show minimal expression on normal BCPs. Such markers include CD73, CD86, CD123 and CD304. Overexpression of these markers indicates immunophenotypic abnormalities. Such abnormalities are usually present in blasts even at the time of diagnosis and are considered as LAIPs.

Step-5 MRD detection:

- The presence of any two immunophenotypic abnormalities, including deviation from normal or LAIPs or both, is required to label MRD in the given sample [64].
- A combination of such well-charted gates/boundaries of normal is then used as a template/semi-automated algorithm to analyze a given sample. The number of MRD events is gated using a combination of markers, which provides the best separation of residual leukemic blasts from normal BCPs, and the percentage of MRD out of all singlet cells is calculated.
- This MRD level is further corrected for MRD in singlet total nucleated cells as described below. An example of positive MRD has been shown in Fig. 2.

Step-6 MRD calculation and correction for total nucleated cells:

- An additional page or dotplots are included in the template to analyze an FCS file with Syto13 dye, CD19, and CD45 (Fig. 2). Nucleated cells are isolated using Syto-13 positive expression with FSC vs. Syto13 dotplot, followed by exclusion of doublets and isolation of CD19-positive events. The percentage of Syto13+CD19-positive cells in total nucleated cells from Syto13 tube is determined. Similar dotplots are created for the main MRD FSC file, except for the first dotplot for Syto-13 positive cells. Link the gates isolating singlet events and subsequent CD19-positive events from the Syto13 tube and main MRD tube. It ensures the exact shape and position of gates related to dotplots from both tubes, avoiding subjectivity.
- A correction factor (CF) for CD19+ cells in nucleated cells is calculated as a ratio of the percentage of Syto-13+CD19-positive

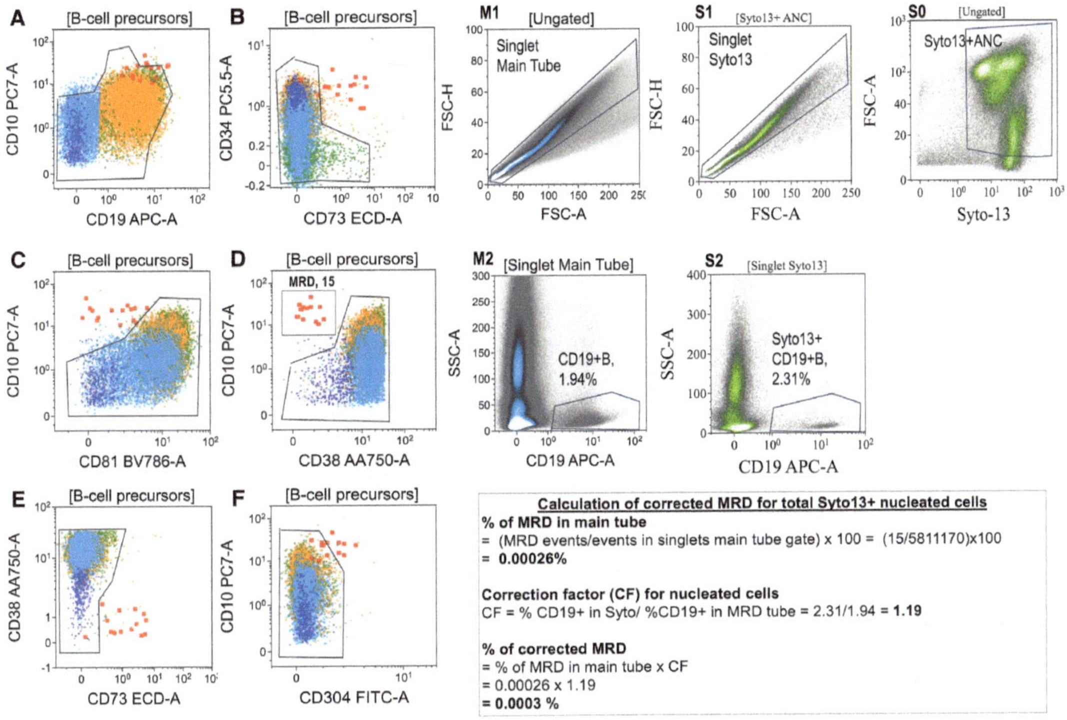

Fig. 2 B-ALL MRD calculation. Dotplots **A–F** illustrate an example of low-level B-ALL MRD highlighted with red dots. These cells show moderate to bright expression of CD10 and CD19 (dotplot-**A**), bright CD34 and moderate CD73 (dotplot-**B**), aberrant downregulation of CD38 and CD81, and overexpression of CD304, thus clearly deviating from BCPs. Initially, MRD percentages are calculated using MRD events (15 events) from total singlet cells (5,811,170 events) as "0.00026%." The final MRD in total nucleated cells is determined by multiplying this initial MRD value by the correction factor (CF) determined for total nucleated cells. CF was determined as the ratio of CD19+ B cells in Syto-tube gated with "Syto13+CD19+B" (derived in density plots **S1–S3**) to the CD19+ B cells in the main MRD tube gated with "CD19+B" (derived in density plots **M1–M2**). In this example, the CF was calculated as 1.19. Thus, the corrected final MRD was calculated as "0.00026% × 1.19 = 0.0003%."

cells from the Syto-13 tube to CD19-positive cells from the main MRD tube. The final (corrected) level of MRD is determined by multiplying the MRD percentages in the main tube by the correction factor for nucleated cells.

4.3 CD19-Agnostic Gating in Post-Anti-CD19 Therapy MRD Assessment

The increasing use of anti-CD19 agents (e.g., blinatumomab, CD19-directed CAR-T cells) frequently results in CD19 loss or downregulation on leukemic B-blasts, necessitating CD19-independent gating strategies for flow cytometric MRD (FCM-MRD) detection. Consequently, antibody panels need to be modified to incorporate alternative B-cell markers. As given in Table 2, our post-antiCD19 B-ALL MRD panel supplements conventional markers with CD22, CD24, and CD81 for B-cell identification. However, these markers require careful interpretation due to their expression in non-B-cell populations:

- CD22 can be expressed on basophils and mast cells.
- CD24 is present on granulocytes, which may lose SSC and appear in the mononuclear cell (MNC) region upon degranulation.

We included CD66b+CD64 to exclude granulocytes and monocytes, and HLA-DR and CD33 to exclude basophils and mast cells in the post-anti-CD19 therapy B-ALL MRD panel [50, 65]. This avoids the false positivity and background interference by CD24+ low SSC granulocytes and non-specifically bound monocytes.

Note: Special attention must be given to distinguishing CD19-negative normal B-cell precursors (BCPs), such as ELP-I and BCP-0, to avoid false-positive MRD detection.

4.3.1 Clinical Relevance of Low-Level MRD Post-AntiCD19 Therapy

Emerging evidence suggests that any detectable MRD, even at low levels, may hold significant prognostic value in patients receiving anti-CD19 therapies [66, 67]. Modern post-antiCD19 FCM-BMRD protocols achieve a sensitivity of 0.0005% (5×10^{-6}). Our modified gating strategy for posttherapy MRD assessment is demonstrated in Fig. 3.

4.4 Gating Approach for B-ALL MRD Post-Anti-CD19 Therapy Analysis

Additional gating steps:

- As illustrated in Fig. 3, following the exclusion of doublets and debris as described in the post-conventional chemotherapy B-ALL MRD approach, mononuclear cells (MNCs) are selectively gated by excluding granulocytes and monocytes using a CD66b/CD14 versus SSC dotplot.
- Subsequently, a Boolean gate ("Merge 22/24/81") is constructed using OR logic from CD22 versus CD24 and CD22 versus CD81 dotplots to include all events positive for CD22, CD24, or CD81 while excluding negative populations.
- This Boolean gate is then applied to the next two dotplots, where a secondary Boolean gate ("Merge 33 ×CD22/10/34") is generated to include only events positive for CD34, CD10, or HLA-DR with dim to negative CD33 expression (based on CD34 vs. CD33, CD10 vs. CD33, and CD22 vs. HLA-DR dotplots).
- Although this secondary gate eliminates most non-B cells, it may still include CD22-positive basophils, mast cells, and pDC precursors, which require additional exclusion steps to eliminate. These remaining non-B-cell populations are removed through further gating strategies: debris is excluded using SSC

Table 2

B-ALL MRD—post-anti-CD19 targeted therapy antibody panel

Tube-1: sytodye tube															
Fluiroc-hrome	**BUV737**	**BUV395**	**BV786**	**BV711**	**BV650**	**BV605**	**BV510**	**BV421**	**FITC/ Vio-bright B515**	**PE**	**ECD/ PECF594**	**PC5.5**	**PC7**	**APC**	**APC AF_700**	**APC AF_750/ APC-VIO770**
Antibody	X	X	X	X	X	X	X	X	Syto13	CD22	X	X	X	x	CD45	
Clone									Green-fluorescent	REA340					J.33	
Cat No									S7575	130-123-245					A79390	
Company									Invitrogen	MB					BC	
Tube-2: MRD main tube																
Antibody	**CD304**	**CD19**	**CD38**	**CD33**	**HLA-DR**	**CD86**	**CD20**	**CD123**	**CD66b+ CD64**	**CD22**	**CD73**	**CD24**	**CD10**	**CD34**	**CD45**	**CD81**
Clone	U21-1283	HIB19	HIT2	WM53	G46-6	BU63	2H7	9F5	G10F5+	REA340	AD2	ALB9	ALB1	581	J.33	REA513
Cat No	749093	569107	563964	563171	564231	747529	563067	562 517	564679+	130-123-245	562 817	B23133	A46527	IM2472U	A79390	130-130-060
Company	BD	BD	BD	BD	BD	BD	BD	BD	BD	MB	BD	BC	BC	BC	BC	MB

BC Beckman Coulter Ltd., *BD* Becton Dickinson Biosciences, *Cat No* catalog number, *MB* Miltenyi Biotec

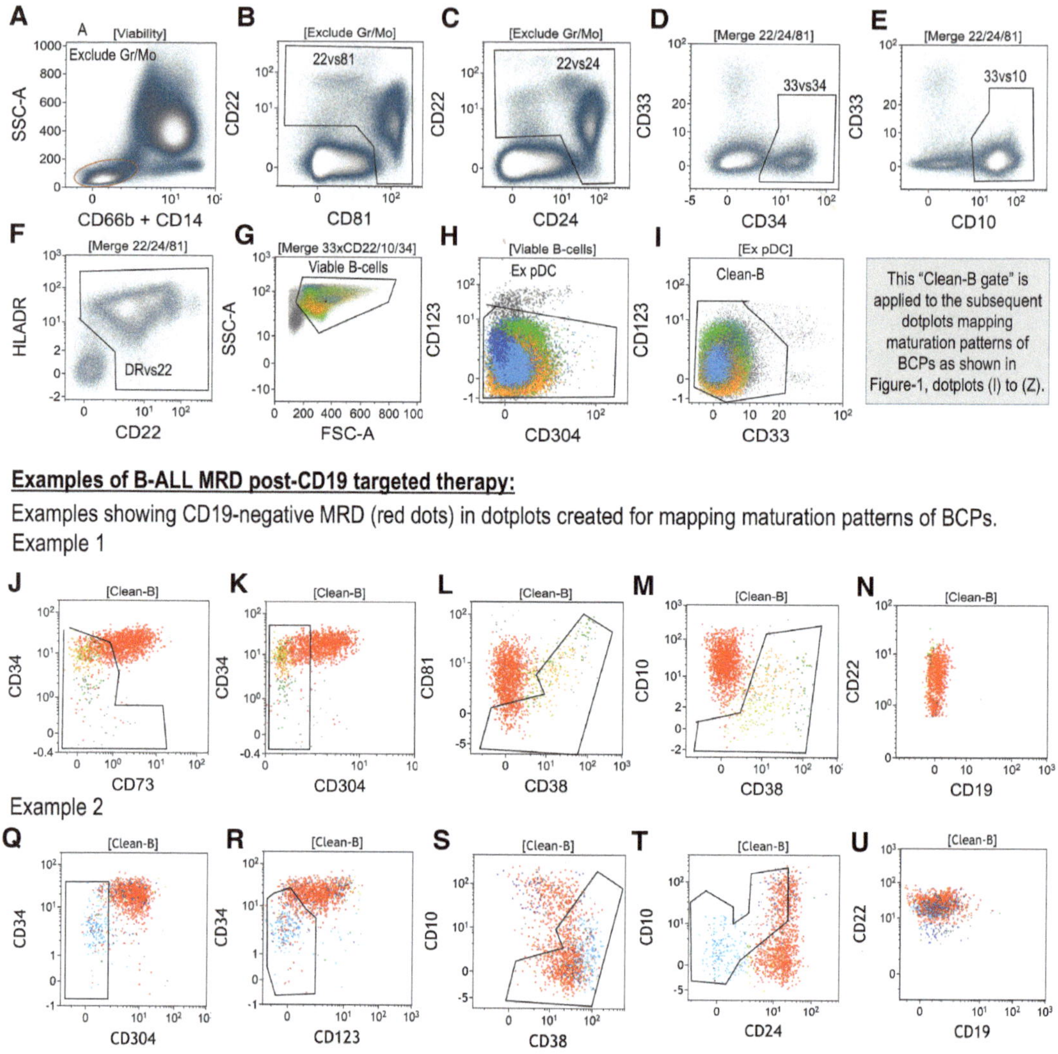

Fig. 3 Post-anti-CD19 targeted therapy MRD Strategy. Given potential CD19 downregulation post-antiCD19 targeted therapy, B-cells are isolated using a CD19-independent gating strategy. Mononuclear cells (MNCs) are gated after excluding granulocytes and monocytes using CD66b/CD14 vs. SSC (dotplot-**A**). A Boolean gate "Merge 22/24/81" is constructed using "OR" logic from CD22 vs. CD81 (dotplot-**B**). and CD22 vs. CD24 (dotplot-**C**) plots. A secondary Boolean gate "Merge 33 × CD22/10/34" includes events from CD34 vs. CD33 (dotplot-**D**), CD10 vs. CD33 (dotplot-**E**), and CD22 vs. HLA-DR plots (dotplot-**F**). Additional steps remove debris, plasmacytoid dendritic cells, and basophils (dotplots **G–I**), yielding a refined "clean B-cell" population. This "Clean-B gate" is applied to the subsequent dotplots mapping maturation patterns of BCPs as shown in Fig. 1, (dotplots **G–Z**). These cells are then assessed for MRD through deviation-from-normal patterns using a standardized B-lineage maturation template. Dotplots **J–N** and **Q–U** demonstrated two B-ALL MRD-positive examples with negative CD19 expression (red dots)

versus FSC parameters, while plasmacytoid dendritic cell precursors, mast cells, and basophils are discriminated based on their bright CD123 and CD33 expression, along with heterogeneous CD304 co-expression.

- The resulting "clean B-cell" population is then evaluated for residual disease by assessing deviation-from-normal patterns using a standardized B-lineage maturation template as previously described.
- MRD is detected and calculated as described above.

4.5 Key Considerations

- The most critical MRD assessments post-conventional chemotherapy occur at the end of induction (EOI; day 28–35) and end of consolidation (EOC; day 78–82) MRD [4, 6, 14, 45, 46, 68]. Additional protocol-specific checkpoints include Day-8 peripheral blood (COG) and Day-15 bone marrow (St. Jude, European groups) [69]. The standard actionable threshold is ≥ 0.01%, with bone marrow preferred over peripheral blood due to 1–3 log greater sensitivity [70]. MRD also informs prognosis in patients receiving Blinatumomab or CAR-T cells, where lower cut-offs may be clinically meaningful [66, 67].
- Coustan-Smith et al. conducted systematic studies (later validated by multiple groups) identifying >20 differentially expressed markers that distinguish leukemic blasts from normal B-cell precursors [71]. Clinically validated markers for MRD monitoring include: CD73, CD86, and CD304, overexpressed in 60%–77%, 50%–60%, and 42%–59% of diagnostic cases, respectively, with higher prevalence at MRD timepoints (CD73: 90%–95%; CD86: 65%–71%; CD304: 40%–41%) [72, 73] CD123, CD58, and CD66c, differentially expressed in 65%–85%, 55%–94%, and 36%–46% of cases [73], with CD123, CD304, and CD66c linked to *BCR::ABL1*. Other markers that have been reported in various studies include: CD9, CD13, CD33, CD44, CD45RA, CD49f, CD58, CD81, CD200, etc.
- *Caution:* Some "leukemia-associated" markers (e.g., CD73, CD86) show weak physiological expression in specific BCP subsets (e.g., CD10+CD34– BCP2 vs. CD10+CD34+ BCP1). Therefore, interpretation should rely on composite immunophenotypic patterns rather than isolated markers [48].
- Awareness of regenerative and low-level mimics is critical, particularly in high-sensitivity assays where rare events become more prominent. Normal BCPs, especially regenerative BCP1 (hematogones), show shifts characterized by CD34/CD10 downregulation and CD20 upregulation compared to treatment-naïve marrow. These changes are most pronounced post-intensive therapy and diminish during consolidation and

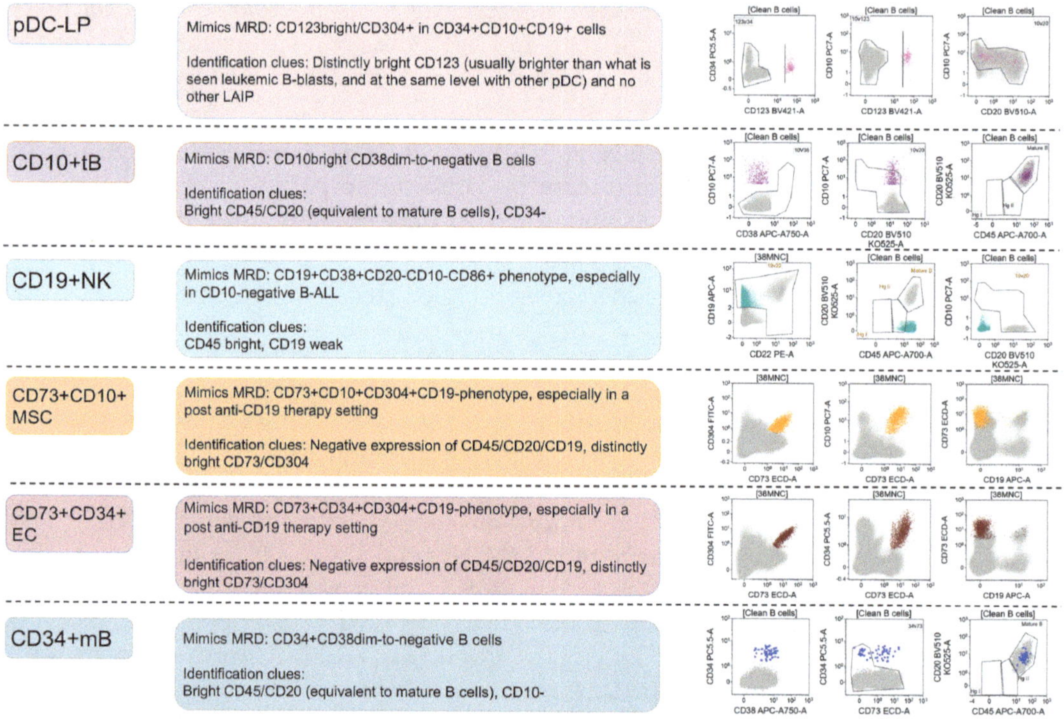

Fig. 4 Immunophenotypic mimics of B-lymphoid blasts in flow cytometric MRD (B-MRD) analysis. This figure presents representative dot plots of recurrently encountered low-level cellular populations that can mimic residual leukemic B-blasts during high-sensitivity flow cytometric MRD evaluation and clues to distinguish them from true MRD. These immunophenotypically distinct, yet potentially confounding, populations are color-coded and annotated for reference. These include: pDC-LP (pink dots): Plasmacytoid dendritic cell-like precursors showing partial B-lineage features; CD10$^+$ tB cells (deep purple dots): Transitional B-cells with strong CD10 expression, sometimes overlapping with B-lymphoblast phenotype; CD19$^+$ NK cells (cyan dots): Natural killer cells aberrantly expressing CD19, potentially misclassified as B-lineage events; MSC (orange dots): Mesenchymal stromal/stem cells occasionally detected in bone marrow aspirates; EC (deep brown dots): Endothelial cells with distinct immunophenotypic signatures; CD34$^+$ mB cells (bright blue dots): CD34–positive mature B-cells, a rare but recognized subset with potential overlap in blast gate regions

follow-up. Figure 4 summarizes common mimics and their distinguishing features. Establishing mimic-aware templates and algorithms—preferably using regenerative marrows from negative T-ALL/AML MRD samples—can significantly reduce interpretation errors in B-ALL MRD analysis.

- A sensitivity of 10^{-4} can typically be achieved using a standardized FCM BMRD panel with 0.5–1 million cells analyzed. However, Stow et al. showed that MRD levels between 10^{-4} and 10^{-5} (below the clinically actionable cutoff) at the end of induction therapy correlated with a higher cumulative relapse risk compared to MRD-negative cases ($<10^{-5}$) [74]. In routine practice, a lower limit of detection (LOD)/lower limit of quantification (LLOQ) is preferable, as some bone marrow samples may be hemodiluted, underrepresenting residual blasts.

- The Euroflow consortium and our team (TMC) have developed highly sensitive FCM-BMRD assays capable of reaching $<10^{-5}$, comparable to PCR-based methods, when four to five million cells are acquired [48, 75]. Key factors for achieving this include standardized processing (e.g., bulk erythrocyte lysis, instrument settings) and using the "first-pull" (0.5–1 mL BM) for MRD evaluation. With such optimized assays, a LOD of 2×10^{-6} and LLOQ of 1×10^{-5} can be attained with ~4–5 million events. The Euroflow consortium reported 98% concordance with PCR-based methods using this approach.

5 T-ALL MRD

Detection of T-ALL MRD requires an understanding of physiological T-cell development and is primarily based on systematically excluding normal T/NK subsets that share overlapping features with leukemic blasts. The earliest T-cell precursors (double-negative stage) typically express CD2, cytoplasmic CD3, CD7, CD34, CD99, and TdT, but lack CD4, CD5, CD8, and CD48. Progression involves co-expression of CD4 and CD8 (double-positive cells) with acquisition of CD1a, characteristic of cortical thymocytes. Terminal maturation results in single-positive CD4 or CD8 T cells with surface CD3 expression [76–81]. Presence of cells with a thymic (immature) immunophenotype in peripheral blood or bone marrow is considered abnormal as immature T cells are normally restricted to the thymus.

Notably, cytoplasmic CD3 expression in the absence of surface CD3 is also seen in subsets of normal T/NK lineage cells, creating a major challenge for MRD detection. Accurate separation of leukemic blasts from these normal counterparts is essential, as phenotypic overlap may compromise both sensitivity and specificity. Therefore, MRD antibody panels must be carefully optimized to combine intracellular and surface markers, enhancing their discriminatory potential in post-treatment samples.

Methods: Reagents, sample collection, cell counting, red cell lysis, and surface staining are performed using the same protocol described above. T-ALL MRD involves intracellular CD3 expression analysis and additional steps required for intracellular CD3 staining, as described above.

5.1 Antibody Panel: T-ALL MRD

Several multiparametric flow cytometry strategies have been developed to overcome these challenges. The Children's Oncology Group (COG) employs a two-tube, 12-marker panel including CD2, CD3,

CD4, CD5, CD7, CD8, CD34, CD45, and CD99 [49]. The EuroFlow consortium uses a 12-color, two-tube approach with a common 8-marker backbone (CD16+NKp80, cytoplasmic CD3, surface CD3, CD5, CD7, CD34, CD45, CD45RA), supplemented with CD6, CD38, and TdT. Tube-1 incorporates CD117, CD13, and CD33, while Tube-2 includes CD2, CD4, CD8, and CD99 [82]. At Tata Memorial Centre (TMC), a 13-color (14 antibodies), single-tube assay has been standardized (Table 3), which includes:

- T-lineage-specific markers: intracytoplasmic CD3 (cyCD3) and surface CD3 (sCD3)
- Pan-T cell markers: CD2, CD5, and CD7
- Set of maturation and differentiation markers: CD1a, CD2, CD4, CD8, CD38, CD48, and CD45, which help delineate normal T-cell ontogeny
- Immaturity markers: TdT, CD1a, CD99, CD34, and CD10
- Exclusion of NK cell markers: CD16 and CD56

5.2 Gating Approach and Analysis

The analytical strategy for T-MRD detection primarily involves following the major steps (illustrated in Fig. 5).

- Exclude doublets and debris using FSC-A vs. FSC-H and SSC-A vs. FSC-A dotplots. Selectively mononuclear cells (MNCs) by excluding CD38-dim to negative granulocytes on a CD38 vs. SSC dotplot.
- Isolate T/NK-cells using CD7, a pan-T cell marker. $CD7^{+}$ cells typically include mature T-cells, NK cells, a subset of regenerating myeloid precursor cells with weak CD7 expression, a subset of $CD7^{+}$ plasmacytoid dendritic cells (pDCs), and residual leukemic blasts in MRD-positive samples.
- Exclude mature T-cells using surface CD3 expression (sCD3 vs. CD7 dotplot), NK cells using $CD16^{+}/CD56^{+}$ and CD48 expression (CD48 vs. CD16/CD56 dotplot), and weakly $CD7^{+}$ myeloid blasts and pDCs by isolating only cytoplasmic $CD3^{+}$ events (since these two subsets lack cyCD3) on a CD7 vs. cyCD3 dotplot.
- Further exclude immature $CD16^{-}/CD56^{-}$ NK cells using additional NK-cell markers such as CD94 vs. CD161. (Caution: rare cases of T-ALL may express CD161; therefore, we confirm CD5 expression before excluding such events.)
- Similarly, exclude mature T-cells that have lost surface CD3 during processing by evaluating mutually exclusive CD4 and CD8 expression patterns on a CD4 vs. CD8 dotplot. This dotplot also highlights aberrant populations with double-negative or double-positive CD4/CD8 expression.

Table 3
T-ALL MRD antibody panel

Tube-1: sytodye tube													
Fluirochrome	**BV786**	**BV650**	**BV605**	**BV510**	**BV421**	**FITC/ Vio-bright B515**	**PE**	**ECD/ PECF594**	**PC5.5**	**PC7**	**APC**	**APC AF_700**	**APC AF_750**
Antibody	X	X	X	X	X	Syto13	CD7	X	X	X	X	CD45	
Clone						Green-Fluorescent	8H8.1					J.33	
Cat No						S7575	IM1429U					A79390	
Company						Invitrogen	BC					BC	
Tube-2: MRD main tube													
Antibody	**CD161**	**CD94**	**CD3**	**CD16 + CD56**	**CD48**	**CD8**	**CD7**	**cyto CD3**	**CD34**	**CD5**	**CD4**	**CD45**	**CD38**
Clone	HP-3G10	HP-3D9	UCHT1	3G8/ NCAM16.2	TU145	B9.11	8H8.1	UCHT1	8G12	BL1a	13B8.2	J.33	LS198-4-3
Cat No	748281	743951	742623	563830/ 563041	562718	A 07756	IM1429U	A07748	347203	A51075	IM2468	A79390	B49200
Company	BD	BD	BD	BD	BD	BC	BC	BC	BD	BC	BC	BC	BC

BC Beckman Coulter Ltd., *BD* Becton Dickinson Biosciences, *Cat No* catalog number, *MB* Miltenyi Biotec

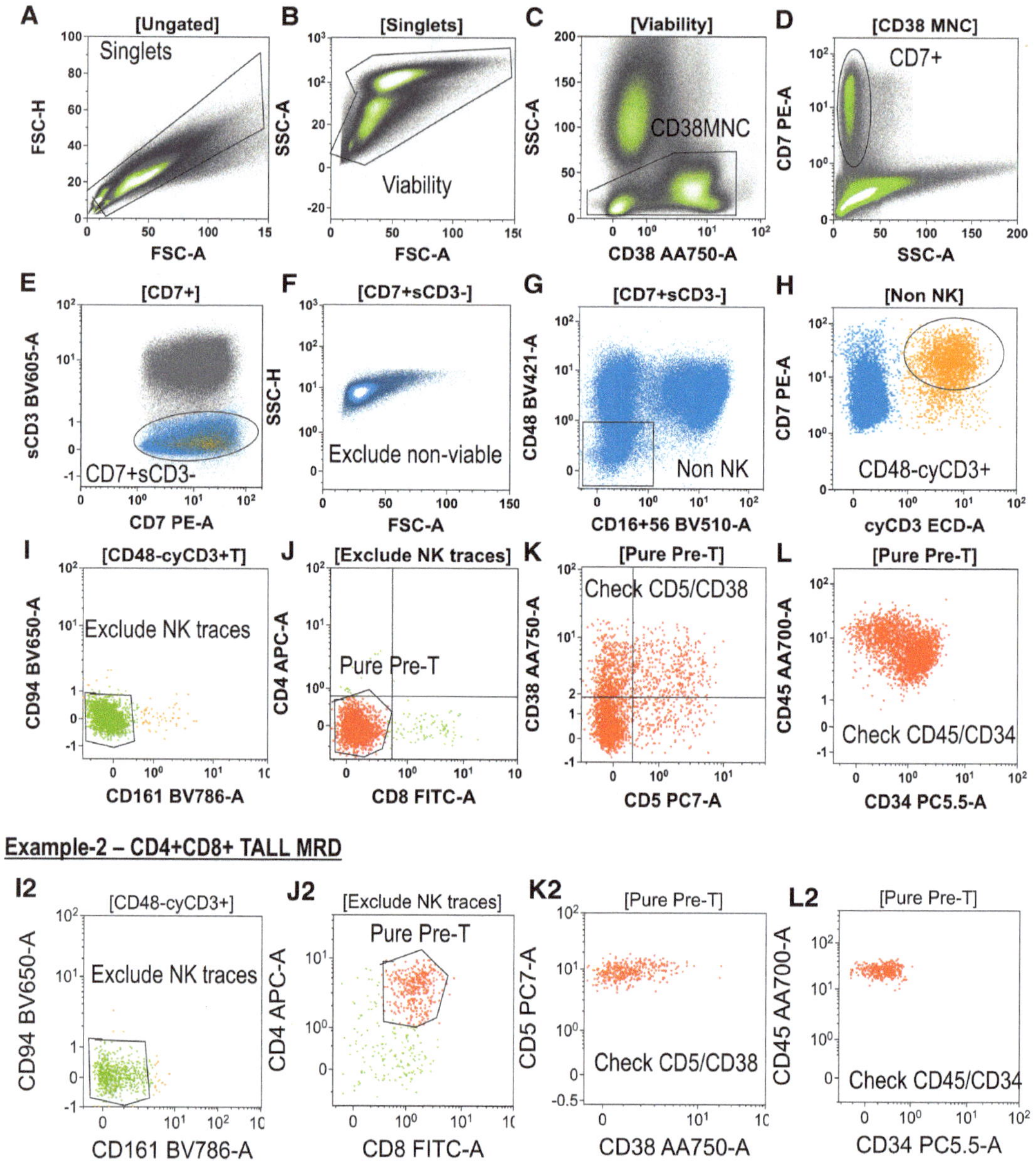

Fig. 5 Gating strategy for T-lymphoblastic leukemia (T-ALL) measurable residual disease (MRD). This figure illustrates the flow cytometric gating that begins with gating singlets (dotplot-**A**) and viable cells (dotplot-**B**), followed by selection of mononuclear cells based on CD38/SSC (dotplot-**C**). CD7+/low SSC cells are gated to enrich for T-lineage cells (dotplot-**D**). Mature surface CD3+ T-cells are excluded, and the rest of the CD7+ cells are gated (dotplot-**E**). If needed, a secondary viability gate may be applied (dotplot-**F**). NK cells present in the CD7+sCD3- cells are excluded using CD16/CD56 and CD48 (dotplot-**G**). Remaining cells are refined as CD48-/cytoCD3+ (dotplot-**H**) with further NK exclusion via CD94 and CD161 (dotplot-**I**). Final confirmation of MRD (red dots) confirmed with expression of CD4 vs CD8 (double-negative or double-positive or only CD4 or CD8 positive expression; dotplot-**J**), CD5 vs CD38 (dotplot-**K**), and CD45 vs CD34 (dotplot-**L**) to assess maturation and detect aberrant phenotypes. Dotplots **I2**–**L2** illustrate another example of CD4+/CD8+ double-positive T-ALL MRD

- After removing residual NK and mature T-cells, assess the suspicious T-cell population (double-negative or double-positive CD4/CD8) for CD5 and CD38 expression.
- CD7+ T-cells with sCD3 negative, cyCD3-positive, CD5+, with or without CD38 expression, confirm the immature nature of T-cells. Additionally, evaluate CD34 and weak CD45 expression, since a subset of T-ALL cases shows reduced CD45 and CD34 expression for further confirmation.
- Blood or BM samples do not show immature T-cells, and hence these immature T-cells indicate residual disease (MRD). It also provides CD38 expression levels on MRD events, which are relevant for anti-CD38 targeted immunotherapy.

5.3 Key Considerations

- Markers such as CD1a, CD99, TdT, and CD34, known for T-cell immaturity, can aid in MRD detection [28, 83]. However, CD1a and CD34 are expressed only in a subset of T-ALL cases, while CD99 and TdT are often limited by post-therapeutic downregulation, reducing their reliability during MRD monitoring across all samples [49, 84]. TdT is technically challenging due to nuclear localization and is absent in up to 10% of T-ALL patients [49, 83, 85].
- Recently, loss of CD48 has emerged as a useful discriminator, as CD48 is moderately expressed on mature T-cells and NK cells but is typically absent on T-ALL blasts [28].
- Incorporation of additional NK-cell markers, such as NKp80, can assist in excluding $CD56^{-}/CD16^{-}$ early NK cells. Caution is warranted, however, since NKp80 may also be expressed on blasts at diagnosis; thus, knowledge of the diagnostic immunophenotype of T-ALL blasts is essential for accurate interpretation. Similarly, a subset of cases with early T-cell precursor ALL (ETP-ALL) may demonstrate partial or heterogeneous CD56 expression. In these situations, $CD56^{+}$ cells should be excluded carefully, taking into account co-expression of other NK-associated markers such as CD48, CD94, CD161, or NKp80.
- In rare cases, T-ALL blasts may express surface CD3 (sCD3), usually with weak intensity. Such $sCD3^{+}$ events should not be excluded directly; instead, sCD3 expression must be interpreted in combination with negative CD48 and aberrant CD4/CD8 loss or co-expression for reliable MRD detection. Figure 5 provides representative examples of T-ALL MRD.
- Similar to B-ALL, in T-ALL, EOI and EOC assessments using ≥0.01% are common, though some protocols use ≥0.1% thresholds [86–93]. MRD testing may extend to weeks 10–16 or later. Protocol-specific variations—such as EOC focus (ICiCLe, GMALL), sequential testing in Modified BFM, week-22

positivity in GIMEMA, or differing EOI/EOC thresholds in PALG and PATHEMA—highlight the need for adaptable MRD strategies [87–89, 92].

6 AML MRD

Similar to ALL MRD, multiple studies demonstrated the prognostic value for disease-free survival (DFS), cumulative incidence of relapse, and overall survival (OS) at any detectable levels [5, 7, 10, 17, 19, 21, 24, 35, 44, 54, 94]. However, the standard threshold for defining MFC-MRD positivity in clinical decision-making for AML is defined at ≥0.1% of CD45+ cells [35]. One of the technical considerations for detecting MRD with lower thresholds (0.01-0.001%) is a requirement of acquisition of ≥2 million events and adequately validated AML MRD assay. Popular time intervals for AML MRD Assessment include [5, 21, 35, 94].

- Post-induction: predicts chemosensitivity; guides consolidation intensity
  At the end of the first induction cycle (day 14–21) and second induction cycle (day 28–42)
- Post-consolidation: Before maintenance therapy (identifies patients needing alternative strategies)
- Transplant settings: guides relapse prevention interventions

Pretransplant (defines baseline risk) and day +30, +60, and + 100 post-transplant.

6.1 Antibody Panel: AML MRD

AML MRD assessment is more complex compared to ALL, as AML leukemic cells must be discriminated from diverse myeloid progenitors that differentiate into granulocytic, monocytic, erythroid, dendritic, basophil, and mast cell lineages [35, 59, 60, 62]. Hence, in addition to the ELN-recommended backbone markers (CD45, CD34, CD38, CD117, HLA-DR), the AML MRD panel should include lineage-specific markers to identify normal myeloid maturation pathways [59, 62, 95]. These include:

- Granulocytic precursors: CD13, CD15, CD33
- Monocytic precursors: CD64, CD33, CD14, CD36
- Erythroid precursors: CD36, CD71
- Plasmacytoid dendritic cells, basophils, and mast cell precursors: HLA-DR, CD123, CD203c

Deviations—such as asynchronous expression or abnormal antigen intensities—can be detected using the "different-from-normal" (DfN) approach. Additionally, lymphoid markers often expressed in AML (e.g., CD7, CD19, CD56) should be included to capture

LAIPs along with other markers (CD2, CD4, CD5, CD11b, CD25), enhancing sensitivity.

Overexpression of CD33, CD123, and aberrant CD7/CD56 in CD34+CD38− indicates leukemic stem cells (abnormal hematopoietic stem cells, also referred to as leukemic stem cells (LSC)) [96–98]. Further aberrancies can be detected using CD45RA, CD52, CD54, CD96, CD97, CD366 (Tim-3), and CD371 (CLL-1) in committed progenitors and stem cells [76–78, 98, 99]. The ELN recommends a minimum eight-color MRD panel, while expanded ≥10–12 color panels increase sensitivity by enabling simultaneous detection of aberrant expression patterns, and allow incorporation of diagnostic immunophenotype-based customized antibodies, improving assay accuracy [35, 62]. We use a single 16-color AML MRD panel (Table 4). We also provided an example of 12 and 13-color panels that can be used based on the instrument available (Table 5).

Methods: Reagents, sample collection, cell counting, red cell lysis, and surface staining are performed using the same protocol described above.

6.2 Gating Approach to AML MRD

AML MRD analysis also involves two principal approaches, i.e., LAIP and DfN approaches [35, 59, 61, 62, 79]. The LAIP approach critically depends on the diagnostic immunophenotype of AML blasts and involves aberrant expression of lymphoid markers (e.g., CD2, CD7, CD19, and CD56). The presence of significant alterations in the expression of progenitor markers (e.g., CD34, CD38, CD117, HLA-DR) in diagnostic samples can also be considered as LAIPs for MRD assessment.

The DfN approach detects MRD by recognizing deviations from normal antigenic maturation in hematopoietic progenitors and does not depend on the diagnostic immunophenotype. It relies on a comprehensive antibody panel covering normal myeloid maturation toward granulocytic, monocytic, dendritic, erythroid, B-lymphoid, and basophil/mast cell lineages. Early hematopoietic stem cells (HSCs) typically show $CD34^{+}$, weak/negative CD38, relatively higher CD45, weak HLA-DR, variable CD117, weak CD13/CD33, and lack lymphoid (CD2, CD5, CD7, CD56) or lineage-specific markers (CD11b, CD15, CD36, CD64, CD45RA, CD371, etc.). These stem cells further show differentiation towards the committed progenitors of granulocytic, monocytic, erythroid, dendritic, and basophil lineages. Deviations from these well-defined maturation sequences are considered aberrant and indicate residual disease.

6.2.1 Gating Strategy (Fig. 6)

- The initial gating strategy for AML MRD analysis also follows the same principles involving doublet exclusion using FSC-H vs. FSC-A properties, removal of debris with FSC vs. SSC properties, and isolation of mononuclear cells using CD38 vs. SSC.

Table 4

AML MRD antibody panel (16 color)

Fluiro chrome	BUV737	BUV395	BV786	BV711	BV650	BV605	BV510	BV421	FITC/ BB515	PE	PECF594	PC5.5	PC7/ PEV IO-770_	APC	APC AF_700	APC AF_750
Antibody	CD64	CD56	CD117	CD19	CD14	CD36	HLA-DR	CD 123	CD15	CD13	CD7	CD33	CD203c	CD34	CD45	CD38
Clone	10.1	NCAM16.2	104D2	SJ25C1	M5E2	CLB IVC7	L243	9F5	80H5	WM16	M-T701	D3HL 60.251	E-NPP3	581	J.33	LS198-4.3
Cat No	612776	563554	313237	563036	563419	745161	307 646	562 517	B36298	555394	562541	A70198	A66906	IM24 72U	A79390	B49200
Company	BD	BD	BL	BD	BD	BD	BL	BD	BC	BD	BD	BC	BC	BC	BC	BC

BC Beckman Coulter Ltd., *BD* Becton Dickinson Biosciences, *BL* BioLegend, *Cat No* catalog number, *MB* Miltenyi Biotec

Table 5

Proposed 12- and 13-color AML MRD antibody panel

Tube no.	Fluoro-chrome -1	Fluoro-chrome -2	Fluoro-chrome-3	Fluoro-chrome -4	Fluoro-chrome -5	Fluoro-chrome -6	Fluoro-chrome -7	Fluoro-chrome -8	Fluoro-chrome -9	Fluoro-chrome -10	Fluoro-chrome -11	Fluoro-chrome -12	Fluoro-chrome -13	Fluoro-chrome -14	Fluoro-chrome -15	Fluoro-chrome-16
	Backbone markers					Myeloid lineage markers		Myelomonocytic lineage markers			Leukemia-associated or cross-lineage markers					
12-color antibody panel																
MRD 1	HLA-DR	CD34	CD117	CD45	CD38	CD33	CD13	CD15	CD64 or DI[a]	CD7	CD123	CD71 or DI	X	X	X	X
MRD 2	HLA-DR	CD34	CD117	CD45	CD38	CD33	CD14	CD64	CD36	CD19	CD56	CD203c or DI[a]	X	X	X	X
13-color antibody panel																
MRD 1	HLA-DR	CD34	CD117	CD45	CD38	CD33	CD13	CD15	CD64 or DI[a]	CD36	CD7	CD123	CD203c or DI[a]	X	X	X
MRD 2	HLA-DR	CD34	CD117	CD45	CD38	CD33	CD371	CD14	CD64	CD45RA	CD19	CD56	CD71 or DI[a]	X	X	X

Abbreviation: DI-drop in antibody The antibody panel can be designed by selecting appropriate antibody-fluorochrome combinations based on the user's experience, instrument configuration, and published literature CD203c can be helpful in AML cases with positive CD117 but negative HLADR expression at diagnosis [a]One can add any antibody marker at this place based on the diagnostic immunophenotype of AML. For example, if blasts were CD11b positive at diagnosis, it can be added instead of CD71 or CD203c or if blasts were positive for CD2 or CD5 or CD25, it can be added at this empty space

- It is followed by the enrichment of progenitor cells excluding CD45-bright lymphocytes and monocytes using "Progen gate," and mapping the maturation sequences of myeloid progenitors.
- Divide progenitor cells into three compartments—[1] $CD34^{+}$ [2], $CD117^{+}/CD34^{-}$, and [3] $CD34^{-}/CD117^{-}$ (dotplot C). Select first—CD34+ compartment and exclude B-cell precursors using Boolean gates resulting from combining gates CD19/CD33 or CD19/CD117, and isolate myeloid/erythroid progenitors (dotplots D1–D2).
- Remove remnants of $CD14^{+}CD36^{+}$ monocytes using the "Clean MyBL" gate (dotplots D3), and analyze these purified progenitors for abnormal maturation patterns across different relevant antibody combinations with reference to normal maturation patterns (Fig. 6, G-Box).
- Similarly, evaluate $CD117^{+}/CD34^{-}$ compartment progenitors for aberrancies using deviation from normal maturation patterns with reference to normal maturation patterns (Fig. 6, F-Box).
- Finally, analyze the $CD117^{-}/CD34^{-}$ compartment. Exclude lymphocytes with negative expression of CD13 and CD33 with "$CD13^{-}/CD33^{-}$" gate and basophils/plasmacytoid dendritic cells with "exclude Baso/pDC" gate and look for abnormal cell population after denoting normal myeloid/erythroid progenitors (Fig. 6, E-Box).
- Blast/Progenitor cells are characterized by weak CD45 expression + low SSC, expression of immature markers (CD34, CD117), and absence of maturation markers (CD11b, CD14, strong CD15).

6.3 Key Considerations

- The compartment-based strategy allows targeted assessment of progenitor subsets according to diagnostic leukemia-associated immunophenotypes (LAIPs), such as AML blasts lacking CD34 or CD117 expression. This approach facilitates focused evaluation of relevant compartments (e.g., analysis of the $CD117^{+}$ $CD34^{-}$ subset in CD34-negative AML) while simultaneously monitoring maturation trajectories through established differentiation pathways (Fig. 6). Examples of MRD cases across compartments are shown in Fig. 7.
- One should note that although LAIPs defined at diagnosis provide a framework for MRD monitoring, post-therapy immunophenotypic shifts—resulting from clonal evolution or phenotypic plasticity—may lead to false negatives. Other limitations of the LAIP-based approach include the unavailability of diagnostic immunophenotypes in referral cases, and CD34+ progenitors may show weak expression of CD7/CD56 in regenerating BM samples, leading to false positive results [62, 79, 80]. Regenerating erythroid progenitors usually have

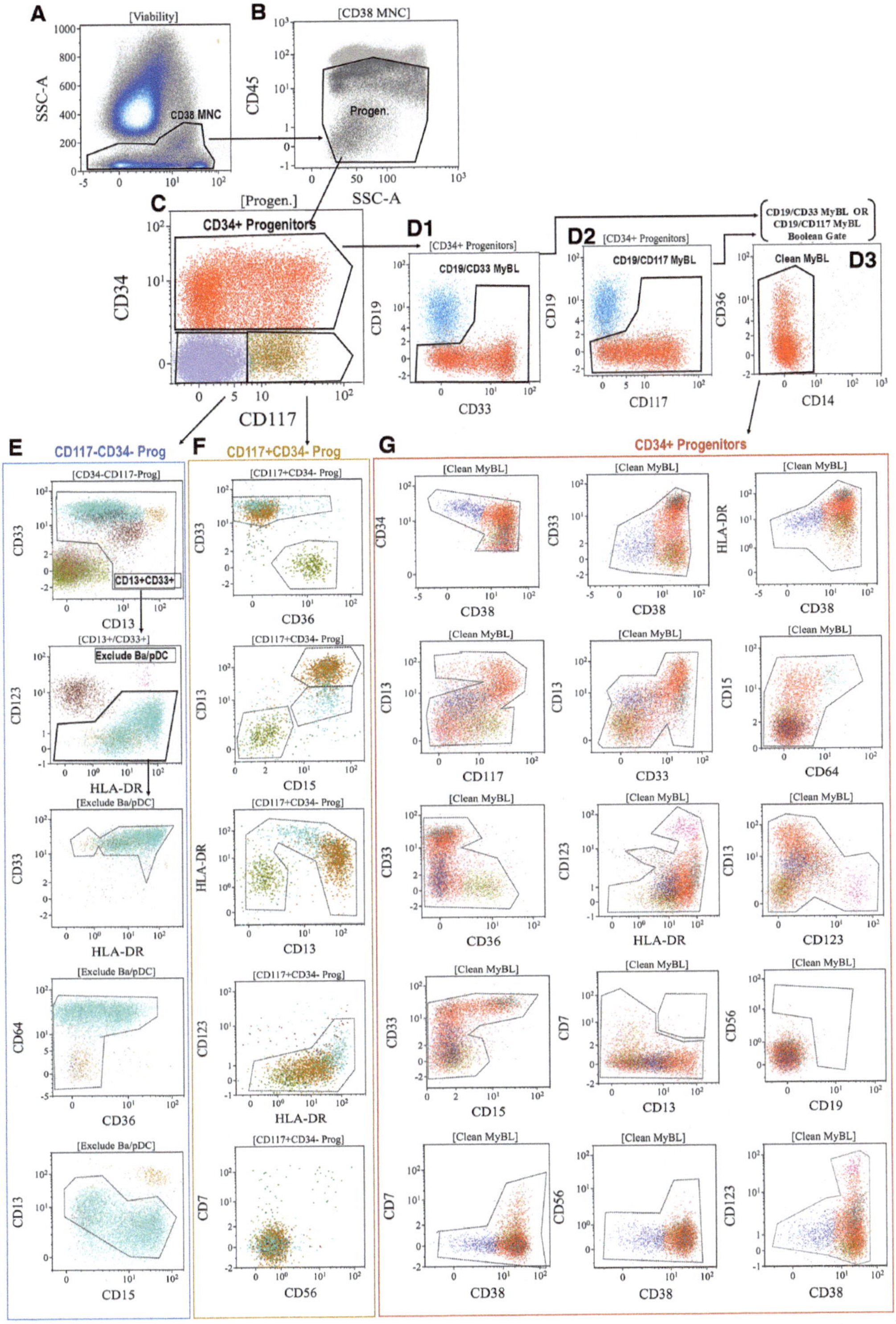
A
[Viability]
SSC-A
CD38 MNC
B
[CD38 MNC]
CD45
Progen.
SSC-A
C
[Progen.]
CD34+ Progenitors
CD34
CD117
D1
[CD34+ Progenitors]
CD19/CD33 MyBL
CD19
CD33
D2
[CD34+ Progenitors]
CD19/CD117 MyBL
CD19
CD117
CD19/CD33 MyBL OR
CD19/CD117 MyBL
Boolean Gate
Clean MyBL
D3
CD36
CD14
E
CD117-CD34- Prog
[CD34-CD117-Prog]
CD33
CD13
CD13+CD33+
[CD13+/CD33+]
Exclude Ba/pDC
CD123
HLA-DR
[Exclude Ba/pDC]
CD33
HLA-DR
[Exclude Ba/pDC]
CD64
CD36
[Exclude Ba/pDC]
CD13
CD15
F
CD117+CD34- Prog
[CD117+CD34- Prog]
CD33
CD36
[CD117+CD34- Prog]
CD13
CD15
[CD117+CD34- Prog]
HLA-DR
CD13
[CD117+CD34- Prog]
CD123
HLA-DR
[CD117+CD34- Prog]
CD7
CD56
G
CD34+ Progenitors
[Clean MyBL]
CD34
CD38
CD33
CD38
HLA-DR
CD38
CD13
CD117
CD13
CD33
CD15
CD64
CD33
CD36
CD123
HLA-DR
CD13
CD123
CD33
CD15
CD7
CD13
CD56
CD19
CD7
CD38
CD56
CD38
CD123
CD38

more CD7 expression than myeloid progenitors. Restricted expression of CD13 or CD33 results in clustered maturation patterns during hematopoietic recovery [47, 59, 62, 79, 80].

- Sensitivity optimization: Acquisition of ≥1 million cells is recommended for reliable 0.01% sensitivity [35, 62].
- Panel design: ≥8-color panels are now standard, with EuroFlow recommendations providing harmonized approaches [35, 62].
- Data analysis: Use of reference databases or predefined templates for normal regeneration patterns improves DfN accuracy
- The European Leukemia Net (ELN) working group defines MFC-MRD positivity as ≥0.1% of CD45+ cells [35], but, numerous studies demonstrate that MRD levels <0.1% (down to 0.001% or any detectable levels) retain prognostic value for disease-free survival (DFS), relapse and overall survival (OS) [5, 7, 10, 17, 19, 21, 24, 35, 44, 54, 94]. Detecting low-level MRD (0.01%–0.001%) requires ≥2 million events and validated assays. Consensus for BM sampling timepoints includes post-induction (cycle 1: day 14–21; cycle 2: day 28–42), postconsolidation (before maintenance), and in transplant settings (pretransplant baseline; day +30, +60, +100 post-transplant) [5, 21, 35, 94].

7 Multiple Myeloma MRD

Similar to acute leukemia, several studies have established the major role of MRD-status in the prognostication, response assessment, and treatment modification in MM, especially for patients treated with autologous stem cell transplant (ASCT) [81, 100–116]. The International myeloma working group (IMWG) has included flow cytometric MRD-negative status using next-generation flow (NGF) in bone marrow as a criterion to define stringent CR (sCR) [117]. NGF-MRD or high-sensitivity MFC MRD has been shown to be an equally sensitive technique compared to NGS MRD, with the

Fig. 6 Gating strategy for AML measurable residual disease (MRD) analysis. Mononuclear cells (MNCs) are identified by excluding granulocytes (CD38 vs. SSC, dotplot-**A** and then gated to enrich for progenitor cells by excluding CD45-bright lymphocytes and monocytes (Progen gate, dotplot-**B**). In dotplot-**C**, progenitor subsets are classified into: (1) $CD34^+$, (2) $CD117^+/CD34^-$, and (3) $CD34^-/CD117^-$ compartments. $CD34^+$ B-cell precursors were excluded using Boolean gates combining CD19/CD33 (dotplot-**D1**) or CD19/CD117 (dotplot-**D2**) to isolate non-lymphoid $CD34^+$ progenitors. A "Clean MyBL" gate was created by removing $CD14^+CD36^+$ monocytes (dotplot-**D3**), and these cells were analyzed for abnormal maturation patterns, the combinations of various markers, as shown in dotplots included in "G-Box", based on normal reference patterns defined with gates. Similarly, $CD117^+/CD34^-$ progenitors are further assessed for aberrancies using deviation from normal maturation patterns indicated with gates shown in dotplots included in "F-Box", and $CD117^-/CD34^-$ progenitors are studied after excluding $CD13^-/CD33^-$ cells and basophils/plasmacytoid dendritic cells as shown in dotplots included in "E-Box"

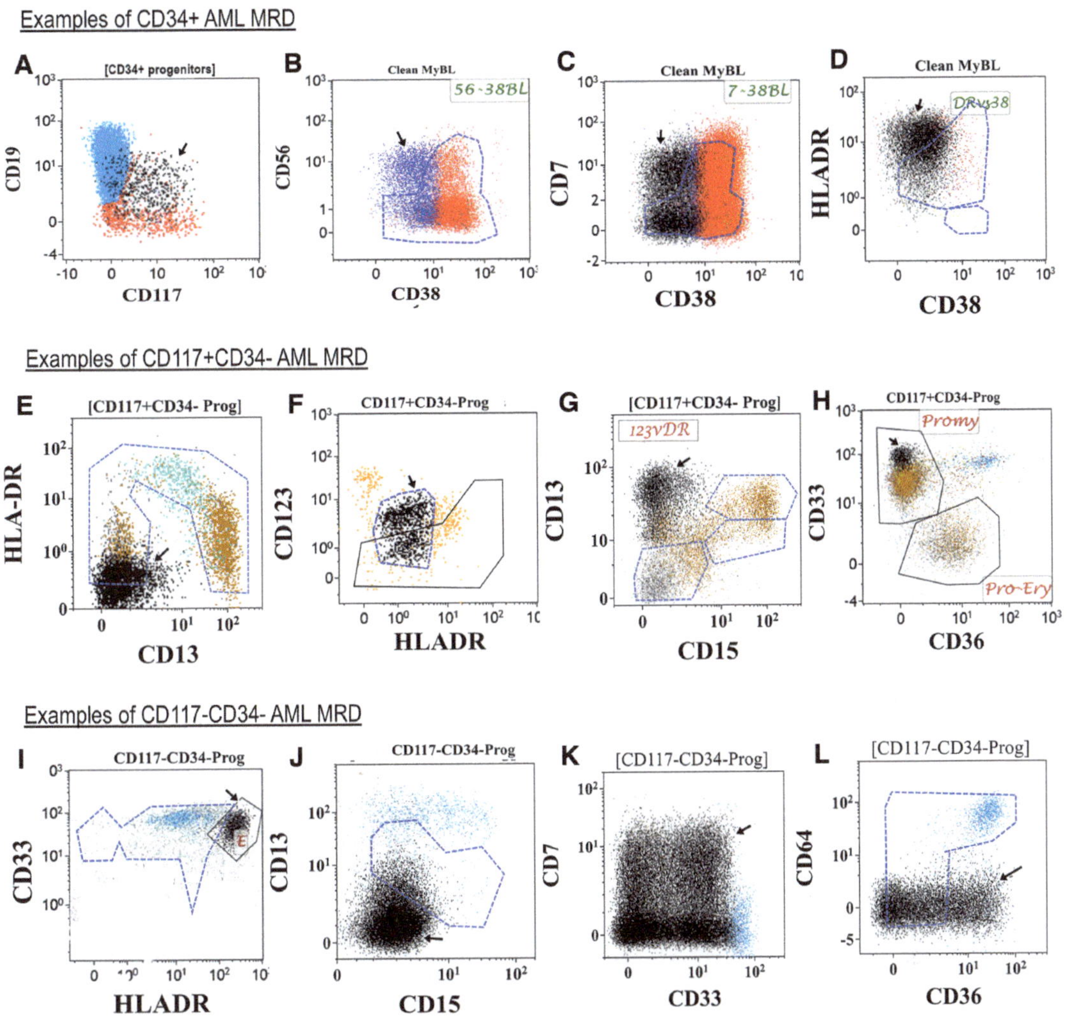

Fig. 7 Illustrative examples of AML measurable residual disease (MRD) detection across progenitor compartments. MRD populations are shown from three compartments in Fig. 6. Each dotplot represents MRD from a different patient. MRD in the CD34$^+$ progenitors region (dotplots **A–D**), CD117$^+$/CD34$^-$ progenitors (dotplots **E–H**), and CD117$^-$/CD34$^-$ progenitors (dotplots **I–L**). Residual leukemic cells are marked with black dots in all plots, except in **B**, where MRD is highlighted in blue color dots to indicate aberrant CD56 expression (dotplot-**B**) in CD38$^-$ stem cell populations

advantages of greater applicability, less time-consuming, and knowledge of the degree of hemodilution [118–122].

7.1 Antibody Panel—MM MRD

Similar to other MRD assays, MM MRD assay panel designing involves isolation of plasma cells (PCs), differentiating abnormal PCs (aPCs) from normal PCs (nPCs), and confirmation with clonality establishment. Traditionally, plasma cells are identified with expression of CD138 and strong expression of CD38. However, since the introduction of anti-CD38 therapies, conventional CD38 is no longer available for identifying PCs, which has led to the inclusion of additional markers for gating plasma cells, such as CD229, CD319, CD54, intracellular CD38 [123]. The differentiation of aPCs from nPCS is done using underexpression of CD19, CD27, CD45, and CD81 and overexpression of CD56 and CD117 [122, 124]. The clonality is confirmed with restricted expression of one of the intracellular light chains (Table 6).

Methods: Reagents, sample collection, cell counting, red cell lysis, and surface staining are performed using the same protocol described above. MM MRD involves intracellular light chain expression analysis and additional steps required for intracellular kappa and lambda light chain staining, as described above.

7.2 Gating Approach (Fig. 8)

1. As described earlier, doublets and debris are excluded using light scatter characteristics—FSC and SSC.
2. Evaluate CD38 expression on monocytes and lymphocytes gated using CD45 vs. SSC dotplot as an internal quality control for CD38 expression. It helps to verify the possibility of daratumumab therapy in the absence of adequate clinical and treatment details.
3. Plasma cells are gated using a Boolean gate "Initial PC" gate as a result of "CD138 vs. CD38 gate" OR "CD38 vs. CD45 gate" OR "CD138 vs. CD45 gate" function. This "Initial PC" gate is broader and includes other hematopoietic cells, such as a few monocytes (resulting from the non-specific binding of CD138 and NK cells with moderate CD38 expression). These non-PCs are removed, and PCs are further purified using the "additional PC" gate in a dot plot of CD54 vs. CD319+ CD229 (CD38 vs. CD319+CD229), after removal of CD54 and CD229+CD319 negative events. The PCs are further purified using "KL-gate," which includes only cytoKappa or cytoLambda positive events and removes cytoplasmic light chain negative events. The cytoKappa-positive cells are given a cobalt blue color, and the cytoLambda-positive cells are given an orange color. This color coding helps track cytoKappa or cytoLambda positive cells in subsequent dotplots (caution: the cytoplasmic light chain negative events need to be included in a patient with non-producing myeloma, which is extremely rare).
4. Mature B-cell contamination is removed after excluding bright CD19, CD20, and CD45 positive cells.
5. Aberrent overexpression of CD56, CD117, and CD20 and loss of CD19, CD27, CD45, and CD81 is studied by applying "Viable PC" gate to a combination of different markers such as CD19 vs CD20, CD56, CD45, CD27, CD81, and CD117; CD27 vs CD45, CD56, CD81, and CD117, CD56 vs CD81 and CD117.

Table 6

MM MRD antibody panel

Tube-1: MRD main tube													
Fluiro chrome	**BV786**	**BV650**	**BV605**	**BV510**	**BV421**	**FITC**	**PE**	**PECF594**	**PC5.5**	**PC7**	**APC**	**APC AF_700**	**APC AF_750**
Antibody	CD117	CD54	CD38	CD56	CD138	cyto Kappa	cyto Lambda	CD81	CD27	CD19	CD229+ CD319	CD45	CD20
Clone	104D2	M-L13	HB7	HCD56	MI15	Poly-Goat	Poly-Goat	JS81	1A4CD27	J3-119	HLy9.25 / 235614	J.33	B9E9(HRC20)
Cat No	313237	743049	562665	318340	562935	C 15623	C15189	565428	B21444	IM3628	17-2299-42 / 564338	A71117	A66331
Company	BL	BD	BD	BL	BD	BC	BC	BD	BC	BC	Ebio / BD_	BC	BC

BC Beckman Coulter Ltd., *BD* Becton Dickinson Biosciences, *BL* BioLegend, *Cat No.* catalog number, *MB* Miltenyi Biotec

6. The aPCs are gated using the best combination, suggesting abnormal antigen expression, and applied to the cytoKappa vs. cytoLambda dotplot to confirm the clonality through light chain restriction. We use the ratio of ≥9:1 for clonality confirmation.
7. MRD levels are calculated as a percentage of clonal aPC gated in total singlet viable cells. In addition, the percentage of total PCs in total singlet viable cells is also calculated.
8. Additionally, parameters that help in the assessment of hemodilution in the given sample are evaluated (shown in Fig. 8), including 1] percentages of hematogones using "CD19+ CD81+CD20dim to negative, CD45 dim and CD38 moderate positive expression, 2] CD117 bright positive and CD45 moderate positive mast cells, 3] CD117 dim to moderate positive and CD45 weak positive myeloid/erythroid precursor cells, and 4] CD45 negative low SSC erythroid cells [124, 125].

7.3 Key Considerations

- This is a high-sensitivity MRD assay with LOD of 10 events and LLOQ of 30 events and a sensitivity of 0.0001%. To reach such sensitivity, we acquire at least 10 million viable cells, and considering up to 70% cell loss during processing, we take a sample volume that has approximately thirty million cells, calculated using an automated cell counter.
- In multiple myeloma, peripheral blood (PB) MRD has emerged as a valuable prognostic tool, outperforming conventional markers such as M-protein and free light chain assays [126]. Achieving a sensitivity of 0.0001% typically requires 10–12 mL of blood (~10 million cells). While PB-MRD does not replace bone marrow (BM) MRD, it detects positivity in about one-third to one-half of BM-MRD–positive cases [115]. Importantly, PB-MRD offers a minimally invasive method that is more sensitive for predicting early relapse or progression and can obviate the need for painful BM procedures in PB-MRD—positive patients, given the high concordance with BM positivity.
- Plasma cells can possess high SSC characteristics; hence, do not exclude the cells with high SSC from plasma cell gating.
- If excessive cell loss occurs during cytoplasmic processing, acquire half the cells after surface staining alone and the other half with cytoplasmic light-chain staining. The two data files can be merged for analysis, as backbone markers are identical. Once abnormal plasma cells are identified with light-chain restriction, MRD can be calculated from the surface-stained tube, minimizing bias from cytoplasmic processing—related cell loss.

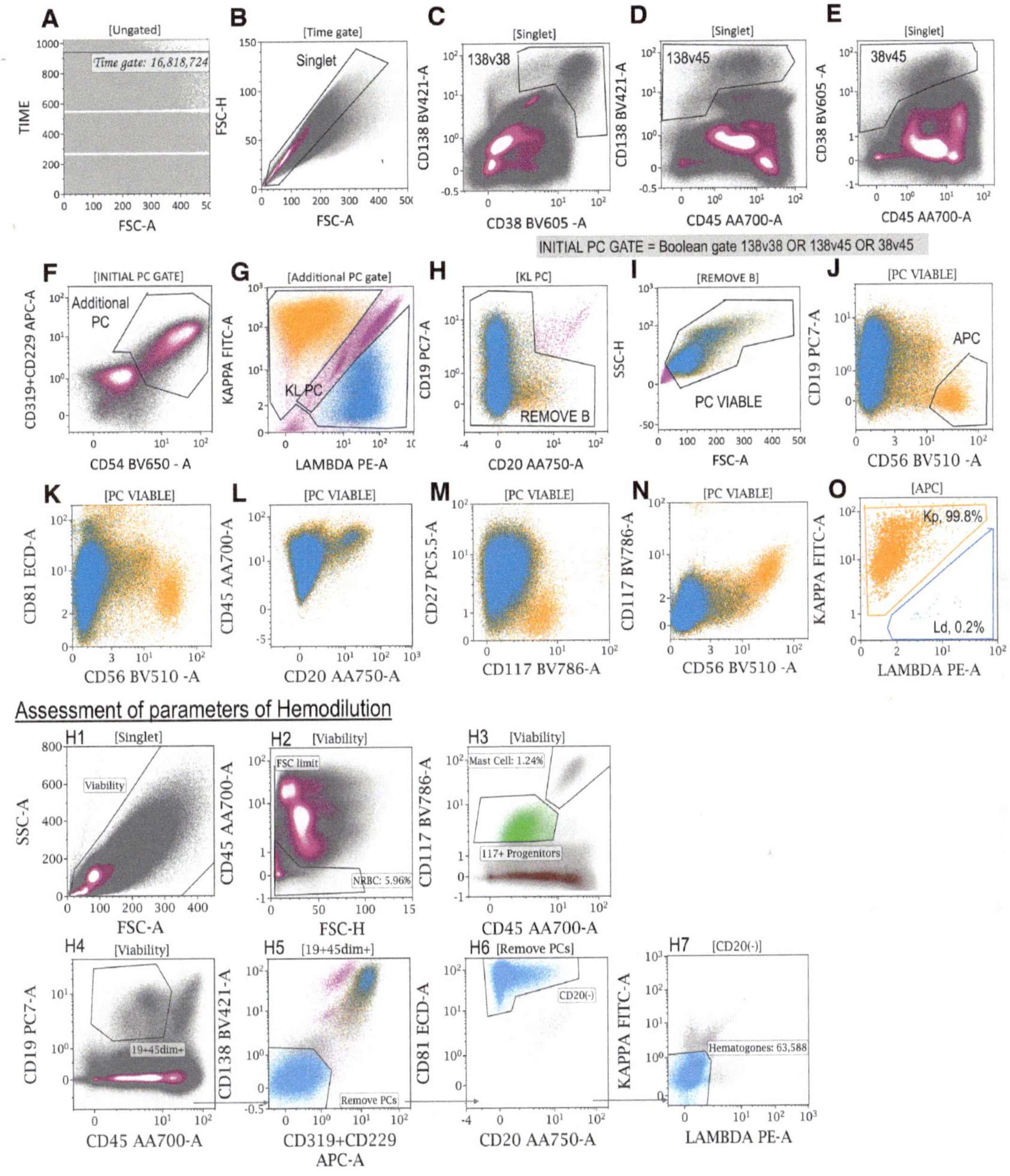

8 CLL MRD

Among mature lymphoproliferative diseases, MRD in CLL has been studied extensively, and its clinical value is well established [127–132]. It is a robust prognostic and predictive biomarker in chronic lymphocytic leukemia (CLL). MRD kinetics in CLL are therapy dependent; for example, time-limited regimens show linear progression-free survival (PFS) improvement with each log MRD depletion at the end of treatment (EoT), whereas BTK inhibitors induce modest early reductions followed by stable or slowly declining levels [133–136]. BCL2 inhibitors achieve rapid depletion within 6–12 months, and therapeutic antibodies preferentially clear circulating disease [133, 135]. Stratification demonstrates that only deep MRD negativity (<0.01%) confers durable remission, while rising peripheral blood (PB) MRD ≥1-log increase often precedes clinical progression [132, 135, 137–139]. Wide therapeutic options improved survival outcomes in CLL patients with longer complete remissions, making MRD assessment a vital tool to monitor these patients. PB monitoring offers a practical alternative to bone marrow (BM), with PB MRD levels generally equating to one log higher BM MRD levels [139]. Recommended practice is to confirm PB negativity at 3 months and BM at 6 months, with sustained dual negativity defining time to MRD negativity and potentially obviating invasive BM testing [139].

8.1 Antibody Panel—CLL MRD

MFC plays a crucial role in the diagnosis of CLL/SLL and immunophenotypic markers like co-expression of CD19 and CD5, dim to negative CD20, expression of CD200 and /or CD23, weak surface light chain restriction (Kappa, and Lambda) is an immunophenotypic signature of typical CLL according to ERIC (European Research Initiative on CLL) /ESCCA (European Society for Clinical Cell Analysis) groups [4]. Thus, CLL MRD antibody panel includes CD19 and CD45 to isolate mature B-cells, CD5, CD20, and CD200 (or CD23) for the identification of abnormal B-cells and kappa/ lambda light chains for clonality confirmation. Additional markers recommended for MRD monitoring in CLL include markers like CD43, CD79b, CD81, CD200, and CD160. CD3 is also usually used to exclude the doublets of B and T lymphocytes to avoid false positivity. CD40 can be added as an additional gating marker in

Fig. 8 Gating strategy for highly sensitive measurable residual disease (MRD) in multiple myeloma (MM) analysis. This figure illustrates the flow cytometric gating that begins with gating singlets (dotplot-**B**) following exclusion of possible noise using Time parameter analysis (dotplot-**A**). Initially, plasma cells were isolated using a broad Boolean gate "INITIAL PC GATE" constructed by combining events gated with "138v38" (dotplot-**C**), "138v45" (dotplot-**D**), and "38v45" (dotplot-**E**) using "OR" logic function. Plasma cells are further purified using the "Additional PC" gate using positive coexpression of CD319+CD229 and CD54 (dotplot-**F**). This gate still includes some non-PC cells, which are removed using their negative expression for both intracellular light chain expression (dotplot-**G**) using the "KLPC" gate. If a few mature B-cell or doublets of PCs and mature B-cells are present, they are removed using the "REMOV B" gate (dotplot-**H**), and finally, clean viable PCs are gated using a refined "PC VIABLE" gate (dotplot-**I**). This "PC VIABLE" gate is applied to the subsequent dotplots to study the aberrant loss or gain markers included in the panel to differentiate abnormal PCs (APC) from normal PCs (dotplots **J–N**). APCs are gated by applying the "APC" gate to the dotplot with any combination of markers that provide the best separation of abnormal PCs. MRD is confirmed by confirming the clonal nature of APC by studying the intracellular light chain (kappa or lambda) restriction. Dotplots **H1–H7** illustrate the assessment of other hematopoietic cells (CD45-negative nucleated red cells, mast cells, CD117+ myeloid/erythroid progenitors, and hematogones) that serve as parameters of hemodilution assessment

patients receiving anti-CD19 therapy. We use an 11-color antibody panel given in Table 7.

Methods: Reagents, sample collection, cell counting, red cell lysis, and surface staining are performed using the same protocol described above.

8.2 Gating Approach (Fig. 9)

- As described earlier, the initial gating involves doublet exclusion using FSC-H vs. FSC-A properties and removal of debris with FSC vs. SSC properties.
- A broad lymphocyte gate "CD45 Lym" is applied in CD45 vs. SSC dotplot to include CD45 moderate to bright and low SSC lymphocyte predominant events.

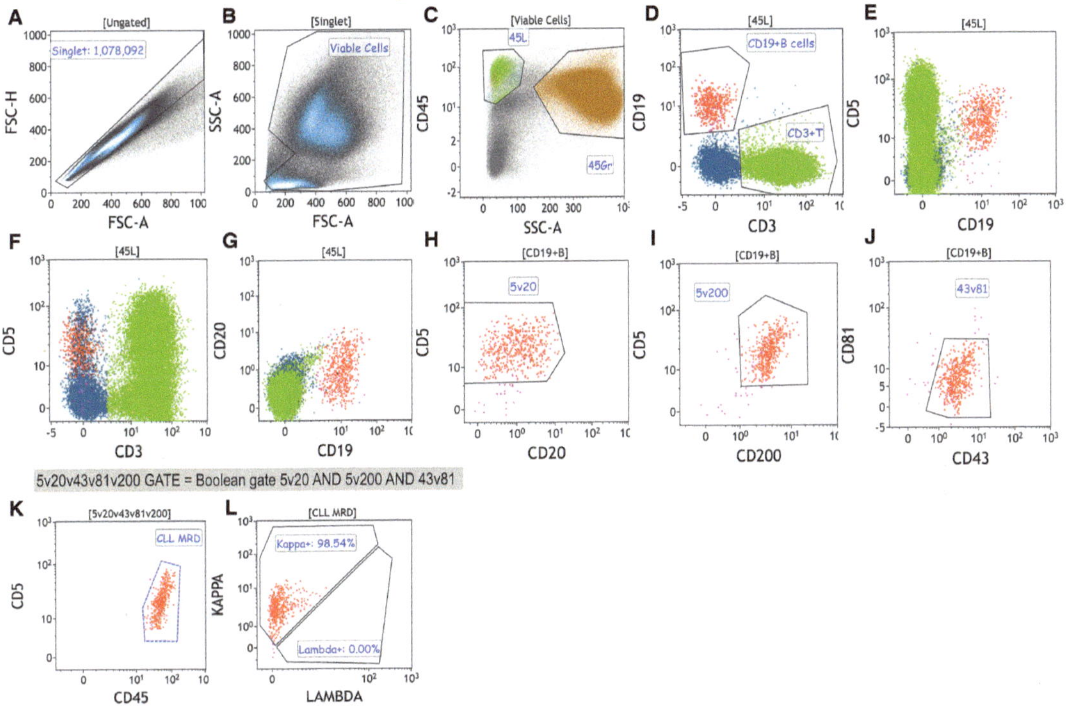

Fig. 9 Gating strategy for highly sensitive measurable residual disease (MRD) in chronic lymphocytic leukemia (CLL) analysis. This figure illustrates the flow cytometric gating approach that begins with gating singlets (dotplot-**A**) and viable cells (dotplot-**B**). Initially, lymphoid cells were isolated using a broad gate "CD45L" with SSC versus CD45 (dotplot-**C**) and applied to dotplots **D–G** for studying B and T-lymphoid cells. B-cells are gated and highlighted with red dots, and T-cells with light green dots. B-cells isolated with "CD19+B cell" gate are studied for the aberrant expression of CD5 and weak to negative expression of CD20 (dotplot-**H**), moderate homogenous expression of CD200 (dotplot-**I**), and weak to negative expression of CD81 and moderate expression of CD43 (dotplot-**J**). A Boolean gate "5v20v43v81v200" is constructed by combining events gated with "5v20" (dotplot-**H**), "5v200" (dotplot-**I**), and "43v81" (dotplot-**J**) using "AND" logic function. Abnormal B-cells are further purified using the "CLL MRD" gate using positive expression of CD5 and moderate to bright expression of CD45. MRD is confirmed by confirming the clonal nature of abnormal B-cells by studying the light chain (kappa or lambda) restriction (dotplot-**L**)

Table 7
CLL MRD antibody panel

Tube-1: MRD main tube											
Fluoro-chrome	**BV605**	**BV510**	**BV421**	**FITC**	**PE**	**PECF594**	**PC5.5**	**PC7**	**APC**	**APC AF_700**	**APC VIO 770**
Antibody	CD43	CD20	CD200	Kappa	Lambda	CD19	CD38	CD5	CD3	CD 45	CD81
Clone	1G10	2H7	MRC OX-104	GOAT Poly	GOAT Poly	HIB19	LS198.4.3	BL1a	UCHT1	J.33	REA513
Cat No	563378	563067	564114	C15623	C15189	562294	A70205	A51075 / A21690	561811	A71117	130-130-060
Company	BD	BD	BD	BC	BC	BD	BC	BC	BD	BC	MB

BC Beckman Coulter Ltd., *BD* Becton Dickinson Biosciences, *Cat No* catalog number, *MB* Miltenyi Biotec

- Cells in CD45 gate are reviewed in three dotplots: CD3 vs. CD19, CD5 vs. CD19, and CD3 vs. CD5 to check overlap between B and T cells.
- CD19-positive but CD3-negative B cells are gated using a "CD19+B" gate, and T cells are gated with "CD3+T" gate.
- CLL cells are identified with aberrant expression of CD5, weak to negative CD20 and moderate CD200, moderate CD43 and weak CD81 after applying "CD19+B" gate to a combination of different markers such as CD5 vs. CD20, CD5 vs. CD200, and CD43 vs. CD81.
- A Boolean gate of combination of gates "5v20 AND 5v200 AND 43v81" and CD5 vs. CD45 dotplots. All abnormal cells were gated using "CLL MRD" gate, which is applied to the Kappa vs Lambda dotplot to confirm the clonality through light chain restriction. We use the light chain ratio of ≥9:1 for clonality confirmation.
- MRD levels are calculated as a percentage of CLL cells gated with "CLL MRD gate" in total singlet viable cells in bone marrow and in total CD45+ WBCs in the blood sample.

8.3 Key Considerations

- It should be noted that BTK inhibitor therapy, such as Ibrutinib, downregulates the immunosuppressive molecule CD200, leading to its negative expression [140]. Therefore, one should be cautious when analyzing the results post-ibrutinib therapy. In such cases, CD23 can be added as an alternative marker.
- Based on the standardized definitions established by the ERIC group in Leukemia (2021), the MRD status is a tiered classification system. An undetectable MRD (uMRD) status at a given sensitivity level (e.g., uMRD4) is defined by finding fewer than one CLL cell per specified number of total normal cells acquired (e.g., <1/10,000). Conversely, a detectable MRD (dMRD) status at the next level (e.g., dMRD4) is assigned when any CLL cells are found, but the count remains below the threshold for the higher uMRD tier, indicating persistent, low-level disease. This precise stratification is critical for interpreting treatment depth and patient prognosis.
- CLL MRD analysis in BM remains the gold standard for confirming undetectable MRD (uMRD) in clinical trials due to its superior sensitivity. However, PB testing is a valuable, less invasive tool for screening and monitoring; the detection of MRD in PB obviates the need for a BM aspiration. A sustained or gradual decrease in PB MRD is observed with effective therapy, while a significant rise (>1-log) in PB MRD levels can serve as an early biomarker, pre-empting clinical disease progression in a majority of cases.
- The existence and magnitude of a compartment effect between peripheral blood (PB) and bone marrow (BM) in CLL are

treatment-dependent. PB monitoring offers a practical alternative to bone marrow (BM), with PB MRD levels generally equating to one log higher BM MRD levels, though this varies significantly by therapy. For example, alemtuzumab exhibits a large (2-log) discrepancy, while targeted agents like ibrutinib show minimal effect. Consequently, BM assessment remains necessary to confirm undetectable MRD for therapies with a known significant compartment effect, though a high PB MRD level (>1%) renders a BM aspirate redundant.

9 MRD Monitoring in B-Cell Lymphomas Other NHLs

Flow cytometric MRD monitoring in B-cell lymphomas other than CLL and in T-cell non-Hodgkin lymphomas (T-NHL) remains largely restricted to clinical trials using qPCR/NGS-based approaches and has not been studied extensively using MFC [141–143]. Available evidence indicates prognostic relevance comparable to CLL, with lower MRD levels correlating with improved outcomes; however, studies are limited and subtype-specific [142–144]. For example, MRD has been shown to predict relapse risk in hairy cell leukemia, mantle cell lymphoma, follicular lymphoma, and diffuse large B-cell lymphoma in selective settings [145–148]. Despite these promising data, the evidence base remains insufficient for standardized clinical application of MFC-MRD in other NHL. MRD assays for these NHL entities are not included in this chapter due to limited scope.

10 Quality Control (QC) and Quality Indicators

Quality control in flow cytometry-based MRD assessment should include the following points [62, 80, 149, 150].

10.1 Instrument Quality Control

With daily calibration beads such as CS&T beads, Rainbow beads, flow check beads, etc., as per the manufacturer's recommendations.

Reagent Quality Assurance: It includes validation of new reagent lots against established controls and use of pre-validated antibody cocktails to minimize pipetting variability, ensure lot-to-lot consistency, and reduce inter-operator variability.

10.2 MRD analysis Template Development, Verification, and Maintenance [2,61,62,79,80]

Template Establishment: Minimum 20 control bone marrow samples required, which should preferably include ten normal non-leukemic samples and ten post-chemotherapy regenerating samples, possibly samples spanning relevant age groups.

- Template verification: demonstration of ≤10% coefficient of variation for marker expression intensities and verification of sensitivity to 0.01% (10^{-4}) aberrant populations [82].

- Template maintenance/revalidation: it includes monthly re-evaluation of templates using control samples, quarterly template updates using control samples, and updated regeneration patterns [90].
- Competency assurance: It is done through personnel training, regular participation in MRD-specific workshops and continuing medical education (CME) programs, and annual inter/intra-laboratory competency assessments [151–154].
- External quality assessment: it includes enrollment in international EQA programs (e.g., UK NEQAS) or regular performance review with peer laboratories [151–156].

10.3 Additional QC Measures

Sample Processing Controls can be performed through:

- Assessment of cell viability using viability dyes (e.g., 7-AAD, DAPI), especially in transported samples [62, 80].
- One of the critical considerations is hemodilution of BM samples, causing false-negative results [3, 35, 48, 62]. It is recommended to use first-pull aspirates (1–2 mL) to minimize the risk of hemodilution and to acquire at least 1×10^6 viable events per tube to achieve optimal sensitivity. However, acquisition of $\geq 4–5 \times 10^6$ viable cells can further improve the MRD detection, allowing next-generation flow cytometry in ALL to reach <0.001% sensitivity [3, 48–50, 157].
- Hemodilution assessment via either marrow representing cellular elements such as mast cells, erythroid precursors, myeloid/lymphoid precursors, etc., or monitoring cells predominant in blood and scant in marrow, e.g., mature granulocytes (CD16+) [53, 62, 158–160].
- One of the key quality indicators of MRD assays is the limit of detection (LOD) and lower limit of quantitation (LLOQ), which defines assay sensitivity and precision [161]. The LOD is the lowest MRD level distinguishable from background, requiring a signal above negative controls and the LLOQ denotes the lowest MRD level quantifiable with acceptable accuracy and precision (CV <25%) [61, 162, 163]. In flow cytometry, LOD is typically determined by the number of abnormal cells detected within a large population of normal cells, corresponding to a sensitivity of $10^{-5}–10^{-6}$ (1 in 100,000–1 in 1,000,000), and LLOQ usually requires higher cell counts than the LOD for reproducibility [2, 48–50, 61, 79]. Both parameters are influenced by sample quality, antibody panel design, instrument sensitivity, and data analysis methods.
  In ALL, CLL, and MM, flow cytometry-based MRD assays typically reach an LOD of 10–20 events ($10^{-4}–10^{-5}$) and an LLOQ of 40–50 events ($10^{-3}–10^{-4}$) due to clear separation of leukemic from normal counterpart [3, 28, 48, 49, 80, 150, 164, 165].

- In AML, defining LOD/LLOQ is challenging due to immunophenotypic heterogeneity and overlap with normal progenitors [2, 61, 62]. The immunophenotypic overlap of AML blasts with normal myeloid progenitor cells influences the degree of LAIPs that distinguish MRD from normal progenitors. Consequently, the LOD and LLOQ established for one specific immunophenotype may not be applicable to AML cases with differing LAIPs. The ELN working group recommends assay- and panel-specific validation, with technical guidelines suggesting ≥20 LAIP-associated events for LOD and ≥100 events for LLOQ [35, 62]. Future studies may refine immunophenotype-specific thresholds to improve clinical applicability.

Data Analysis QC can be performed by monitoring the following parameters [152–154]

- Blind re-analysis of 10% random cases
- Quarterly review of false positive/negative rates
- Maintenance of an internal MRD case registry for longitudinal follow-up

10.4 Computational Requirements for High-Sensitivity Assays

High-sensitivity assays generate large FCS files (300–700 GB), necessitating high-performance computing workstations or laptops. Recommended configurations include at least 16 GB RAM, a dedicated graphics card (RTX 3050 or higher with >4 GB VRAM), and SSD storage.

11 Future Directions and Advancements

Emerging computational strategies have enhanced flow cytometry-based MRD detection, notably through database-guided analysis using Infinicyt software, which applies an integrated Analytical Framework leveraging principal component analysis and reference database comparisons or machine learning algorithms [166–174].

While conventional 8–13 color panels face limitations in marker diversity and subclone detection, advanced platforms such as full spectrum flow cytometry (FSFC) and mass cytometry (CyTOF) allow simultaneous assessment of ≥18 markers per tube, improving fluorochrome resolution and enabling precise identification of rare malignant populations [175–178]. Achieving assay sensitivity up to 2×10^{-6} (0.0002%) by acquiring up to 10 million cells underscores the need for standardized high-parameter panels, multicenter harmonization, and automated analytics [3, 48]. Integrating computational tools with traditional methods enhances sensitivity, reproducibility, and clinical utility, with hybrid models—combining pattern recognition, quantitative metrics, and expert oversight—offering promising avenues for ultra-sensitive MRD evaluation.

References

1. Dekker SE, Rea D, Cayuela J-M, Arnhardt I, Leonard J, Heuser M (2023) Using measurable residual disease to optimize management of AML, ALL, and chronic myeloid leukemia. Am Soc Clin Oncol Educ Book 43:e390010
2. Tembhare PR (2023) Monitoring measurable/minimal residual disease in acute myeloid leukemia: multiparametric flow cytometry-based approach. Indian J Med Paediatr Oncol 44(06):554–565
3. van Dongen JJ, van der Velden VH, Brüggemann M, Orfao A (2015) Minimal residual disease diagnostics in acute lymphoblastic leukemia: need for sensitive, fast, and standardized technologies. Blood 125(26):3996–4009
4. Borowitz MJ, Devidas M, Hunger SP, Bowman WP, Carroll AJ, Carroll WL et al (2008) Clinical significance of minimal residual disease in childhood acute lymphoblastic leukemia and its relationship to other prognostic factors: a Children's Oncology Group study. Blood 111(12):5477–5485
5. Short NJ, Zhou S, Fu C, Berry DA, Walter RB, Freeman SD et al (2020) Association of measurable residual disease with survival outcomes in patients with acute myeloid leukemia: a systematic review and meta-analysis. JAMA Oncol 6(12):1890–1899
6. Berry DA, Zhou S, Higley H, Mukundan L, Fu S, Reaman GH et al (2017) Association of minimal residual disease with clinical outcome in pediatric and adult acute lymphoblastic leukemia: a meta-analysis. JAMA Oncol 3(7):e170580
7. Hourigan CS, Gale RP, Gormley NJ, Ossenkoppele GJ, Walter RB (2017) Measurable residual disease testing in acute myeloid leukaemia. Leukemia 31(7):1482–1490
8. Moors I, Vandepoele K, Philippé J, Deeren D, Selleslag D, Breems D et al (2019) Clinical implications of measurable residual disease in AML: review of current evidence. Crit Rev Oncol Hematol 133:142–148
9. Döhner H, Wei AH, Appelbaum FR, Craddock C, DiNardo CD, Dombret H et al (2022) Diagnosis and management of AML in adults: 2022 recommendations from an international expert panel on behalf of the ELN. Blood 140(12):1345–1377
10. Inaba H, Coustan-Smith E, Cao X, Pounds SB, Shurtleff SA, Wang KY et al (2012) Comparative analysis of different approaches to measure treatment response in acute myeloid leukemia. J Clin Oncol 30(29):3625–3632
11. Loken MR, Alonzo TA, Pardo L, Gerbing RB, Raimondi SC, Hirsch BA et al (2012) Residual disease detected by multidimensional flow cytometry signifies high relapse risk in patients with de novo acute myeloid leukemia: a report from Children's Oncology Group. Blood 120(8):1581–1588
12. Brodersen LE, Gerbing RB, Pardo ML, Alonzo TA, Paine D, Fritschle W et al (2020) Morphologic remission status is limited compared to ΔN flow cytometry: a Children's Oncology Group AAML0531 report. Blood Ad 4(20):5050–5061
13. Schuurhuis GJ, Heuser M, Freeman S, Bene MC, Buccisano F, Cloos J et al (2018) Minimal/measurable residual disease in AML: a consensus document from the European LeukemiaNet MRD Working Party. Blood 131(12):1275–1291
14. Pui CH, Pei D, Raimondi SC, Coustan-Smith E, Jeha S, Cheng C et al (2017) Clinical impact of minimal residual disease in children with different subtypes of acute lymphoblastic leukemia treated with response-adapted therapy. Leukemia 31(2):333–339
15. Tembhare PR, Narula G, Khanka T, Ghogale S, Chatterjee G, Patkar NV et al (2020) Post-induction measurable residual disease using multicolor flow cytometry is strongly predictive of inferior clinical outcome in the real-life management of childhood T-cell acute lymphoblastic leukemia: a study of 256 patients. Front Oncol 10:577
16. Winter SS, Dunsmore KP, Devidas M, Wood BL, Esiashvili N, Chen Z et al (2018) Improved survival for children and young adults with T-lineage acute lymphoblastic leukemia: results from the children's oncology group AALL0434 methotrexate randomization. J Clin Oncol 36(29):2926–2934
17. Freeman SD, Virgo P, Couzens S, Grimwade D, Russell N, Hills RK et al (2013) Prognostic relevance of treatment response measured by flow cytometric residual disease detection in older patients with acute myeloid leukemia. J Clin 31(32):4123–4131
18. Karol SE, Coustan-Smith E, Pounds S, Wang L, Inaba H, Ribeiro RC et al (2023) Clinical impact of minimal residual disease in blood and bone marrow of children with acute myeloid leukemia. Blood Adv 7(14):3651–3657
19. Langebrake C, Creutzig U, Dworzak M, Hrusak O, Mejstrikova E, Griesinger F et al (2006) Residual disease monitoring in childhood acute myeloid leukemia by multiparameter flow cytometry: the MRD-AML-BFM Study Group. J Clin 24(22):3686–3692
20. Borowitz MJ, Wood BL, Devidas M, Loh ML, Raetz EA, Salzer WL et al (2015) Prognostic significance of minimal residual disease in

high risk B-ALL: a report from Children's Oncology Group study AALL0232. Blood 126(8):964–971
21. Buldini B, Rizzati F, Masetti R, Fagioli F, Menna G, Micalizzi C et al (2017) Prognostic significance of flow-cytometry evaluation of minimal residual disease in children with acute myeloid leukaemia treated according to the AIEOP-AML 2002/01 study protocol. Bri J Haematol 177(1):116–126
22. Weng XQ, Shen Y, Sheng Y, Chen B, Wang JH, Li JM et al (2013) Prognostic significance of monitoring leukemia-associated immunophenotypes by eight-color flow cytometry in adult B-acute lymphoblastic leukemia. Blood Cancer J 3(8):e133
23. Chen X, Cherian S (2021) Role of minimal residual disease testing in acute myeloid leukemia. Clin Lab Med 41(3):467–483
24. Vedula RS, Lindsley RC (2017) Measurement of residual disease in acute myeloid leukemia. Curr Hematol Malig Rep 12(6):574–581
25. Ribera JM, Oriol A, Morgades M, Montesinos P, Sarrà J, González-Campos J et al (2014) Treatment of high-risk Philadelphia chromosome-negative acute lymphoblastic leukemia in adolescents and adults according to early cytologic response and minimal residual disease after consolidation assessed by flow cytometry: final results of the PETHEMA ALL-AR-03 trial. J Clin 32(15):1595–1604
26. Popov A, Henze G, Roumiantseva J, Budanov O, Belevtsev M, Verzhbitskaya T et al (2023) A simple procedure to identify children with B-lineage acute lymphoblastic leukemia who can be successfully treated with low or moderate intensity: Sequential versus single-point minimal residual disease measurement. Pediat Blood Cancer 70(6):e30295
27. Platzbecker U, Middeke JM, Sockel K, Herbst R, Wolf D, Baldus CD et al (2018) Measurable residual disease-guided treatment with azacitidine to prevent haematological relapse in patients with myelodysplastic syndrome and acute myeloid leukaemia (RELAZA2): an open-label, multicentre, phase 2 trial. Lancet Oncol 19(12):1668–1679
28. Reiterová M, Kohlscheen S, Maglia O, Sala S, Schumich A, Maurer-Granofszky M et al (2025) Flow-cytometric MRD detection in pediatric T-ALL: a multicenter AIEOP-BFM consensus-based guided standardized approach. Clin Chem Lab Med 63(7):1419–1426
29. Zhang W, Cai J, Wang X, Ma Y, Zhu X, Yu J et al (2025) Prognostic and therapeutic implications of measurable residual disease levels during remission induction of childhood ALL. Blood 145(17):1890–902
30. Appelbaum FR (2018) Hematopoietic cell transplantation as treatment of patients with acute myeloid leukemia with measurable residual disease after consolidation therapy. Best Pract Res Clin Haematol 31(4):405–409
31. Ceppi F, Rizzati F, Colombini A, Conter V, Cazzaniga G (2021) Utilizing the prognostic impact of minimal residual disease in treatment decisions for pediatric acute lymphoblastic leukemia. Expert Rev Hematol 14(9):795–807
32. Pettit K, Stock W, Walter RB (2016) Incorporating measurable ('minimal') residual disease-directed treatment strategies to optimize outcomes in adults with acute myeloid leukemia. Leuk Lymphoma 57(7):1527–1533
33. Scrideli CA, Assumpção JG, Ganazza MA, Araújo M, Toledo SR, Lee ML et al (2009) A simplified minimal residual disease polymerase chain reaction method at early treatment points can stratify children with acute lymphoblastic leukemia into good and poor outcome groups. Haematologica 94(6):781–789
34. Vega-García N, Perez-Jaume S, Esperanza-Cebollada E, Vicente-Garcés C, Torrebadell M, Jiménez-Velasco A et al (2020) Measurable residual disease assessed by flow-cytometry is a stable prognostic factor for pediatric T-cell acute lymphoblastic leukemia in consecutive SEHOP protocols whereas the impact of oncogenetics depends on treatment. Front Pediatr 8:614521
35. Heuser M, Freeman SD, Ossenkoppele GJ, Buccisano F, Hourigan CS, Ngai LL et al (2021) 2021 Update on MRD in acute myeloid leukemia: a consensus document from the European LeukemiaNet MRD Working Party. Blood 138(26):2753–2767
36. Xu D, Yang Y, Yin Z, Tu S, Nie D, Li Y et al (2023) Risk-directed therapy based on genetics and MRD improves the outcomes of AML1-ETO-positive AML patients, a multi-center prospective cohort study. Blood Cancer J 13(1):168
37. Gui G, Hourigan CS (2022) Measurable residual disease assessment as a surrogate marker in new drug development in acute myeloid leukemia. Cancer J 28(1):73–77
38. Meddi E, Savi A, Moretti F, Mallegni F, Palmieri R, Paterno G et al (2023) Measurable residual disease (MRD) as a surrogate efficacy-response biomarker in AML. Inte J MolSciences 24(4):3062
39. Health Quality Ontario. Minimal residual disease evaluation in childhood acute lymphoblastic leukemia: a clinical evidence review. Ont Health Technol Assess Ser 2016;16(7):1–52. PMID: 27099643; PMCID: PMC4808716
40. Bradstock KF, Janossy G, Tidman N, Papageorgiou ES, Prentice HG, Willoughby

M et al (1981) Immunological monitoring of residual disease in treated thymic acute lymphoblastic leukaemia. Leuk Res 5(4-5):301–309
41. van Dongen JJ, Hooijkaas H, Comans-Bitter M, Hählen K, de Klein A, van Zanen GE et al (1985) Human bone marrow cells positive for terminal deoxynucleotidyl transferase (TdT), HLA-DR, and a T cell marker may represent prothymocytes. J Immunol 135(5):3144–3150
42. Cheson BD, Cassileth PA, Head DR, Schiffer CA, Bennett JM, Bloomfield CD et al (1990) Report of the National Cancer Institute-sponsored workshop on definitions of diagnosis and response in acute myeloid leukemia. J Clin Oncol 8(5):813–819
43. Hagenbeek A, Martens AC (1991) Minimal residual disease in acute leukaemia: preclinical studies in a relevant rat model (BNML). Baillieres Clin Haematol 4(3):609–635
44. Zhao Z, Lan J (2024) Detection methods and prognosis implications of measurable residual disease in acute myeloid leukemia. Ann Hematol 103(12):4869–4881
45. Ramos Elbal E, Fuster JL, Campillo JA, Galera AM, Cortés MB, Llinares ME et al (2024) Measurable residual disease study through three different methods can anticipate relapse and guide early interventions in childhood acute lymphoblastic leukemia. Clin Transl Ooncol 26(1):278–287
46. Saygin C, Cannova J, Stock W, Muffly L (2022) Measurable residual disease in acute lymphoblastic leukemia: methods and clinical context in adult patients. Haematologica 107(12):2783–2793
47. Zhou Y, Wood BL (2017) Methods of detection of measurable residual disease in AML. Curr Hhematol Malig Rep 12(6):557–567
48. Tembhare PR, Subramanian Pg PG, Ghogale S, Chatterjee G, Patkar NV, Gupta A et al (2020) A high-sensitivity 10-color flow cytometric minimal residual disease assay in b-lymphoblastic leukemia/lymphoma can easily achieve the sensitivity of 2-in-10(6) and is superior to standard minimal residual disease assay: a study of 622 patients. . Cytometry B Clin Cytom 98(1):57–67
49. Tembhare PR, Chatterjee G, Khanka T, Ghogale S, Badrinath Y, Deshpande N et al (2021) Eleven-marker 10-color flow cytometric assessment of measurable residual disease for T-cell acute lymphoblastic leukemia using an approach of exclusion. Cytometry B Clin Cytom 100(4):421–433
50. Chatterjee G, Dhende P, Raj S, Shetty V, Ghogale S, Deshpande N et al (2024) 15-color highly sensitive flow cytometry assay for post anti-CD19 targeted therapy (anti-CD19-CAR-T and blinatumomab) measurable residual disease assessment in B-lymphoblastic leukemia/lymphoma: real-world applicability and challenges. Eur J Haematol 112(1):122–136
51. Dillon R, Hills R, Freeman S, Potter N, Jovanovic J, Ivey A et al (2020) Molecular MRD status and outcome after transplantation in NPM1-mutated AML. Blood 135(9):680–688
52. Jacobsohn DA, Loken MR, Fei M, Adams A, Brodersen LE, Logan BR et al (2018) Outcomes of measurable residual disease in pediatric acute myeloid leukemia before and after hematopoietic stem cell transplant: validation of difference from normal flow cytometry with chimerism studies and wilms tumor 1 gene expression. Biol Bblood Marrow Transplant 24(10):2040–2046
53. Óskarsson J, Rögnvaldsson S, Thorsteinsdottir S, Aspelund T, Gunnarsson SB, Hákonardóttir GK et al (2023) Determining hemodilution in diagnostic bone marrow aspirated samples in plasma cell disorders by next-generation flow cytometry: proposal for a bone marrow quality index. Blood Cancer J 13(1):177
54. Patkar N, Kakirde C, Shaikh AF, Salve R, Bhanshe P, Chatterjee G et al (2021) Clinical impact of panel-based error-corrected next generation sequencing versus flow cytometry to detect measurable residual disease (MRD) in acute myeloid leukemia (AML). Leukemia 35(5):1392–1404
55. Singh J, Gorniak M, Grigoriadis G, Westerman D, McBean M, Venn N et al (2022) Correlation between a 10-color flow cytometric measurable residual disease (MRD) analysis and molecular MRD in adult B-acute lymphoblastic leukemia. Cytometry B Clin Cytom 102(2):115–122
56. van der Linde R, Gatt PN, Smith S, Fernandez MA, Vaughan L, Blyth E et al (2023) Measurable residual disease (MRD) by flow cytometry in adult B-acute lymphoblastic leukaemia (B-ALL) and acute myeloid leukaemia (AML): correlation with molecular MRD testing and clinical outcome at one year. Cancers 15(20):5064
57. Buccisano F, Palmieri R, Piciocchi A, Arena V, Maurillo L, Del Principe MI et al (2022) Clinical relevance of an objective flow cytometry approach based on limit of detection and limit of quantification for measurable residual disease assessment in acute myeloid leukemia. A post-hoc analysis of the GIMEMA AML1310 trial. Haematologica 107(12):2823–2833
58. Paiva B, Vidriales MB, Sempere A, Tarín F, Colado E, Benavente C et al (2021) Impact

of measurable residual disease by decentralized flow cytometry: a PETHEMA real-world study in 1076 patients with acute myeloid leukemia. Leukemia 35(8):2358–2870
59. PR T (2023) Monitoring measurable/minimal residual disease in acute myeloid leukemia: multiparametric flow cytometry-based approach. Ind J Med Paediatr Oncol 44(6):554–565
60. Roshal M, Gao Q (2025) Flow cytometry evaluation of acute myeloid leukemia minimal residual disease based on an understanding of the normal maturation patterns in the blast compartments. Am J Clin Pathol 163(5):775–793
61. Tettero JM, Dakappagari N, Heidinga ME, Oussoren-Brockhoff Y, Hanekamp D, Pahuja A et al (2023) Analytical assay validation for acute myeloid leukemia measurable residual disease assessment by multiparametric flow cytometry. Cytometry B Clin Cytom 104(6):426–439
62. Tettero JM, Freeman S, Buecklein V, Venditti A, Maurillo L, Kern W et al (2022) Technical aspects of flow cytometry-based measurable residual disease quantification in acute myeloid leukemia: experience of the European LeukemiaNet MRD Working Party. HemaSphere 6(1):e676
63. https://euroflow.org/protocols/
64. Bendall SC, Davis KL, Amir el AD, Tadmor MD, Simonds EF, Chen TJ et al (2014) Single-cell trajectory detection uncovers progression and regulatory coordination in human B cell development. Cell 157(3):714–725
65. Cherian S, Miller V, McCullouch V, Dougherty K, Fromm JR, Wood BL (2018) A novel flow cytometric assay for detection of residual disease in patients with B-lymphoblastic leukemia/lymphoma post anti-CD19 therapy. Cytometry B Clin Cytom 94(1):112–120
66. Nguyen D, Jabbour E, Jain N, Karrar O, Nasnas C, Haddad FG et al (2024) Clearance of very low levels of measurable residual disease with blinatumomab significantly improves outcomes in B-cell acute lymphoblastic leukemia. Blood 144(Suppl 1):1465
67. Pulsipher MA, Han X, Maude SL, Laetsch TW, Qayed M, Rives S et al (2022) Next-generation sequencing of minimal residual disease for predicting relapse after tisagenlecleucel in children and young adults with acute lymphoblastic leukemia. Blood Cancer Discov 3(1):66–81
68. Bassan R, Brüggemann M, Radcliffe HS, Hartfield E, Kreuzbauer G, Wetten S (2019) A systematic literature review and meta-analysis of minimal residual disease as a prognostic indicator in adult B-cell acute lymphoblastic leukemia. Haematologica 104(10):2028–2039
69. Campana D, Pui CH (2017) Minimal residual disease-guided therapy in childhood acute lymphoblastic leukemia. Blood 129(14):1913–1918
70. Kotrova M, Volland A, Kehden B, Trautmann H, Ritgen M, Wäsch R et al (2020) Comparison of minimal residual disease levels in bone marrow and peripheral blood in adult acute lymphoblastic leukemia. Leukemia 34(4):1154–1157
71. Coustan-Smith E, Song G, Clark C, Key L, Liu P, Mehrpooya M et al (2011) New markers for minimal residual disease detection in acute lymphoblastic leukemia. Blood 117(23):6267–6276
72. Sędek Ł, Theunissen P, Sobral da Costa E, van der Sluijs-Gelling A, Mejstrikova E, Gaipa G et al (2019) Differential expression of CD73, CD86 and CD304 in normal vs. leukemic B-cell precursors and their utility as stable minimal residual disease markers in childhood B-cell precursor acute lymphoblastic leukemia. J Immunol Methods 475:112429
73. Tembhare PR, Ghogale S, Ghatwai N, Badrinath Y, Kunder N, Patkar NV et al (2018) Evaluation of new markers for minimal residual disease monitoring in B-cell precursor acute lymphoblastic leukemia: CD73 and CD86 are the most relevant new markers to increase the efficacy of MRD assay. Cytom B Clin CCytom 94(1):100–111
74. Stow P, Key L, Chen X, Pan Q, Neale GA, Coustan-Smith E et al (2010) Clinical significance of low levels of minimal residual disease at the end of remission induction therapy in childhood acute lymphoblastic leukemia. Blood 115(23):4657–4663
75. Theunissen P, Mejstrikova E, Sedek L, van der Sluijs-Gelling AJ, Gaipa G, Bartels M et al (2017) Standardized flow cytometry for highly sensitive MRD measurements in B-cell acute lymphoblastic leukemia. Blood 129(3):347–357
76. Das N, Panda D, Gajendra S, Gupta R, Thakral D, Kaur G et al (2024) Immunophenotypic characterization of leukemic stem cells in acute myeloid leukemia using single tube 10-colour panel by multiparametric flow cytometry: deciphering the spectrum, complexity and immunophenotypic heterogeneity. Int J Lab Hematol 46(4):646–656
77. Al-Mawali A, Gillis D, Lewis I (2016) Immunoprofiling of leukemic stem cells CD34+/CD38-/CD123+ delineate FLT3/ITD-positive clones. J Hematol Oncol 9(1):61
78. Khaldoyanidi SK, Hindoyan A, Stein A, Subklewe M (2022) Leukemic stem cells as a target for eliminating acute myeloid leukemia: gaps in translational research. Crit Rev Oncol Hematol 175:103710

79. Wood BL (2020) Acute myeloid leukemia minimal residual disease detection: the difference from normal approach. Curr Protoc Cytom 93(1):e73
80. Chen X, Wood BL (2017) Monitoring minimal residual disease in acute leukemia: Technical challenges and interpretive complexities. Blood Rev 31(2):63–75
81. Yoroidaka T, Yamashita T, Murata R, Yoshihara K, Yoshihara S, Ueda M et al (2023) Measurable residual disease assessment using next-generation flow in patients with relapsed and refractory multiple myeloma treated with a combination of carfilzomib, lenalidomide, and dexamethasone. Anticancer Re 43(1):157–165
82. Sedek L, Soriano B, Słota Ł, Brüggemann M, Lasia A, Kohlscheen S et al (2024) A novel 12-color 2-tube next generation flow approach enables sensitive detection of measurable residual disease in virtually all T-cell acute lymphoblastic leukemia patients. Blood 144(Suppl 1):4218
83. Dworzak MN, Fröschl G, Printz D, Zen LD, Gaipa G, Ratei R et al (2004) CD99 expression in T-lineage ALL: implications for flow cytometric detection of minimal residual disease. Leukemia 18(4):703–708
84. Roshal M, Fromm JR, Winter S, Dunsmore K, Wood BL (2010) Immaturity associated antigens are lost during induction for T cell lymphoblastic leukemia: implications for minimal residual disease detection. Cytom B Clin Cytom 78(3):139–146
85. Zhou Y, Fan X, Routbort M, Cameron Yin C, Singh R, Bueso-Ramos C et al (2013) Absence of terminal deoxynucleotidyl transferase expression identifies a subset of high-risk adult T-lymphoblastic leukemia/lymphoma. Mod Pathol 26(10):1338–1345
86. Dunsmore KP, Winter SS, Devidas M, Wood BL, Esiashvili N, Chen Z et al (2020) Children's oncology group AALL0434: a phase III randomized clinical trial testing nelarabine in newly diagnosed t-cell acute lymphoblastic leukemia. J Clin Oncol 38(28):3282–3293
87. Das N, Banavali S, Bakhshi S, Trehan A, Radhakrishnan V, Seth R et al (2022) Protocol for ICiCLe-ALL-14 (InPOG-ALL-15-01): a prospective, risk stratified, randomised, multicentre, open label, controlled therapeutic trial for newly diagnosed childhood acute lymphoblastic leukaemia in India. Trials 23(1):102
88. Aziz Z, Zahid M, Mahmood R, Maqbool S (1997) Modified BFM protocol for childhood acute lymphoblastic leukemia: a retrospective analysis. Med Pediatr Oncol 28(1):48–53
89. Gökbuget N, Hoelzer D, Arnold R, Böhme A, Bartram CR, Freund M et al (2000) Treatment of Adult ALL according to protocols of the German Multicenter Study Group for Adult ALL (GMALL). Hematol Oncol Clin N Am 14(6):1307–25. ix
90. Rijneveld AW, van der Holt B, de Weerdt O, Biemond BJ, van de Loosdrecht AA, van der Wagen LE et al (2022) Clofarabine added to intensive treatment in adult patients with newly diagnosed ALL: the HOVON-100 trial. Blood Adv 6(4):1115–1125
91. Teachey DT, O'Connor D (2020) How I treat newly diagnosed T-cell acute lymphoblastic leukemia and T-cell lymphoblastic lymphoma in children. Blood 135(3):159–166
92. Ribera JM, Ortega JJ, Oriol A, Bastida P, Calvo C, Pérez-Hurtado JM et al (2007) Comparison of intensive chemotherapy, allogeneic, or autologous stem-cell transplantation as postremission treatment for children with very high risk acute lymphoblastic leukemia: PETHEMA ALL-93 Trial. J Clin Oncol 25(1):16–24
93. Huguet F, Chevret S, Leguay T, Thomas X, Boissel N, Escoffre-Barbe M et al (2018) Intensified therapy of acute lymphoblastic leukemia in adults: report of the randomized GRAALL-2005 clinical trial. J Clin Oncol 36(24):2514–2523
94. Buckley SA, Wood BL, Othus M, Hourigan CS, Ustun C, Linden MA et al (2017) Minimal residual disease prior to allogeneic hematopoietic cell transplantation in acute myeloid leukemia: a meta-analysis. Haematologica 102(5):865–873
95. Grimwade D, Freeman SD (2014) Defining minimal residual disease in acute myeloid leukemia: which platforms are ready for "prime time"?. Blood 124(23):3345–3355
96. Bruserud Ø, Aasebø E, Hernandez-Valladares M, Tsykunova G, Reikvam H (2017) Therapeutic targeting of leukemic stem cells in acute myeloid leukemia - the biological background for possible strategies. Expert Opin Drug Discov 12(10):1053–1065
97. Cloos J, Harris JR, Janssen J, Kelder A, Huang F, Sijm G, et al (2018) Comprehensive protocol to sample and process bone marrow for measuring measurable residual disease and leukemic stem cells in acute myeloid leukemia. J Vis Exp. (133):56386
98. Wouters R, Cucchi D, Kaspers GJ, Schuurhuis GJ, Cloos J (2014) Relevance of leukemic stem cells in acute myeloid leukemia: heterogeneity and influence on disease monitoring, prognosis and treatment design. Exp Rev Hematol 7(6):791–805

99. Heo SK, Noh EK, Ju LJ, Sung JY, Jeong YK, Cheon J et al (2020) CD45(dim)CD34(+) CD38(-)CD133(+) cells have the potential as leukemic stem cells in acute myeloid leukemia. BMC Cancer 20(1):285
100. Wijnands C, Noori S, Donk N, VanDuijn MM, Jacobs JFM (2023) Advances in minimal residual disease monitoring in multiple myeloma. Crit Rev Clin Lab Sci 60(7):518–534
101. Noori S, Wijnands C, Langerhorst P, Bonifay V, Stingl C, Touzeau C et al (2023) Dynamic monitoring of myeloma minimal residual disease with targeted mass spectrometry. Blood Cancer J 13(1):30
102. San-Miguel J, Avet-Loiseau H, Paiva B, Kumar S, Dimopoulos MA, Facon T et al (2022) Sustained minimal residual disease negativity in newly diagnosed multiple myeloma and the impact of daratumumab in MAIA and ALCYONE. Blood 139(4):492–501
103. Derman BA, Kansagra A, Zonder J, Stefka AT, Grinblatt DL, Anderson JrLD et al (2022) Elotuzumab and weekly carfilzomib, lenalidomide, and dexamethasone in patients with newly diagnosed multiple myeloma without transplant intent: a phase 2 measurable residual disease-adapted study. JAMA Oncol 8(9):1278–1286
104. de Tute RM, Pawlyn C, Cairns DA, Davies FE, Menzies T, Rawstron A et al (2022) Minimal residual disease after autologous stem-cell transplant for patients with myeloma: prognostic significance and the impact of lenalidomide maintenance and molecular risk. J Clin Oncol 40(25):2889–2900
105. Costa LJ, Chhabra S, Medvedova E, Dholaria BR, Schmidt TM, Godby KN et al (2022) Daratumumab, carfilzomib, lenalidomide, and dexamethasone with minimal residual disease response-adapted therapy in newly diagnosed multiple myeloma. J Clin Oncol 40(25):2901–2912
106. Cavo M, San-Miguel J, Usmani SZ, Weisel K, Dimopoulos MA, Avet-Loiseau H et al (2022) Prognostic value of minimal residual disease negativity in myeloma: combined analysis of POLLUX, CASTOR, ALCYONE, and MAIA. Blood 139(6):835–844
107. Yee AJ, Raje N (2021) Minimal residual disease in multiple myeloma: why, when, where. Hematology Am Soc Hematol Educ Program 2021(1):37–45
108. Oliva S, Bruinink DHO, Rihova L, D'Agostino M, Pantani L, Capra A et al (2021) Minimal residual disease assessment by multiparameter flow cytometry in transplant-eligible myeloma in the EMN02/HOVON 95 MM trial. Blood Cancer J 11(6):106
109. Soh KT, Tario Jr.JD, Hahn TE, Hillengass J, McCarthy PL, Wallace PK (2020) Methodological considerations for the high sensitivity detection of multiple myeloma measurable residual disease. Cytometry B Clin Cytom 98(2):161–173
110. Paiva B, Puig N, Cedena MT, Rosiñol L, Cordón L, Vidriales MB et al (2020) Measurable residual disease by next-generation flow cytometry in multiple myeloma. J Clin Oncol 38(8):784–792
111. Martinez-Lopez J, Wong SW, Shah N, Bahri N, Zhou K, Sheng Y et al (2020) Clinical value of measurable residual disease testing for assessing depth, duration, and direction of response in multiple myeloma. Blood Adv 4(14):3295–3301
112. Munshi NC, Avet-Loiseau H, Rawstron AC, Owen RG, Child JA, Thakurta A et al (2017) Association of minimal residual disease with superior survival outcomes in patients with multiple myeloma: a meta-analysis. JAMA Oncol 3(1):28–35
113. Munshi NC, Avet-Loiseau H, Anderson KC, Neri P, Paiva B, Samur M et al (2020) A large meta-analysis establishes the role of MRD negativity in long-term survival outcomes in patients with multiple myeloma. Blood Adv 4(23):5988–5999
114. Szalat R, Anderson K, Munshi N (2024) Role of minimal residual disease assessment in multiple myeloma. Haematologica 109(7):2049–2059
116. Tzasta A, Wijnands C, Baalman K, van Gool AJ, Zweegman S, Jacobs JFM (2015) Advances in multiple myeloma blood-based monitoring and its clinical applications. Crit Rev Clin Lab Sci 62(7):491–509
115. Paiva B, Shi Q, Puig N, Cedena MT, Orfao A, Durie BGM et al (2025) Opportunities and challenges for MRD assessment in the clinical management of multiple myeloma. Nature Rev Clin Oncol 22(6):424–438
117. Kumar S, Paiva B, Anderson KC, Durie B, Landgren O, Moreau P et al (2016) International Myeloma Working Group consensus criteria for response and minimal residual disease assessment in multiple myeloma. Lancet Oncol 17(8):e328–e46.
118. Zhou M, Chen Y, Gong Y, Zhu M, Cen J, Pan J et al (2024) Evaluation of next-generation sequencing versus next-generation flow cytometry for minimal-residual-disease detection in Chinese patients with multiple myeloma. Discover Oncol 15(1):78
119. Yoroidaka T, Takamatsu H, Urushihara R, Itagaki M, Yoshihara S, Sato K et al (2025) Prognostic value of minimal residual disease detected by EuroFlow next-generation flow cytometry and next-generation sequencing in patients with multiple myeloma achieving complete response and receiving lenalidomide

maintenance after autotransplant: a prospective comparison study. Haematologica 110(9):2160–2170
120. Medina A, Puig N, Flores-Montero J, Jimenez C, Sarasquete ME, Garcia-Alvarez M et al (2020) Comparison of next-generation sequencing (NGS) and next-generation flow (NGF) for minimal residual disease (MRD) assessment in multiple myeloma. Blood Cancer J 10(10):108
121. Sriram H, Kunjachan F, Khanka T, Gawai S, Ghogale S, Deshpande N et al (2022) Expression levels and patterns of B-cell maturation antigen in newly diagnosed and relapsed multiple myeloma patients from Indian subcontinent. Cytometry B Clin Cytom 102(6):462–470
122. Chatterjee G, Gujral S, Subramanian PG, Tembhare PR (2017) Clinical relevance of multicolour flow cytometry in plasma cell disorders. Indian J Hematol Blood Transfus 33(3):303–315
123. Tembhare PR, Ghogale S, Tauro W, Badrinath Y, Deshpande N, Kedia S et al (2018) Evaluation of CD229 as a new alternative plasma cell gating marker in the flow cytometric immunophenotyping of monoclonal gammopathies. Cytometry B Clin Cytom 94(3):509–519
124. Soh KT, Came N, Otteson GE, Jevremovic D, Shi M, Olteanu H et al (2022) Evaluation of multiple myeloma measurable residual disease by high sensitivity flow cytometry: an international harmonized approach for data analysis. Cytom B Clin Cytom 102(2):88–106
125. Puig N, Flores-Montero J, Burgos L, Cedena MT, Cordón L, Pérez JJ et al (2021) Reference values to assess hemodilution and warn of potential false-negative minimal residual disease results in myeloma. Cancers 213(19):4924
126. Tembhare PR, Sriram H, Khanka T, Gawai S, Bagal B, Ghogale SG et al (2024) Circulating tumor plasma cells and peripheral blood measurable residual disease assessment in multiple myeloma patients not planned for upfront transplant. HemaSphere 8(4):e63
127. Simon F, Ligtvoet R, Robrecht S, Cramer P, Kutsch N, Fürstenau M et al (2025) End point surrogacy in first-line chronic lymphocytic leukemia. J Clin Oncol 43(4):381–91.
128. Greenberg MR, Lucido T, Singh K, Rhodes JM (2025) The impact of minimal residual disease measurement in the management of chronic lymphocytic leukemia. Cancers 17(10):1708
129. Rios-Olais FA, McGary AK, Tsang M, Almader-Douglas D, Leis JF, Buras MR et al (2024) Measurable residual disease and clinical outcomes in chronic lymphocytic leukemia: a systematic review and meta-analysis. JAMA Oncol 10(9):1221–1227
130. Munir T, Cairns DA, Bloor A, Allsup D, Cwynarski K, Pettitt A et al (2024) Chronic lymphocytic leukemia therapy guided by measurable residual disease. N Engl J Med 390(4):326–337
131. Rios-Olais FA, Hilal T (2023) Measurable residual disease in chronic lymphocytic leukemia: current understanding and evolving role in clinical practice. Curr Treat Option On 24(8):907–928
132. Fisher A, Goradia H, Martinez-Calle N, Patten P, Munir T (2023) The evolving use of measurable residual disease in chronic lymphocytic leukemia clinical trials. Front Oncol 13:1130617
133. Wierda WG, Allan JN, Siddiqi T, Kipps TJ, Opat S, Tedeschi A et al (2021) Ibrutinib plus venetoclax for first-line treatment of chronic lymphocytic leukemia: primary analysis results from the minimal residual disease cohort of the randomized phase II CAPTIVATE study. J Clin Oncol 39(34):3853–3865
134. Uchiyama T, Yokoyama A, Aoki S (2020) Measurable residual disease in the treatment of chronic lymphocytic leukemia. J Clin Exp Hematopathol 60(4):138–145
135. Kater AP, Seymour JF, Hillmen P, Eichhorst B, Langerak AW, Owen C et al (2019) Fixed duration of venetoclax-rituximab in relapsed/refractory chronic lymphocytic leukemia eradicates minimal residual disease and prolongs survival: post-treatment follow-up of the MURANO phase III study. J Clin Oncol 37(4):269–277
136. Ghia P, Rawstron A (2018) Minimal residual disease analysis in chronic lymphocytic leukemia: a way for achieving more personalized treatments. Leukemia 32(6):1307–1316
137. Kater AP, Levin MD, Dubois J, Kersting S, Enggaard L, Veldhuis GJ et al (2022) Minimal residual disease-guided stop and start of venetoclax plus ibrutinib for patients with relapsed or refractory chronic lymphocytic leukaemia (HOVON141/VISION): primary analysis of an open-label, randomised, phase 2 trial. Lancet Oncol 23(6):818–828
138. Rawstron AC, Fazi C, Agathangelidis A, Villamor N, Letestu R, Nomdedeu J et al (2016) A complementary role of multiparameter flow cytometry and high-throughput sequencing for minimal residual disease detection in chronic lymphocytic leukemia: an European Research Initiative on CLL study. Leukemia 30(4):929–936
139. Rawstron A (2025) Measurable/minimal residual disease: what method and what cut-off? Hematol Oncol Clin N Am 39(5):917–934

140. Long M, Beckwith K, Do P, Mundy BL, Gordon A, Lehman AM et al (2017) Ibrutinib treatment improves T cell number and function in CLL patients. J Clin Investig 127(8):3052–3064
141. Chandhok NS, Sekeres MA (2025) Measurable residual disease in hematologic malignancies: a biomarker in search of a standard. EClinicalMedicine 86:103348
142. Galimberti S, Genuardi E, Mazziotta F, Iovino L, Morabito F, Grassi S et al (2019) The minimal residual disease in non-Hodgkin's lymphomas: from the laboratory to the clinical practice. Front Oncol 9:528
143. Chase ML, Armand P (2018) Minimal residual disease in non-Hodgkin lymphoma - current applications and future directions. Br J Haematol 180(2):177–188
144. Pott C, Jurinovic V, Trotman J, Kehden B, Unterhalt M, Herold M et al (2024) Minimal residual disease status predicts outcome in patients with previously untreated follicular lymphoma: a prospective analysis of the phase III GALLIUM Study. Clin Oncol 42(5):550–561
145. Ravandi F, Kreitman RJ, Tiacci E, Andritsos L, Banerji V, Barrientos JC et al (2022) Consensus opinion from an international group of experts on measurable residual disease in hairy cell leukemia. Blood Cancer Jl 12(12):165
146. Epstein-Peterson ZD, Drill E, Aypar U, Batlevi CL, Caron P, Dogan A et al (2024) Immunochemotherapy plus lenalidomide for high-risk mantle cell lymphoma with measurable residual disease evaluation. Haematologica 109(4):1149–1162
147. Cheminant M, Derrieux C, Touzart A, Schmit S, Grenier A, Trinquand A et al (2016) Minimal residual disease monitoring by 8-color flow cytometry in mantle cell lymphoma: an EU-MCL and LYSA study. Haematologica 101(3):336–345
148. Chovancová J, Bernard T, Stehlíková O, Šálek D, Janíková A, Mayer J et al (2015) Detection of minimal residual disease in mantle cell lymphoma-establishment of novel eight-color flow cytometry approach. Cytometry Clin Cytom 88(2):92–100
149. DiGiuseppe JA, Wood BL (2019) Applications of flow cytometric immunophenotyping in the diagnosis and posttreatment monitoring of b and t lymphoblastic leukemia/lymphoma. Cytometry B Clin Cytom 96(4):256–265
150. Wood BL (2013) Flow cytometric monitoring of residual disease in acute leukemia. Methods Mol Biol 999:123–136
151. Scott SD, Fletcher M, Whitehouse H, Whitby L, Yuan C, Mazzucchelli S et al (2019) Assessment of plasma cell myeloma minimal residual disease testing by flow cytometry in an international inter-laboratory study: Is it ready for primetime use?. Cytometry B, Clin Cytom 96(3):201–208
152. Björklund E, Matinlauri I, Tierens A, Axelsson S, Forestier E, Jacobsson S et al (2009) Quality control of flow cytometry data analysis for evaluation of minimal residual disease in bone marrow from acute leukemia patients during treatment. J Pediatr Hematol Oncol 31(6):406–415
153. Greig B (2019) Quality control of immunophenotyping. Methods Mol Biol 2032:227–279
154. Maurer-Granofszky M, Schumich A, Buldini B, Gaipa G, Kappelmayer J, Mejstrikova E et al (2021) An extensive quality control and quality assurance (QC/QA) program significantly improves inter-laboratory concordance rates of flow-cytometric minimal residual disease assessment in acute lymphoblastic leukemia: an I-BFM-FLOW-network report. Cancers 13(23):6148
155. Bras AE, Matarraz S, Nierkens S, Fernández P, Philippé J, Aanei CM et al (2022) Quality assessment of a large multi-center flow cytometric dataset of acute myeloid leukemia patients-a euroflow study. Cancers;14(8):2011
156. Kalina T, Flores-Montero J, Lecrevisse Q, Pedreira CE, van der Velden VH, Novakova M et al (2015) Quality assessment program for EuroFlow protocols: summary results of four-year (2010-2013) quality assurance rounds. Cytometry A 87(2):145–156
157. Kalina T, Flores-Montero J, van der Velden VHJ, Martin-Ayuso M, Böttcher S, Ritgen M et al (2012) EuroFlow standardization of flow cytometer instrument settings and immunophenotyping protocols. Leukemia 26(9):1986–2010
158. Hoffmann J, Thrun MC, Röhnert MA, von Bonin M, Oelschlägel U, Neubauer A et al (2023) Identification of critical hemodilution by artificial intelligence in bone marrow assessed for minimal residual disease analysis in acute myeloid leukemia: the Cinderella method. Cytometry A 103(4):304–312
159. Tettero JM, Heidinga ME, Mocking TR, Fransen G, Kelder A, Scholten WJ et al (2024) Impact of hemodilution on flow cytometry based measurable residual disease assessment in acute myeloid leukemia. Leukemia 38(3):630–639
160. Vigliotta I, Armuzzi S, Barone M, Solli V, Pistis I, Borsi E et al (2022) The ALLgorithMM: how to define the hemodilution of bone marrow samples in lymphoproliferative diseases. Front Oncol 12:1001048

161. Armbruster DA, Pry T (2008) Limit of blank, limit of detection and limit of quantitation. Clin Biochem Re 29(Suppl 1):S49–52
162. https://www.cytometry.org/web/modules/Module_25.pdf
163. Clinical and Laboratory Standards Institute (2004). Protocols for determination of limits of detection and limits of quantitation AGW, PA USA: CLSI; CLSI document EP17
164. Dworzak MN, Gaipa G, Ratei R, Veltroni M, Schumich A, Maglia O et al (2008) Standardization of flow cytometric minimal residual disease evaluation in acute lymphoblastic leukemia: multicentric assessment is feasible. Cytometry B Clin Cytom 74(6):331–340
165. Arroz M, Came N, Lin P, Chen W, Yuan C, Lagoo A et al (2016) Consensus guidelines on plasma cell myeloma minimal residual disease analysis and reporting. Cytometry B Clin Cytom 90(1):31–39
166. Aanei CM, Veyrat-Masson R, Selicean C, Marian M, Rigollet L, Trifa AP et al (2021) Database-guided analysis for immunophenotypic diagnosis and follow-up of acute myeloid leukemia with recurrent genetic abnormalities. Front Oncol 11:746951
167. Piñero P, Morillas M, Gutierrez N, Barragán E, Such E, Breña J et al (2022) Identification of leukemia-associated immunophenotypes by databaseguided flow cytometry provides a highly sensitive and reproducible strategy for the study of measurable residual disease in acute myeloblastic leukemia. Cancers 14(16):4010
168. Reiter M, Diem M, Schumich A, Maurer-Granofszky M, Karawajew L, Rossi JG et al (2019) Automated flow cytometric MRD assessment in childhood acute B- Lymphoblastic leukemia using supervised machine learning. Cytometry A 95(9):966–975
169. Wödlinger M, Reiter M, Weijler L, Maurer-Granofszky M, Schumich A, Sajaroff EO et al (2022) Automated identification of cell populations in flow cytometry data with transformers. Comput Biol Med 144:105314
170. Verbeek MWC, Rodríguez BS, Sedek L, Laqua A, Buracchi C, Buysse M et al (2024) Minimal residual disease assessment in B-cell precursor acute lymphoblastic leukemia by semi-automated identification of normal hematopoietic cells: a EuroFlow study. Cytometry PB Clin Cytom 106(4):252–263
171. Nguyen PC, Nguyen V, Baldwin K, Kankanige Y, Blombery P, Came N et al (2023) Computational flow cytometry provides accurate assessment of measurable residual disease in chronic lymphocytic leukaemia. Br J Haematol 202(4):760–770
172. Duetz C, Bachas C, Westers TM, van de Loosdrecht AA (2020) Computational analysis of flow cytometry data in hematological malignancies: future clinical practice?. Curr Opin Oncol 32(2):162–169
173. Ko BS, Wang YF, Li JL, Li CC, Weng PF, Hsu SC et al (2018) Clinically validated machine learning algorithm for detecting residual diseases with multicolor flow cytometry analysis in acute myeloid leukemia and myelodysplastic syndrome. EBioMedicine. 37:91–100
174. Shopsowitz KE, Liu L, Setiadi A, Al-Bakri M, Vercauteren S (2022) Machine learning optimized multiparameter radar plots for B-cell acute lymphoblastic leukemia minimal residual disease analysis. Cytometry B Clin Cytom 102(5):342–352
175. Soh KT, Conway A, Liu X, Wallace PK (2022) Development of a 27-color panel for the detection of measurable residual disease in patients diagnosed with acute myeloid leukemia. Cytometry A 101(11):970–983
176. Maag AH, Swanton H, Kull M, Vegi NM, Feuring M (2023) Immunophenotypical profiling of myeloid neoplasms with erythroid predominance using mass cytometry (CyTOF). Cytometry A 103(7):551–9562
177. Fokken H, Waclawski J, Kattre N, Kloos A, Müller S, Ettinger M et al (2024) A 19-color single-tube full spectrum flow cytometry assay for the detection of measurable residual disease in acute myeloid leukemia. Cytometry A 105(3):181–195
178. Matthes T (2024) Phenotypic analysis of hematopoietic stem and progenitor cell populations in acute myeloid leukemia based on spectral flow cytometry, a 20-color panel, and unsupervised learning algorithms. Int J Mol Sci 25(5):2847

Check for updates

Chapter 13

Immunophenotyping of Cellular Therapy Products Against Oncogenic and Viral Targets

Christopher A. Lazarski, Anushree A. Datar, Patrick J. Hanley and Sandeep K. Srivastava

Abstract

Flow cytometric analysis can provide valuable information on the viability, phenotype, and therapeutic potential of cell therapy products using a minimal amount of cell numbers. In this chapter, we describe methods for the analysis of clinical products that incorporate extra considerations within the protocols to help maintain consistency and confidence of staining. This can include methods for evaluating the transduction efficiency for T cells expressing chimeric antigen receptors (CAR-T), characterizing the frequency of diverse lymphocyte populations within polyclonal cell cultures and postinfusion research samples, and measuring the activity or exhaustion in response to antigen-specific stimulation. This chapter also includes recommendations for storage and management of staining materials, frequency of compensation analysis, and reference standards.

Keywords Chimeric antigen receptors T cell (CAR-T cell), Virus-specific antigens (VST cells), Multicolor flow cytometry, Sequential gating, T cell subsets

1 Introduction

Accurate enumeration of cell populations is critical when evaluating the safety and potency of cellular products cultured in vitro for extended periods of time.

Adoptive immunotherapy, which utilizes immune effector cells engineered ex vivo to express chimeric antigen receptors (CARs) that facilitate the elimination of cancer cells, has emerged as a groundbreaking method in cancer treatment. The CD19 antigen found on B cells was the first clinically approved target for CAR-T cell therapy, and treatment led to impressive remission rates in patients with precursor B-cell acute lymphoblastic leukemia (B-ALL), diffuse large B-cell lymphoma (DLBCL), primary mediastinal B-cell lymphoma (PMBCL), and mantle cell lymphoma (MCL) [1].

Angélique Biancotto and Pradeep K. Dagur (eds.), *Immunophenotyping*, Methods in Molecular Biology, vol. 3087, https://doi.org/10.1007/978-1-0716-5607-5_13,

In addition to CAR-T cell therapy, polyclonal T cell therapy is being investigated as treatment for patients with relapsed acute lymphoblastic leukemia (ALL), acute myeloid leukemia (AML), and diffuse intrinsic pontine glioma (DIPG) [2–4], as well as for intractable viral infections common after bone marrow transplant, including cytomegalovirus (CMV), Epstein–Barr virus (EBV), and adenovirus [5, 6]. Polyclonal T cell therapy differs from CAR-T cells, as polyclonal cells recognize the cognate tumor antigen or viral antigen as peptides in association with MHC Class I and MHC Class II via their endogenously expressed T cell receptor, and not through transduction of an engineered construct. By expanding these cells with peptides specific to tumor-associated antigens (TAA-T cells) or viral-specific antigens (VST cells), the cellular product can contain a mix of CD4+ helper T cells and CD8+ cytotoxic T cells, which react with patient-specific HLA molecules.

In the absence of firm guidelines from regulatory bodies regarding GMP flow cytometry, consensus and sharing of best practices within the cell therapy community can address potential issues with product quality and efficacy. As cellular products for therapy become more complex and their medical use intensifies, the importance of properly performed flow cytometry will only continue to grow [7]. Methods to assess these parameters, such as CAR T cell subpopulations, exhaustion, senescence, and differentiations, must be highly standardized, robust, and transferable. Multiparametric immunophenotyping by flow cytometry facilitates fast, comprehensive, and primarily antibody-based routine analysis of progenitor and immune cells in peripheral blood, bone marrow, or cerebrospinal fluid in both healthy individuals and patients with leukemia and lymphoma [8].

In this chapter, we have examined and illustrated the principles to be applied in the analysis of cellular products for therapeutic use. Central to all these examples is the practice of good flow cytometry, including the use of sequential gating, doublet exclusion, vital dyes, and setting meaningful analysis gates.

2 Materials

2.1 Fluorescent Conjugated Monoclonal Abs (mAb) for Surface Staining of CAR-T cell (Table 1)

1. Stem cell-like memory T (Tscm) staining buffer (TSS): 2 mL of 25% Human Serum Albumin to 48 mL of Hank's balanced salt solution (HBSS), final concentration of Human Serum Albumin is 1%.
2. Pierce™ Recombinant Biotinylated Protein L: ThermoFisher scientific product number 29997, 0.5 mg/vial, dissolved in 10 mL HBSS, final concentration will be 50 μg/mL, aliquot into 80 μl and freeze at −80°C, and secondary antibody Streptavidin PE.
3. Falcon® 5 mL Round Bottom Polystyrene Test Tube, BD FACSymphony™ Cell Analyze, and Flow Jo for data analysis.

2.2 Therapeutic T Cell Staining Reagents

1. FACS Buffer: 1× PBS supplemented with 5 mm EDTA (store at room temperature), 1× PBS store at room temperature.
2. Live/Dead fixable near infrared dead cell stain kit for 633 nm excitation and Live/Dead Fixable Aqua Dead Cell Stain Kit for 405 nm excitation.
3. FC receptor block.
4. Monoclonal antibodies:
   - 4.1. TNK lineage: CD8 FITC (clone REA734); CD16 PE (clone REA423); CD56 PE (clone REA196); CD3 PerCP Vio700 (clone REA613); CD4 PE Vio770 (clone REA623); CD45 APC (clone REA747).
   - 4.2. DC mono lineage: CD83 FITC (clone REA714); CD19 PE (clone REA675); CD3 PerCP Vio700 (clone REA613); CD14 PE Vio770 (clone REA599); CD45 APC (clone REA747).
   - 4.3. TCR memory: CCR7 FITC (clone REA546); CD45RO PE (clone REA611); CD3 PerCP Vio700 (clone REA613); TCRγδ PE Vio770 (clone REA591); TCRα β APC (clone REA652); CD95 APC Vio770 (clone REA738).
   - 4.4. Pre-processing phenotype: CD4 VioBlue (clone REA623); CD45 VioGreen (clone REA747); CD14 FITC (clone TUK4); CD19 PE (clone REA675); CD3 PerCP Vio700 (clone REA613); CD56 PE Vio770 (clone REA196); CD83 APC (clone HB15); CD8 APC Vio770 (clone REA734).
   - 4.5. Pre-processing memory: CD62 VioBlue (clone 145/15); TCRαβ VioGreen (clone REA652); CCR7 FITC (clone REA546sssss); CD45RO PE (clone REA611); CD3 PerCP Vio700 (clone REA613); CD95 APC (clone DX2); TCRγδ APC Vio770 (clone REA591).
5. 5 mL U-bottom FACS tubes, 15 mL V-bottom centrifuge tubes, and 1.5 mL black opaque microcentrifuge tubes.
6. CD-Chex Plus lyophilized staining control (Streck), ACK Lysis buffer (store at room temperature).

3 Methods

3.1 CAR-T Manufacturing and Flow Staining of CAR-T Product

The manufacturing process for CAR-T cells involved stimulating T cells with interleukin-2 (IL-2) and CD3/CD28 paramagnetic beads in a culture plate. This was followed by a 2-day period of lentiviral transduction (CD19/CD22 CAR) and expansion. On days 7–9, the cells were harvested, concentrated, and analyzed via flow cytometry to determine the phenotype of the final T cell product and its subtypes, ensuring they met the final product release criteria [9].

3.1.1 Compensation Set Up for Flow

1. Add one drop of compensation bead to the designated fluorescent tag tube with 50 μL of TSS buffer.
2. Vortex or flick to mix, then incubate at 4°C for 20 min in the dark.
3. The tubes are now ready for acquisition to set up compensation.

3.1.2 Fluorescence Minus One (FMO) Staining with Cell

Beads are not used to incorporate viability dyes into the compensation matrix. Instead, cellular material is stained separately in tubes with cells, and these samples are acquired on the cytometer simultaneously with bead acquisition. Prepare another set of designated fluorescent tubes with cells (1×10^6) as FMO controls. Follow the extracellular staining protocol as discussed in the next section.

Key points:

a. Add 1×10^6 cells per tube precisely in 5 mL U-bottom FACS tubes.

b. Always keep TSS buffer or samples on wet ice or at 4°C

c. For the last wash before adding antibodies, always use a vacuum to discard the remaining buffer.

d. Make antibody cocktails (volume of antibody as shown in Table 1) with TSS buffer (50 μL).

3.1.3 Extracellular Flow Staining

1. Prepare 1×10^6 cells per tube.
2. Set the centrifuge to $400 \times g$, 5 min.
3. Adjust the volume to approximately 4 mL and centrifuge.
4. Discard the supernatant and wash the cells with 3 mL TSS buffer, pipetting up and down before centrifugation.
5. Repeat Steps 3 and 4.
6. Vacuum the buffer without disturbing the cell pallets.
7. Add 50 μl of TSS to each tube.
8. Add 20 μl Protein L to each tube except the unstained tube to stain the CAR-T antigen. Incubate at 4°C for 20 min.
9. Repeat Steps 3–6.
10. Add 100 μl of FACS antibodies cocktail and incubate at 4°C for 20 min. Avoid light.
11. Repeat Steps 3–6.
12. Add 400 μl TSS buffer and pipette up and down to resuspend the cells. Keep tubes cool and avoid light.

Table 1
Flow antibody details for multicolor CAR-T phenotype

S.N.	Flurochrome	Antibody	Clone	Company	Catalog number	Volume to use
1	BV650	CD45RA	HI100	BD Biosciences	563963	5 μL
2	PERCP-CY5.5	KLRG1	2F1	Biolegend	138418	5 μL
3	PE	CAR				
4		7AAD		BD Biosciences	51-68981E	5 μL
5	PECY7	PD1	EH12.2H7	Biolegend	329918	2.5 μL
6	APC	CCR7	150503	R&D Systems	FAB197A-100	10 μL
7	APC-CY7	CD4	RPA-T4	Biolegend	300518	1 μL
8	E450	LAG3	3DS223H	eBioscience	48-2239-42	2 μL
9	BV605	CD8	SK1	BD Biosciences	564116	5 μL
10	FITC	CD57	HCD57	Biolegend	322306	1 μL
11	BV786	CD137	4B4-1	BD Biosciences	741000	5 μL
12	BV711	TIM3	F38-2E2	Biolegend	345024	5 μL
13	BUV496	CD3	UCHT1	BD Biosciences	564809	5 μL

PERCP-CY5.5 peridinin chlorophyll protein-cyanine 5.5 tandem, *PE* phycoerythrin, *7-AAD* 7-aminoactinomycin D, *PE–Cy7* phycoerythrin–cyanine 7 tandem, *APC* allophycocyanin–cyanine, *APC-Cy7* allophycocyanin–cyanine H7 tandem, *E450* eFluor 450, *BV605* brilliant violet 605, *BV786, BV711, BV650, FITC* fluorescein isothiocyanate, *BUV496* brilliant ultraviolet 496

13. Analysis: for best results, analyze the cells on the flow cytometer as soon as possible, and for the best results, perform the analysis on the same day. For extended storage (up to 16–18 h) and greater flexibility in planning cytometer time, fixed the cells in 1%–4% paraformaldehyde to prevent deterioration before storage.

3.1.4 Detection and Data Analysis

- **Compensation Considerations:**

1. Enough events must be collected to make a statistically significant determination of spillovers. About 5000–10,000 events for both the positive and negative populations is ideal, but less can be used if necessary.
2. Performed instrument calibration/standardization procedures using a calibration bead and according to the established SOP of the laboratory before acquiring compensation or sample tubes.

3. Run an unstained cell sample for autofluorescence control. Adjust FSC (forward scatter) and SSC (side scatter) detector settings so that cells of interest are displayed on scale and can be gated as desired.
4. While gating on the cells of interest, adjust fluorescent detector settings so that the autofluorescence background is roughly within the first decade of the log scale of the fluorescence intensity histogram. Compare autofluorescence control (unstained cells) with stained cell positive controls to ensure that the stained cells are on scale for each parameter.
5. Set compensation by running beads stained individually with each antibody-fluorochrome conjugate (single-stain).
6. Next, set compensation by running cells stained individually with each antibody-fluorochrome conjugate (single-stain).
7. Finally, check the 13-color-stained cell population. The previous steps should have compensated the signals sufficiently, so no further adjustment is necessary for this sample.
8. Start acquiring control, sample data, and save.
9. Run freshly stained compensation controls for each experiment. Compensation settings may need to be readjusted for each multi-color staining experiment. Therefore, for each multi-color experiment, single- and dual-color staining controls are needed so that Steps 3–7 can be repeated. To minimize the amount of adjustment between sets of reagents, use the brightest staining reagent/cells of each fluorochrome for initial compensation setup.

- **CART Cell Phenotype Gating Strategy to Analyze the 13-Color-Antibody Flow Panel**

1. Defined the CAR-T cell phenotype to identify and most commonly used markers for T cell transduction efficiency (TE), subpopulations, subsets, exhaustion, senescence, differentiation, and stimulation/activation markers.
2. A detailed analysis of fluorochromes and antibodies from different providers was conducted to compare and optimize reagent combinations for our instrument's characteristics, ensuring successful 13-color flow cytometry. This systematic review addressed issues related to sampling, preparation, instrument settings, spillover and compensation, specific and nonspecific labeling, reagent performance, and general principles of panel construction.

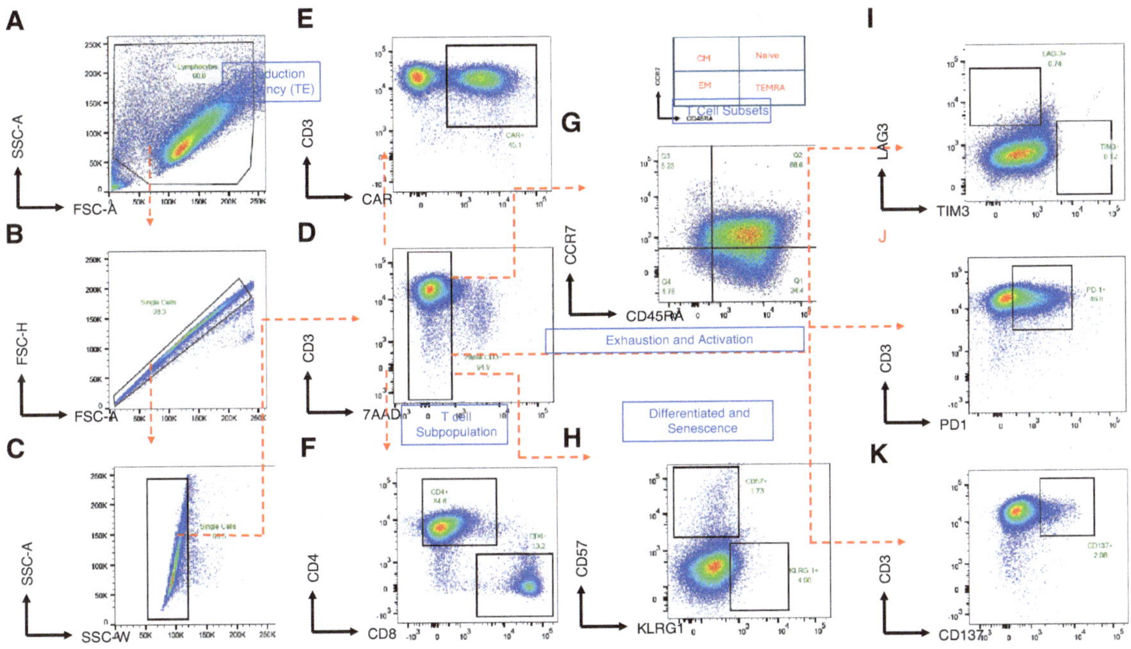

Fig. 1 CART cell phenotype gating strategy to analyze the 13-color-antibody flow panel. Final CAR-T product harvested on day-9 was stained with 13-antibody and acquired on 15-color BD Symphony flow cytometry. **A**, Gated lymphocytes to exclude cell debris on FSC-A vs. SSC-A. **B** and **C**, Lymphocytes were gated for singlets, by examining FSC-A vs. FSC-H (or FSC-W) or SSC-A vs. SSC-H (or SSC-W), with the latter being more sensitive for gating out doublets. **D**, Gated singlets to CD3 vs. 7AAD for total viable CD3+ T cells. **E**, On viable CD3+ T cells, gated CAR-positive T cell for Transduction efficiency (TE). **F**, the percentage subpopulation of CD4+ or CD8+ cells on Gate viable CD3+ T cells. **G**, the percentage of subsets of T cells on gated viable CD3+ T cells. **H**, Gated CD57 and KLRG1 for differentiation and senescence population on viable CD3+ T cells. **I–K**, On viable CD3+ T cells gated LAG3, TIM3, and PD1 for exhaustion and CD137 for activation population in final CAR-T cell products

3. The developed panel is displayed in Fig. 1. The gating strategy is based on the principles of removing noise signals from air bubbles and disturbing debris, as well as clumps of multiple cells, as these can skew the final results. This is achieved by examining FSC-A vs. FSC-H (or FSC-W) or SSC-A vs. SSC-H (or SSC-W), with the latter being more sensitive for gating out doublets (Fig. 1B, C).

4. Once single cells of interest are isolated in our analysis, we first gate for live CD3+ cells by excluding dead cells stained with 7AAD to determine the percentage of T cells (Fig. 1D) and then gate for CD19 CAR-positive T cells using biotinylated protein L and secondary antibody streptavidin PE for the transduction efficiency in the final CAR-T product (Fig. 1E).

5. Dead cells can be sticky and may pick up dye nonspecifically, leading to inaccurate results and false conclusions. The final T cell percentage and transduction efficiency (TE) vary with each submitted protocol. For this study, the final percentage of CD3+ cells was greater than 70%, and the percentage of TE was greater than 5%, as per the release criteria submitted during the IND submission to the FDA.
6. For the T cell subpopulation, we gated CD4 on the y-axis and vs. CD8 on the x-axis, on viable CD3+ T cells (Fig. 1F).
7. Staining of final CAR-T cells with antibodies to CD45RA and CCR7 revealed 4 subsets of CD3+ T cells, which include Naïve T, central memory, effector memory, and terminal effector memory T cell [10].
8. In the subsets gating on CD3+ T cells, the naive T (TNaive cells are CD45RA+CCR7, central memory (TCM) cells, which are CD45RA−CCR7+, CD45RA−CCR7− effector memory (TEM) cells, and CD45RA+CCR7− terminal effector memory T cells (TEMRA cell) (Fig. 1G).

- **Gating Strategy to Define the Differentiated and Senescence CAR T Cells**

1. Gate CD3+ cells and then gate D57 on the y-axis and KLRG1 on x-axis to define the T cell senescence and differentiation.
2. Gate CD3+ cells and then gate CD57 on the y-axis and KLRG1 on the x-axis to define the T cell senescence and differentiation. Gated CD3+ CD57+ T cell subsets were identified as a senescent cell, whereas gated CD3+ KLRG1+ T cell subsets identified terminally differentiated cells. (Fig. 1H)

- **Gating Strategy to Define the CAR T Cells Exhaustion and Activation Markers:**

1. Gating strategy to define the CAR T cells exhaustion and activation markers. Selected PD1, Tim3, and LAG3 on gated viable CD3+ T cells to identify the exhausted T cells. (Fig. 1I and J)
2. Select CD137 on gated viable CD3+ T cells to identify the activated T cells. (Fig. 1K).
3. All analyses can be performed on the CD3+ T-cell compartment, with gating on CD4+ and CD8+ subsets to determine the percentage expression of TE markers, functional subsets, exhaustion markers, senescence markers, differentiation states, and stimulation/activation markers.

3.2 Multiparameter Staining of Mixed Lymphocyte Populations for Clinical Release of In Vitro GMP Cultured Product

3.2.1 Preparation of Staining Reagents

1. For Live/Dead Vital Dyes, bring one vial of fluorescent reactive dye (Component A) to room temperature and one vial of anhydrous DMSO (Component B).
2. Add 50 μL of DMSO to reconstitute the lyophilized Live/Dead Vital Dye. Vortex to mix well and visually confirm the resuspension of the dye. Store the reconstituted stock dye mix at −20°C and reuse for up to 6 months post-reconstitution. To create a working solution, add 1 μL of reconstituted stock solution to 500 μL of PBS at 1:500 ratio.
3. For antibody preparations, label microcentrifuge tubes with the name of the corresponding antibody mix. Pipette tips are changed between antibodies to avoid carryover. Store the antibody preparation at 4°C or on ice. Discard any remaining antibody mix. Vortex tubes and briefly spin down in a tabletop microcentrifuge for 10 sec.
4. Add 1 μL of each antibody per sample.
5. Add 1 μL to account for CD-Chex Plus control.
6. Add 1 μL to account for pipetting error.
7. For Live/Dead Vital Dyes, bring one vial of fluorescent reactive dye (Component A) to room temperature and one vial of anhydrous DMSO (Component B).
8. Add 50 μL of DMSO to reconstitute the lyophilized Live/Dead Vital Dye. Vortex to mix well and visually confirm the resuspension of the dye. Store the reconstituted stock dye mix at −20°C and reuse for up to 6 months post reconstitution. To create a working solution, add 1 μL of reconstituted stock solution to 500 μL of PBS at 1:500 ratio.
9. For antibody preparations, label microcentrifuge tubes with the name of the corresponding antibody mix. Pipette tips are changed between antibodies to avoid carryover. Store the antibody preparation at 4°C or on ice. Discard any remaining antibody mix. Vortex tubes and briefly spin down in a tabletop microcentrifuge for 10 sec.
10. Add 1 μL of each antibody per sample.
11. Add 1 μL to account for CD-Chex Plus control.
12. Add 1 μL to account for pipetting error.

13. Warm CD-Chex Plus stock vial to room temperature on a rocking platform for 30 min.

3.2.2 Preparation of Fluorescence Minus One Control (FMOs)

1. FMOs are necessary for the proper placement of positive vs. negative gates in multicolor flow cytometry. FMOs measure the background intensity within a specific channel due to both autofluorescence and spillover from other fluorescent markers. FMOs are created from the sample to be investigated, and each FMO uses all the fluorescent stains except the one that is to be investigated. FMO controls should be created for each staining panel to be used.
2. To create the staining master mix for each FMO.
3. Add 1 μL of each antibody except the antibody that is the target of the FMO. For example, to make the tube for FMO FITC for the TNK lineage stain, the staining mix would contain CD16 PE, CD56 PE, CD3 PerCP Vio700, CD4 PE Vio770, and CD45 APC. The cells in question would also receive viability staining for Live/Dead Near IR.
4. To create the staining mix for the FMO Live/Dead Near IR tube, add 1 μL of each antibody. However, the cells in the FMO Live/Dead tube will not be stained with Live/Dead Near IR.
5. Label 5 mL FACS tubes with FMO and with each fluorescent marker to measure (e.g., FMO FITC, FMO PE, FMO APC, etc.). Label a 5 mL tube as unstained.
6. Count cells and resuspend at 2×10^6 cells per mL in 1× PBS.
7. Aliquot 100 μL of cell suspension (2×10^5 cells) into each tube.
8. Add 100 μL of Live/Dead Near IR working solution to FMO FITC.
9. Add 100 μL of Live/Dead Near IR working solution to FMO FITC, FMO PE, FMO PerCP Vio700, FMO PE Vio770, FMO APC.
10. Add 100 μL of 1× PBS to FMO Live/Dead Near IR and the unstained tube.
11. Incubate for 20 min in the dark at room temperature.
12. Add 2 mL of FACS Buffer to each tube and spin down at 400 x*g* for 5 min at room temperature.
13. Decant supernatant and resuspend cell pellets in 100 μL of FACS Buffer.
14. Add 10 μL of Fc Receptor Block to all tubes except the unstained tube. Incubate in the dark at room temperature for 5 min.

15. Add the antibody mix to each of their respective tubes. Incubate cells in the dark at room temperature for 25 min.
16. Wash cells by adding 2 mL FACS Buffer and spin down at 400 xg for 5 min at room temperature. Decant the supernatant and repeat the wash on cell pellet.
17. Decant supernatant and resuspend cell pellets in 200 μL of FACS Buffer. Set tubes aside at 4°C or on ice until ready to run.

3.2.3 Staining of CD-Chex Plus

1. Label 5 mL FACS tubes with CD-Chex and with TNK, DC Mono, and unstained labels.
2. Aliquot 100 μL of pre-warmed CD Chex to each tube and add 10 μL of Fc Receptor Block to all tubes except the unstained tube. Incubate in the dark for 5 min.
3. Add the antibody mix to each of the respective tubes and incubate in the dark at room temperature for 15 min.
4. Add 2 mL ACK Lysis Buffer to each tube and incubate in the dark at room temperature for 30 min.
5. Add 2 mL of FACS Buffer to each tube and spin down at 400 x*g* for 5 min at room temperature.
6. Decant supernatant and resuspend the cell pellet in 2 mL of FACS Buffer as a final wash. Spin the downtube at 400 x*g* for 5 min at room temperature.
7. Decant supernatant and resuspend the cell pellet in 200 μL. Set tubes aside at 4°C or on ice until ready to run.

3.2.4 Staining of Patient Samples (Not Including Whole Blood)

1. Label 5 mL tubes with sample ID and with TNK, DC Mono, TCR, Exhaust, and unstained labels.
2. Count cells and resuspend at 2×10^6 cells per mL in 1× PBS.
3. Aliquot 100 μL of cell suspension (2×10^5 cells) into each tube.
4. Add 100 μL of Live/Dead Near IR working solution to TNK and DC Mono tubes. Add 100 μL of 1× PBS to TCR, Exhaust, and unstained tubes.
5. Incubate for 20 minu in the dark at room temperature.
6. Add 2 mL of FACS Buffer to each tube and spin down at 400 × *g* for 5 min at room temperature.
7. Decant supernatant and resuspend cell pellets in 100 μL of FACS Buffer.
8. Add 10 μL of Fc Receptor Block to all tubes except the unstained tube. Incubate in the dark at room temperature for 5 min.

9. Add the antibody mix to each of their respective tubes. Incubate cells in the dark at room temperature for 25 min.
10. Wash cells by adding 2 mL FACS Buffer and spin down at 400 × *g* for 5 min at room temperature. Decant the supernatant and repeat the wash on cell pellet.
11. Decant supernatant and resuspend cell pellets in 200 μL of FACS Buffer. Set tubes aside at 4°C or on ice until ready to run.

3.2.5 Preprocessing Staining of Whole Blood Samples

1. Label 5 mL tubes with sample ID and with Pheno, Memory, and unstained labels.
2. Aliquot 100 μL of whole blood into each tube.
3. Add 10 μL of Fc Receptor Block to all tubes except the unstained tube. Incubate in the dark at room temperature for 5 min.
4. Add the antibody mix to each of their respective tubes and incubate in the dark at room temperature for 15 min.
5. Add 3 mL of ACK Lysis buffer and incubate in the dark at room temperature for 12 min.
6. Add 1 mL of FACS Buffer and spin at 400 x *g* for 5 min.
7. Decant supernatant and wash cell pellets again with 2 mL of FACS Buffer. Spin at 400 x *g* for 5 min.
8. Decant supernatant and resuspend in 200 μL of FACS Buffer. Set tubes aside at 4°C or on ice until ready to run.

3.2.6 Staining of Cell Therapy Products in Preparation for Clinical Freezing

1. Label 5 mL tubes with sample ID and with TNK, DC Mono, TCR, and unstained labels.
2. Count cells and resuspend at 2×10^6 cells per mL in 1× PBS.
3. Aliquot 100 μL of cell suspension (2×10^5 cells) into each tube.
4. Add 100 μL of Live/Dead Near IR working solution to TNK and DC Mono tubes. Add 100 μL of 1× PBS to TCR and unstained tubes.
5. Incubate for 20 min in the dark at room temperature.
6. Add 2 mL of FACS Buffer to each tube and spin down at 400 × *g* for 5 min at room temperature.
7. Decant supernatant and resuspend cell pellets in 100 μL of FACS Buffer.
8. Add 10 μL of Fc Receptor Block to all tubes except the unstained tube. Incubate in the dark at room temperature for 5 min.
9. Add the antibody mix to each of their respective tubes. Incubate cells in the dark at room temperature for 25 min.

10. Wash cells by adding 2 mL FACS Buffer and spin down at 400 × *g* for 5 min at room temperature. Decant supernatant and repeat wash on cell pellet.
11. Decant supernatant and resuspend cell pellets in 200 µL of FACS Buffer. Set tubes aside at 4°C or on ice until ready to run.

3.2.7 Instrument Preparation and Sample Collection

1. Flow cytometry instruments should have regular QC checks performed to demonstrate that the instrument is functioning within specifications.
2. Single-color controls and unstained controls can be collected to create a compensation matrix for the samples on the same day as the samples are collected. A previously run compensation can only be used if the approach has been validated prior to use.
3. Briefly vortex the samples prior to collection to ensure mixing of the samples.
4. Settings for voltage gains should remain constant for all samples collected, and all samples should be collected on the same day to avoid loss due to cell death.
5. Dot plots and histograms can be populated within the workspace to confirm the presence of the populations of interest during sample collection.
6. Data should be immediately saved and stored on a secure drive.

3.2.8 Gating Strategy for Phenotypic Identification of T Cells, B Cells, Monocytes, NK Cells, and Dendritic Cells

1. The population we want to determine using the gating strategy includes T-cells (CD3+), T helper cells (CD3+CD4+), T-cytotoxic cells (CD3−+CD8+), NK cells (CD3−CD16+ CD56+), B Cells (CD19+), Monocytes (CD14+), DCs (CD3−CD83+), and T cell receptors (CD3+TCRab+ and CD3+TCRgd+).
2. For accurate gating, utilize controls including an unstained sample and Fluorescence Minus One Controls (FMOs). These controls illustrate where unstained samples, unaffected by any antibodies, appear on the dot plot and create a visual reference pinpointing a positively stained, marked by a specific antibody. Stained cells expressing the phenotype of interest will exhibit a distinct population shift when compared to the unstained sample or FMOs.
3. When analyzing specific phenotypes, identification of a population expressing two antibodies may require gating for a double-positive population. This can be achieved by either creating quadrants or boxing the double-positive population. Using the quadrant system, the population gets divided into four distinct populations (Fig. 2A)

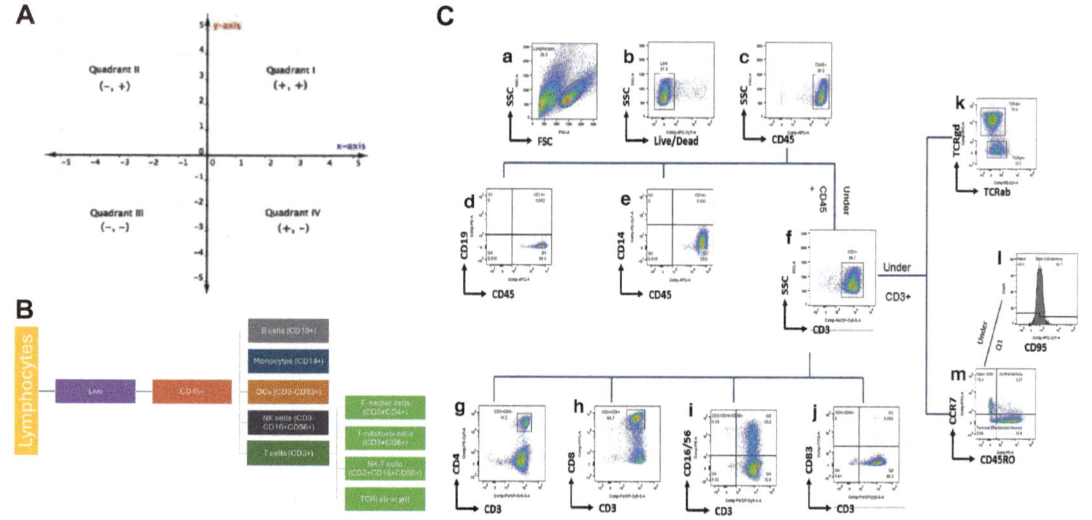

Fig. 2 **A**, Figure illustrates the quadrant system used for data analysis when using two unique antibodies to gate for different populations. The quadrant system divides the plot into four distinct quarters, labeled as Quadrant I (top right), Quadrant II (top left), Quadrant III (bottom left), and Quadrant IV (bottom right). **B**, A schematic overview of the basic gating strategy and the different T cell subsets and cells of interest. **C**, TAA-T cell phenotype gating strategy to analyze the six color-antibody flow panel: Final TAA-T product harvested on day 28 was stained with six color-antibody flow panels described above and acquired on 6 color BD Canto II. (a) Gated lymphocytes to exclude cell debris on FSC-A vs SSC-A. (b) Lymphocytes were gated for viability by gating for the Live/Dead channel against SSC, with the negative population representing the total percentage of live cells. (c) Gated live cells for the CD45 population to capture total viable CD45+ cells. (d and e) Using CD45 on FSC, gated for B cells and monocytes by changing the SSC to CD19 and CD14, respectively. (f) Gated for the total viable percentage of CD3+ T cells under the CD45 gate. (g) Gated for the percentage subsets of CD4 T cells on gated viable CD3 cells. (h) Gated for the percentage subsets of CD8 T cells on gated viable CD3 cells. (i) Gated for the percentage of NK cells on gated viable CD3 cells. (j) Gated for the percentage of Dendritic cells on gated viable CD3 cells. (k) Gated to differentiate the product for TCRab and TCRgd under total viable CD3+ T cells. (l) Gated to differentiate the different memory subsets present in the product under total viable CD3+ T cells. (m) To further gate and distinguish populations for memory subset and stem cell memory subset, cells from Q1 of gate (l) were evaluated using CD95

   a. Q1—x-axis negative, y-axis positive
   b. Q2—x-axis positive, y-axis positive (double positive)
   c. Q3—x-axis positive, y-axis negative
   d. Q4—x-axis negative, y-axis negative (double negative)

4. For gating populations from CD Chex Plus control samples, our workflow begins by creating three leukocyte gates based on Forward scatter vs. side scatter—these gates include lymphocyte, monocyte, and granulocyte.
5. Create a polygon gate to select CD45+ cells by side scatter, and within this gate select out specific lymphocyte populations by phenotypic markers.

6. CD Chex Plus lists specific values per lot for CD3+, CD3+ CD4+, CD3+ CD8+, CD16+/CD56+ CD3−, CD19+ cells under the CD45+ lymphocyte gate, and CD14+ cells are found under the CD45+ monocyte gate (Fig. 2B).
7. Compare the frequency of lymphocytes, monocytes, and granulocytes within CD Chex Plus control samples with the CD Chex Plus QC Lot information and the frequency of specific stained lymphocyte and monocyte subsets. Frequencies should fall within the range listed for the specific Lot QC information, with any deviations possibly requiring a variance to be written.
8. For gating lymphocyte populations in patient samples and cell therapy products, our workflow begins gating for T cells and their subsets by initially selecting the "lymphocytes" population based on the FSC vs. SSC scale to encompass all cell subsets. (Fig. 2C)
9. To refine our population of interest, apply gating on the Live/Dead stain to exclude dead cells. Define the "Live" population as the negative population on the dot plot to establish the second gate.
10. Within the live cell population, gate for "CD45" on the x-axis while maintaining SSC on the y-axis. The CD45+ population serves as a parent gate encompassing T cells, B cells, NK cells, monocytes, and DCs.
11. With CD45 on the x-axis, switch the y-axis to "CD19." Identifying the double-positive CD45+CD19+ population provides the percentage of B cells.
12. Adjusting the y-axis to "CD14," identify the double-positive CD45+CD14+ population to determine the percentage of monocytes.
13. For gating T cell subsets and NK cells, further gating is conducted within the CD45 gate. The gated populations represent the percentage of CD45 cells expressing our target phenotype.
14. Within CD45+ positive cells, change the x-axis to "CD3" while keeping SSC on the y-axis. The positive population reveals the percentage of T cells present. Subsequently, by maintaining the x-axis on CD3 and altering the y-axis for different phenotypes, determine the subset percentages of T cells.
15. Maintain the x-axis on CD3; switch the y-axis from SSC to "CD4." The double-positive population CD3+CD4+ indicates the percentage of T helper cells.

16. Maintain the x-axis on CD3 and switch the y-axis from SSC to "CD8." The double-positive population CD3+CD8+ indicates the percentage of T cytotoxic cells.
17. Maintain the x-axis on CD3 and switch the y-axis from SSC to "CD16/CD56." CD3−CD16+CD56+ identifies the percentage of NK cells, while CD3+CD16+CD56+ identifies the percentage of NK-T cells.
18. Maintain the x-axis on CD3 and switch the y-axis from SSC to "CD83." CD3−CD83+ population indicates the percentage of DCs present in the product.
19. To identify the two distinct TCR populations, gate within the CD3+ population for TCRab on the x-axis and TCRgd on the y-axis of lymphocytes.
20. Next, we gate on viable cells by selecting Live/Dead Near IR low cells vs. cells staining brightly for Live/Dead Near IR (dead/dying).
21. Cell frequencies and PDF layouts of all phenotypic gates are recorded in a separate written document to be reviewed and signed off by QC staff.

4 Notes

1. Select the correct compensation bead that binds the species or reagent and can be excited by the appropriate laser. Do not wash the compensation bead, as washing could result in the loss of stained beads [7].
2. The absolute compensation values depend not only on the spectral overlap between two fluorophores but also on the PMT voltage. Remember that the default CST settings may not be optimal for your sample. Instead, use stained cells to adjust the PMT voltages specifically for your experiment. To maintain balanced PMT voltages, it is crucial to build a balanced panel [7].
3. To measure the amount of spectral spillover in each channel, single-stained cells, or antibody-binding beads, each labeled with one of the fluorescent reagents used in the panel, are separately collected at the beginning of acquisition in order to apply compensation to the subsequent samples [1].
4. The expression of CD57 and KLRG1 T-cells displays key features of cellular senescence, including proliferation and terminally differentiated T cells. CD57−KLRG1+ subset retained the ability to proliferate and secrete IFN-γ in response to cognate antigen, the CD57+KLRG1+ subset failed to proliferate

and displayed only trace IFN-γ secretion, suggesting that these were terminally differentiated senescent T-cells. [11].

5. Expression of programmed cell death 1 (PD-1), T-cell immunoglobulin domain and mucin domain protein 3 (TIM-3), and lymphocyte activation gene 3 (LAG-3) are hallmarks of T cell exhaustion, and exhausted T cells are functionally distinct from T effector and T memory cells [1]. Similarly, CD137 (4-1BB), a member of the tumor necrosis factor receptor superfamily, is an inducible costimulatory receptor expressed on activated T cells [12].

References

1. Blache U, Weiss R, Boldt A, Kapinsky M, Blaudszun AR, Quaiser A, Pohl A, Miloud T, Burgaud M, Vucinic V, Platzbecker U, Sack U, Fricke S, Koehl U (2021) Advanced flow cytometry assays for immune monitoring of car-t cell applications. Front Immunol 12:658314. https://doi.org/10.3389/fimmu.2021.658314
2. Dave H, Terpilowski M, Mai M, Toner K, Grant M, Stanojevic M, Lazarski C, Shibli A, Bien SA, Maglo P, Hoq F, Schore R, Glenn M, Hu B, Hanley PJ, Ambinder R, Bollard CM (2022) Tumor-associated antigen–specific T cells with nivolumab are safe and persist in vivo in relapsed/refractory Hodgkin lymphoma. Blood Adv 6(2):473–485. https://doi.org/10.1182/bloodadvances.2021005343
3. Hont AB, Cruz CR, Ulrey R, O'Brien B, Stanojevic M, Datar A, Albihani S, Saunders D, Hanajiri R, Panchapakesan K, Darko S, Banerjee P, Fortiz MF, Hoq F, Lang H, Wang Y, Hanley PJ, Dome JS, Bollard CM, Meany HJ (2019) Immunotherapy of relapsed and refractory solid tumors with ex vivo expanded multi-tumor associated antigen specific cytotoxic T lymphocytes: a phase I study. J Clin Oncol 37(26):2349–2359. https://doi.org/10.1200/JCO.19.00177
4. Kinoshita H, Cooke KR, Grant M, Stanojevic M, Cruz CR, Keller M, Fortiz MF, Hoq F, Lang H, Barrett AJ, Liang H, Tanna J, Zhang N, Shibli A, Datar A, Fulton K, Kukadiya D, Zhang A, Williams KM, Dave H, Dome JS, Jacobsohn D, Hanley PJ, Jones RJ, Bollard CM (2022) Outcome of donor-derived TAA-T cell therapy in patients with high-risk or relapsed acute leukemia post allogeneic BMT. Blood Adv 6(8):2520–2534. https://doi.org/10.1182/bloodadvances.2021006831
5. Keller MD, Bollard CM (2020) Virus-specific T-cell therapies for patients with primary immune deficiency. Blood 135(9):620–628. https://doi.org/10.1182/blood.2019000924
6. Bollard CM, Gottschalk S, Leen AM, Weiss H, Straathof KC, Carrum G, Khalil M, Wu MF, Huls MH, Chang CC, Gresik MV, Gee AP, Brenner MK, Rooney CM, Heslop HE (2007) Complete responses of relapsed lymphoma following genetic modification of tumor-antigen presenting cells and T-lymphocyte transfer. Blood 110(8):2838–2845. https://doi.org/10.1182/blood-2007-05-091280
7. Campbell JDM, Fraser AR (2018) Flow cytometric assays for identity, safety and potency of cellular therapies. CytometryB Clin Cytom 94B:725–735. https://doi.org/10.1002/cyto.21735
8. Peinelt A, Bremm M, Kreyenberg H, Cappel C, Banisharif-Dehkordi J, Erben S, Rettinger E, Jarisch A, Meisel R, Schlegel PG, Beck O, Bug G, Klusmann JH, Klingebiel T, Huenecke S, Bader P (2022) Monitoring of circulating CAR T cells: validation of a flow cytometric assay, cellular kinetics, and phenotype analysis following tisagenlecleucel. Front Immunol 13:830773. https://doi.org/10.3389/fimmu.2022.830773
9. Panch SR, Srivastava SK, Elavia N, McManus A, Liu S, Jin P, Highfill SL, Li X, Dagur P, Kochenderfer JN, Fry TJ, Mackall CL, Lee D, Shah NN, Stroncek DF (2019) Effect of cryopreservation on autologous chimeric antigen receptor T cell characteristics. Mol Ther 27(7):1275–1285. https://doi.org/10.1016/j.ymthe.2019.05.015
10. Zhang DKY, Adu-Berchie K, Iyer S, Liu Y, Cieri N, Brockman JM, Neuberg D, Wu CJ, Mooney DJ (2023) Enhancing CAR-T cell functionality in a patient-specific manner. Nat Commun 14(1):506. https://doi.org/10.1038/s41467-023-36126-7
11. Kared H, Martelli S, Ng TP, Pender SL, Larbi A (2016) CD57 in human natural killer cells and

T-lymphocytes. Cancer Immunol Immunother 65(4):441–452. https://doi.org/10.1007/s00262-016-1803-z

12. Otano I, Azpilikueta A, Glez-Vaz J, Alvarez M, Medina-Echeverz J, Cortés-Domínguez I, Ortiz-de-Solorzano C, Ellmark P, Fritzell S, Hernandez-Hoyos G, Nelson MH, Ochoa MC, Bolaños E, Cuculescu D, Jaúregui P, Sanchez-Gregorio S, Etxeberria I, Rodriguez-Ruiz ME, Sanmamed MF, Teijeira Á, Berraondo P, Melero I (2021) CD137 (4-1BB) costimulation of CD8+ T cells is more potent when provided in cis than in trans with respect to CD3-TCR stimulation. Nat Commun 12(1):7296. https://doi.org/10.1038/s41467-021-27613-w

Part III

Technologies

Check for updates

Chapter 14

Guidelines for Gating Flow Cytometry Data for Immunological Assays

Janet Staats, Anagha Divekar, J. Philip McCoy and Holden T. Maecker

Abstract

"Gating" refers to the selection of successive sub-populations of cells for analysis in flow cytometry. While automated analysis algorithms have seen a greatly increased use, there is still heavy reliance on manual gating, based on expert knowledge of cell characteristics. This can lead to variability in how gates are applied, even between individuals experienced in the field. For clinical assays, guidelines often exist, and software may automate the gating process. But there are no comprehensive guidelines across the various types of immunological assays performed in research settings using flow cytometry. Here, we attempt to provide such guidelines, focused on the most general and pervasive types of gates, why they are important, and what recommendations can be made regarding their use. We do so through the display of example data, collected by academic, government, and industry representatives. These guidelines should be of value to both novice and experienced flow cytometrists analyzing a wide variety of immunological assays.

Keywords Flow cytometry, Gating, Analysis, Immunophenotyping

1 Introduction

The analysis of flow or mass cytometry data uses a process commonly called "gating," whereby populations of cells are virtually separated from one another. This is accomplished through the use of specialized flow cytometry software programs to identify subsets within a sample of particles or cells and to assess the expression levels of specific markers of interest on those selected subsets, without confounding data from other cellular subsets. In this context, some authors may refer to an "analysis region" rather than a "gate," which formally refers to physical sorting of cells, but we will not distinguish the two here.

Various aspects of a cell subset affect the ease with which that subset can either be manually gated or identified by a clustering algorithm. These include the relative size and abundance of the subset, its separation from other populations, and the number of distinguishing markers for that subset used in the staining panel.

Angélique Biancotto and Pradeep K. Dagur (eds.), *Immunophenotyping*, Methods in Molecular Biology, vol. 3087, https://doi.org/10.1007/978-1-0716-5607-5_14,

Most software programs used for flow cytometry analysis allow for gates to be drawn in one or two dimensions (i.e., on histograms or on two-parameter plots). Two-dimensional gates may be rectangular, elliptical, or irregular polygons, and the plots on which they are drawn may depict contours, density, or individual events (dot plots). In theory, gates could be drawn in more than two dimensions at once, but this is practically not feasible for humans. Hence, subsets that are distinguished by a series of markers are usually identified by sequential 1- or 2-dimensional gates (known as Boolean hierarchical gating). These are usually Boolean "and" gates (sequentially drilling down on smaller subsets), but can also involve "or" and/or "not" gates.

It is worth noting that the limitation of visualizing in two dimensions at a time means that we are always missing some information when performing manual gating. Consider the case of two-dimensional data for which there are subpopulations expressing neither, one, or both of the two markers in question. Examination of histograms for each marker will not necessarily resolve the proportions of these subpopulations accurately, while a two-dimensional plot will. If we extend this concept to highly multidimensional data, it becomes clear that we are not getting a complete picture of the data by only examining two dimensions at a time. For this reason, where possible, analysis of high-dimensional data should probably include both automated analysis (which can "see" in many dimensions) as well as manual gating.

Because of the vast number of different assays that can be performed, flow cytometry has developed into a widely used technology in a number of scientific endeavors. These applications can be roughly divided into four major areas: clinical/diagnostic, clinical/translational research, biopharma/drug discovery, and basic research. While these four areas use a common technology, each has nuances that relate to the specific applications. Thus, when guidelines for flow cytometry are established, some elements might be common to all applications, while others might apply to one or a few of the applications. Perhaps for this reason, there are few general guidelines on how to place gates or analysis regions that are applicable to the vast majority of flow cytometry analyses.

As one might expect, the clinical environment has led the way in efforts to establish practice guidelines for flow cytometry, in large part due to the implementation of CLIA-88 (Clinical Laboratory Improvement Act of 1988). These laws were enacted to assure reliable, high-quality, accurate, and safe testing in all clinical laboratories. The laws encourage a standard of practice, as well as proficiency testing and inspections. CLIA, in turn, led directly to the issuance of guidelines for the performance of flow cytometry in the clinical environment by the College of American Pathologists (CAP) and the Clinical Laboratory Standards Institute (CLSI) (formerly NCCLS). These guidelines deal with a variety of issues, ranging from lot testing to instrument setup to establishing standard operating procedures (SOPs) for nearly every aspect of the laboratory.

In clinical and translational research environments, there have been a number of assay-specific guidelines that have been established. Perhaps best known are the guidelines for CD4 T cell counting in the context of HIV disease. One of the first of these was the guidelines published by the US Centers for Disease Control and Prevention (CDC) in 1992, with supplements and revisions to these guidelines issued in 1994, 1997, and 2003 [1–4]. Similar guidelines for enumeration of CD4 T lymphocytes have been put forward by the World Health Organization [5].

Guidelines for additional clinical tests have been published and have gained wide acceptance. Among these are guidelines for enumeration of CD34+ hematopoietic stem cells (proposed by the International Society of Hematotherapy and Graft Engineering [6]) and guidelines for detecting paroxysmal nocturnal hemoglobinuria (PNH) [7]. These include specific directions for how the gating is to be performed for these assays.

While the clinical arena is heavily regulated with enforceable guidelines, the remaining areas in which flow cytometry is used are far less regulated. Except for some network clinical trial work, for which GCLP compliance is required, most often adherence to guidelines in research environments is voluntary rather than mandated. For biopharmaceutical and translational research cytometry, there have been guidelines established to promote high-quality practices, although these pertain more to general laboratory practice rather than specific flow cytometry issues. In the realm of translational research, several guidelines have been proposed that would also be applicable to basic research. Examples of these include guidelines for intracellular cytokine staining, proposed by the Cancer Immunotherapy Consortium [8], and guidelines for HLA-peptide multimer assays proposed by the Cancer Vaccine Consortium [9]. Of these, only the former included details concerning how the gating should be performed.

Basic research using flow cytometry is the area in which the least has been done to standardize or harmonize how assays are performed. This is due to a number of factors, including the vast array of assays that are performed by flow cytometry in basic research, the fiercely independent nature of many researchers, and the lack of ability to enforce compliance with guidelines in this area. Nomura et al. demonstrated that gating of flow cytometric data was a significant factor in inter-laboratory variations [10], a finding that was later confirmed by Finak et al. [11]. The latter study also explored the advantages offered by an automated gating strategy.

It is worth noting that the complexity of flow cytometry experiments has increased from one-to-four colors in the 1980's to up to 40–50 parameters today [12]. This has been led most recently by the rise of spectral cytometry, which uses full-spectrum measurement followed by mathematical deconvolution of signals, rather than narrow-band detection channels and compensation for spillover into those channels [13]. The invention of mass cytometry, or CyTOF, has also led to datasets with up to 50 or more parameters [14]. While aspects of gating specific to CyTOF are detailed elsewhere [15], the increasing number of parameters places even more importance on the gating process. Gating strategies become more complex with many more options, and quality control of factors such as spillover correction becomes ever more important.

To deal with this complexity, recent years have seen an increased use of algorithms to help automate the gating process (reviewed in [16, 17]). These can range from tools embedded within manual gating software that help define the edges of a selected cell population to supervised approaches for analyzing a complete data set to unsupervised algorithms that perform dimensionality reduction and/or clustering of cell subsets. The latter include viSNE [18], UMAP [19], FlowSOM [20], and many others. Yet, even these algorithms perform much better (i.e., provide more interpretable output) when data are initially gated on at least live, single cells. Furthermore, with increasing complexity and decreasing supervision comes the prospect that the automated method may produce spurious or undesirable results, or at least will not match the manual gating results of recognized experts. As such, automated gating tends to still require some manual confirmation.

Gating is subjective; substantial variations in gating can be observed between different analysts. In addition to operator expertise, this is in part dependent on the quality of staining. Many factors influence the observed fluorescence intensity of the samples, as summarized in Table 1. While most of these factors have been reviewed previously, it is generally assumed that the operator will make appropriate adjustments for these factors in terms of gating. Ideally, however, these factors should be first minimized through technical means (e.g., experimental design and careful adherence to protocol and to best practices).

An important aspect of hierarchical gating is knowledge of the expression profile of a population of interest across many different markers. Hierarchical gates can include both positive and negative markers, and the sequential order of the hierarchical gates can influence the ability to clearly resolve the final population. Therefore, accurate and robust gating should be based on a strategy that involves an optimal sequence of marker-based gates, rather than just a single terminal gate based on one or two markers.

The process of defining an optimal gating hierarchy may be iterative and should be considered already at the time of panel design. The hierarchy should be verified on test data (e.g., controls and/or pilot samples), and training on its use may be required for large, multi-site projects. In such projects, a shared template will be required

Table 1
Factors affecting population resolution in flow cytometry

Root level	Factor	References
Biological	Donor variability in expression levels, background	[21]
	Sample integrity (shipping, cryopreservation/thawing, viability, etc.)	[22–24]
	Aggregates	[25]
Instrument	Optical configuration (lasers, laser power, filters)	[26]
	Daily performance, including laser delays, voltage calibration	[27]
	Hydrodynamics (air bubbles, clogs)	[28, 29]
Reagents	Antibody clones, titers, and lot-to-lot variability	[30, 31]
	Panel design (fluorochrome brightness and spillover)	[32]
	Reagent degradation, especially tandem dyes	[32]
	Staining protocol, including activation, fixation, permeabilization, intracellular vs. cell-surface staining, washing	[31]
Data transformation	Accuracy of compensation for fluorescence spillover	[33, 34]
	Scaling (logarithmic vs. biexponential, valley artifact)	[35, 36]

along with instructions on its implementation. One suggestion is to use a set of relevant existing data files as a training tool and share these across operators and sites participating in a given study. Finally, the application of the template to each set of data should ideally be checked by more than one user to achieve consensus.

What should the gating strategy be for a particular type of experiment? That is a question impossible to fully answer here, given the breadth of immunological experiments in flow cytometry. However, some general principles can be stated. A hierarchy of gates usually follows an order somewhat as follows:

1. A CD45 threshold, applied during data acquisition (this is relevant mainly for samples of mixed immune and nonimmune cells)
2. A time gate, or at least examination of the time parameter to determine if time gating is needed to exclude regions of unstable flow rate during cell acquisition
3. A single cell gate, usually on forward and/or side light scatter height vs. width or height vs. area
4. A live/dead gate and/or exclusion gate for populations not to be analyzed (i.e., a "dump channel")

5. A scatter gate, to remove debris and identify populations of interest, like lymphocytes, monocytes, or granulocytes
6. Specific population gates, based on lineage markers in a sequential hierarchy (e.g., CD3 followed by CD4 vs. CD8), and often including negative markers as well

With this hierarchy in mind, we address each of these six categories of gates in sequential sections of the Guidelines below. We also discuss data display as it relates to setting gates, including compensation and scale transformations. We describe how various types of controls can be used to set gates, the units to report from gates, and other special considerations. Within these sections, we also explain data displays that are used for quality control, such as "back gating" to verify gate placement and "$n \times n$" plot matrices to check compensation.

2 Guidelines

2.1 Acquisition Threshold or Trigger Gate

2.1.1 Definition

When acquiring samples using a flow cytometer, the cells of interest are usually in a suspension that contains many other cells or particles. Each cell or particle is measured by the cytometer as an event in the form of a pulse; some events represent signals of interest and some are noise. An operator may set a limit for the type of pulse that is acceptable to be included in the resulting data file; this is called an acquisition (detection) threshold or trigger gate. Trigger gates are set prior to acquisition as a means of excluding events of noninterest or noise. The most common type of trigger gate is used to exclude debris based on size, measured using forward scatter. Using forward scatter, the trigger gate is set to exclude events that are too small to be considered cells. Another type of commonly used trigger gate is based on CD45 expression, where CD45 is a pan-leukocyte marker (see Fig. 1). Setting the trigger gate on CD45+ cells will essentially exclude all nonleukocyte events, cleaning downstream gating and improving the accuracy of downstream subset enumeration.

Note that on some cytometers, a ***storage gate*** can be set in addition to an acquisition threshold or trigger. Care should be taken that the storage gate, as well as the acquisition gate, includes all populations of potential interest for later analysis. For example, a storage gate set on lymphocytes (using forward vs. side scatter) would preclude downstream analysis of monocytes.

2.1.2 How Data is Affected

Trigger gates have a "Goldilocks" zone. A low or no-trigger gate leads to excess noise and reduces overall signal-to-noise. But when trigger gates are set too high, they partially or completely exclude the

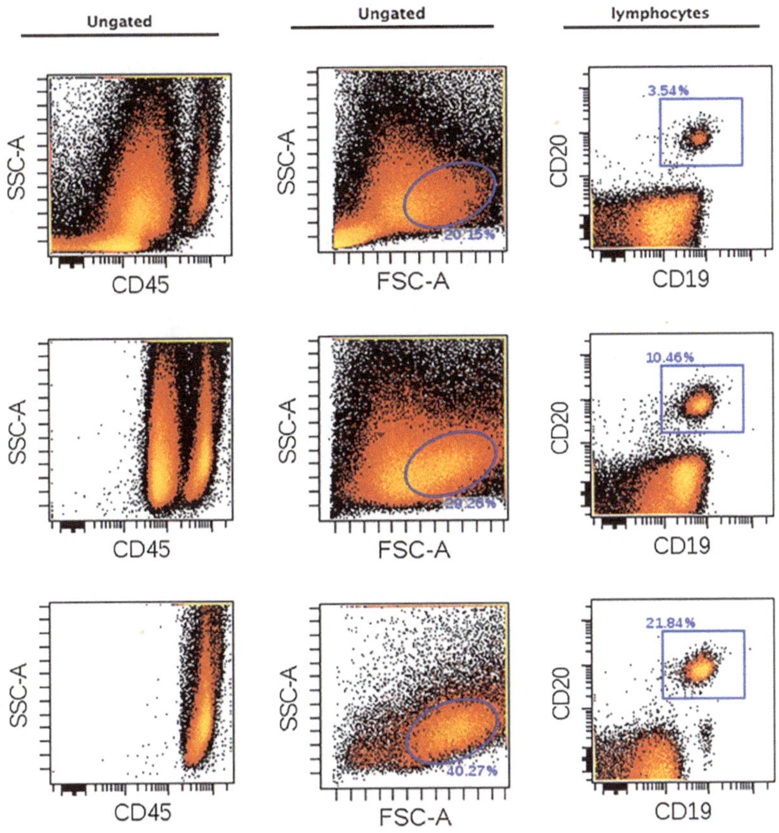

Fig. 1 A CD45 trigger gate (threshold) is used to exclude nonleukocytes. In a sample heavily contaminated with erythrocytes, a CD45 threshold can help to exclude these cells and "clean up" the downstream gates. In this example, the same erythrocyte-contaminated sample was run with no (top row), low (middle row), or high (bottom row) threshold on CD45. Left column: CD45 vs. SSC plots show the huge contribution of CD45– events without a high enough threshold. Middle column: FSC vs. SSC plots show the progressive difficulty of identifying the lymphocyte population without sufficiently high CD45 triggering. Right column: Enumeration of B cells from the lymphocyte gate. Even with identically placed FSC vs. SSC lymphocyte gates, the B cell enumeration is inaccurate without a sufficiently high CD45 trigger, as the denominator is diluted by contaminating CD45– events

target population of interest and decrease the overall signal. When applied properly, trigger gates effectively remove unwanted events or cells, thus decreasing noise and improving the overall signal-to-noise.

2.1.3 Conclusion

Threshold gating is a very effective tool for reducing noise. To ensure the removal of noise and not signal, setting trigger gates properly requires both prior knowledge of the sample and the target population of interest.

2.2 Time Gate

2.2.1 Definition

Time was first introduced as a flow cytometry parameter to measure kinetic responses [37]. Subsequently, in 1987, James Watson first described the concept of using time as a quality control (QC) parameter [28]. The principle of flow cytometry is based on regulating the differential pressure between sample and sheath fluid to establish a laminar flow, enabling a single cell to pass in front of the laser for interrogation. The steps involved with acquiring a sample influence the fluid dynamics throughout the course of acquisition. When a sample is first introduced into the cytometer, a burst of pressure may cause a high density of events to pass in front of the laser during the first few seconds. After the initial sample introduction, the flow rate stabilizes and becomes laminar. Nearing the end of acquisition, the sample volume is greatly reduced, which may influence the fluid dynamics again. During acquisition, fluidic changes may be observed as either an increase or a decrease in the event rate. The most frequent culprits are air bubbles or clogs. For these reasons, the time parameter may be used during analysis to exclude those events that were acquired during periods of nonlaminar flow. This is called a time gate. For manual analysis, a time gate is created by visualizing time versus scatter or fluorescence. Alternatively, there are automated software options, such as FlowClean [29] or FlowAI [38], that may be used to effectively remove anomalies, such as those caused by changes in the event rate, from an FCS file. Of note, the time parameter may be displayed differently based on the software being used for acquisition or analysis.

2.2.2 How Data is Affected

Anomalies in the flow rate result in wider CVs and/or shifts in scatter and fluorescence parameters (see Fig. 2). The effect of nonlaminar flow on the resulting data may be modest or severe, depending upon the degree of fluidic instability and the integrity of any subsequent quality control gates, such as singlet and fluorescence gates. In the most severe case, an entire data file may be unacceptable for analysis due to unstable fluidics that occurred throughout the time of acquisition.

2.2.3 Conclusion

The time parameter is an effective QC tool for identifying temporal changes in fluid dynamics that occurred throughout sample acquisition. Gating on "good" acquisition interval(s) can improve data quality.

2.3 Singlet Gates

2.3.1 Definition

Aggregates consist of multiple cells or particles; they form for many reasons, including cells clumping together due to dying/death, adherent properties, antigen presentation, and also high sample concentration, high flow rate, and even dye–dye aggregation. When running clumped samples with many aggregates, it is important to

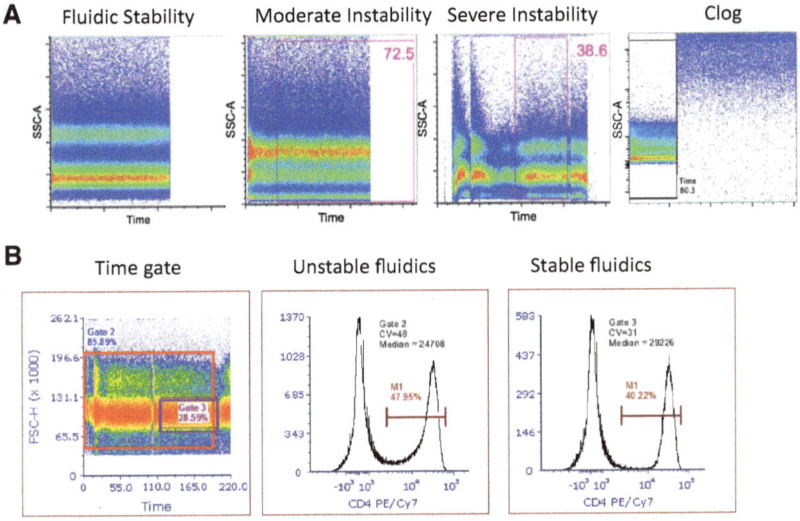

Fig. 2 Use of time gates. **A**, Time as a QC parameter. Time versus SSC-A is used to visualize the temporal changes in fluid dynamics that occur throughout the sample acquisition. *Left to right: first plot* illustrates an ideal time parameter plot. The burst of sample at the beginning, observed as the red or higher-density, lower-scatter population, has an initially wider SSC CV that rapidly resolves into a tight SSC CV. The SSC CV remains the same throughout the remainder of acquisition and slightly decreases near the end of acquisition, as the sample rate decreases (red turns to orange). There are no skips or jumps in the number of events relative to time and no shifting of SSC over time. *Second plot*: moderate nonlaminar flow. A more noticeable sample burst at the beginning of acquisition with a wide SSC-A CV. SSC-A CV changes throughout the acquisition in an oscillating pattern, caused by swirling fluidics. The time gate, indicated in pink, is set to include the most stable portion of the events acquired. *Third plot:* This illustrates nonstable fluidics that occurred throughout the entire sample acquisition. Small clogs are indicated by skips or jumps in time, where the event rate is greatly reduced, followed by bursts of events when the clog breaks. Changes in SSC-A CV and shifting of SSC-A are prominent throughout the acquisition. The time gate, in pink, is placed around the most stable area, although no area in this example is ideal. While a significant portion of events is excluded due to unstable fluidics, the data included in the time gate represents those events with the highest integrity. *Fourth plot:* This illustrates a stable clog that occurred a short time into the sample acquisition. **B**, Time gating affects CV of downstream gates. Human PBMC were stained with CD4 PE-Cy7. The leftmost plot is ungated; the middle and right histograms are gated on FSC/SSC singlets. If placement of the time gate includes sections of the plot where nonlaminar flow exists, then the CV, as well as median of the CD4+ cell peak and the frequency of CD4+ cells is affected. Inclusion of events from the unstable fluidics results in a wider CV and higher percent positive cells, and exclusion of these events results in a tighter peak with lower CV and higher fluorescence intensity

filter the samples prior to acquisition to remove as many aggregates as possible. When filtering is insufficient to remove all aggregates, they may be further removed using singlet gating.

To understand how singlet gates work, it's important to understand how signals are detected by a flow cytometer. When a particle passes in front of the laser, the cytometer measures the resulting photons of light in the form of a pulse. Digital cytometers are capable of collecting three values that define a pulse: ***Width*** is a measurement that represents the time a cell passes in front of the laser. Width is proportional to the size of the cell or particle and is unaffected by PMT voltage settings [39]. ***Height*** is the measurement of the intensity of the signal (maximum fluorescence, size, or scatter) and is affected by PMT voltages. ***Area*** is not measured; it is derived using both width and height measurements and is only slightly affected by PMT voltage, owing to its relationship to both width and height. Area scaling is used in some digital instruments to equalize height and area.

Pulses can be distorted by coincidence when two events (a doublet) pass in front of the laser simultaneously. This distortion results in a two-fold stretching of the area and width, while the height remains unchanged. During analysis, singlet gating uses the distortion aggregates impose on pulse width and area to distinguish singlets from aggregates. Singlet gating is performed using either forward scatter (FSC) or side scatter (SSC) parameters. It is important to remember that FSC is collected using a diode, unless you have an instrument specially equipped with FSC PMT, and SSC is collected using a PMT. Since PMTs are more sensitive than diodes, a better resolution between singlets and aggregates may be visualized using the SSC over FSC parameter. As a best practice, both FSC and SSC singlet gates would be used sequentially to remove coincidental events occurring in different orientations.

There are two common approaches to singlet gating; one is dependent upon area scaling, and one is not: area versus height and width versus height. For area vs. height singlet gating, coincidence is identified by the events for which there is an increase in area without a corresponding increase in height. Thus, cells along the diagonal are singlets and should be included in the singlet gate. Use of area versus height singlet gating requires the area scaling to be properly adjusted on your cytometer using cells; otherwise, there is a risk for gating errors. Maximum resolution between singlets and aggregates is observed using width versus height, as both are independently measured values and are not affected by area scaling [40]. For width versus height singlet gating, coincidence is identified by an increase in width, while height remains unchanged. Irrespective of the pulse value used to establish singlet gates, debris may be readily distinguished from cells in the lower corner and may be excluded using the same gate as the aggregates. Examples of singlet gates are shown in Fig. 3.

2.3.2 How Data is Affected

It is important to remove aggregates during data analysis of immunophenotyping data, as these can give false double-positive events. As an example, if a cluster of cells includes B (CD19+CD3−) and T (CD19−CD3+) cells and if these are not removed from the data analysis, then the operator may report the presence of a cell population that is CD3+CD19+ (a T-B cell), which does not exist. Applying doublet discrimination helps remove these oddities from the data.

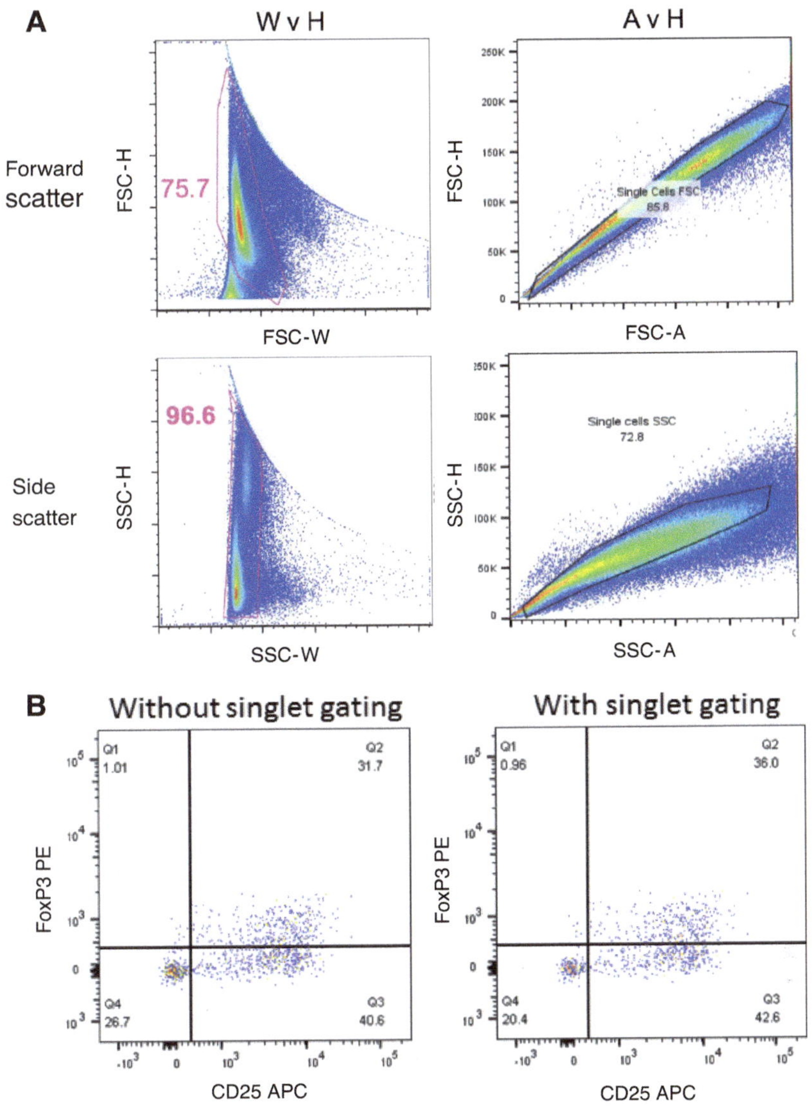

Fig. 3 Singlet gating. **A**, Width vs. height and area versus height singlet gates using FSC and SSC. Whole blood was subjected to erythrocyte lysis, stained, and acquired by flow cytometry. *Top row*: forward scatter parameter is used to create singlet gates. *Bottom row*: side scatter is used to create singlet gates. *Left column*: width (-W) versus height (-H) plots are used for singlet gating and to exclude debris in the lower corner. Aggregates are those events for which there is an increase in -W relative to -H. *Right column*: area (-A) versus -H is used for gating singlets; aggregates are those events for which -A increases disproportionately to -H. **B**, Exclusion of doublets can alter the frequency of gated populations. Whole blood was stained with a cocktail containing CD3, CD4, CD127, CD25, and Foxp3. The left image depicts the frequency of Treg cells without doublet exclusion, and right panel shows the frequency of Treg cells with singlet gating or doublet exclusion. Note the increase in frequency of CD25+Foxp3+ cells following gating on singlets. Singlet gating was performed using SSC-A vs. SSC-H, followed by FSC-A vs. FSC-H

2.3.3 Conclusion

In the context of immunophenotyping, singlet gates are necessary for accurate identification of cells that are co-expressing two markers or double positives. Once singlet gating is applied in some instances, this will "clean up" data and thus result in loss of a certain cell population, whereas in other cases it may result in enrichment of a cell type due to loss of nonspecific populations. There are, however, circumstances under which one might want to evaluate aggregates, and for these assays, the use of singlet gating would not be applicable [41].

2.4 Exclusion Gates

2.4.1 Definition

In the most general sense, an exclusion gate is any gate that removes unwanted cell populations, before positive gating on markers of interest. Practically, this is often useful to prevent the interference of these excluded populations on the downstream analysis. Besides time and singlet gates already discussed above, two common variations of such exclusion gates are viability gates and lineage marker cocktails (often called a "dump channel"). Viability is most frequently assessed by the fact that live cells have intact membranes and thus exclude dyes that would otherwise enter the cell and bind, for example, DNA. Propidium iodide and 7-AAD are examples of such dyes. Alternatively, agents that react with free amine [24] or thiol groups can be used, taking advantage of the fact that relatively few such reactive groups are present on the cell surface, but many more are available in the interior of cells. Many dyes are available in this category, with different fluorescent properties, such that they can be combined with almost any combination of fluorochrome-labeled antibodies. A major advantage of these dyes is that they form a covalent bond with cellular proteins, and thus do not diffuse away if cells are fixed, permeabilized, or otherwise processed after application of the viability stain. For this reason, they are often referred to as "fixable" viability dyes.

2.4.2 How Data is Affected

Dead cells tend to bind antibodies and/or fluorochromes nonspecifically, and as such, interfere with the quantitation of cells identified with positive markers alone. This is particularly important when quantifying rare populations like antigen-responsive T cells (Fig. 4). In such assays, dead cells may add to the "background" or unstimulated response, as well as potentially obscuring the distinction between positive and negative populations.

A lineage cocktail uses markers for one or more cell lineages that are not of interest for downstream analysis. For example, B, T, NK, and monocyte markers are commonly combined in a lineage cocktail for the downstream gating of dendritic cells (DCs), which bear none of these markers. This exclusion improves the ability to perform positive gating on DCs (commonly gated as lineage—HLA-DR+). But other applications are also possible. Identification of antigen-specific T cells with certain reagents may be improved by first gating out monocytes and B cells, which express Fc receptors. Samples that have

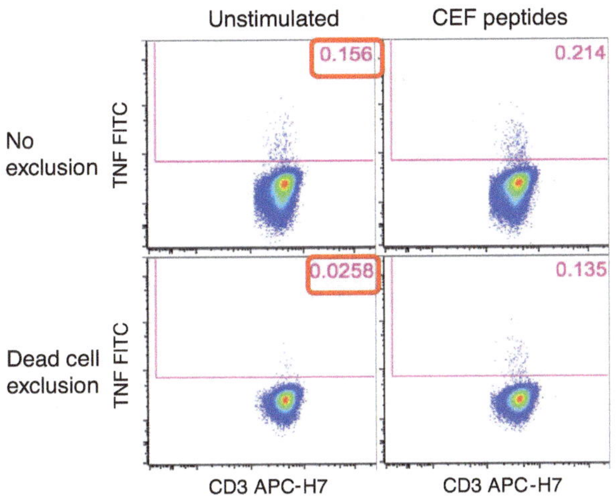

Fig. 4 Excluding dying/dead cells reduces noise. PBMCs were thawed, stimulated with either no specific antigen or CEF peptide mix (optimal 8–9mers for CMV, EBV, and influenza) in the presence of protein transport inhibitors, permeabilized, stained, and acquired. For all plots, CD3 versus TNF are shown after gating on CD8+ T cells. For this ICS assay, a positive response is defined as a response that is at least 2× background and greater than 0.10%, and values are reported after background subtraction. *Left column*: unstimulated. *Right column*: CEF stimulated. *Top row*: The background-subtracted CEF response in the absence of a dead cell exclusion gate is 0.058%, or no response. *Bottom row*: When the same FCS file as was used in the top row is analyzed using a dead cell exclusion gate, the background is reduced, and the net response is 0.109% and thus positive

erythrocyte contamination may benefit from exclusion gating on CD235a, an erythrocyte-specific marker. And use of a platelet-specific antibody like CD62P can be used to remove platelet-bound lymphocytes and thereby clean up intracellular cytokine staining of antigen-responsive T cells [42].

To some extent, lineage cocktails also help to remove dead cells via their nonspecific binding; but they are not a perfect surrogate for a viability dye, and can serve other functions as noted above. Of course, a viability dye and a lineage cocktail can also be combined in a single channel.

2.4.3 Conclusion

Exclusion gates reduce noise, thus enhancing the overall signal-to-noise. Rare populations are more accurately quantitated with exclusion of dead cells and/or irrelevant cell lineages.

2.5 Scatter Gates

2.5.1 Definition

Fluorescence flow cytometry typically measures forward and right-angle (side) scatter of the laser light used to excite fluorochromes on the particles of interest. These two parameters contain information on cell size and granularity, respectively. As such, they are frequently used as initial gates on immune cell populations, such as lymphocytes, monocytes, and granulocytes.

2.5.2 How Data is Affected

While a population such as lymphocytes may appear to have a relatively homogeneous scatter profile, there are in fact subtle differences in the scatter profile of different lymphocyte subsets. Most dramatic is the increased forward scatter of proliferating lymphocytes (blasts); a tight lymphocyte gate thus misses most of the proliferating cells (Fig. 5A). But even populations such as B cells or CD8+ T cells can have subtle shifts in their scatter relative to the rest of the lymphocytes. The result is that the placement of the lymphocyte gate can have small but potentially significant effects on the relative percentages of different subsets reported within that gate (Fig. 5B).

2.5.3 Conclusion

Scatter gates should be used with caution and checked by backgating to avoid unintended exclusion of relevant cells (recovery) and inclusion of irrelevant cells (purity). Backgating refers to the visualization of a final gated population in all prior gating steps (generally with those prior gates removed). This allows one to see the effect of each parent gate in the hierarchy, and can quickly identify parental gates that create undesirable effects (for example, a too-tight scatter gate that cuts through a specific downstream subpopulation of interest).

2.6 Specific Population Gates

2.6.1 Definition

A specific population gate defines a cell population, such as CD4+ T cells, B cells, or monocytes, focusing further subset identification

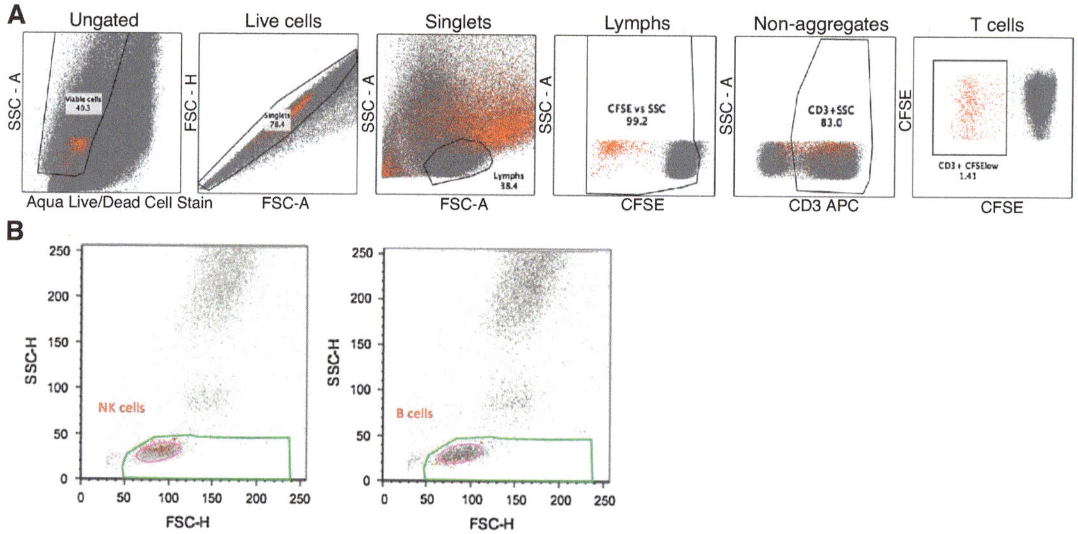

Fig. 5 Differential scatter profiles of specific lymphocyte populations. **A**, Proliferating cells. In a CFSE proliferation experiment, backgating on CFSE-low cells (red) shows their position in other gated dimensions. Note the scatter gate (third plot from left) misses almost all of the proliferating cells, which are higher in both FSC and SSC dimensions. **B**, Lymphocyte subsets. NK cells (left panel) and B cells (right panel) were gated separately using specific markers from the green, liberal scatter gate for lymphocytes and then projected back into the scatter dimensions as red dots. Note that the pink (tight) scatter gate for lymphocytes would exclude a subset of both NK and B cells because of their biased and/or diffuse scatter profile relative to most lymphocytes

within those populations. The size, placement, and order of such gates can influence the reported results in ways that may not be obvious to an inexperienced operator.

2.6.2 How Data is Affected

Knowledge of the expression properties of specific subsets of interest can be useful for setting a parental population gate. For example, CD3 may be used as a parental gate prior to analysis of T cell subsets such as γδ T cells. By examining a dot plot of CD3 vs SSC, one can identify the γδ T cells as CD3high relative to the rest of the CD3+ population (Fig. 6A). Knowing this, care should be taken not to exclude CD3high cells when enumerating γδ T cells as a subset thereof.

Another example in T cell gating concerns the simple discrimination of CD4+ and CD8+ populations. Since these are largely mutually exclusive populations, it may seem irrelevant whether one gates on CD3+CD4+ and CD3+CD8+ populations, or rather sequentially gates on CD3+ and then CD4 vs. CD8. However, the presence of a small but variable population of CD4+CD8+ T cells means that these two strategies will not give identical results. In fact,

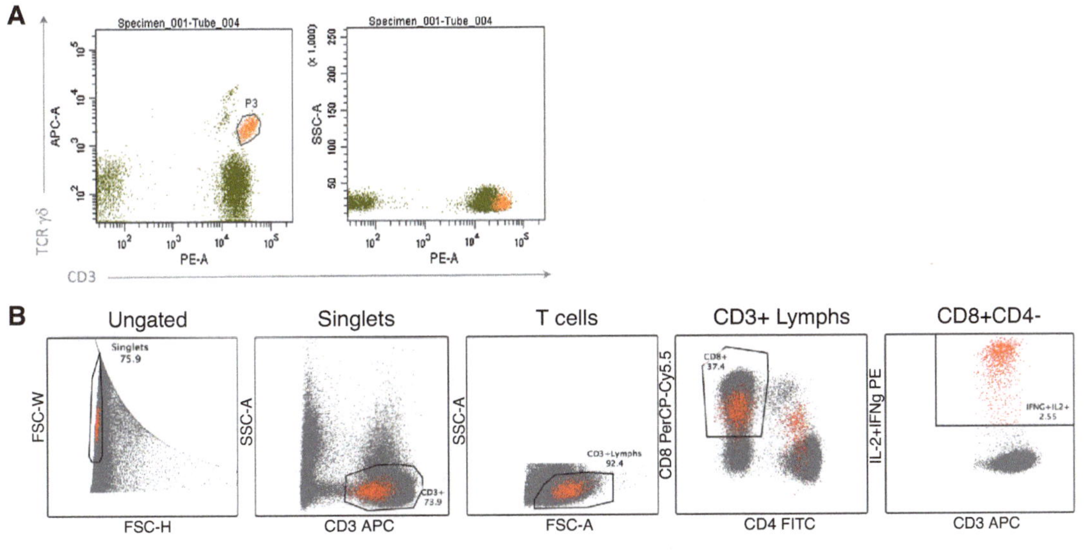

Fig. 6 **A**, γδT cells are CD3-high. Human lysed-washed blood was stained with CD3 and TCRγδ antibodies. Orange population indicates TCRγδ cells on a plot of CD3 vs. TCRγδ (left) and CD3 bright staining on the TCRγδ population depicted in CD3 vs. SSC plot (right image). **B**, Cytokine-positive CD8 T cells are CD8-low and CD3-low. Backgating of IL-2+IFNγ+ CD8 T cells (red) shows their position in all other gated dimensions. Note their shifted position in CD8 and CD3 dimensions relative to noncytokine-producing cells, and how these gates were intentionally drawn to include dim CD3+ and CD8+ cells. Drawing the cytokine gate against CD3 is also helpful to see if the cytokine-positive population is being truncated by the CD3 gating

the former strategy will double-count these CD4+CD8+ events as belonging to both populations, while the latter method allows their true discrimination.

Still another example with regard to T cell gating is the down-modulation of CD3, CD4, and CD8 on activated T cells. When performing stimulation assays, the gates for these markers need to be sufficiently "loose" to avoid cutting off potentially small subpopulations of activated (and thus responding) T cells that have partially down-modulated these markers (Fig. 6B). While intracellular staining for CD3, CD4, and CD8 minimizes the downmodulation, there is still generally a downward shift of activated cells for these markers, as seen in the figure.

Rare event gates are another source of potential variability. For example, gates on cytokine+ cells responding to specific antigen, or MHC-tetramer+ T cells, require consistent placement relative to the negative population. If the gate is drawn too close to the negative population, there will be increased and variable "background" created, from cells that are on the edge of the negative population but become counted as positive. Conversely, if the gate is drawn too far away from the negative population, dim positives will be missed. In this regard, a gating control (e.g., unstimulated cells for cytokine assays or an FMO for MHC-tetramer staining) may be useful for setting the boundary of the positive gate.

2.6.3 Conclusion

Knowledge of the staining properties of subsets within an initial population gate is important for accurate and consistent downstream enumeration. Backgating (see Figs. 5A and 6B) is a useful technique for checking all prior gates in a hierarchy, to be sure they are not unintentionally excluding populations of interest. A gating control can also be useful, particularly for rare and/or dim populations.

2.7 Data Display

2.7.1 Compensation

Definition

The premise of flow cytometry is that one marker is tagged with one fluorescent dye, and the photons generated by the dye are measured in a single detector. Spectral overlap occurs when the fluorescence emission from one dye emits photons into the fluorescence emission spectrum of a neighboring detector. In a multi-color panel, spectral overlap from multiple dyes may be observed in a single detector. Electronic subtraction is applied to correct the interference caused by the spectral overlap in each channel; this is called compensation. A compensation matrix is generated to correct the spectral overlap for each dye versus all of the other dyes in the panel, respectively. The goal of compensation is to specifically detect fluorescence from each dye in its own respective detector. Compensation does not change the raw data, it merely changes how the data are displayed. By reducing the apparent interference from neighboring fluorophores using compensation, the markers tagged with those fluorophores are displayed with more separation, enhancing the appearance of distinct populations.

How Data is Affected

During analysis, where visualizing a clear separation of populations is essential, an error in compensation may lead to an erroneous classification of cells, including false positives or false negatives (Fig. 7). It is therefore important to assess the accuracy of compensation by examining biaxial plots of all combinations of fluorescence channels before the application of gating on fluorescence channels. Typically, an N × N set of plots is generated (Fig. 8A), with appropriate transformation (see Sect. 2.7.2 below), where N is the number of fluorescence channels in the experiment. Some flow cytometry software allows the automatic creation of such a plot layout. The user can then judge the accuracy of compensation by the degree to which populations that should be orthogonal to each other for particular markers do in fact align orthogonally. Under- or over-compensation can be easily identified as an artificially curved edge, or "hook," to such populations. Correction of compensation errors should first investigate the single-color controls and gates set on positive and negative populations in these controls. Recalculation of compensation in the analysis software can then be undertaken. In the worst case, manual adjustment of compensation can be done (Fig. 8B), but this should only be undertaken by highly experienced users and as a last resort, since it is inherently biased and can change the data in unintended ways.

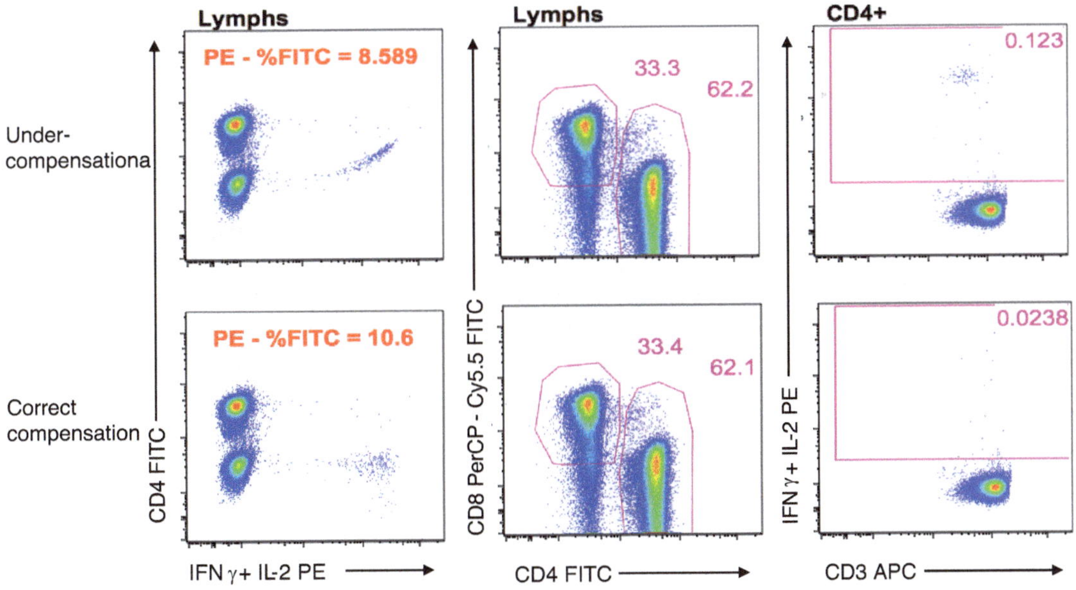

Fig. 7 Undercompensation leads to a false positive response in an ICS assay. Undercompensation between PE (IFNγ+IL-2) and FITC (CD4) created the appearance of CD4+ IFNγ+IL-2+ cells, when in fact, with correct compensation, these cells were CD4−. In this case, the operator used software to create a compensation matrix but neglected to verify the accuracy of the compensation matrix by visual inspection of the data (e.g., using an N × N plot matrix as described in the text)

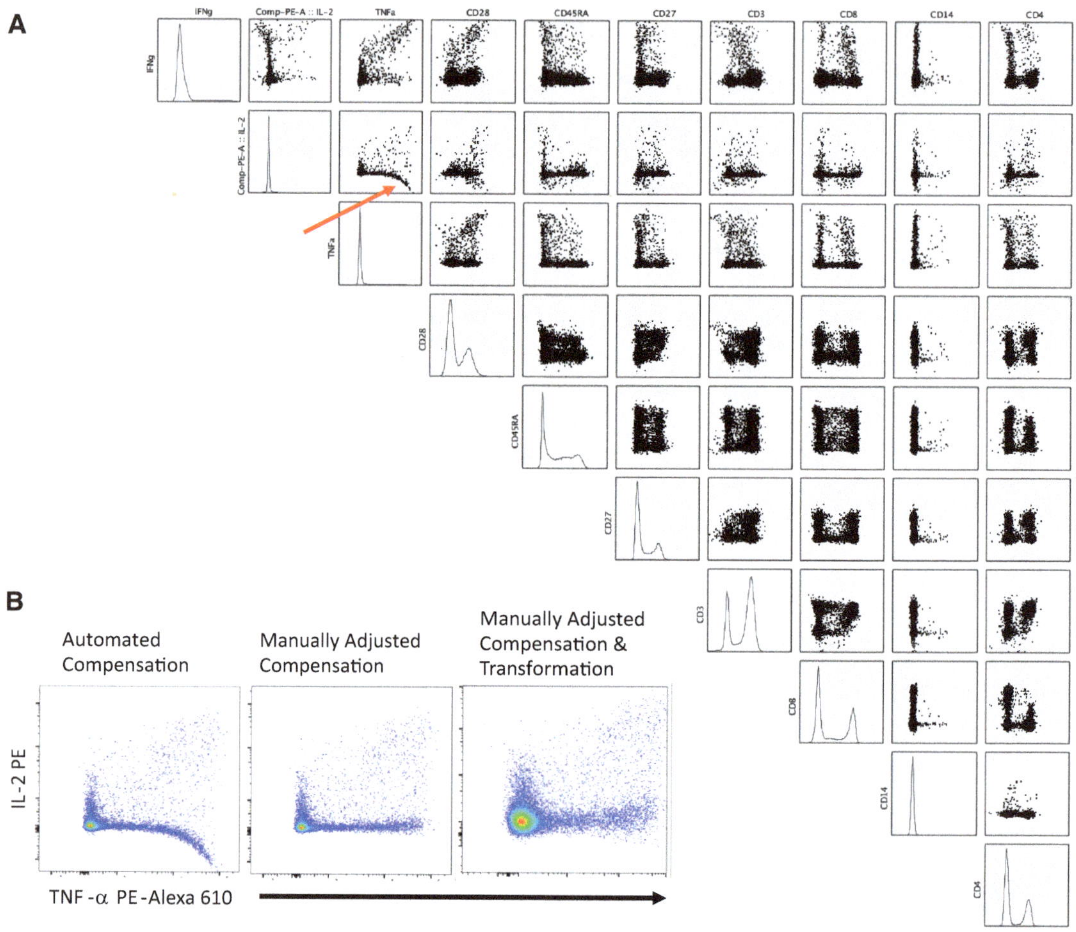

Fig. 8 **A**, N×N dot plots for compensation checking. Some software programs allow automated generation of a series of dot plots showing each parameter combination in the experiment. When viewed with a high-level gate, such as live singlets, this view can allow visualization of over- or under-compensation, as curved or "hooked" populations that should be orthogonal. The red arrow indicates apparent over-compensation. **B**, Correction of the over-compensation in IL-2 vs. TNF parameters. Middle plot shows visually corrected population alignment, while the left plot shows further scale transformation to better visualize the cells around zero. While incorrect compensation will adversely affect downstream results, manual compensation correction can affect other parameters in unintended ways. As a result, it should only be done by an expert user and as a last resort

Conclusion

It is critical to appropriately compensate cytometry data prior to analysis. To ensure proper compensation, operators should manually verify compensation for accuracy, looking for "hooks" created by over- or under-compensation, as well as sub-negative populations resulting from over-compensation with a neighboring detector. An easy way to review compensation is by visualizing a compensation layout containing N×N dot plots.

2.7.2 Transformation

Definition

The range of signals in flow cytometry can span very negative values (particularly after compensation) to very high values (10^4 or greater). Given the approximately log-normal distribution of many populations in flow data, a linear scale will severely distort populations and make inefficient use of the display space. However, a simple logarithmic scale is unable to display zero or negative values. Therefore, transformations have emerged that preserve the ability to display negative and zero values while approximating a logarithmic scale in the positive space. These have variably been referred to as logicle, biexponential, or VLog scales [35, 43, 44]. All current flow cytometry software offers a version of such transformation, often with automated application in a data-dependent fashion.

How Data is Affected

Choosing an optimal transformation is important for a number of reasons. First, visualizing the effects of compensation requires a scale that smoothly transitions through zero and displays negative data. Under- or over-compensation can then be accurately judged based on how orthogonally the edges of populations align in the negative to low positive range. Second, the choice of transformation is important to the efficient use of the graphical display space. If a majority of the axis length is devoted to high values (e.g., $>10^2$), and if the data consists entirely of dim events, then the transformation is not appropriate. Conversely, if a majority of the axis length is devoted to negative values, and high positive populations are compressed and difficult to distinguish, the transformation is not appropriate. Finally, an inappropriate transformation can create artifacts such as "splitting" of a negative population into what appears to be two populations ("valley artifact"). This particular artifact is a result of displaying too many decades of logarithmic scaling at the low end of the log scale (usually below 10 or 1). Since judgment of positive and negative expression is often a goal of flow cytometry, such artifacts can clearly be an impediment to interpretation.

Conclusion

Choosing the right transformation is critical for accurate visualization of compensation, for efficient use of display space, and for complete visualization of negative populations without introducing the artificial appearance of breaks in the data.

2.8 Analytical Controls

2.8.1 Unstained Cells

An unstained cell sample is often used for the calculation of compensation, relative to single positive samples for each fluorochrome. However, this is not strictly required for performing compensation, if negative populations are included in each single-color control. An unstained control is, however, still useful for judging the PMT voltage gain in each sample, since it represents the position of the most negative possible population. It is generally not useful for determining gate placement on stained samples, since it doesn't account for either spillover from other fluorescent channels or nonspecific binding in the channel of interest.

2.8.2 Isotype Controls

Isotype controls are useful when deciding gate placement in flow cytometry. These controls are made of the same isotype as the antibody of interest and have the same fluorescent dye attached to them. In specific instances, such as when monocytes are being stained with an antibody tagged with tandem dyes such as PE-Cy7, APC-Cy7, APC-Fire 750, APC-H7, or PE-CF594, use of an isotype control antibody is extremely helpful. Monocytes tend to nonspecifically bind to dyes such as Cy7, Fire 750, and H7. Figure 9 shows the effect

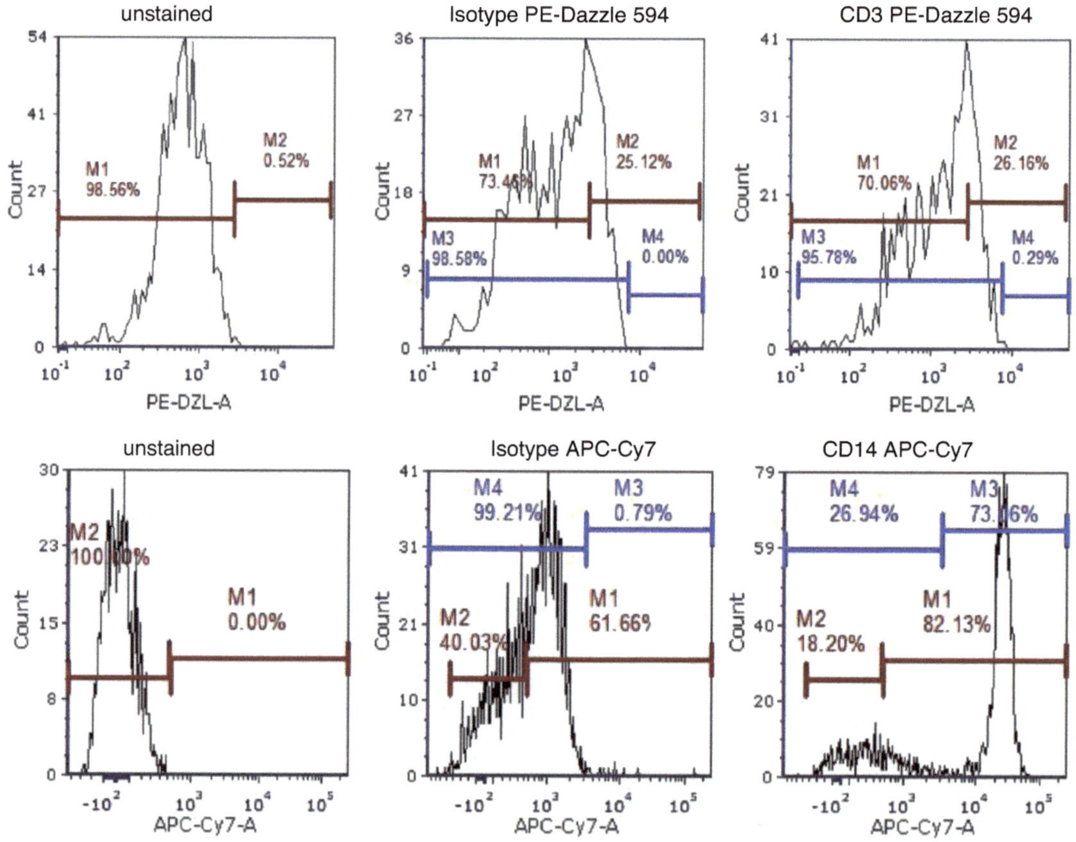

Fig. 9 Tandem dyes nonspecifically stain monocytes, and thus, isotype should be used to set marker gate/s appropriately. Top row: human lysed washed blood was not stained or stained with isotype control PE/Dazzle 594 or CD3 PE/Dazzle 594. Histograms are gated on FSC and SSC Singlets, followed by monocyte gate (scatter-based). Red marker gates (M1 and M2) indicate gate placement using unstained cells (leftmost panel) and blue marker gates (M3 and M4) depict the frequency of CD3-positive cells based on gate placement with isotype control (middle panel). If care is not taken in interpreting data based on unstained cells, then users may report monocytes to be CD3+. Bottom row: a similar case using APC/Cy7. Here, we show that gates based on the isotype control are more appropriate for the positive monocyte stain, CD14

on tandem dye binding to isotype controls. Some manufacturers formulate their antibodies with reagents to reduce this background, and others may sell buffers to overcome these issues (for example, Monocyte Blocker from BioLegend).

2.8.3 Fluorescence Minus One Controls (FMOs)

In the absence of situations like those shown in Fig. 9, the major background signals in high-dimensional flow cytometry tend to come from spillover rather than nonspecific binding. As such, an FMO control, which leaves out a single antibody conjugate but maintains all other stains in the panel, is often more useful for setting gates than an isotype control in the absence of other stains.

2.8.4 Combined Isotype and FMO Controls

A control that leaves out one antibody conjugate but replaces it with an isotype control antibody can be considered a "combined" control. With such a control, one attempts to combine the benefits of FMO (accounting for spillover) and isotype controls (accounting for nonspecific binding). While this, in theory, may be the "best of both worlds," it is worth mentioning that there are still drawbacks to the use of isotype controls that are not alleviated by combining it with FMO. Specifically, the particular isotype control antibody may or may not have similar nonspecific binding properties as the test antibody [45]; minimally, the protein concentration of the isotype control should be matched to the test antibody to at least approximate those properties.

2.8.5 Biological Controls, e.g., Unstimulated Cells

For functional readouts like cytokines in intracellular cytokine staining (ICS), the best gating control may be an internal biological control, e.g., unstimulated cells from the same donor. Even so, it can be difficult to decide exactly where to place the cytokine or functional marker gate relative to the negative population. Placing the gate too close to the negative population increases the background of false positive events, but placing it too far away can reduce the readout of true positives. This becomes especially critical when dealing with rare events such as antigen-specific responses. Guidelines for this have been published, both for manual gating [8] and for gate selection based on a statistical measure [46]. For manual gating, it is generally advised to set the gate above the "halo" of cells at the edge of the negative population, since increased background, which may be sample-specific, will be the most detrimental to the detection of low-level responses.

Biological controls can also be used as a way to check staining across batches of samples. When control marker staining is shifted in one batch vs. others, there may be a technical error in the staining protocol. This can indicate that a specific batch's data may need to be interpreted with increased caution, if not discarded (Fig. 10).

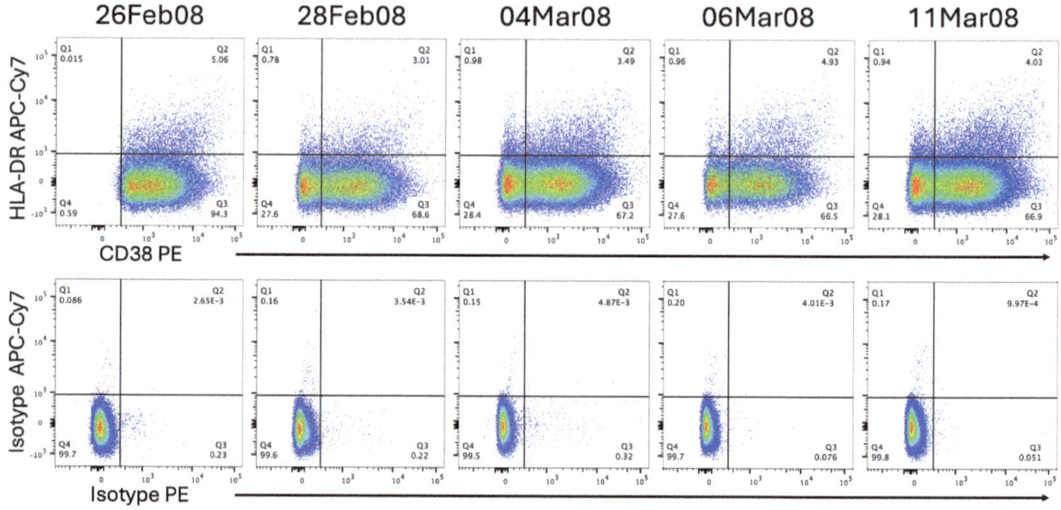

Fig. 10 A biological control for batch-specific quality control and gate setting. Thawed PBMCs from a leukopak batch of control cells are stained with CD38 and HLA-DR (top row), with an isotype control (bottom row) used to set the gate. This becomes erroneous in the left-most batch (26Feb08), for which CD38 staining is artifactually shifted to the right. Data from this batch needs to be reviewed and interpreted with caution, if not discarded

2.9 Flow Cytometry Units

It is worth noting that the readout of a gated population can be expressed in different units. Most gating software by default reports percent-of-parent statistics, i.e., the percentage of cells in the parent population that are contained in the gate of interest. However, this may not always be the best readout to report.

Consider the case of lymphocytes, which have major subpopulations of B, T, and NK cells. If an individual's T cell count increases, the percentage of lymphocytes that are B and NK cells will necessarily decrease, even if their absolute frequency is unchanged. In order to avoid the effect of such reciprocal relationships, one can report absolute cell counts (per microliter of blood). However, this requires a volumetric measurement or use of counting beads. It can be done most accurately by single-platform methods (where the counting beads or volumetric measurement is done in the same flow cytometry assay as the cell population staining). Alternatively, a dual platform method can be used, relating the results of a counting assay to the flow cytometry staining. For example, total lymphocyte count can be measured by a complete blood count with differential (CBCD). The total lymphocyte count can then be multiplied by the percentage of lymphocytes gated as B, T, and NK cells, to derive absolute counts of these three populations. Because of variations in the way lymphocytes are defined in the CBCD versus flow cytometry platform, however, this will necessarily be less accurate than single-platform counting. Still, both methods can be made relatively reproducible [47].

Short of reporting absolute counts, one can also report population percentages, but using a higher-level ancestor gate. For example, each gated population in a PBMC sample could be reported as a percent of all live cells (after exclusion of aggregates). While this is not the same as absolute counting, it does diminish the effect of changes in reciprocal populations. It is also easily done using the reported percent-of-parent statistics by simply multiplying percentages of each ancestor gate up to the top-level gate being reported. Some software programs can provide such reporting automatically by choosing the desired ancestral gate from which to report percentages.

Finally, it is possible to report not the frequency or percentage of cells in a gate, but rather their mean or median fluorescence intensity (MFI). This is particularly useful when reporting data on markers that may not have distinct positive and negative populations, but whose distribution represents a continuum ("smear") or for which shifts in expression level are the biologically important readout. Examples include CD80 and CD86 expression on dendritic cells, whose intensity relates to cell maturation/activation. Rather than measuring the percent of "positive" cells using a potentially arbitrary cutoff, it makes more sense to report MFI of the entire population. However, when comparing MFI across experiments, day-to-day variations in staining and instrument setup will confound such results. Calibration of relative MFI to measured equivalent soluble fluorochrome (MESF), using beads of standardized intensity, is recommended in such cases [48].

2.10 Special Considerations

Flow cytometry experiments and, therefore, associated results can be affected by artifacts in the assay, which are not considered or known prior to performance of a particular assay. These artifacts can include samples with too few events or difficult-to-discern rare populations, or shifts in a cell population from one sample to another. Visual inspection of gates on all samples is therefore critical before finalizing the gating. In determining whether to move a gate for a specific sample, one should attempt to maintain a fair and unbiased approach, i.e., keeping gates as similar as possible except where clear shifts in a population warrant changing the gate. Samples to be directly compared, such as stimulated and unstimulated conditions from the same donor, need to have the exact same gating applied.

When gating a large study, or when multiple operators and/or sites are involved, it becomes necessary to create a gating template. This is a document with default gates provided; that is the starting point for analysis of each batch of samples. It represents not only the hierarchy of all gates to be applied but also a "best guess" of where those gates should lie. Movement of gates from the template position should only be undertaken when necessary, and with the caveats described above in mind. In general, central analysis by a single operator (or at least verified by a single operator) is more consistent than separate analysis by different individuals or sites [11, 49].

Some types of artifacts can be anticipated and possibly avoided, including many of the factors listed in Table 1. Some variables that should be controlled to avoid excessive population shifts include: reagent lot control, consistent washing between wells or tubes, and avoiding degradation of problematic tandem dyes that can be idiosyncratic between samples. It is, of course, always preferable to avoid artifacts that would complicate gating in the first place, but this may not always be perfect. An example of a "preventable" population shift is that caused by brilliant dye interaction, as shown in Fig. 11.

Finally, a piece of idiosyncratic advice regarding gate placement relative to x- and y-axes. Even with appropriate transformation and data display, there are almost always events that are below the axis for negative populations. As such, gates on the negative population need to be placed ***over*** the relevant axis to include events that are below that axis. Depending on the plot size and visual resolution, this difference may be insidious; two gates may look essentially identical, but one includes events below the axis and one does not. Experienced cytometrists, therefore, typically draw a gate and then drag it well beyond the axis in order to avoid this issue.

3 Discussion/Conclusions

In research and clinical trial settings, the application of gates to flow cytometry data has been done largely in the absence of wide-ranging consensus or written guidelines. This is surprising given the pervasiveness of flow cytometry in immunology research and immune monitoring settings. The guidelines set forth here were initially drafted by a small group of experienced flow cytometrists from academic, industrial, and government labs. They were further vetted

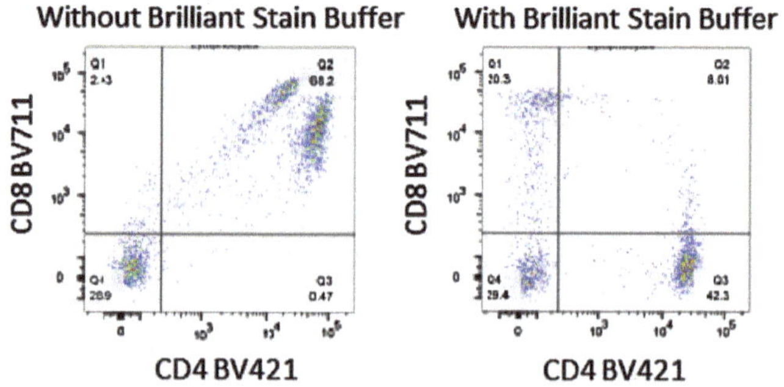

Fig. 11 Interaction between fluorophores may affect populations. Human lysed whole blood was stained with CD4 Brilliant Violet 421™ and CD8 Brilliant Violet 711™ in the absence (left) or presence (right) of Brilliant Stain Buffer™. Note the loss of mutually exclusive expression of CD4 and CD8 in the left plot

by presentation at the Federation of Clinical Immunology Societies (FOCIS) 2017 and 2018 annual meetings. The result is an attempt at a consensus document on the highest-level principles for gating flow cytometry data in immunological assays.

The guidelines presented here mainly focus on types of gates that are (or should be) common to most immunological assays. These include acquisition thresholds (or triggers), time gates, singlet gates, exclusion gates, and scatter gates. Examples of how the definition of these gates can affect downstream data highlight their general usefulness. We also discuss specific population gates, which tend to be more assay-centric, for example, T cell and rare event gates in the setting of in vitro-stimulated and/or antigen-specific T cell assays. Finally, we discuss the quality control of gating. This can include the use of "back-gating" to identify the effect of each gate in a gating hierarchy, allowing for optimization of the recovery of a desired population. It can also include the review of compensation for artifacts that cause downstream gates to report inaccurate results. Finally, the application of gates to multiple conditions and donors is discussed, including the need to visually inspect all files, creating gates that can be applied across samples while still allowing for donor-specific corrections if needed.

Much of the advice presented here can be broadly summarized as follows: (1) *All relevant parameters in the data file should be used to most purely and completely define populations of interest.* These parameters include time, which can give clues to artifacts such as clogs or bubbles in the data acquisition. They also include the height, area, and/or width of the light scatter parameters, which can be used in combination to exclude cell aggregates. Expression of one or more parameters may be used to exclude irrelevant events (exclusion gating), in addition to parameters used to include relevant events. And backgating can be useful to judge the adequacy of individual gates in a hierarchy, including light scatter gates, with regard to their unbiased selection of the population of interest. Importantly, the fact that a parameter is not used in the gating scheme does not mean that it may not affect the gated results, so visual inspection of all parameters in the data is recommended.

(2) *Once defined for a single sample, gates should be checked to ensure fair application across all samples.* This may require adjusting gates between donors, but it should generally not entail changes in the gate among samples or conditions to be compared within the same donor. Attempts should be made to avoid gates that slice arbitrarily through an obvious cluster of cells, including cases where part of a cell cluster lies below the axis of a histogram or dot plot. In some instances, an otherwise homogeneous cell population will display a continuous distribution of intensity for a given parameter. It may then be more sensible to report the mean (or median) intensity of

the population, rather than arbitrarily gating a positive and negative subset for that parameter and reporting percent positive or negative.

The eventual goal of these guidelines is not only accuracy in defining populations of interest, but also reproducibility between experiments, between operators, and between donors and cohorts. This is particularly important in longitudinal and/or multi-site studies. In these cases, it is best to have all gating verified by the same personnel, even if initially performed by different personnel. Further, it is advisable to continually refer to historically gated data to ensure reproducible treatment relative to contemporary samples for studies that are analyzed over a period of time.

In this document, we have tried to illustrate the most overarching principles that apply to gating flow cytometry data from immunological assays. There may be exceptions to these principles, but they are likely to be few. There are also undoubtedly situations where one can avoid some of these recommendations without peril. But it is also likely that following most or all of the recommendations increases the confidence that artifacts and false conclusions will be avoided. Furthermore, there are certainly additional and more detailed recommendations that could be added to these guidelines. But with increasingly detailed recommendations comes the likelihood that they will not apply as broadly across different situations and assays. So we have mostly avoided too much specificity in these guidelines, in favor of exposing the high-level concepts behind the gating strategies discussed. We hope that the wide dispersion of these concepts leads to an overall higher level of accuracy and reproducibility of flow cytometry gating.

Acknowledgments

The authors thank the U.S. National Heart, Lung, and Blood Institute (NHLBI), External Quality Assurance Program Oversight Laboratory (EQAPOL) consortium, Duke Immune Profiling Core, and Jennifer Enzor for providing data and analysis examples, and the FOCIS Immunophenotyping Course participants for helpful suggestions and vetting of these guidelines.

References

1. (1992) Guidelines for the performance of CD4+ T-cell determinations in persons with human immunodeficiency virus infection. MMWR Recomm Rep 41:1–17
2. (1994) 1994 revised guidelines for the performance of CD4+ T-cell determinations in persons with human immunodeficiency virus (HIV) infections. Centers for Disease Control and Prevention. MMWR Recomm Rep 43:1–21
3. (1997) 1997 revised guidelines for performing CD4+ T-cell determinations in persons infected with human immunodeficiency virus (HIV). Centers for Disease Control and Prevention. MMWR Recomm Rep 46:1–29
4. Mandy FF, Nicholson JKA, McDougal JS et al (2003) Guidelines for performing single-platform absolute CD4+ T-cell determinations with CD45 gating for persons infected with

human immunodeficiency virus. Centers for Disease Control and Prevention. MMWR Recomm Rep 52:1–13
5. Health OW (2007) Laboratory guidelines for enumerating CD4 T lymphocytes in the context of HIV/AIDS. World Health Organization Regional Office for South-East Asia, New Delhi
6. Sutherland DR, Anderson L, Keeney M et al (1996) The ISHAGE guidelines for CD34+ cell determination by flow cytometry. International Society of Hematotherapy and Graft Engineering. J Hematother 5:213–226
7. Illingworth A, Marinov I, Sutherland DR et al (2018) ICCS/ESCCA consensus guidelines to detect GPI-deficient cells in paroxysmal nocturnal hemoglobinuria (PNH) and related disorders part 3 – data analysis, reporting and case studies. Cytometry B Clin Cytom 94:49–66
8. McNeil LK, Price L, Britten CM et al (2013) A harmonized approach to intracellular cytokine staining gating: Results from an international multiconsortia proficiency panel conducted by the Cancer Immunotherapy Consortium (CIC/CRI). Cytometry A 83:728–738
9. Britten CM, Janetzki S, Ben-Porat L et al (2009) Harmonization guidelines for HLA-peptide multimer assays derived from results of a large scale international proficiency panel of the Cancer Vaccine Consortium. Cancer Immunol Immunother 58:1701–1713
10. Nomura L, Maino VC, Maecker HT (2008) Standardization and optimization of multiparameter intracellular cytokine staining. Cytometry A 73:984–991
11. Finak G, Langweiler M, Jaimes M et al (2016) Standardizing flow cytometry immunophenotyping analysis from the Human ImmunoPhenotyping Consortium. Sci Rep 6:20686
12. Park LM, Lannigan J, Jaimes MC (2020) OMIP-069: forty-color full spectrum flow cytometry panel for deep immunophenotyping of major cell subsets in human peripheral blood. Cytometry A 97:1044–1051
13. Sanders CK, Mourant JR (2013) Advantages of full spectrum flow cytometry. J Biomed Opt 18:037004
14. Tanner SD, Baranov VI, Ornatsky OI et al (2013) An introduction to mass cytometry: fundamentals and applications. Cancer Immunol Immunother 62:955–965
15. Olsen LR, Leipold MD, Pedersen CB et al (2019) The anatomy of single cell mass cytometry data. Cytometry A 95:156–172
16. Verschoor CP, Lelic A, Bramson JL et al (2015) An introduction to automated flow cytometry gating tools and their implementation. Front Immunol 6:380
17. Mair F, Hartmann FJ, Mrdjen D et al (2016) The end of gating? An introduction to automated analysis of high dimensional cytometry data. Eur J Immunol 46:34–43
18. Amir ED, Davis KL, Tadmor MD et al (2013) viSNE enables visualization of high dimensional single-cell data and reveals phenotypic heterogeneity of leukemia. Nat Biotechnol 31:545–552
19. Becht E, McInnes L, Healy J et al (2018) Dimensionality reduction for visualizing single-cell data using UMAP. Nat Biotechnol 37:38–44
20. Van Gassen S, Callebaut B, Van Helden MJ et al (2015) FlowSOM: using self-organizing maps for visualization and interpretation of cytometry data. Cytometry A 87:636–645
21. Burel JG, Qian Y, Lindestam Arlehamn C et al (2017) An integrated workflow to assess technical and biological variability of cell population frequencies in human peripheral blood by flow cytometry. J Immunol 198:1748–1758
22. Disis ML, Rosa C, dela, Goodell V et al (2006) Maximizing the retention of antigen specific lymphocyte function after cryopreservation. J Immunol Methods 308:13–18
23. Posevitz-Fejfár A, Posevitz V, Gross CC et al (2014) Effects of blood transportation on human peripheral mononuclear cell yield, phenotype and function: implications for immune cell biobanking. PLoS ONE 9:e115920
24. Perfetto SP, Chattopadhyay PK, Lamoreaux L et al (2006) Amine reactive dyes: an effective tool to discriminate live and dead cells in polychromatic flow cytometry. J Immunol Methods 313:199–208
25. Shankey TV, Rabinovitch PS, Bagwell B et al (1993) Guidelines for implementation of clinical DNA cytometry. International Society for Analytical Cytology. Cytometry 14:472–477
26. Perfetto SP, Roederer M (2007) Increased immunofluorescence sensitivity using 532 nm laser excitation. Cytometry A 71:73–79
27. Maecker HT, Trotter J (2006) Flow cytometry controls, instrument setup, and the determination of positivity. Cytometry A 69:1037–1042
28. Watson JV (1987) Time, a quality-control parameter in flow cytometry. Cytometry 8:646–649
29. Fletez-Brant K, Špidlen J, Brinkman RR et al (2016) flowClean: automated identification and removal of fluorescence anomalies in flow cytometry data. Cytometry A 89:461–471
30. Maecker HT, McCoy JP, Nussenblatt R (2012) Standardizing immunophenotyping for the human immunology project. Nat Rev Immunol 12:191–200
31. Lamoreaux L, Roederer M, Koup R (2006) Intracellular cytokine optimization and standard operating procedure. Nat Protoc 1:1507–1516

32. Maecker HT, Frey T, Nomura LE et al (2004) Selecting fluorochrome conjugates for maximum sensitivity. Cytometry A 62:169–173
33. Roederer M (2001) Spectral compensation for flow cytometry: visualization artifacts, limitations, and caveats. Cytometry 45:194–205
34. Chevrier S, Crowell HL, Zanotelli VRT et al (2018) Compensation of signal spillover in suspension and imaging mass cytometry. Cell Syst 6:612–620.e5
35. Finak G, Perez J-M, Weng A et al (2010) Optimizing transformations for automated, high throughput analysis of flow cytometry data. BMC Bioinform 11: 546
36. Novo D, Wood J (2008) Flow cytometry histograms: transformations, resolution, and display. Cytometry A 73:685–692
37. Martin JC, Swartzendruber DE (1980) Time: a new parameter for kinetic measurements in flow cytometry. Science 207:199–201
38. Monaco G, Chen H, Poidinger M et al (2016) flowAI: automatic and interactive anomaly discerning tools for flow cytometry data. Bioinformatics 32:2473–2480
39. Hoffman RA (2009) Pulse width for particle sizing. Curr Protoc Cytom Chapter 1:Unit 1.23
40. Wersto RP, Chrest FJ, Leary JF et al (2001) Doublet discrimination in DNA cell-cycle analysis. Cytometry 46:296–306
41. Furman MI, Barnard MR, Krueger LA et al (2001) Circulating monocyte-platelet aggregates are an early marker of acute myocardial infarction. J Am Coll Cardiol 38:1002–1006
42. Nomura LE, Walker JM, Maecker HT (2000) Optimization of whole blood antigen-specific cytokine assays for CD4(+) T cells. Cytometry 40:60–68
43. Parks DR, Roederer M, Moore WA (2006) A new "Logicle" display method avoids deceptive effects of logarithmic scaling for low signals and compensated data. Cytometry A 69:541–551
44. Bagwell CB, Hill BL, Herbert DJ et al (2016) Sometimes simpler is better: VLog, a general but easy-to-implement log-like transform for cytometry. Cytometry A 89:1097–1105
45. Andersen MN, Al-Karradi SNH, Kragstrup TW et al (2016) Elimination of erroneous results in flow cytometry caused by antibody binding to Fc receptors on human monocytes and macrophages. Cytometry A 89:1001–1009
46. Richards AJ, Staats J, Enzor J et al (2014) Setting objective thresholds for rare event detection in flow cytometry. J Immunol Methods 409:54–61
47. Hultin LE, Chow M, Jamieson BD et al (2010) Comparison of interlaboratory variation in absolute T-cell counts by single-platform and optimized dual-platform methods. Cytometry B Clin Cytom 78:194–200
48. Vogt RF, Cross GD, Henderson LO et al (1989) Model system evaluating fluorescein-labeled microbeads as internal standards to calibrate fluorescence intensity on flow cytometers. Cytometry 10:294–302
49. Maecker HT, Rinfret A, D'Souza P et al (2005) Standardization of cytokine flow cytometry assays. BMC Immunol 6:13

Check for updates

Chapter 15

Quality Control of Immunophenotyping

Bruce Greig

Abstract

Applications of flow cytometry and immunophenotyping offer precise and accurate means for providing information used to both diagnose and monitor disease; they serve as a standard platform for many research endeavors that study discrete populations of biological entities. The proper use of this highly sophisticated technology requires daily and ongoing monitoring of both the instrument and the methodology. Best practices for this begin with quality control (QC) procedures designed to set up and monitor the reagents used, the instrument performance, and the results to ensure that they are working properly both on the day of use and over time. If the results of those QC procedures are outside of acceptable, then recording the corrective action taken must also be included in the quality control records. Quality assurance (QA) is a way to know that the three phases of testing, namely, pre-analytic, analytic, and post-analytic procedures, are being followed. This chapter describes the procedures used to assess quality control as it pertains to flow cytometry and immunophenotyping in all three phases of testing.

Keywords Quality control, Quality assurance, Flow cytometry, Levy-Jennings, Immunophenotyping, Monoclonal antibodies, Immunofluorescence, Instrument optimization, Color compensation, Quantitative controls, Qualitative controls, FMO controls, Proficiency testing, Method validation, Accuracy, Precision, Key performance indicators

1 Introduction

Quality Control is defined as a set of procedures performed each day for the continuous and immediate monitoring of laboratory work to confirm whether the results are reliable and can be used to diagnose, monitor, or ascertain the presence of disease in clinical applications, as well as produce results that are accurate, precise, and reproducible in a research or clinical setting.

Why perform quality control in immunophenotyping? The reasons are as follows:

1. Ensures that all measurands meet a standard
2. It monitors day-to-day consistency and reliability
3. It provides confidence in results

Angélique Biancotto and Pradeep K. Dagur (eds.), *Immunophenotyping*, Methods in Molecular Biology, vol. 3087, https://doi.org/10.1007/978-1-0716-5607-5_15,

4. It ensures reproducibility
5. It addresses regulatory requirements

Quality control (QC) and quality assurance (QA) as they pertain to immunophenotyping are a broad topic that covers many different components, each of which is used so that the results of a flow cytometry assay are reliable. QC is what you do to ensure your experiment or assay will do what it is supposed to do at each step, from setup to acquisition to analysis to interpretation to the final answer. QA is the ongoing monitoring of all quality control procedures that reflect the overall acceptability of the results. QC procedures should demonstrate that everything being used to produce results is in proper working order prior to running patient samples in the clinical lab or biological specimens used in a research lab.

The results and records of the QC procedures on both the instrument and reagents used by the laboratory should be readily available for inspection purposes [1, 2] as well as provide a means for long-term monitoring to detect trends and exceptions. Flow cytometry laboratories that perform clinical testing must meet and maintain several regulatory requirements found on the *College of American Pathologists (CAP) Flow Cytometry Inspection Checklist*. Fig. 1 [3] is a partial list of these requirements.

Flow cytometers used in immunophenotyping perform several different measurements and assurance of how well it is working is tantamount to being confident in the results it produces. Once the samples have been acquired, additional analytic steps are necessary

FLO.23275	**QC – Antibody Validation**
FLO.23737	**QC – Reagents/ Stain**
FLO.23800	**QC - Single/Dual Platform Tests**
FLO.23925	**QC Range Verification**
FLO.24230	**QC Corrective Action**
FLO.24250	**QC Testing of Controls**
FLO.24300	**QC Confirmation of Acceptability**
FLO.24475	**Monthly QC Review**
FLO.24650	**Comparability of Instrument/Method**
FLO.25100	**Instrument Function Checks**
FLO.25150	**Optical Alignment**
FLO.25250	**Instrument/Equipment Service Records**
FLO.25300	**Instrument/Equipment Maintenance Review**
FLO.30250	**Fluorescent Beads**
FLO.30260	**Color Compensation Settings**
FLO.30270	**Laser Current**
FLO.30280	**Calibration/Laser Performance**

Fig. 1 Quality control requirements of the CAP flow cytometry inspection checklist [3]

to ensure the interpretation of the results is accurate. All these considerations can be put into three categories:

1. Pre-analytic quality control: quality control as it pertains to the procedures taken to prepare an assay, including specimen requirements and handling, as well as quality assessment of all reagents used to set up the assay.
2. Analytic quality control: quality control as it pertains to the test systems, including instrumentation validation and verification of performance, maintenance with function checks and controls procedures with numbers and types of control materials, and frequency of monitoring to ensure everything is operating properly.
3. Post-analytic quality control: quality control as it pertains to the results generated by an immunophenotyping assay that ensures that they are accurate and precise. Ongoing quality assessment by subscribing to external proficiency testing surveys as well as using key performance indicators (KPI) is considered a best practice of post-analytic quality control.

Each of these steps will be described in this chapter, along with diagrams, and references, and Notes for procedures and further explanation.

2 Materials

1. QC beads, generally sourced from the instrument manufacturer or other commercial vendor
   a. Alignment/performance check beads, sometimes referred to as PQC beads
   b. Compensation setup beads
2. PBS Buffer with or without albumin (see Notes section for full recipe instructions)
3. Monoclonal antibodies (various vendors)
4. 7-AAD viability dye (see Notes section for full recipe instructions)
5. Ammonium chloride RBC lysing reagent (see Notes section for full recipe instructions)
6. Quantitative commercial controls used in immunophenotyping procedures
   a. Normal control: at or within the clinical application range of normal values
   b. Abnormal control: either below or above the clinical application range of normal values

7. Qualitative controls: for tests that use descriptive results, i.e., "positive," "negative," "present," "absent" and/or "not detected" or "below detectable limits"
8. Miscellaneous laboratory equipment
   a. Refrigerators and Freezers
   b. Centrifuges
   c. Water baths and incubators
   d. Biological safety hoods

3 Methods

3.1 Pre-analytic Quality Control Procedures

3.1.1 The Procedure Manual

The procedure manual, whether hard copy or electronic, is a fundamental laboratory document that is the first pre-analytic step in Quality Control of immunophenotyping. It not only provides directions for the procedures used in the laboratory but should also include any information required to perform all quality assurance steps of immunophenotyping experiment. The procedures should be organized in such a way that they can be easily followed by the laboratory personnel. There are many formats found for building a procedure and a procedure manual. Based on the requirements found on the College of American Pathologists (CAP) All Common Checklist [4], the following elements should be included in a test procedure:

- The principle and clinical significance
- Requirements for specimen collection, handling storage, and preservation, labeling, transportation, processing, and criteria for specimen acceptability and rejection
- Step-by-step performance of the procedure, including test calculations and interpretation of results
- Preparation of solutions, controls, reagents, and other materials used in testing
- The analytic measurement range for test results, if applicable
- Control procedures
- Corrective action to take when control results fail to meet the laboratory's criteria for acceptability
- Limitations in the test methodology, including interfering substances
- Reference intervals (normal values)
- Critical/life-threatening test results, if applicable

- Pertinent literature references
- The laboratory's system for entering results in the patient record and reporting patient results, including, when appropriate, the procedure for reporting critical results
- A description of the course of action to take if a system becomes inoperable

3.1.2 Specimen Requirements

Proper specimen collection and timely processing are critical pre-analytical steps required for performing immunophenotyping procedures and are therefore considered as to be part of the quality assurance aspect of quality control. Specimens used in flow cytometry testing can be obtained from several sources, and in the clinical laboratory, these may include peripheral blood, bone marrow aspirates, solid tissue biopsies, fine needle aspirates, and several types of body fluids such as CSF, pleural fluid, synovial fluid, and vitreous fluid. In a research setting, this list may also include frozen tissue, animal samples, plant samples, seawater, and a variety of other sources. Each specimen has specific handling requirements that should be fully described in the procedures of the laboratory. The following list includes the most widely used specimen types and their specific collection requirements for immunophenotyping.

3.1.3 Peripheral Blood and Bone Marrow Aspirates

All blood or bone marrow samples must be in an anticoagulant if the assay is for testing either white blood cells (WBCs) and/or red blood cells (RBCs) [5]. The most used anticoagulants for immunophenotyping include EDTA, (Sodium) Heparin, and ACD (acid citrate dextrose). All are designed to prevent clotting while preserving WBCs; however, there are some advantages and disadvantages to each, and they are discussed below. The time between collection and testing can be critical, depending on the anticoagulant used. For the times listed in this section, the laboratory should do its own validation of each anticoagulant to determine what is acceptable.

1. Ethylenediaminetetraacetic acid (EDTA)preserves lymphocyte viability up to 72 h if refrigerated, 48 h at room temperature. This is the most common anticoagulant used in the clinical laboratory for testing lymphocyte subsets, since it can also be used for performing white blood cell counts and differentials necessary for calculating the absolute lymphocyte counts. Granulocytes, however, may begin to deteriorate in less than 8 h due to the chelation of the calcium in the plasma by the EDTA, which makes this anticoagulant less desirable for flow assays that focus on myeloid lineage populations. These tubes usually have a purple stopper [6].
2. Sodium heparin prevents the formation of fibrin from fibrinogen and thrombin activation, thus preventing clotting of the sample. It may cause some RBCs to clump and, in some cases,

impart a blue background on Wright–Giemsa stains, so it is not suitable for hematology assays [7]; however, it works quite well on bone marrow aspirates and is the anticoagulant of choice for these samples. Samples in heparin are generally good for up to 48 h if refrigerated. These tubes usually have a light green stopper.

3. Acid Citrate Dextrose (ACD) is the same anticoagulant used to collect whole blood donations for transfusion. It has many qualities that make it highly suitable for longer and more stabilized specimens and is more conducive to maintaining cell integrity and viability than either EDTA or heparin. However, it is a liquid additive and thus will make absolute counts inaccurate since it changes the volume of whatever is added to it, so it should not be used for either flow cytometry single-platform absolute counts or hematology quantitative results. Samples in ACD are good for up to 48 h at room temperature and up to 4 days if refrigerated, making it much preferred in laboratories that are not open all day, every week. These tubes usually have a pale-yellow stopper.

3.1.4 Solid Tissue Biopsies and Fine Needle Aspirates

Tissue samples submitted for flow cytometry testing may be from many different sources. In a Hematopathology setting, this would likely include a lymph node biopsy, soft tissue from an organ such as the gut, and even skin punches submitted by a dermatologist. If the sample is submitted from the same facility as the lab doing the testing, in most cases, it will be reasonably fresh and intact. Care should be taken to ensure good viability by using a transport media with the tissue, such as RPMI with or without albumin, that has nutrients and buffers that greatly improve the likelihood of maintaining viability. Using other media, like plain PBS, is discouraged because the samples will deteriorate much more quickly compared to a tissue transport media like RPMI. The use of any fixative in the media is not acceptable since this will be toxic to the tissue and prohibit any possibility of doing a flow cytometry assay, which requires viable cells. Some laboratories use frozen tissue banks that allow the lab to save rare or study samples in a way that they are not damaged by the subfreezing temperature. Previously frozen samples must be thawed and washed before processing into a single cell suspension. Follow published procedures that describe the steps necessary to gently bring the sample back to room temperature prior to setting up the samples for staining. Making a single cell Suspension of the solid tissue can be achieved in several different ways including:

1. Grinding the sample with a mortar and pestle and rinsing with RPMI. The ease of use makes this a routine technique; however, the physical trauma to the samples may reduce both the viability and the quantity of recovered cells.

2. Using a tuberculin-sized needle that is filled with RPMI and repeatedly injecting it into the tissue in a petri dish to extract the cells of interest, leaving behind the specimen connective tissue. The cell suspension can then be washed with RPMI and concentrated prior to staining.
3. Adding an enzyme solution, such as collagenase, that digests the connective tissue, leaving behind the cells of interest. Though effective in practice, there may be higher cell damage and lower viability using this procedure [8].
4. Using mechanical devices that disaggregate the connective tissue, such as the Becton Dickinson Medimachine™ [9]. Small pieces of the tissue are put into a special cup called a Medicon™ which is then inserted into the Medimachine. RPMI is added to the Medicon, the cup is capped, and the machine is turned on. A spindle turns a raised metal lip that gradually forces the tissue through a bladed mesh. This process strips away the connective tissue, leaving a single cell suspension in the RPMI at the bottom of the Medicon cup. The RPMI, which now includes the separated cells, is then withdrawn from the cup using a syringe, filtered, and then processed for staining. Some laboratories perform cell counts after these steps to ensure adequate staining based on concentrations that are appropriate to use with previously made antibody cocktails (a cocktail is two or more antibodies in the same tube).

3.1.5 Fine Needle Aspirates (FNA)

FNAs are frequently submitted for flow cytometry testing in the context of Hematopathology procedures. Collecting an FNA is a much less invasive procedure compared to obtaining a tissue biopsy. The samples are collected by using a very thin needle (22 gauge or smaller) that is inserted into an area of a tissue in question, like a lymph node or other soft tissue. The cells are then aspirated into the syringe and added to a tube containing transport media such as RPMI. Since the cells are already in suspension, very little manipulation of the sample is necessary prior to using it in an immunophenotyping procedure. It can then be washed, concentrated, and resuspended in RPMI prior to staining. These samples may have some peripheral blood contamination that happens during collection and thus will need to include an RBC-lysing step as part of the staining procedure.

3.1.6 Body Fluids and Cerebrospinal Fluid (CSF)

Body fluids submitted for immunophenotyping can be collected from several different sources. These may include peritoneal, bronchial, pleural, synovial, pericardial, vitreous, and cerebrospinal fluids. In most situations, the reason for testing is to find and identify an underlying malignancy such as lymphoma or leukemia. Body fluids collected the same day will usually maintain viability if transported quickly and refrigerated. CSF is collected using a laparoscopic procedure, and even though these specimens are sent to the lab with no additional media in the tube, it is highly recommended that the sample be resuspended in fresh RPMI as soon as possible to prevent the rapid deterioration of any cells present. Studies have shown that CSF with RPMI added may have a much higher rate of recovery compared to samples that do not include a transport media [10].

3.2 Pre-analytic Quality Control Procedures

3.2.1 Reagent Quality Control

Buffers and lysing solutions: all reagents used in an immunophenotyping experiment must be tested for efficacy prior to being used in a procedure [11]. For example, PBS (phosphate buffered saline) and RBC lysing agents such as ammonium chloride (NH_4CL), can be tested quite easily using a sample type that is commonly submitted such as peripheral blood. Set up a sample using the new reagents, then acquire it on the instrument and review the results for acceptability. Is the light scatter okay and can you see differences in the major populations present? Are the RBCs lysed as expected? If the immunophenotyping procedures include reagents used for intracellular staining, the performance of those reagents should also be tested prior to putting them use for specimen testing. Record the results of the QC in a Reagent QC log to track performance. Include the lot number if applicable, the specimen used and whether the results were as expected, usually by either "pass" or "fail" depending on the criteria included in the procedure. As with all QC records, the supervisor or designee should review the results monthly.

3.2.2 Antibody Quality Control

Primary reagents like monoclonal antibodies require a more thorough control process than other test materials [12]. Quality control of antibodies should include the following procedures:

1. Antibody specificity: does the antibody react with the intended antigen [13]? This aspect of antibody QC requires specific specimen types to prove the efficacy of the reagent. For example, if CD33 is being tested, the control material should include myeloid lineage cells such as granulocytes and monocytes, since that antigen is fully expressed on those cells. Lymphocytes in the sample serve as a negative control. Other antibody control material includes:
    a. Commercial quality controls: these are preserved whole blood specimens that include a list of expected results, both qualitative and, in some cases, quantitative, for several antibodies in the package insert of the product being used. These controls are generally used for daily testing and verification of a multicolor antibody panel, also known as a cocktail (a cocktail is two or more antibodies in the same

tube) and are required to be run daily for clinical applications like peripheral blood lymphocyte subsets or stem cell quantitation assays. Commercial controls will generally have known qualitative and quantitative values for most of the antibodies used for immunophenotyping in a clinical laboratory. Two levels of the controls should be used each day the test is performed on patient samples and must pass before the testing can be reported. One control with expected quantitative ranges found normally and a second control that is either above or below the normal range is required. Running both levels verifies the analytic method range (AMR) established for the method during initial validation of the procedure (see Fig. 2 for an example).

b. In-house controls: other control samples that can be used to check antibody specificity and reactivity may be sourced from previously tested material that has been confirmed to be positive/negative for the antibody, depending on the populations present in the sample. If alternate testing is performed in the laboratory, such as Immunohistochemistry (IHC), the flow cytometry antibody results can be compared to the results of the alternate assay, such as a tissue section stained for CD20. However, this choice may only offer limited comparisons since, in many cases, the number of antibodies used in flow cytometry applications may be greater than those found on the Histology test menu. A second source of control material is previously tested samples that have included the reagents used in the cocktail

Phenotype^{a,e}	WBC Type^{d}	% Positive Cell^{a,b}	Expected Range^{f}	Absolute Number^{a,c}	Expected Range^{f}
CD2^{+}	Lymphocytes	85.0	73.0 - 97.0	-	-
CD3^{+e}	Lymphocytes	78.4	68.4 - 88.4	1625	1275 - 1975
CD3^{+}/CD4^{+e}	Lymphocytes	51.1	45.1 - 57.1	1059	879 - 1239
CD5^{+}	Lymphocytes	74.6	64.6 - 84.6	-	-
CD3^{+}/CD8^{+e}	Lymphocytes	25.3	19.3 - 31.3	524	354 - 694
CD3^{-}/CD16^{+}	Lymphocytes	9.2	4.2 - 14.2	-	-
CD3^{-}/CD16^{+}•CD56^{+}	Lymphocytes	10.5	5.5 - 15.5	218	73 - 363
CD19^{+}	Lymphocytes	11.2	6.2 - 16.2	232	82 - 382
CD20^{+}	Lymphocytes	11.1	6.1 - 16.1	-	-
CD3^{-}/CD56^{+}	Lymphocytes	8.6	3.6 - 13.6	-	-
HLA • DR^{+}	Lymphocytes	15.8	9.8 - 21.8	-	-
CD14^{+}	Monocytes	85.0	70.0 - 100.0	-	-
CD45^{+}	Leukocyte Gate	---------------- CD45 is to be used for gating purposes only.^{g} ----------			
CD34^{+}	Progenitor Cells	0.04	0.01 - 0.07	2.8	0.6 - 5.0
Lymphocyte^{c}		31.4	21.4 - 41.4	2074	1674 - 2474
Monocyte		8.4	3.4 - 13.4	-	-
Granulocyte		57.1	47.1 - 67.1	-	-
Absolute^{b} WBC^{c} (x 10^3/μL)		-	-	6.6	6.1 - 7.1

Fig. 2 Commercial control package insert: example of a package insert from a whole blood commercial control listing reactivity and expected quantitative values for antibodies used in immunophenotyping [14]

to be QCed or individual antibodies themselves. Does the cocktail or individual antibody reactivity agree with the previously tested material? The College of American Pathologists (CAP) also offers samples that can be used for a limited number of other rare antibodies, such as CD103 and CD1a. Some laboratories with sub-zero freezers may cryo-preserve samples that have previously been tested for quality control of new shipments/lot numbers of antibodies.

c. Proficiency testing samples: if the laboratory participates in an external proficiency testing program such as the CAP Survey, there may be leftover samples that are still adequate for quality control testing purposes that can be used for routine and/or rare antibodies. Returned results of these surveys can then be compared to the antibodies being tested for quality in the laboratory. This may not be practical if there is a considerable time delay before the phenotype of the testing material is known, unless the lab has some way of preserving the cells.

3.2.3 Antibody Concentration and the Stain Index

Most antibody manufacturers provide directions that include the amount of antibody to use for samples typically submitted for immunophenotyping. In some instances, knowing the concentration of the sample will be necessary to ensure saturation staining [15, 16]. If a sample has a high cell concentration, then the recommended amount of antibody/test may not be enough to ensure adequate staining. In other cases, the amount of antibody recommended may be several times more than is necessary. In that scenario, the excess antibody may impact the staining by increasing the background signal, resulting in lower resolution of the antibody. Choosing the antibody concentration to use can be determined by calculating the stain index (SI). The SI is the ratio of the difference between the positive and negative populations divided by two times the standard deviation (SD) of the negative population (see Fig. 3) [17, 18].

The stain index can be used to compare different antibodies, and it can also be used to compare different concentrations of the same antibody. In some cases, the SI will improve using a lower concentration than what is recommended since the background signal width may be smaller than it is using a higher concentration. The higher the SI, the better the resolution between the positive and negative populations. Different SIs can often be seen with different antibody concentrations. The difference can be due to one of two reasons: a lower intensity of the positive peak if the concentration is too low or a wider SD of the negative signal due to excess unbound antibody when the concentration is too high. Titrating the antibody is useful for determining the optimal concentration to use in the experiment. Different concentrations may have little effect on the overall intensity of the positive peak; however, the background/negative signal width

$$\text{Stain Index} = \frac{MFI_{pos} - MFI_{neg}}{2 \times SD}$$

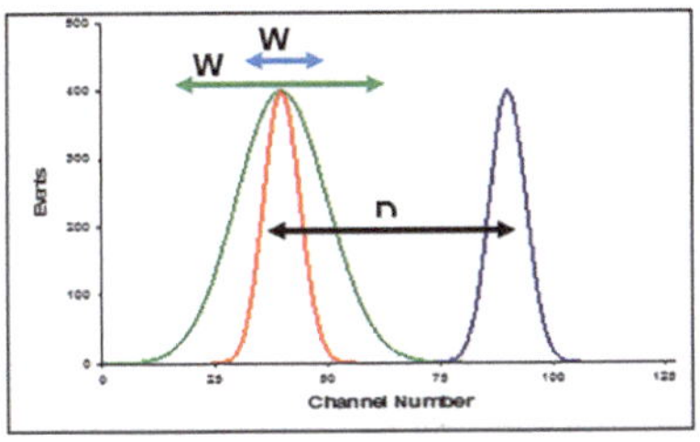

Fig. 3 The stain index (SI) calculation [16, 17]. The SI is the ratio of the difference between the positive mean fluorescence intensity (MFI) and negative mean fluorescence intensity (MFI) populations divided by two times the standard deviation (SD) of the negative population. The SI is used to compare fluorescence intensity of one antibody vs. another or when comparing the concentration of one antibody vs. a different dilution (titer) of the same antibody

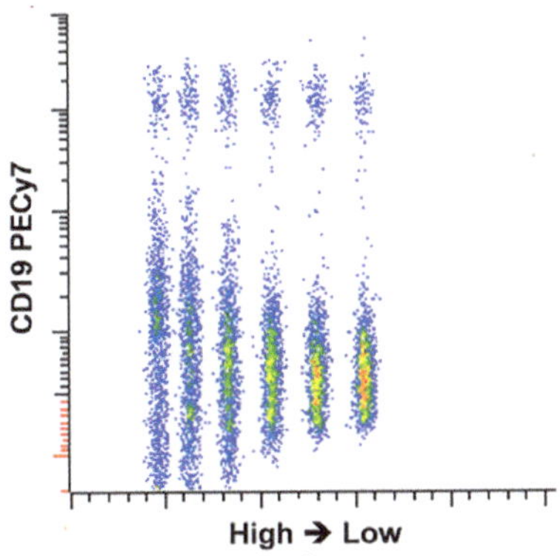

Fig. 4 Antibody concentration vs. CD19 PEcy7 expression. Over the course of different titrations or dilutions, the sample stain index (SI) improves when using lower concentrations of the antibody. There is no noticeable loss of CD19 intensity (Y-axis); however, the background signal is much greater at higher concentrations of the antibody and thus has a lower stain index. Ideally, the standard deviation of the negative signal should be somewhere near that of unstained autofluorescence

decreases because of less unbound antibody and thus a better SI is observed. An example of how the optimal antibody concentration can be calculated is seen in Fig. 4.

The titration procedure compares serial dilutions of the antibody after calculating the stain index of each dilution. The results are then graphed using the SI vs. the concentration of the antibody. Determining the best titer of the antibody should be included in the Quality Control procedure for any new lot number or new shipment prior to using it in an assay. It is okay to use a different amount than what is recommended by the manufacturer if you can document that it still performs as intended (see Fig. 5 as an example).

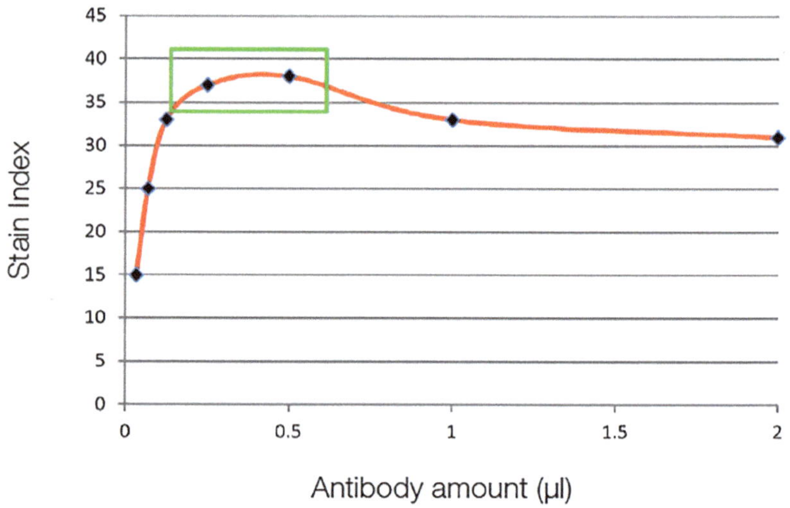

Fig. 5 Antibody concentration vs. stain index. As the antibody concentration increases, the stain index (SI) increases until it levels off and shows little improvement. The optimal concentration to use should be somewhere close to the value of the stain index that begins to cease changing; in the example shown, approximately 0.5 µl/test

See the Notes section for Instructions on the Antibody Titration Procedure.

3.2.4 Multi-Antibody Cocktails

An antibody cocktail is defined as whenever two or more antibodies are combined to make a single or multiuse reagent. As part of the panel design, the antibody cocktail is constructed using combinations that are used for a variety of immunophenotyping applications. There are several advantages to using cocktails. They can save time when setting up multiple experiments and are likely to lower the likelihood of pipetting errors that could happen if the single-use cocktail was made up each time the experiment was performed. Another advantage of using cocktails is that you can standardize delivery volumes. This makes test setups easier since many single reagents use different volumes, and in a multicolor cocktail, this can be very labor-intensive and prone to error. The cocktail can be made in advance and for enough tests to last a few days or weeks. When should antibody cocktails be QCed?

1. Initial QC testing: any cocktail must be proven to be accurate before using it for testing/reporting purposes. Initial quality testing of the cocktail should include comparing the in-use cocktail to the new cocktail by setting both up with a fresh sample. Results should agree for both reactivity and quantitation if applicable. Shelf life of the cocktail must be determined as part of the initial assay validation. Each lab should determine its own pass/fail requirements.

2. Ongoing testing cocktail performance should be documented each day it is used on samples submitted for immunophenotyping. One way to do this is for the person interpreting the results to use written pass/fail criteria to determine if the cocktail is acceptable or has questionable results. Do the antibodies in the cocktail react with expected populations present? Is it negative on the populations it should not react with? Can you distinguish normal vs. abnormal? If the cocktail is used over several days, it should be tested each week using positive and negative controls or previously tested samples. Loss of any antibody reactivity or a definitive change in intensity means that the cocktail should be discarded. Criteria for acceptability should be included in the QC procedure. Document the results of the quality control and keep records of the testing.

3.2.5 Antibody Cocktail Setup Procedure (example)

A 4-color cocktail setup procedure can be found in the Notes section. The example shown in the recipe is for a 10-test, 4-color cocktail for T-lymphocyte subset testing using titrated antibodies and buffer that is aliquoted using 100 μl cocktail/test. Use a previously determined titer amount for each antibody in the cocktail and add enough buffer to create a standardized aliquot for testing, for example, 100 μl/test. The additional volume of the buffer will also allow better mixing of the cocktail and the specimen.

3.2.6 Conclusion, Pre-Analytic Quality Control

Every experiment in immunophenotyping uses specific samples from a wide range of sources. Each laboratory should develop specific sample requirements that include the instructions for collection, the media used to transport the sample, and the storage and conditions required for testing. In addition, the use of any reagents to perform the test must be checked for suitability using controls that confirm each of the proper preparation steps of the assay works as intended. Records of all quality controls must be kept in a concise way that allows easy retrieval for inspection purposes as well as to monitor performance over time. After the successful completion of these pre-analytical steps, the immunophenotyping testing can now proceed to the analytic phase.

3.3 Analytic Quality Control Procedures

3.3.1 Instrument Quality Control

Quality control of the immunophenotyping assay continues with consideration of the instruments used to run the assay, such as a flow cytometer. These very sensitive and highly adaptable platforms are rather simple in terms of how they work. Basically, a fluidics system consisting of either buffered saline or deionized water provides a means for the sample to be introduced into a flow cell chamber. The flow cell has a very specific shape that, through the principle of hydrodynamic focusing, takes the cells or other materials in a random mixture and lines them up into a single file by forcing them through a small opening as they exit the flow cell. Upon exiting the flow cell, the sample stream is bisected by one or more light sources, typically lasers with monochromatic emissions. The intersection of the stained cell fluorochromes with the laser produces several specific emissions with different wavelengths that are then directed by fiber optic cables or mirrors and filters into electronic detectors commonly known as photomultiplier tubes (PMTs). The PMTs then convert those signals into a digital format for data analysis using software programs.

Despite the simplicity of its operation, the quality of the results can be greatly affected by the performance of any or all of these flow cytometer components: fluidics, optics, and electronics. Daily instrument maintenance and quality control are paramount to assuring reliable results. This section of Immunophenotyping Quality Control will review and describe the steps to perform on the instrument prior to running test samples. See Notes section: Instrument Quality Control Workflow, a step-by-step guide for setting up a flow cytometer for immunophenotyping assays. An explanation of each step in the analytic QC of the instrument is found in the following section.

3.3.2 Instrument Quality Control: Flow Cytometer Startup and Maintenance

1. Fluidics system startup: the flow cytometer fluidics system is truly the lifeblood of the instrument. Regular maintenance of the fluidics system is tantamount to proper operation and helps ensure high-quality results. Prior to turning on the instrument, check the levels of the fluidic containers. They are usually found either on a fluidics cart or other storage rack near the instrument. This includes the sheath fluid tank along with cleaning solutions, or DI water, etc., that may be a part of the whole fluidics system. If the instrument has a waste container, dispose of anything from the last time it was used. Be sure to add bleach or some other decontamination agent to the container to minimize the danger of any hazardous waste that may result from using biological samples. Most flow cytometers have built-in fluidics startup programs that initially provide a means for the sheath fluid and other solutions to be primed and distributed throughout the system, including the flow cell, and ultimately into the waste container. As part of the startup, a series of priming and flushing cycles takes place and that all the fluidics lines are free of air bubbles. Sometimes, when the instrument is shut down for several hours, bubbles can come out of the sheath fluid solution still in the instrument, much like what is observed on the side of a glass of water left standing overnight. If the bubbles are not adequately removed as part of the fluidics startup, they may cling to the inside of the flow cell and create signal focusing issues, much like any other obstruction in the fluidics system.

2. Fluidics clean cycle: samples routinely run on the instrument, such as tissue cell suspensions and bone marrow specimens, may have residual debris or fibrin or other non-intact cell residue that can build up on the inside of the sample intake, fluidics tubing, or the flow cell. After starting up the instrument, the fluidics system should be put through a short cleaning cycle that includes running dilute bleach (10%) for 5 min followed by a DiH_2O sample for 5 min. Running a series of cleaning tubes helps to remove the buildup and prevent clogs or other obstructions that might be present in the fluidics system. Some labs prefer to do this cleaning sequence one or more times a day or after running several samples in succession. A good rule-of-thumb is that for every 1 min you run dilute bleach, it should be followed for the same amount of time by running a tube of DiH_20. This is so the bleach is adequately removed from the fluidics system. A full system clean of 60 min (30/30) or more should be performed at least once a month to more thoroughly clean the fluidics lines and flow cell chamber. Be careful to detach any filters in the fluidics lines since the bleach can cause the filter material to break down and potentially clog the system.
3. Fluidics function checks as part of bead QC: as mentioned before, issues with fluidics are probably the most common cause of instrument-related problems. Just about every instrument has a quality control program that uses latex beads to measure and standardize each parameter of the instrument operation. Usually, performance issues are first indicated by out-of-range bead QC results, including diminished resolution of either or both light scatter parameters, Forward Scatter (FSC) and Side-Scatter (SSC). A well-maintained instrument that has been set up and optimized should demonstrate that the FSC/SSC dot plot of a lysed normal blood sample has at least three distinct WBC populations corresponding to the lymphocytes, monocytes, and granulocytes. Whereas, if the instrument has population resolution issues seen by light scatter due to fluidics problems, the populations tend to drift downward toward the origin, and the populations appear condensed or smeared compared to distinct, fully expanded clusters (see Fig. 6).

   If debris builds up over time in the fluidics lines or the sample line, or if the sample stream is obstructed in some way, the Forward Scatter Channel (FSC) signal is usually the first to demonstrate this problem. There will either be a gradual loss of the signal, resulting in a lower location on Forward Scatter of a population or an increase in the detector gain if you are using PMT settings that are auto-adjusted by the bead QC software to meet an expected target channel range for each detector. Eventually, these changes to the detector settings will exceed the defined limits of the bead QC program and the results will have one or more parameter failures.

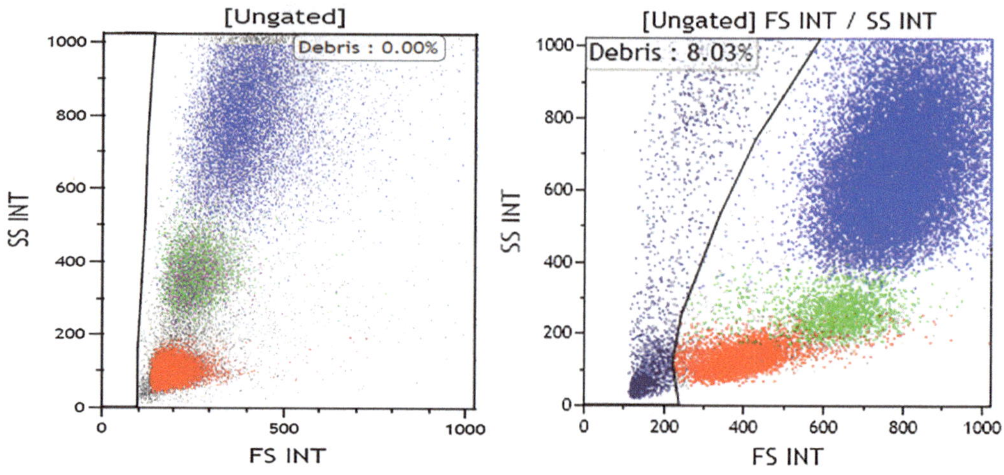

Fig. 6 Poor light scatter resolution vs. good light scatter resolution. Left dot plot: RBC-lysed peripheral blood sample with lymphocyte (red), monocyte (green), and granulocyte (blue) WBC populations with suboptimal FSC/SSC signals due to possible fluidics issues. Right dot plot: same instrument with improved resolution and separation of all three populations after cleaning the fluidics and purging the system of bubbles

4. Laser performance: the solid-state lasers found on most flow cytometers are considerably more dependable and less prone to failure than the high-power water-cooled lasers that were used on many of the research-type instruments years ago. However, they still need time to warm up and equilibrate. Usually, the time required to run the fluidics start-up cycle and cleaning steps is typically long enough for the lasers to warm up. In most cases, the instrument will have an indication of when it is ready to be used following the warm-up time for the laser.

5. Flow cell, optics, and electronics QC Using Commercial Beads: Following instrument startup, priming the fluidics, initial cleaning, and laser warm-up, the flow cytometer must then be checked for proper operation. This is typically accomplished using latex beads or other materials, usually sourced from an instrument vendor, to ascertain the overall function of the fluidics, electronics, and optical alignment. These quality control checks are where the "quality" of Quality Assurance begins in flow cytometry. The CAP checklist requires that performance of function checks on all instruments follow a defined schedule [19]. For results from a flow cytometer to be reliable and accurate, the setup and monitoring of the flow cytometer performance should be checked every time the instrument is started up each day. Some labs may require this as often as once each shift if the instrument is used continuously.

   To ensure that the flow cytometer is fully operational, and each event signal is meeting a set standard, most instrument manufacturers use bead QC programs that test all the operational aspects of the flow cytometer. After establishing initial baseline settings on new bead lots, the performance is then monitored on an ongoing basis each day the instrument is used. The QC beads are generally and are polychromatic, and thus the signals can be detected across all the PMT and light scatter detectors. The beads have different sizes and fluorescent intensities so that linearity and sensitivity can be measured along with PMT performance. If, for instance, the PMT voltage required to place the bead signal in the baseline channel is outside the software limits of acceptability, then that parameter will be flagged as failing (see Fig. 7).

6. Focal point/alignment check: the alignment of the laser to the flow cell stream is critical and has a direct impact on the quality of the results. Most flow cytometers have one or more laser light sources that are directed by a series of lenses and prisms to a fine focal point at the exit point of the flow cell. If the laser or the flow cell is not properly aligned or focused, then the resolution of one or more signals may be suboptimal, resulting in lower sensitivity. An instrument with good alignment means that the laser light bisecting the sample stream is precise and finely tuned. A bright, sharp signal provides better resolution and sensitivity compared to a signal that is not quite as precise (see Fig. 8).

   Running QC beads each day does not actually change the alignment of the light with the flow cell, but provides a way of knowing just how the performance compares to the baseline standard established earlier. Performing an actual alignment of the instrument laser or flow cell is not something done routinely, especially in a clinical setting. That procedure is typically performed by a service engineer as part of an initial setup of the instrument or as part of regularly scheduled preventative

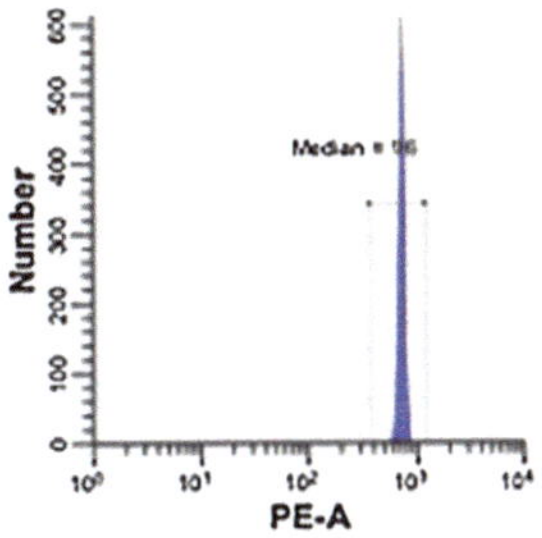

Fig. 7 Performance quality control (PQC) beads signal daily. The bead signals are placed in a specific channel of each detector by the PQC software, which automatically adjusts the detector gain/voltage to match the baseline target value range. If the voltage required is outside the acceptable range, the software will flag the results for review

Fig. 8 Optical alignment and fine focusing. Good optical alignment equates to fine focusing, which can provide a sharper, clearer signal that gives better resolution and higher sensitivity (right image) compared to an instrument with possible alignment issues (left image)

maintenance, or whenever a major component like a laser or a PMT is replaced. Some research flow cytometers may allow the user to change or adjust the optical alignment, but this is the exception, not the rule, since in most instruments the flow cell itself may be semi-permanently attached to the optical bench.

7. PMT performance: besides checking the alignment of the laser and the flow cell, the QC beads are indicators of PMT performance. After setting up a baseline range for each signal, the daily performance of the QC beads is measured against the results of the baseline measurements. Each of these performance checks provides information about the detector efficiency, sensitivity, resolution, precision, linearity, background (noise) level, and laser delay. By definition:

   a. Detector efficiency: a measurement of how well the detector is working. Ideally, it is the maximum signal output with minimum PMT voltage. A failing PMT will have a loss of efficiency over time and subsequently require more voltage to put the emitted signal within the program's baseline range of the scale. An increase in PMT voltage may then result in a higher background signal and subsequent loss of sensitivity.

   b. Sensitivity is the ability to distinguish true positive signals from true negative signals. One of the ways to determine PMT sensitivity is to run polychromatic beads and calculate the stain index (SI). The SI equation uses the fluorescence intensity of the positive peak vs. the negative peak and can be calculated using this equation:

      SI = MFI of positive bead − MFI of the unlabeled bead/2 × SD of unlabeled bead.

(SI = stain index, MFI = mean fluorescence intensity, SD = standard deviation).

The standard deviation (SD) of the negative peak reflects the width of the background signal. Some fluorochromes have a broad negative signal while others may have one that is much narrower and thus have a lower SD and, in many cases, a higher SI.

The SI calculation on some instruments is performed by the bead QC software, but it can also be calculated on a spreadsheet if you know the results to use in the equation. Knowing the stain index of each detector and monitoring it on at least a weekly basis is highly recommended. Each laboratory should establish an acceptable range, usually ± 2 SD of the mean SI for each PMT. If, over time, the bead sensitivity decreases, it may be indicative of instrument issues such as faulty fluidics, dirty optics, electronic noise, or even a failing laser. If the trend continues over the course of multiple days, the service engineer should be notified if minor troubleshooting does not resolve the issue.

c. Signal resolution is the ability to detect discrete signals in multiple results where the differences are very small. Good resolution may usually result in higher sensitivity. One way of determining how well the instrument can resolve multiple populations with small differences is to run beads that have varying intensities. There should be good separation between each of the populations throughout the range of the detector, including very weak signals vs. negative signals and very bright signals being on scale and not crushed against the far axis. If the signals are not clearly differentiated, it may indicate an alignment/focal point problem, often seen when there are fluidics issues, especially if it happens in more than one channel. Cleaning the flow cell using a manufacturer-recommended cleaning solution can oftentimes improve the quality of the results (see Fig. 9).

d. Precision is the ability to detect the same signal repeatedly within a defined range of the baseline. The measurement of precision is the coefficient of variation (CV) of multiple results. High CV's mean poor precision, which negatively impacts the reproducibility of your results.

e. Linearity is the property of sequential results to be detected in a straight line over time. If the detector has good linearity, a signal that has twice the output of the first signal should then be 2× brighter than the first signal and so on for subsequent results with increasing positive peaks. This is very important for experiments that measure DNA ploidy, where the G2+M population should be 2× the G1/G0

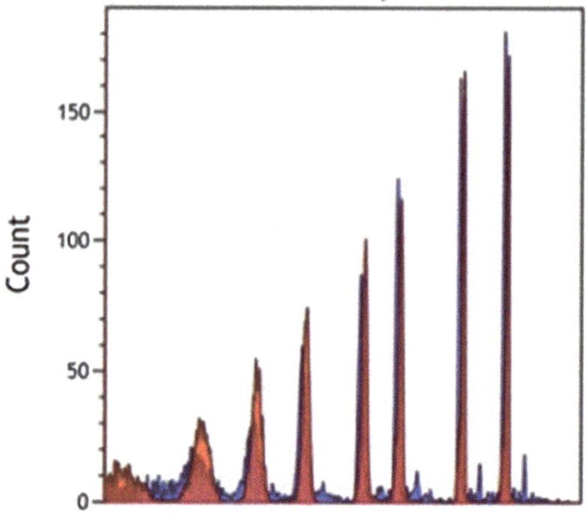

Fig. 9 Multiple fluorescent intensities demonstrating high resolution. A flow cytometer with finely tuned alignment and optimized PMT settings can resolve multiple populations and demonstrate discrete differences in intensity

population [20–22]. If it is not, then it might be misinterpreted as being aneuploid. Good linearity means that the reliability of multiple signals is expectedly proportional.

f. Background level is sometimes referred to as the noise level. It's an extraneous signal that is detected when no sample is being run at all. It is present on all PMT outputs and can vary depending on the detector. Knowing the background signal is important because it may impact the sensitivity if it is higher than the otherwise negative signal produced by the cells being measured. Higher background levels that increase over time are often seen in PMT's or other instrument electronics that are beginning to fail or have high-voltage settings. If routine troubleshooting does not improve the background signal, it might require attention by a service engineer.

g. Laser delay is the time difference between the bead being excited by the first laser, followed by the signal output from the second or subsequent lasers. If the delay is too long, it can result in the possibility of missing or abbreviating the detection of the second signal. Generally, laser delay issues can be traced back to fluidics problems. Restriction or something that slows down the fluidics stream due to obstructions like dirty tubing, or a partially obstructed flow cell, will oftentimes cause laser delay errors. Performing regular instrument fluidics cleaning will usually prevent this problem.

8. Levy–Jennings charts: in addition to the bead QC indicating the overall operation of the instrument on a specific day, it also provides a way to monitor the performance of several parameters over the course of days or weeks to demonstrate reliability and consistency. Most flow cytometers are rather stable from day to day; however, the alignment or precise focal point can drift or deteriorate over time due to any number of reasons, including a fluidics issue, faulty electronics, laser failure, vibrations, alignment issues, or some other unknown cause. A gradual decrease in performance may predict a future failure. The bead QC program usually includes graphical displays of all measured results plotted over time. These are known as Levy–Jennings (LJ) charts, which allow the user to observe or detect changes or trends in performance that can be addressed before they totally fail (see Fig. 10).

If only one detector shows a trend, then the problem may be a faulty PMT. If more than one detector from the same laser shows this problem, it may indicate that the laser performance or alignment is the potential problem. In some cases, the issue may be a damaged fiber optic cable that transmits the signals from the flow cell to the PMTs. This may be due to high laser output that clouds the fiber optic over time. Generally, a problem like this must be addressed by a service engineer. If more than one parameter is out over multiple lasers, then the problem is more likely to be the fluidics or an impending failure of the laser or even the instrument electronics.

QC bead results summary: performing routine maintenance and running QC beads to check fluidics, alignment, and instrument operation is the first step in preparing the flow cytometer to run experiment samples. Each lab should determine how and when to

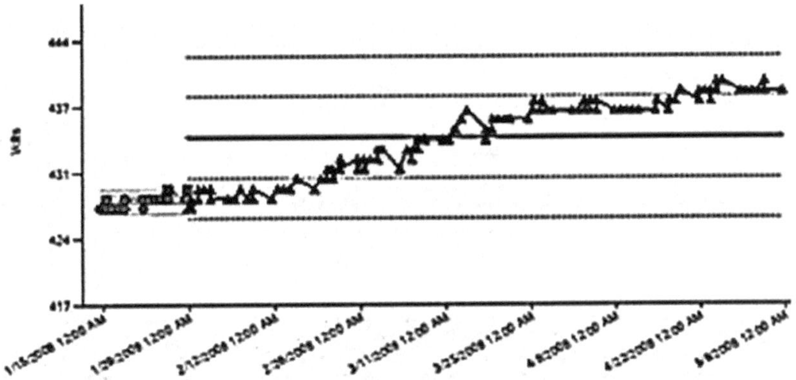

Fig. 10 Levy–Jennings graph of daily PMT QC settings over time. The voltage of the FL detector required to place the bead signal within the baseline specifications is gradually rising over time, which may indicate an impending failure of that PMT or possible alignment adjustments needed

respond to a QC failure. In some cases, performing a system clean of the fluidics and flow cell will often fix the problem, and in other cases, more detailed troubleshooting is required. If the problem continues, then a service call may be necessary to find and fix the issue. Document daily performance results as well as any troubleshooting performed. Successful completion of the performance QC is the first step in the preanalytical QC process. The next step, instrument optimization, is optional since it should be performed initially when setting up an experiment, but only as needed from day to day. That step is described in the next section.

3.3.3 Optimizing the Instrument Settings

Introduction: optimizing the flow cytometer instrument settings is a way to maximum performance or gain the best signal:noise ratio for each detector [23]. The signal:noise ratio is the difference between the median positive signal and the median negative signal. Think of optimizing as fine-tuning the instrument settings as compared to a course adjustment from running only the QC performance beads. Do not confuse optimizing the settings with setting up compensation. Color compensation setup follows optimizing the instrument settings. In the clinical lab, think of this as moving the instrument from being "bead-ready" to one that is "sample-ready." Why? Because in some cases, bead signal intensity will be different from cell signal intensity and using stained cell samples to optimize the bead settings may improve the resolution of the detector to identify discrete populations present. In many cases, these differences may be indicative of an abnormal population. If the default setting is too low, there may be a loss of sensitivity because the weak signal may be too low to detect, an intermediate signal that blends into either the brighter positive signal or overlaps with the negative signal due to lower resolution and separation of all signals present. On the other hand, if the bead PMT setting is too high, the positive signal might only be partially observed if it is pushed up against the far axis. This makes the upper end of the signal look like it is off scale when, instead, it is simply stacked up in the last channel and may be erroneously underestimated visually. See Fig. 11 for examples of PMT settings that are too low, too high, and optimal using a stained cell sample with bright, intermediate, and negative populations present.

How often is it necessary to optimize? Most flow cytometers are relatively stable and give reproducible results day after day, so the frequency of performing optimization really depends on a few specific situations:

- Whenever an instrument is initially installed.
- After any major maintenance, such as replacing a laser, re-aligning the flow cell, changing out a PMT, or changing the configuration of the PMTs.
- Setting up a new experiment/panel

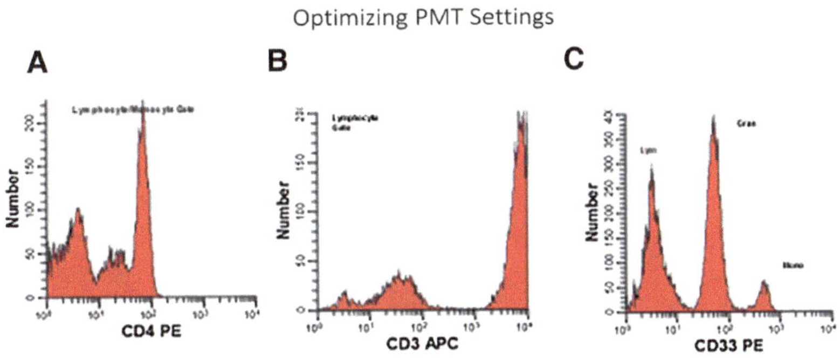

Fig. 11 Optimizing PMT settings. In Histogram (**A**), WBCs are gated on lymphocytes and monocytes stained with CD4. The PMT setting is too low and as a result the intermediate population (monocytes) is difficult to resolve. In Histogram (**B**), the default PMT setting is too high, so CD3+ T cells gated on lymphocytes appear to be off scale when, in fact, they are still on scale; however, the brightest events are stacked up against the right axis and could be visually underestimated. In (**C**), the PMT setting is optimal. A WBC sample stained with CD33 shows clear resolution between the bright (monocytes), intermediate (granulocytes), and negative (lymphocytes) WBC populations

- Setting up an in-use experiment using different fluorochromes in a detector optimized for another fluorochrome. For example, substituting Alexa 488 for FITC in the FL1 channel. This may also be necessary if changing lot numbers of some fluorochromes, especially if it is a tandem.
- Adding new fluorochromes or removing fluorochromes from assays already established.
- If you change the QC bead baseline configuration or set up a baseline using a new lot number of QC beads. If you copy the target channels from one lot to the new lot, it may not be necessary to optimize the settings; however, verify this by running comparisons of the old lot vs. the new lot to confirm. Another good QC practice is to check the settings using a fully stained sample and observing each PMT for optimal performance at least once a month.

1. Optimizing light scatter: forward (FSC) vs. side (SSC) setup
   Begin the optimization procedure by running a cell sample, such as lysed normal whole blood (LWB), using the default settings from the QC performance beads. The first difference you may notice is that the light scatter settings and possibly the threshold will need to be adjusted. Adjust the FSC/SSC light scatter settings until you can see good separation between the lymphocyte, monocyte, and granulocyte populations. Set the threshold to just below the lymphocyte population to mini-

mize the noise and debris events, but still show the entire cluster of lymphocytes present. Keep the granulocyte population on scale; however, take advantage of the entire FSC/SSC area to see each population clearly, like that seen in Fig. 12.

2. PMT optimization
   Optimizing the fluorescence PMTs can be performed in several different ways, and each has advantages and disadvantages, but the purpose of detector optimization is to see an improvement in signal separation between the bright positive, intermediate, and negative populations present. Two of the most popular methods used to optimize the PMT settings are the Unstained Sample Method and the Single-stained Sample method. Note: optimization should be performed with the compensation settings turned off. Compensation setup should only happen after all PMT settings have been completed.

3. PMT optimization methods

   a. Unstained sample method: optimizing the PMTs using the unstained sample method has been used for several years and is relatively easy to perform and not very expensive since no antibodies are required. Set up the instrument, starting with the default PMT settings from the QC performance beads. Acquire an unstained lysed whole blood sample and draw a gate on the lymphocyte population on forward (FSC) vs. side light scatter (SSC). While still acquiring the sample, adjust the PMT settings for each fluorescent parameter in the testing panel until the leading edge of the negative signal goes just beyond the background noise signal. Continue until each PMT has been adjusted as necessary (see Fig. 13).

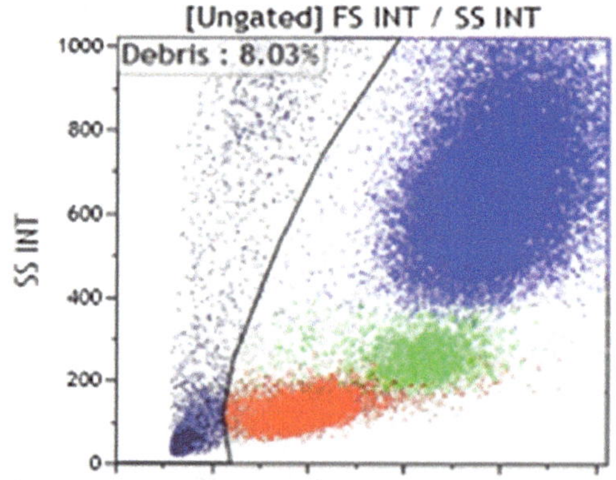

Fig. 12 Forward Scatter (FSC) vs. side Scatter (SSC) of a lysed whole blood sample with high resolution, demonstrating good separation between all three WBC populations with the noise threshold set just below the lymphocytes to allow visualization of the intact cells vs. debris

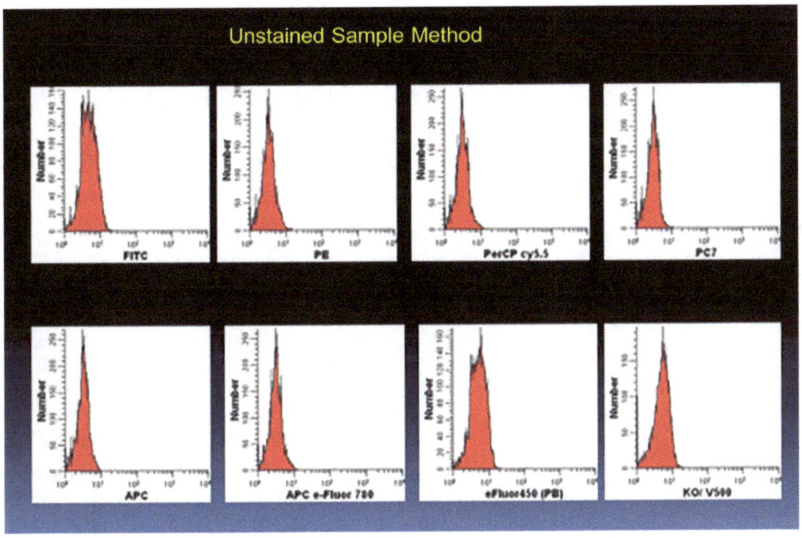

Fig. 13 The unstained sample method for optimizing PMT settings: using an RBC-lysed, whole blood specimen, the PMT settings are adjusted so that the leading edge of the negative signal is just beyond the background/noise signal in each detector

One advantage of this method is that it does not require any antibodies, so it is rather easy to set up. However, a disadvantage to using this method is that without knowing the intensity of a positive result, the signal may be too high and stacked up at the end of the scale or too low, like when a weak positive population cannot be differentiated from a negative or bright positive population. Use the unstained settings as a starting point, but also include a fully stained sample that shows both positive and negative populations to be sure that all the populations are visible on the scale of each PMT.

b. Single-stained sample method: this optimization method uses a panel of lysed whole blood samples that are single-stained with each fluorochrome conjugate used in the experiment. The advantage of this method is being able to visualize the full range of signal intensities of multiple populations: bright, intermediate, and negative for each PMT. For the single-stained samples, choose an antibody that has multiple levels of expression depending on the populations present. Three antibodies that are useful for this method are:

- CD4: gated on lymphocytes and monocytes, stains positive for T-helper lymphocytes (bright), monocytes (intermediate), and is negative for T-Cytotoxic-Suppressors

(CD8+). Granulocytes are also negative for CD4 if the entire WBC population is gated.

- CD33: another good marker for visualizing three or more populations of WBCs. Monocytes are strongly positive, granulocytes stain with an intermediate intensity, and lymphocytes are negative.
- CD8: gated on lymphocytes, stains T-Cytotoxic Suppressor cells very bright, Natural Killer cell lymphocytes with a heterogeneous mid-intensity, and CD4+ T cells are negative. CD8 is one of the brightest antibodies available and gated on lymphocytes, it will express intensity at the highest decade for most samples; however, CD8 may not be the best antibody to use for optimization simply because the intermediate population (NK lymphocytes) is oftentimes not very distinct compared to the intermediate populations observed using CD4 or CD33 (see Fig. 14).

While acquiring the stained sample, adjust the PMT setting so that the bright peaks are on scale, the intermediate peaks can be clearly seen, and the negative population is clearly defined, just past the background signal. This includes seeing a difference between discrete, intermediate populations vs. bright or negative populations. The stained samples chosen to optimize with should include as many of the same fluorochrome conjugates available using the antibodies previously suggested or others that will demonstrate three separate

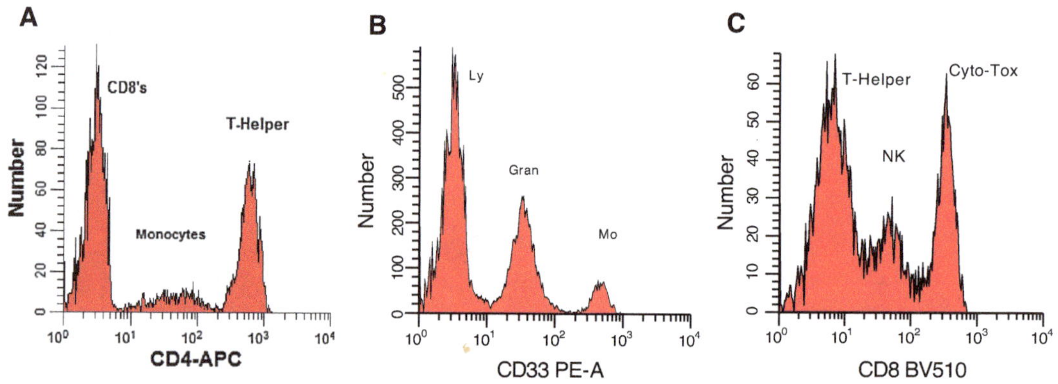

Fig. 14 The stained cells method for optimizing PMTs. Using an RBC-lysed peripheral blood sample, in Histogram (**A**), WBCs gated on lymphocytes and monocytes stained with CD4, show three separate populations that can be clearly visualized after adjusting and optimizing the PMT settings. T-helper cells (CD4+) stain bright, monocytes have an intermediate intensity, and T-Cytotoxic suppressor lymphocytes (CD8+) are negative. In Histogram (**B**), ungated WBCs stained with CD33, good resolution can be visualized after adjusting and optimizing the PMT settings. shows monocytes stain bright, granulocytes intermediate, and lymphocytes are negative. In Histogram (**C**), WBCs stained with CD8 and gated on lymphocytes express bright+ CD8, intermediate positive for NK cells, and negative for T-helper (CD4+) lymphocytes

populations. If an antibody is not found in every color used in the experiment, use a combination panel that includes one or more of the other choices in the missing colors of the primary antibody conjugate. Conversely, use antibodies that are included in the panel, adjusting the detector to see the best separation of positive vs. negative signals that allow discernment of discrete populations while keeping the bright positive signal on scale. See Notes section: Single-stained Sample Optimization Procedure, which describes this method of optimizing the instrument settings.

3.3.4 Color Compensation

Introduction: compensation is the mathematical correction for spectral overlap between different fluorophores. Once the instrument PMT settings are established, the next step in analytic quality control is to set up a compensation matrix. This step must follow the optimization step since no more adjustments should be made to the PMTs; doing so will impact the compensation, which then has to be recalculated. If two or more color conjugates are used in an experiment, especially if they are excited by the same laser, there is a high probability that some of each signal will overlap into the adjacent signal. This overlap must be accounted for and adjusted to ensure signal fidelity, each color independent of the adjacent or overlapping signal. Compensation correction is especially important in distinguishing weakly expressing antigens from negative populations since the bright signal of one fluorochrome might spillover and be detected in the dim region of an adjacent fluorochrome. Even though band-pass filters allow specific wavelengths of light to pass through to the detector, while blocking others, there may still be a signal from one fluorophore overlapping with the signal of another fluorophore. This can mistakenly make a single positive or even negative signal appear to have a weak positive signal (see Fig. 15).

For the sample that is positive for one antibody to not appear as co-expressing the other antibody the spillover signal must be subtracted from the total signal detected. Conversely, if the compensation correction is too high, also known as overcompensation, it could potentially cause a weak dual-positive population to disappear by being drawn into the signal of the adjacent conjugate. See Fig. 16, CD 34 vs. HLA-DR for an example of under- and over-compensation.

The dot plot in the figure is from a clinical flow cytometry leukemia panel. The CD34 vs. HLA-DR appears to have both a dual-positive and a CD34 exclusive population, implying that there are two different blast populations present; however, the dual-positive population is over-compensated. Why does this matter? Because immature cells that are CD34 positive and HLA-DR negative may be mistaken for promyelocytes. Promyelocytic leukemia (PML) is a very dangerous disease that, if not identified early, can in many cases cause disseminated intravascular coagulation (DIC), an internal

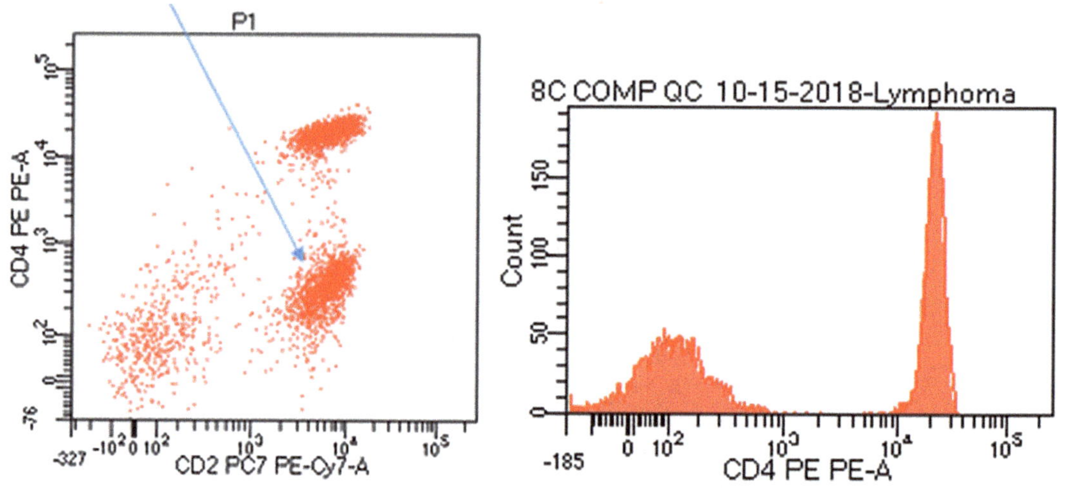

Fig. 15 Lymphocytes stained with CD4 PE vs. CD2 PE-cy7. The left dot plot shows two populations of CD2+ CD4+ cells. The weak CD4+ PE/CD2+ PE-cy7 population is the result of under-compensation of PE-cy7 into the PE channel and should only be positive for CD2

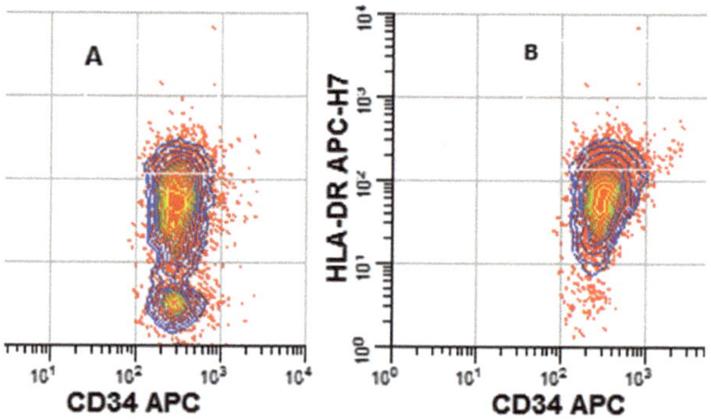

Fig. 16 Different compensation settings of HLA-DR APC-H7 vs. CD34 APC. Dot plot (**A**): overcompensation between HLA-DR APC-H7 and CD34 APC appears to show two separate populations, a dual-positive HLA-DR/CD34 population and a HLA-DR-negative/CD34+ population. Dot plot (**B**): HLA-DR APC-H7 and CD34 APC, correctly compensated, shows a single population that is positive for both CD34 and HLA-DR

bleeding disorder, so distinguishing it from other AMLs is imperative. Correctly compensated, the population in question in this example is, in fact, positive for both CD34 and HLA-DR and was ultimately described as AML, not APL.

1. Setting up compensation on a flow cytometer
   Because of the complexity of setting up compensation for multicolor panels, software compensation programs are used to

Compensation Matrix for an 8-Color Experiment

Signal	into FITC	into PE	into PerCP-cy5.5	into PE-cy7	into APC	into APC-H7	into BV421	into BV 510
FITC	100.00%	14.50%	3.00%	0.20%	0.00%	0.02%	0.00%	3.12%
PE	1.00%	100.00%	22.00%	0.70%	0.02%	0.02%	0.02%	0.03%
PerCP-cy5.5	0.01%	0.00%	100.00%	12.00%	1.61%	7.03%	0.03%	0.05%
PE-cy7	1.00%	5.50%	7.50%	100.00%	0.03%	10.00%	0.03%	1.00%
APC	0.00%	0.11%	1.00%	0.09%	100.00%	8.50%	0.02%	0.03%
APC-H7	0.01%	0.02%	0.06%	0.53%	1.06%	100.00%	0.04%	0.07%
BV421	0.01%	0.02%	0.06%	0.00%	0.00%	0.02%	100.00%	16.50%
BV 510	0.17%	0.16%	0.11%	0.01%	0.00%	0.01%	10.00%	100.00%

Fig. 17 Compensation matrix of an 8-color experiment. Each fluorochrome in the experiment is shown with its respective compensation setting vs. the other fluorochromes of the experiment. The diagonal values are all 100% because a color cannot be compensated against itself. The higher the value, the more overlap each signal has with the opposing signal. When designing an experiment, the use of color-conjugate pairs with very high spectral overlap should be avoided for those markers that are used to identify very discrete dual positive populations, since they might be lost due to the high compensation correction value

create the compensation matrix using a compensation panel that includes both unstained and single-fluorochrome-stained beads or cells, one tube for each conjugate that will be used in the experiment. After acquiring the panel, compensation is then calculated by the compensation software program. A portion of the first signal is subtracted from the total of the overlapping signal, and vice versa for each of the fluorochromes where there is spillover detected. A portion of the signal subtracted from the total signal corrects for any spillover. The calculations applied thoughout the entire panel are called a compensation matrix (see Fig. 17).

Setting up compensation on instruments that only detect four or fewer colors could be done just by looking at one signal vs. another and making the necessary adjustments to correct for any under- or over-compensation of the signals. Newer digital flow cytometers can detect upward of ten or more colors; some research platforms can detect 20 or more. This means calculating the compensation now requires the software to take into consideration every possible combination of fluorochromes. For instance, to set up compensation for a 4-color experiment could require up to 12 calculated values vs. the number for a 10-color experiment, which could be up to 90 different calculated values.

2. Compensation setup considerations
   a. Label-specific compensation: many multicolor panels include the use of tandem and monochrome fluorochromes. In some cases, a panel with more than one tube may use different fluorochromes in different tubes that are detected in the same channel, such as FITC and Alexa 488 or PE-cy5 and PerCP-cy5.5. This requires different compensation settings for each conjugate used also known as label-specific settings.
   b. Most compensation software programs allow the user to include both generic as well as label- or lot-specific antibodies in experiments where there is more than one color being detected in the same PMT in a multi-tube panel. For instance, PE-cy5 and PerCPcy5.5 are both detected using the same PMT; however, the spillover of PE-cy5 into the adjacent PE channel will be different from the spillover of PerCP 5.5 into PE and will therefore need a separate compensation value. This changes the signal/noise ratio along with the compensation needed to account for that difference. In some cases, different lot numbers of the same conjugate, oftentimes seen when using tandem fluorochromes, may require different compensation settings due to intensity difference between the lot numbers. These settings are now considered lot-specific as opposed to label-specific. Label- and lot-specific settings allow both antibodies to be used in the same experiment since the compensation now includes calculations for both fluorochromes.
3. The compensation panel can be set up in a few different ways. Some instrument software compensation programs use single-color beads in several different conjugates along with a non-labeled (unstained) bead. Compensation software programs require a negative/unstained population of some kind, and so an unlabeled bead is first acquired using the PMT voltages previously determined during optimization. This establishes the background signal from which everything beyond this signal is considered positive. Single-colored/stained beads are subsequently acquired, one for each conjugate/lot number in the experiment, with no compensation applied. After all the tubes are acquired, any spillover that might be present in more than one channel, because the single color should only be positive in the channel intended, can then be subtracted from that unintended channel. The compensation program then calculates and applies mathematical compensation to each signal until there is no spillover from the opposite color.
4. The file format used for flow cytometry results is designated ".fcs" as part of each listmode filename. All results acquired as fcs files version 3.0 and above are acquired uncompensated;

Compensation Beads

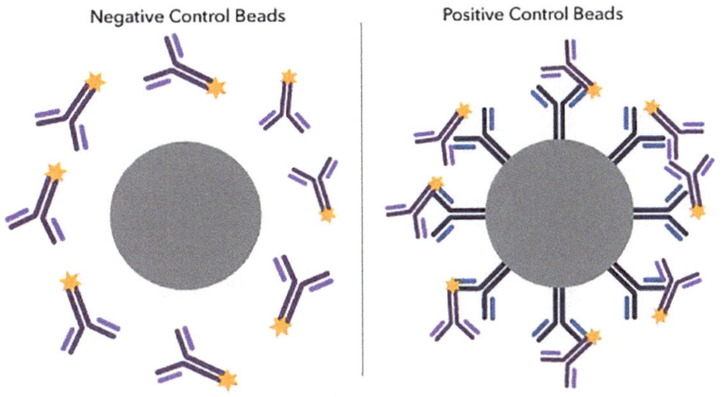

Fig. 18 Compensation beads. Schematic of a negative and positive compensation bead particle highlighting the positive control bead coating with the species-specific antibody (blue), which binds to fluorochrome-conjugated antibodies (purple)

however, the compensation calculation is included as a keywork in the file that can then be applied or turned off, post-acquisition. This allows the user the option of using the calculated compensation values that are included in the file or even adjusting to a more suitable setting if desired.

Compensation can be set up in one of three ways or combinations of the three ways. The first method uses vendor-supplied single-colored beads, mentioned previously, that are surface-coated with the fluorochrome conjugates such as a FITC bead, PE bead, APC bead, etc. Another method is to use beads that are coated with anti-IgG, also known as "capture beads" or "Comp beads." Comp beads can be stained with any antibody conjugate that is used in the experiment. Comp beads include both an unstained bead and another bead, which is surface-coated with anti-mouse IgG-kappa that reacts/stains positive with almost any antibody in the panel, as long as it is an IgG subclass (see Fig. 18).

The advantage of using comp beads is that, unlike cells, they are non-specific and will react with almost any antibody, so that a positive control for each antibody-fluorochrome combination can be used to calculate the compensation. A disadvantage of using comp beads vs. cells is that beads do not always demonstrate the same fluorescence intensity as a stained cell. If the bead signal is very bright or is at the upper end of the scale, turning down the PMT voltage to lower the signal changes the setting you worked hard to optimize. Positive staining cells, on the other hand, may be difficult to come by, especially for rare antigens. However, cells may more closely resemble the signal intensity of those used in the experiment, so a compensation setup using cells may be more accurate than using comp beads or single-color beads. Another disadvantage to using beads is that they will not work with some panel reagents, such as 7-AAD or Syto-6, which are dyes, not antibodies, so substituting antibodies with those fluorochromes may be necessary.

5. Another consideration when setting up compensation using stained cells is deciding what sample types to use. The stained control samples need to be at least as bright as any that will be used in the experiment, since it is the brightest positive signals that can spill over into the adjacent channels. One of the best sample types and antibodies to use in the compensation panel is CD8-stained lymphocytes. CD8 is a high-density antigen on lymphocytes, and thus whatever conjugate is used, CD8 will be as bright or even more brighter than other common antigens. It is also negative on CD4+ T-lymphocytes, so there is both a positive and negative population present for calculating the compensation. If CD8 is not available in every conjugate of the panel, choose one that will have a positive signal intensity of at least the thirrd decade on a log scale. Other antibodies to consider would be either/or both CD33 and CD14. CD33 gated on all WBC's in LWB should demonstrate a bright positive monocyte population, intermediate positive granulocytes, and be negative on lymphocytes. CD14 will be very bright on monocytes and negative on lymphocytes and granulocytes. CD16 and CD15 are other antibodies to consider using to set up compensation. CD16 will be very bright on neutrophils, intermediate positive on NK lymphocytes, and negative on monocytes. Additionally, CD15-stained granulocytes will have a very strong positive signal, and it is negative on lymphocytes. Avoid using cell cultures to set up comp panels unless your experiment uses these cell sources, due to the increased background/noise or auto-fluorescent signal they often have compared to normal WBC's. In most cases, the calculated compensation from the controls will be acceptable for most, if not all, the parameters in the panel. See the Notes section of this chapter for a procedure for setting up compensation (Sect. 4.7).

3.3.5 Compensation Controls

Introduction: Once the compensation is set up, it is important to test the settings on stained cell samples. Cells are preferred over beads since the samples in your experiment will, in most cases, be cell-based. If possible, use these same cells from samples you will be using routinely. Testing the compensation settings can be done in a few different ways [24, 25]. One of the most widely used methods uses a panel called "Fluorescence-Minus-One (FMO)" controls.

1. Fluorescence-Minus-One (FMO) Controls

FMO controls are one of the most widely used ways for checking compensation settings. The advantage of the FMO panel is that it allows the user to observe any compensation issues for all of the conjugates used in the panel. If PE is not included in tube 2, any positive signals in PE must be due to spillover from another detector. The compensation can then be adjusted to remove the unwanted signal. For this procedure, stained cell samples are set up and include tubes for every conjugate found in the experiment, minus 1 color. For example, a 10-color experiment would require ten tubes that include every antibody in your cocktail except one.

Example of an FMO compensation control panel:

a. Tube 1: **(blank)**/FL2/FL3/FL4/FL5/FL6/FL7/FL8/FL9/FL10.

b. Tube 2: FL1/**(blank)**/FL3/FL4/FL5/FL6/FL7/FL8/FL9/FL10.

c. Tube 3, 4, etc., until each color has been omitted once. Continue building the panel for all ten tubes, omitting a different color in each tube.

d. Run the FMO Pane on the flow cytometer using the PMT and compensation settings determined previously.

e. Build a worksheet of 2-parameter dot plots that include the offset or bi-exponential display of each negative/blank channel vs. each of the other positive channels. Bi-exponential dot plots are used to demonstrate both over- and under-compensation since the baseline is set to below the origin. See Fig. 19, which shows the difference between a dot plot with and without an offset/bi-exponential display.

Using mean fluorescence intensity (MFI) values is another way to check for proper compensation. The compensation can then be adjusted until both the negative and positive Y-axis MFI are somewhat similar, as well as visually proportional, as seen by the "fishtail" or "trumpet" pattern of the bright positive population (see Fig. 20).

Once reviewed and adjusted, the compensation settings are ready to be used in the experiment instrument settings.

Despite their utility for checking and adjusting compensation, FMO panels can be complicated, expensive, and time-consuming since they require multiple antibodies and strict attention to making sure each tube has all of the correct markers as well as the correct number of antibodies included in each tube. For instance, a 10-tube FMO panel would require 90 antibodies or nine antibodies in each of the ten tubes to be fully utilized. If a fully stained tube is included, the number of antibodies required for the panel increases to 100.

An alternative, less complicated and less expensive method to using FMO controls is to set up a panel that is just the opposite of

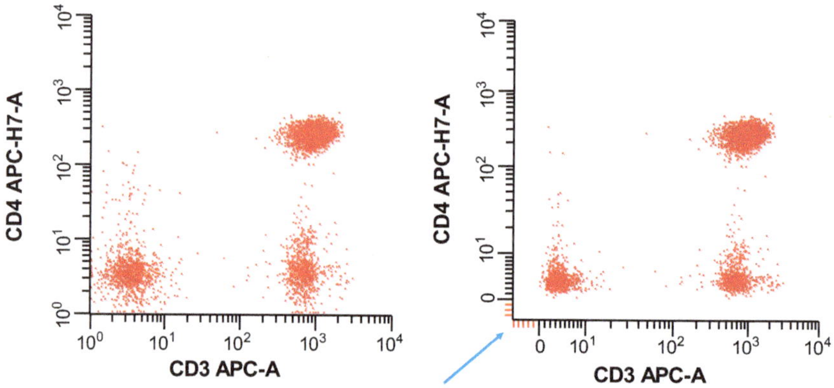

Fig. 19 Log vs. offset/bi-exponential dot plots. The dot plot on the left shows the origin of each axis at the bottom. The dot plot on the right is the same dot plot with the axis offset/bi-exponential setting turned on, allowing visualization of the events both above and below the origin (blue arrow) for demonstrating compensation

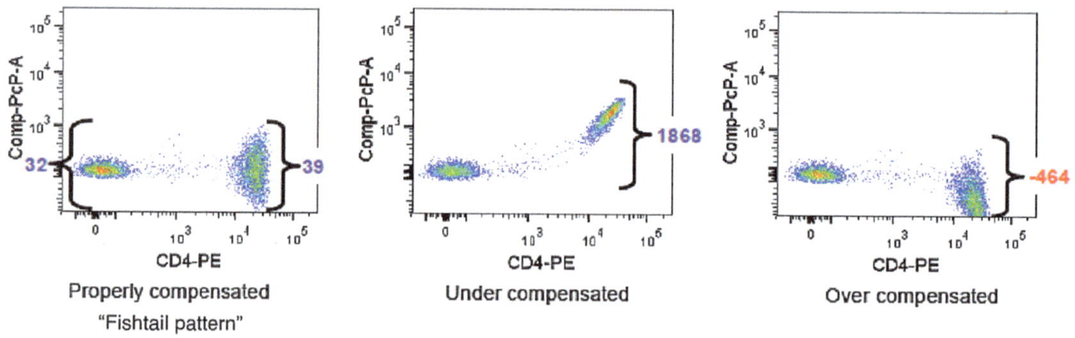

Fig. 20 Examples of compensation using MFI values illustrate what (**A**) correct, (**B**) under- and (**C**) over-compensation looks like using three different settings. The ideal compensation setting (Dot plot **A**) shows the Mean Fluorescence Intensity (MFI) of the positive peak to be very close to the MFI of the negative peak along the Y-axis. This is sometimes referred to as the "fishtail" or "trumpet" pattern. Under-compensation in the middle figure (**B**) shows the MFI of the positive peak to be significantly greater than the negative peak MFI. The MFI of the right dot plot (**C**) shows the positive MFI to be significantly less than the negative peak MFI and thus is overcompensated

FMO, called "Fluorescence-Minus-*All-But*-One (FMABO)". The FMABO panel is set up using multiple single-stained tubes again, but instead of leaving out one antibody/conjugate in each tube, leave out all but 1 so that a positive result is seen in only one channel. Again, build a worksheet and create offset dot plots of every possible combination and compare the positive MFI result in each file to the other unstained/negative parameters. If there is any spillover detected in the empty channel's vs. the single fluorochrome, the compensation settings can then be adjusted to remove spillover or correct for overcompensation. Include an additional fully stained tube as the last tube in the panel that has reactive antibodies for each parameter in the experiment. This is used to demonstrate all of the compensation settings simultaneously and may be useful for making any final adjustments. See the Notes section for an example of the FMABO panel setup.

Compensation Settings Summary: Multicolor flow cytometry experiments are only as good as the compensation settings used to correct for the spectral overlap of all the color-conjugates found in the panel. Once the compensation is set up initially, daily monitoring is required to ensure that those settings are still valid. Minor adjustments can then be made to the compensation to correct for any drift. If, on the other hand, the PMT settings change by more than 10%, or there is major maintenance performed, the compensation setup should be repeated since either of those situations will likely change the spectral overlap of some or all the conjugates being used. An experiment that has been set up on a flow cytometer that is properly quality-controlled, optimized, and color-compensated helps ensure that results are accurate, precise, and dependable.

3.4 Miscellaneous Laboratory Equipment Quality Control

Immunophenotyping laboratories have other instruments besides the flow cytometer that are required for setting up an assay. Examples include refrigerators and freezers, centrifuges, biological flow hoods, and manual or automatic pipettes, to name a few. These devices must be maintained to be sure of proper operation and suitability for use. If the refrigerator or freezer that you store antibodies in has wide fluctuations in temperature over time, it could damage the reagent. A centrifuge has rotors, safety latches, and timers that should be checked regularly to ensure safe and proper operation. Biological flow hoods are safety apparatus that should have regular scheduled maintenance to ensure potential hazardous material and aerosols are contained. Most laboratories use private companies that specialize in maintaining this sort of equipment. Records of regular maintenance of miscellaneous equipment and daily monitoring of the temperatures are considered a good laboratory practice and are required by some laboratory regulation agencies [26].

Analytic quality control conclusion: daily monitoring of the flow cytometer performance, including alignment checks, background and sensitivity limits, fluidics, PMT performance, and laser output, is a standard of best practice in flow cytometry laboratories and may also be required for regulatory compliance. Consistency and reliability on a day-to-day basis define quality assurance. Tracking the results of QC over time may reveal trends that could predict future instrument problems. Understanding how each component of the instrument is performing: fluidics, optics, alignment, and electronics, helps ensure results that are reliable, accurate, and precise. Besides the flow cytometer, keeping maintenance and operation records of other miscellaneous equipment in the laboratory, such as refrigerators, freezers, biohazard cabinets, and centrifuges, is also important for meeting quality assurance best practice standards.

3.5 Post-Analytic Quality Control Procedures

3.5.1 Quantitative Assay Quality Controls

Once the instrument has successfully passed the analytic quality control procedures, the final step of quality control verification is to run sample controls. This includes both quantitative and qualitative controls [27]. For example, clinical laboratories that report quantitative results such as lymphocyte subset percent and absolute counts, or stem cell % and absolute counts, are required to run at least 2 levels of commercial or verified controls each day the test is performed. The controls should include both a normal control as well as an abnormal control or a control that is outside the normal range so that the full range of reportable results is represented. Quantitative commercial controls come with package insert values that must also be verified by the laboratory prior to being put into use. Oftentimes, the control values listed have a wide range of acceptable limits, which in many cases may be too broad to be useful. The laboratory should not only verify those limits but also calculate ranges of their own based on multiple runs. Using the mean ± 2 standard deviations after ten times will usually be enough to generate statistical ranges that satisfy the requirement. The laboratory calculated control ranges should now reflect what the reports on both normal and abnormal results will be, and not be below the lower limit of detection (LLOD) or exceed the analytical method range (AMR).

For those assays in which a commercial control is unavailable, the use of a known positive sample verified by alternative methods or by another laboratory that performs similar testing can be an acceptable substitute. Records of the controls and how the results were calculated and approved are required for verification and acceptability for use.

3.5.2 Qualitative Assay Quality Controls

Qualitative controls, also known as method controls, test not only the instrument but the staining and processing procedures from setup to acquisition to analysis. Every antibody in an experiment should be tested for accuracy prior to being put into use, whether as part of a cocktail or as a standalone reagent. This quality testing was mentioned previously in the Pre-analytic quality control section. Some commercial controls used for quantitative testing also include a list of antibodies that will react with the control for qualitative purposes. Method controls, on the other hand, are those controls used in an experiment that prove the setup, and staining procedures are working as indicated. As also mentioned in the Pre-analytic QC section, material tested using alternate methods such as Histology and Hematology can be used as method controls. If testing material is available from a recent proficiency survey, it too can be used as a method control.

3.5.3 Isotype Controls

Isotype controls are basically antibodies of the same immunoglobulin class as the primary antibody, including both the heavy and light chain molecules, but lacking the specific component of that antibody to react with the target antigen or are directed against an irrelevant target. Isotype controls can serve as method controls and many labs use these to demonstrate background staining or non-specific staining present when using the test antibody. Traditionally, setting a statistical gate on the positive population was based on where the negative signal ended for the isotype control. But there are some inherent limitations to using isotype controls in this way [28, 29]. In some instances, they may not exactly match the target antibody and may react with an unexpected antigen, also known as non-specific binding. This may introduce uncertainty about what is positive vs. what could be non-specific staining. Another aspect of using an isotype control is that it may have a different fluorescence to protein (F/P) ratio vs. the target antibody. If the F/P ratios are different, then the background staining may be different, and now the negative/positive boundary becomes subjective.

There are situations, however, in which the use of an isotype control can be helpful. Immunophenotyping panels designed to study intracellular antigens can be challenging due to the presence of more Fc receptors inside the cell than on the cell surface, which may result in more non-specific binding of the antibodies and thus a higher background signal. An isotype control under these circumstances can be used to know what non-specific staining looks like to more accurately determine the positive population. Another useful application of isotype controls is when a fluorochrome is being used for the first time to see what typical background staining looks like. This provides a starting point for optimizing the signal as part of setting up the instrument.

Rare event detection is another situation in which the isotype control may serve as a useful reference for determining positive vs. negative, especially in those samples that may have weak expression or low density of the antigen being investigated. However, a better strategy is to create a gating scheme using combinations of gates, also known as Boolean logic. This gating approach can be much more specific than single gates alone; for example, the use of the ISHAGE protocol for counting stem cells [29]. As the population of interest is refined by the sequential gates, the analysis may provide a more accurate assessment of determining positive vs. negative results rather than running a separate isotype control.

3.5.4 Post Analytic Quality Controls: Internal/Intra-Assay Controls

Internal/intra-assay controls are method controls that occur naturally in a sample and demonstrate positive vs. negative for each cell type being studied. For example, lysed whole blood WBCs include lymphocyte, monocyte, and granulocyte populations. An antibody

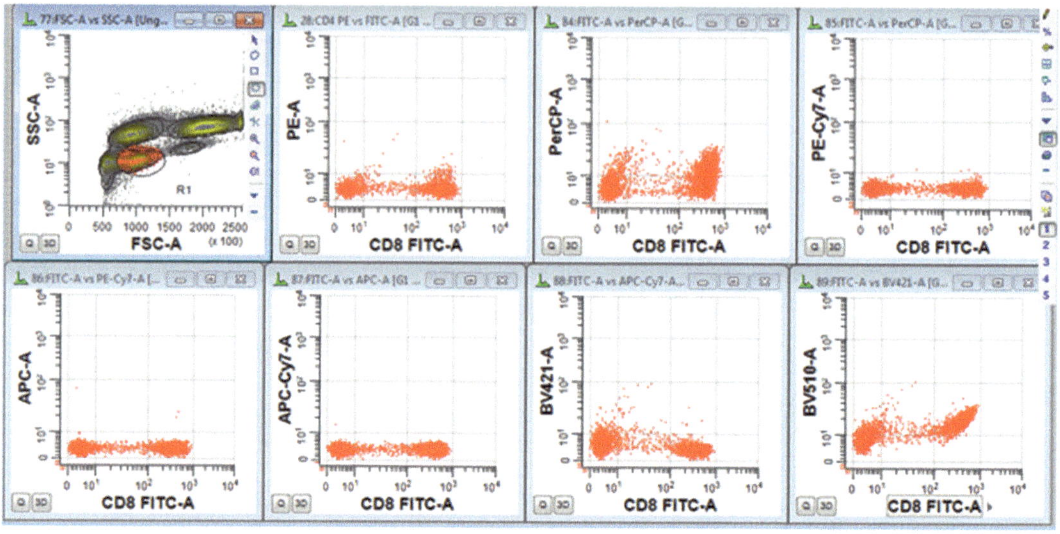

Fig. 21 *FMABO* CD8 FITC control vs. each of the other PMTs used in the experiment. In this example, CD8 FITC vs. each fluorochrome shows good compensation settings for most of the combinations. However, CD8 FITC vs. BV421 is slightly overcompensated, while CD8 vs. BV510 requires additional compensation

like CD33 reacts with monocytes and granulocytes but is negative on lymphocytes. CD19 is positive on B cells but is negative on T and NK cells, as well as monocytes and granulocytes. In each of these examples, there is one or more positive populations and one or more negative populations. In most bone marrow specimens, other examples can be used within the tube to confirm different antibody reactivities, both positive and negative, using this same approach (see Fig. 21).

An advantage of using internal/intra-assay controls is that now the test antibody's reactivity can be described more accurately because the test conditions, F/P ratio, and isotype antigen are all in the same tube compared to using a separate isotype control tube.

Each lab should establish its own criteria for describing the use of internal controls as a method and cocktail quality control check. Abnormal populations typically found in malignancy may aberrantly express antigens or be missing antigens that are expected to be present on normal cells. In these cases, clear descriptions of normal reactivity should be included with each antibody combination used in the experiment as part of the procedure's Interpretation of Results section.

3.5.5 Post Analytic Quality Control Methods: Using Time as a Parameter

Flow cytometer list mode data is stored in a specific format that includes not only a record of every signal or event but also the time point at which that event was detected. This additional measurement is known as the Time parameter. The Time parameter can be a useful quality control indicator to alert the user to problems with the

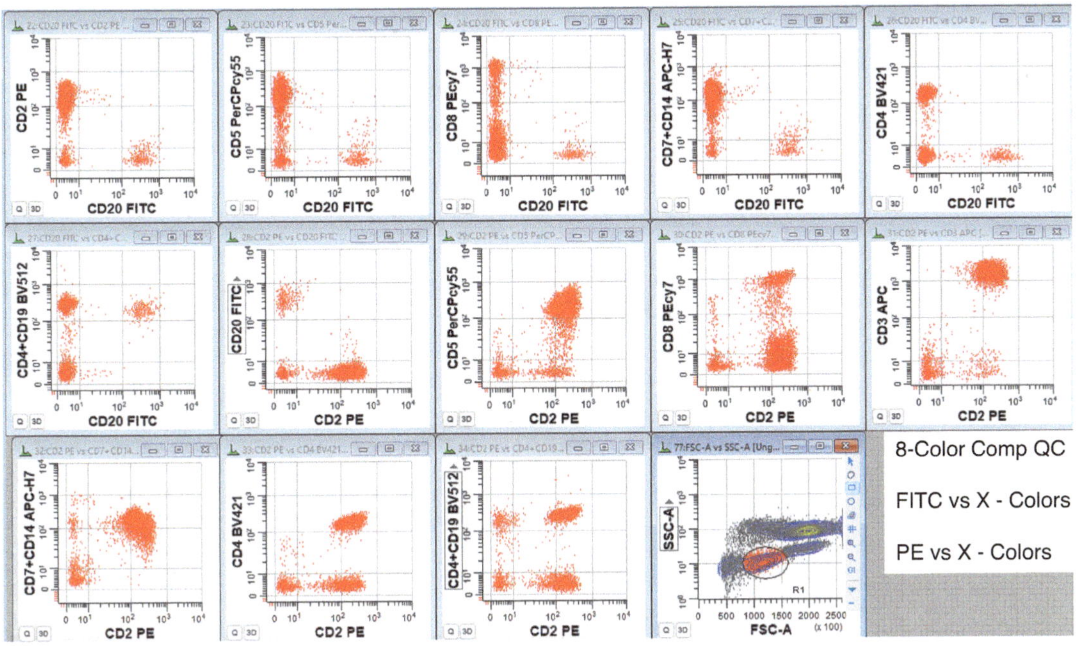

Fig. 22 Example of 8-color compensation. Multiple dot plots of FITC and PE vs. the other seven colors in an 8-color experiment using a cocktail that includes CD8/CD4/CD5/CD2/CD3/CD77/56 or CD16/CD19+CD8. CD19 and CD4 are paired with the same conjugate (BV 510) to observe both co-expression and single staining. This example only shows two of the possible combinations, FITC vs. X and PE vs. X, but additional combinations can be set up by changing the X-axis to a different fluorochrome. This worksheet allows instrument compensation settings to be monitored for any multi-color experiment.

sample acquisition [30]. For example, when an acquisition is either interrupted by something blocking the fluidics or if the sample runs dry and bubbles are created. These situations can result in a loss of signal or the inclusion of bubble events in the list mode file. Bubbles will appear as collected events, especially on light scatter. Looking at a dot plot of time vs. another parameter should normally show a steady sequence of event acquisition, but if there is an interruption or bubbles are detected, the dot plot will demonstrate a noticeable change in the pattern or shift of a parameter. This interruption in the acquisition or inclusion of bubbles can be identified using a Time dot plot. Results before the interruption should look continuous; however, an interruption due to a fluidics issue or introduction of bubble events will be evident on the Time dot plot. The good news is that the results of an otherwise faulty acquisition can be salvaged without having to set up/acquire the sample again by gating out the problem. For example, using Time vs. FL4 or other fluorescence parameter, draw a gate that now includes only the events before the interruption or when air was aspirated. Subsequent dot plots gated on the Time dot plot should be clear of the previously acquired interruption or air bubble events (see Fig. 22).

3.5.6 Post Analytic Quality Control Methods: Singlet Gates

Another issue that can impact the quality of your results is when the sample contains doublets or cellular aggregates. Doublets are when two or more cells, for whatever reason, are bound together as they pass through the flow cell chamber and are counted as one event. Preventing this misidentification can be achieved by drawing a gate on the light scatter that excludes the doublets. This is known as a Singlet gate. Using a singlet gate is an important consideration in immunophenotyping analysis. Doublets can happen when specimens undergo several centrifugation steps, such as preparing bulk lysed samples. Bulk lysing is often used in experiments that look for rare events, especially measurable/minimal residual disease (MRD) cases. To capture these rare events, more specimens are required than would be used under routine circumstances. In a bulk lysing procedure, the use of 1 ml of sample instead of 100 μl increases the number of cells tenfold. Now there are more WBCs, and this is where the problem of doublet/aggregate combinations is likely to occur, especially if there are multiple centrifugation steps during the sample washing and lysing. If doublets are included in the analysis, the results may be misidentified as something abnormal, such as a cell, appearing to co-express both a T-cell antibody and a B-cell antibody when, in fact, it is two separate cells that are a doublet. To eliminate doublets from the analysis, draw a dot plot of Forward Scatter Area (FSC-A) vs. Forward Scatter Height (FSC-H). Please note, some instruments measure FSC width instead of area, and if so, use that signal vs. the FSC-height. Doublets have a larger surface area than single cells and can be seen falling toward the FSC-Area (or Width) axis. Draw a gate around the singlet population found along the diagonal axis of the

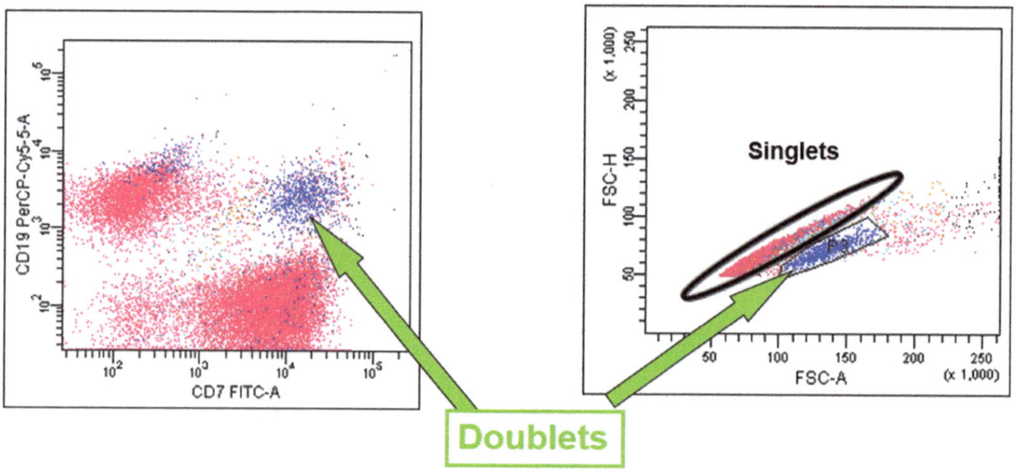

Fig. 23 Intra-assay controls. Antibody reactivity can be assessed using intra-assay controls. Dot plot (**A**) shows lymphocytes (L) in green, monocytes (M) in red, and granulocytes (G) in gold on Forward/Side light scatter. In dot plot (**B**), lymphocyte staining can be used to assess antibody performance since, in the example shown, CD19 is positive for B cells and negative for T cells, monocytes, and granulocytes. CD33 is correctly positive on monocytes (bright) and granulocytes (weak) and is negative on lymphocytes

dot plot to exclude the doublet events (see Fig. 23). This should be the first gate used in the analysis so that subsequent results are only those found within the singlet gate.

3.5.7 Post Analytic Quality Control Methods: The Viability Gate

Immunophenotyping viable cells is imperative because a dead cell can stain non-specifically since the cell membrane is now permeable. Samples received for immunophenotyping, if collected locally or within a day's distance, will usually have high levels of viability, especially if the recommended temperature and transport media are used to deliver the samples. However, in some circumstances, even though all efforts are made to ensure that the sample is viable, the sample, by its very nature of collection, may contain non-viable cells. Apheresis specimens submitted for stem cell counts are a prime example. The mechanical means by which the product is collected, as well as the length of time to perform the apheresis procedure, can result in some portion of the cells dying. Since stem cell counts are typically less than 1% of the overall sample, accurate gating to omit the non-viable cells is crucial to the experiment. Other sample types with the potential for lower viability would include CSF and tissue biopsies, especially if not in appropriate transport media like RPMI. Using a viability dye such as 7-actinomycin D (7-AAD) in the immunophenotyping panel can be very beneficial for identifying the dead cells [31]. These cells typically have a damaged cell membrane, which allows the 7-AAD dye inside the cell and ultimately binds to the disulfide bonds of the DNA molecule in the nucleus. If present, the 7-AAD, being excited by the 488 nm blue laser, emits a signal in the same channel as PerCP (FL3) and thus the cell is considered non-viable. Including a viability gate in the gating scheme is highly recommended and even required if you perform clinical testing on samples over 24 h old. The viability gate can either be used before or after the singlet gate in the sequential gating scheme. See Fig. 24 for an example of how to remove dead cell events from the analysis.

3.6 Post-analytical Quality Assurance Using an External Quality Assessment (EQA)

Regulatory requirements for clinical labs performing immunophenotyping include participating in an external quality assessment program (EQA), also known as proficiency surveys [32]. Laboratory regulatory agencies, such as the College of American Pathologists, require that a laboratory reporting clinical results of any assay participate in an EQA that is designed to test each aspect of the testing and reporting procedure. The laboratory's performance is then compared and graded against other laboratories performing the same testing. Proficiency surveys typically occur 2–3 times per year and will usually include two or more samples with each mailing. Proficiency testing must follow the laboratory's written procedures for all the steps used to test and report patient samples. This not only provides information about the validated procedures but also measures the performance of the instruments used, acceptability of the reagents, and should, over time, even demonstrate potential issues that may not be as obvious on a day-to-day basis. For example, if the returned

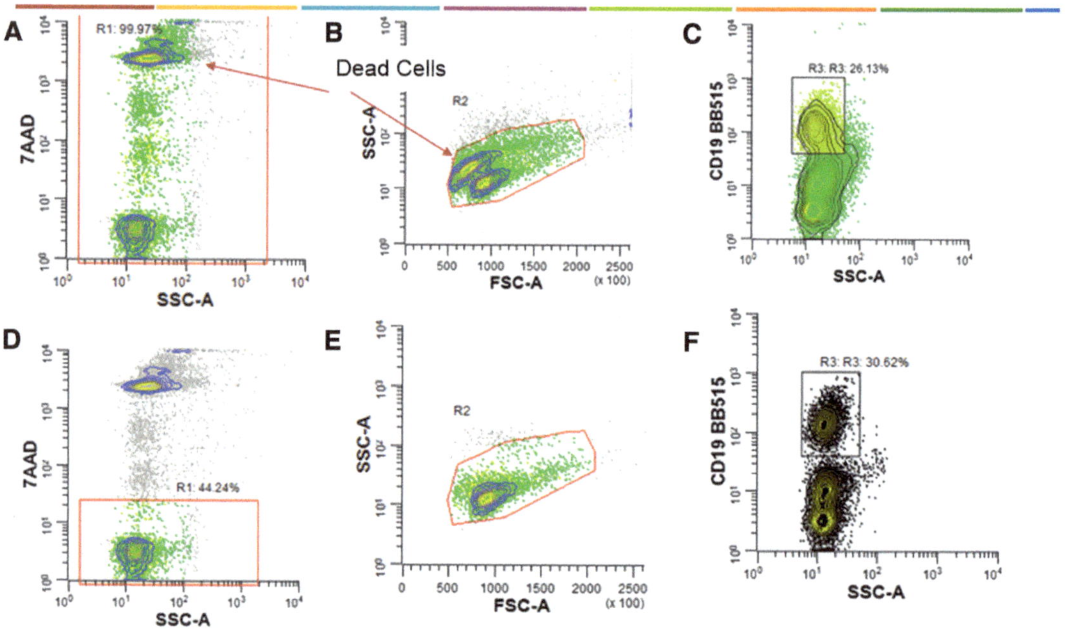

Fig. 24 Using time as a gating parameter. The FSC/SSC dot plot on the left shows what happens when a sample is run dry. This introduces air bubbles, which appear as multiple black dots in the upper area of the Light Scatter dot plot. The middle dot plot, Time vs. CD45, shows a clear break where the bubble events started being acquired. To salvage these results without repeating the sample, a gate is drawn around the "clean" events to the left of the break to omit the bubble events. Subsequent dot plots can be based on this Time gate. For example, the Time-gated FSC/SSC dot plot on the right no longer shows the bubble events

results for CD8 in an immunodeficiency panel are acceptable but consistently trend higher than the mean of the other participants, it may indicate either a reagent or gating issue that might otherwise have gone unnoticed. Three or more returned results that are all above or below the participant mean should be investigated for possible causes of the bias (see Fig. 25).

3.7 Quality Control Records

Quality Control Records: Testing in immunophenotyping clinical applications requires records of performance to be maintained and retained for a certain period [33]. There is a saying that goes "If you did not document what happened, then it did not happen." This is the purpose of quality assurance and quality control records are the proof of that purpose. Each step in the process: pre-analytic, analytic, and post-analytic performance must be documented for acceptability. Best practices in quality control management include having an overall Quality Assurance plan along with the records of procedure and instrument validations, quality control of reagents used for testing, daily instrument performance and control results, assay records of specimen analysis that include the gating diagrams, and any proficiency testing results. For a summary of records that

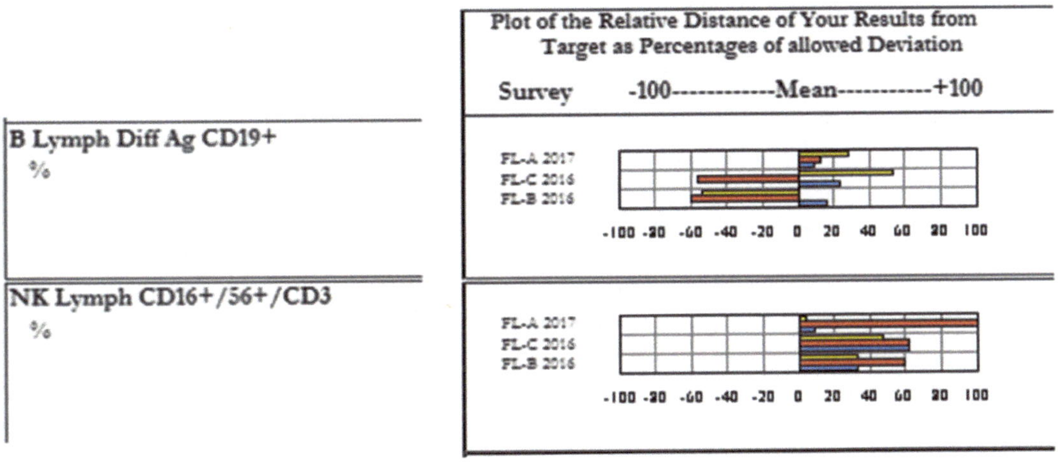

Fig. 25 Using a singlet gate to omit doublet events. The blue population before singlet gating appears to be positive for both CD19, a pan-B cell marker, and CD7, a T cell marker. Normal B cells should be negative for CD7 and normal T cells should be negative for CD19. Using a singlet gate that excludes doublets shows the red populations to be as expected for both CD7 and CD19. Singlet gating is a good practice regardless of the sample type since doublet acquisition may happen quite frequently

should be readily accessible and up to date, see the Notes section of this chapter.

3.8 Key Performance Indicators (KPI)

Key Performance Indicators (KPI) are quantifiable measures over time with a specific objective. Laboratory quality assurance should be considered an ongoing process and KPIs are a useful approach to gauging progress and insights to help understand what that looks like. The laboratory should identify and monitor KPIs with regular reporting of both positive and negative causes and effects that may be used for improvement. Some of the more common KPIs in a laboratory setting are specimen include:

- Turn-around time (TAT): this is useful for understanding the processes from beginning to end and how they can be impacted by delays, miscommunication, wrong specimen types, instrument issues, analytical questions, training or competency assessment, and other factors.
- Troubleshooting or downtime: How often does it happen? Is it specimen or instrument related, or some other root cause? An instrument that has repeated QC failures may require service. What is the response time for service calls? Reccurring problems with an assay may indicate the need to revise a specimen handling procedure, the setup procedure, or identify reagents or test kits that may have quality issues.

- The frequency and variety of repeat testing are a common KPI. How often it happens and tracking the primary reasons for it happening often leads to identifying root causes that, when addressed, may lower the number of repeat tests.
- Specimen issues: includes monitoring the frequency and reasons for unacceptable specimens. Is there proper sample identification and labeling, use of the proper anticoagulant, and acceptable volume? How often are specimens lost or misplaced or damaged in transport? How often and why are specimens unacceptable? These problems can be used to identify systemic issues that may require training or education on the specimen requirement procedures. If samples come from outside locations, there may be a need to do the same education and training to reduce the number of unacceptable specimens or identify other causes for unacceptability.
- Corrected or amended reports: frequency and cause should be documented and tracked as a way to identify areas of improvement to lower this metric.

3.9 Conclusion

Quality control (QC) and quality assurance (QA) in immunophenotyping span all three phases of laboratory testing procedures: pre-analytic, analytic, and post-analytic. As part of an overall quality program, performing and documenting what happens in each of these three phases helps ensure the best outcome of an experiment. Good laboratory practice requires that every aspect of the assay should have measurable methods for ongoing monitoring, including the quality of the reagents, instrument performance, as well as the assay's expected results. Records of QC for each procedure should be well-maintained and up to date, and include the signature of the person or persons who oversee the testing. Monitoring key performance indicators (KPI) is a way to track quality control performance over time.

4 Notes

4.1 Reagent Recipes Used in Immunophenotyping

1. PBS Buffer, pH 7.2, w/sodium azide. Some laboratories include bovine calf serum in their PBS buffer as an additional stabilization agent for the cells. This recipe does not include that additive.
   CAUTION: Sodium azide is an irritant and a poison. Avoid all direct skin contact, wear gloves, and wash hands thoroughly after handling.
   Materials:
   a) PBS, 1×, Mediatech, Inc, Cat. # 21-040-CM

b) Sodium Azide, Fisher Scientific, #S2271-25

Working solution:

a. Add 1 g of sodium azide to 1 l of plain PBS.
b. Mix until dissolved.
c. Check that the pH is between 7.2–7.4.
d. Label and date (date made, and date expired).
e. Store at 2°C–8°C when not in use. Expiration = 3 months.

2. Ammonium chloride (NH4Cl) RBC lysing agent:
Materials:

a. Ammonium chloride, Fisher Scientific, #A661-500
b. Potassium bicarbonate, Sigma Aldrich, 50-180-9064
c. Tetrasodium EDTA, Fisher Scientific, 50-136-8256

Working solution:

a. Place the following into a 1 l of volumetric flask:
8.26 g of ammonium chloride

- 1.0 g of potassium bicarbonate
- 0.037 g (37 mg) tetrasodium EDTA

b. Add deionized water to 1-l line in the flask. Mix all ingredients until dissolved.
c. Check pH: 7.2–7.4. Discard, if >7.4.
d. Dispense into 100 ml bottles. Fill bottles to near the top, so there is no air space*
e. Cap tightly. Put parafilm around the cap to prevent air exposure.
f. Label and date (date made, and date expired). Expiration = 1 week. Store at 4°C until ready to use.

*Note: there should be as little air on top of the lysing solution as possible. Air will cause a decrease in the pH over time and lower the effectiveness of the lyse solution.

3. 7-AAD viability dye
Materials

a. 7-Aminoactinomycin D, Thermo Fisher, Cat.#A-1310, 1 mg.
b. Dimethyl sulfoxide (DMSO), Fisher Scientific, Cat.#BP231-500.
c. 500 ml Flow PBS, stock solution.

7-AAD stock solution:

a. Add 1 ml DMSO to 1.0 mg 7-AAD.

b. Store at room temp/in the dark for up to one month.

7-AAD working dilution

a. Use 5 μl of 7-AAD stock solution/1–2 million cells.

4.2 4-Color T-Cell Subsets Cocktail Recipe for 10 Test Aliquots

Antibody	Volume/test[a]	Antibody total	Buffer volume
CD3 FITC:	10 μl/test	100 μl	650 μl
CD4 PE	10 μl/test	100 μl	
CD8 PerCPcy5.5	10 μl/test	100 μl	
CD45 APC	5 μl/test	50 μl	

[a]Volume/test is based on titration results

4.3 Antibody Titration Procedure

1. Make serial dilutions of the antibody beginning with the package insert recommended volume, followed by dilutions of 1:2, 1:4, 1:8, etc. The diluent used can be a simple PBS buffer with/without calf serum, which is reliable for antibodies that have a high expression such as CD45 and CD8. In some instances, especially very weak expressing antibodies, should be diluted with unlabeled/unconjugated antibody to ensure adequate antibody saturation staining.
2. Include a tube with no antibody (a.k.a. unstained) for comparative purposes.
3. Use the sample type that is typically tested, such as peripheral blood, bone marrow aspirate, or tissues. It is important to have both positive and negative populations in each tube, so cell lines might not be the best choice if no negative populations are typically present in those samples.
4. Treat all samples following standard staining procedures used in your laboratory, including validated incubation times.
5. Run the samples on the flow cytometer with the same instrument settings used for testing purposes.
6. After acquiring the samples, graph the Stain Index vs. the antibody concentration to determine the best overall concentration to use.

4.4 Antibody and Reagent Quality Control Frequency

1. Initially, either new lot and/or new shipment, and after any new reagents or stock cocktails are made.
2. Each day, antibody cocktails and reagents for quantitative assays such as lymphocyte subsets and stem cell panels should be quality controlled using two levels of positive cellular controls that give quantitative results and be within the laboratory-established analytic method range (AMR). One control should be within the normal range and a second (abnormal) control with show results outside the normal range.

3. Each day, appropriate positive and negative staining patterns for each cocktail used should be documented for acceptability by the pathologist or whoever reviews and signs out the results if the cocktails are used for immunophenotyping.
4. Once a week, examples of all cocktails should be verified for efficacy/accuracy/and good compensation by reviewing results from previously submitted cases. The results should include examples of both positive and negative reactivity for all the antibodies used in the cocktail. Records of these reviews should be kept with the quality control records of the lab for historical and inspection purposes. All results in question should be reviewed by the laboratory director for corrective action, if necessary. Unacceptable results should be repeated.
5. As needed, the verification of new tandem or mono-color antibodies, in a cocktail comparing old lot numbers to new lot numbers, should be documented in the New Cocktail QC log.
6. Periodically, examples of compensation verification using fresh samples stained with a multi-color antibody cocktail should be reviewed for compensation accuracy. The verification cocktail should include antibodies that demonstrate both positive and negative reactivity. Acquire the verification tube(s) using current assay PMT and compensation settings. The results should include a worksheet with two-parameter dot plots of the most common and highly compensated fluorochrome pairs. Results should be reviewed for acceptability by either the laboratory supervisor or their designee.

4.5 Instrument Quality Control Workflow

1. Prepare to use the instrument by turning it on and allow sufficient time to warm up the lasers and electronics.
2. Check/refill fluids: sheath tank, cleaning solutions, etc., and empty waste container, add bleach or other neutralizing hazardous waste solution to waste container.
3. Perform scheduled routine maintenance: clean interior of flow cell using an approved medical device cleaning solution, change fluidics filters, once or twice a month, perform an extended whole instrument clean ("long clean").
4. Open Operation Software and initiate instrument fluidics startup if applicable.
5. Perform any periodic maintenance following the manufacturer's guidelines.
6. Prepare QC material: beads, stained cells, etc.
7. Open the Quality Control software program and run beads or some other vendor-suggested material according to the Instrument Setup procedure to check the alignment, sensitivity, precision, and accuracy, electronics/PMT settings, laser

power output, fluidics checks, compensation checks, compensation setup (optional), compensation monitoring (daily), and quantitative commercial controls (if applicable).

8. (Optional step, see step 13 for details) Optimize the instrument settings whenever a new QC bead lot number is put into use, or when there has been major maintenance such as a scheduled PM, laser, or PMT replacement, or laser/flow cell alignment. If you add new parameters to your assay, optimizing those detectors is suggested. If new optimized settings are established, the compensation settings must also be re-established. See Step 13 for details of the compensation setup procedure.
9. Perform compensation monitoring daily using either beads or stained control cells, which includes all fluorochromes used in that day's experiments to assure proper setup for running cell samples. If three or more parameters need adjustment to the compensation settings, then the full Compensation Setup procedure (see Step 14) should be performed using either beads or single-stained cells.
10. Run quantitative commercial controls if those assays are performed in your laboratory. Two levels, including a normal and an abnormal, are required for regulatory purposes. Package insert ranges must be validated before putting into daily use.
11. Record QC results and comment on any out-of-range results. If three or more parameters are out of range or fail specifications or if a single parameter is out three consecutive times, proceed to Troubleshooting. Document any troubleshooting performed and its resolution if the problem is fixed or otherwise addressed.
12. If troubleshooting is required due to exceeding defined limits, document all steps taken to resolve the issue. If results are still out, further action such as contacting the instrument operational support or the vendor service engineer may be required. Document the service performed along with records of the instrument performance after any repairs are made.
13. Optional steps: optimizing the instrument settings
    a. Optimization of the flow cytometer is a procedure for adjusting the instrument settings to achieve maximum performance and is not required daily, but should follow the initial installation and/or any major maintenance, such as a scheduled PM or repair. Optimize the instrument settings using stained controls. A detailed procedure for optimizing the flow cytometer using stained cells can be found in the next Section of the Notes.
    b. If any instrument PMT settings are changed by more than 5volts as part of the performance QC setup (PQC), a new compensation panel should be set up and acquired, but

only after the instrument settings have been optimized. New compensation settings should also be performed:

- After any major changes in PMT settings as part of performance bead QC.
- Whenever a new antibody panel is introduced
- Whenever there are fluorochrome-conjugate changes or additions to an established panel. For example, changing CD45 FITC to CD45 APC, or replacing PC7 with PE-cy7, replacing FITC with Alexa 488, V450 with BV 421, V500 with BV510, APC- R700 with R718, etc.

14. Setting up compensation
    a. Establishing a compensation matrix is required whenever the instrument settings are either established initially or significantly changed. It is not necessary to set it up each day, but it does require monitoring it to ensure that it is appropriate for the panel being used.
    b. The compensation setup procedure uses either single-stained beads or cells with a high-density antibody, such as CD8, gated on lymphocytes. The compensation software program then calculates the amount of compensation to apply to each PMT signal to ensure accurate results.
    c. After the initial setup, compensation should be monitored and demonstrated to work each day the experiment/panel is performed. Some laboratories use a special cocktail that includes all the fluorochromes used in the panels and others monitor compensation post-acquisition using software analysis programs to demonstrate the correct compensation of each pair of fluorochromes.
15. For clinical laboratories that perform quantitative testing, at least two levels of controls with validated ranges must be run every day the flow cytometer is used for those assays. Controls with both normal and abnormal results should demonstrate that the instrument's accuracy and precision limits are still valid.
16. For laboratories that perform qualitative testing, running known positive samples should be included daily to ensure the reliability of the antibodies and cocktails being used.
17. Document all QC results and comment on any results that are out of range and any corrective action taken.
18. Perform scheduled maintenance on any ancillary lab equipment such as centrifuges, fume hoods, water baths, refrigerators and freezers, pipettes, or other miscellaneous laboratory equipment used for sample testing. Temperature-dependent devices should be monitored for maintaining the temperatures

within the acceptable operating range as determined by the laboratory.

19. The laboratory supervisor or designee should review and sign off on all QC results monthly.
20. Records of all QC results and maintenance. Troubleshooting and repairs should be kept for at least 2 years.
21. Proceed with sample testing.

4.6 Instrument Optimization Procedure Using Single-Stained Cells

Set up single tubes using normal samples and stain with one antibody-fluorochrome for each detector on the instrument and add an extra tube for an unstained sample. Follow the routine staining procedure typically used for immunophenotyping. Choose an antibody that demonstrates a positive, intermediate, and negative population. CD4 is a good antibody to use since it is found on normal lymphocytes and monocytes and thus has the three populations needed to determine the optimal setting for the PMTs.

An example of a CD4 8-color optimization panel would include the following tubes:

Tube 1:	Unstained sample
Tube 2:	CD4 FITC
Tube 3:	CD4 PE
Tube 4:	CD4 PerCP cy5.5 or PE-cy5
Tube 5:	CD4 PE-cy7 or PC7
Tube 6:	CD4 APC
Tube 7:	CD4 APC-cy7 or APC-H7
Tube 8:	CD4 BV421 or or V450 or Pacific Blue™
Tube 9:	CD4 BV510 or V500 or Krome Orange™

1. Start with the PMT settings from the performance QC beads, sometimes known as Lyse-Wash settings. Turn off the instrument compensation prior to running the sample. Compensation will be set up following the Optimization procedure.
2. Run the unstained tube and adjust Forward and Side Scatter light signals to display all three WBC populations clearly with a lower threshold just below the debris. Adjust each PMT so that the leading edge of the negative signal extends just beyond the background/noise channel.
3. Next, run the first stained sample tube. If using CD4 draw a gate that includes both the lymphocyte and monocyte populations on light scatter (see Fig. 26).
4. Next, draw either a single-parameter histogram of the fluorochrome being acquired, i.e., CD4 FITC = FL1, etc., or a

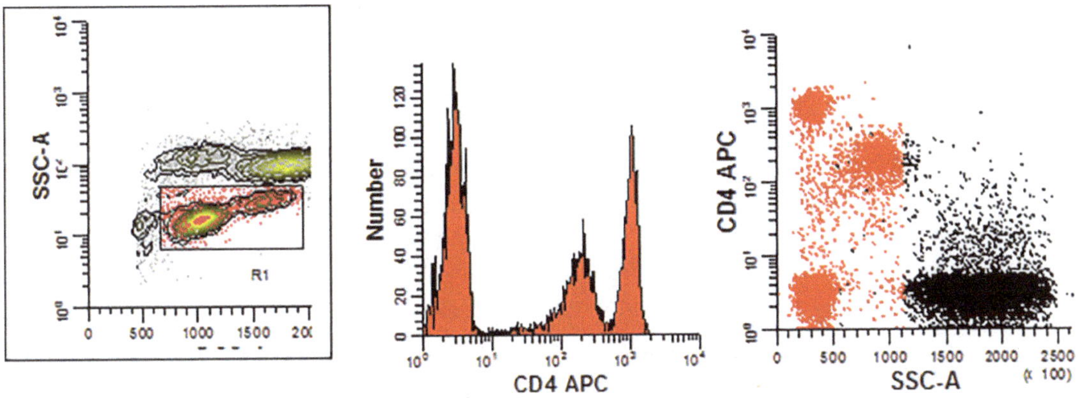

Fig. 26 Using 7-AAD to identify and exclude non-viable cells. 7-AAD emits a signal in approximately the same wavelength as the fluorochrome PerCP when excited by the blue (488 nm) laser. Events in dot plot A have a high number of non-viable cells, which take up 7-AAD, demonstrated by the very bright expression of the 7-AAD on the Y-axis. The nonviable cells can be seen in dot plot B, FSC vs. SSC, as the secondary population on the SSC axis just above the lymphocytes. Dot plot C, SSC vs. CD19, which includes the dead cells, underestimates the actual number of B cells present in the sample. Gating on cells that are 7-AAD negative in dot plot D results in a more accurate count of how many CD19+ B cells are present

two-parameter dot plot of SSC/FL1, FL2, FL3, etc. While running the sample, adjust the PMT voltages for each detector until the three separate populations can be clearly seen, like that of the CD4 histogram and dot plot seen in Fig. 26.

5. Continue running each single-stained tube, adjusting the PMT voltage for each parameter to achieve the best separation between the three populations. At this point, the PMT is optimized.

Save the newly optimized PMT settings. Once these are established, the instrument is now ready to set up the compensation.

4.7 Compensation Setup Procedure Using Stained Cells or Comp Beads

1. Set up single stained controls:
   a. If using stained Comp beads, the antibodies used in the experiment will work in most cases. This method is okay, but it will not always be as indicative of fluorochrome brightness as cells do.
   b. If using the stained cell method, select a fresh whole blood specimen that has a normal WBC and a normal distribution of lymphocytes. Compensation setup tubes should use an antibody found on bright-staining normal WBCs, such as CD8 on lymphocytes, CD14 on monocytes, or CD15 on granulocytes. Bright or highly expressed markers are the most likely to cause spillover into adjacent channels, therefore, the correct compensation setup would be more likely to prevent this from happening.
2. Antibodies and fluorochromes

a. Use one comp sample for each fluorochrome, i.e., CD8 FITC, CD8 PE, etc.
b. For label-specific settings, use one for each tandem lot number.
c. Include an unstained tube, which will be used as a baseline reference.

Below is an example of an 8-color compensation panel that uses CD8 with a different conjugate in each tube.

- Tube 1: unstained sample
- Tube 2: FITC, Alexa 488, BB 515, or 488 nm excitation wavelength conjugate
- Tube 3: PE
- Tube 4: PerCP, PE-cy5, or another yellow/orange conjugate
- Tube 5: PE-cy7, PC7, or other
- Tube 6: APC
- Tube 7: APC-cy7, APC-H7™, or other
- Tube 8: Pacific Blue, V450, BV421, or other
- Tube 9: Krome Orange™, V500, BV512, or other

d. Set up the panel using the same quantity/titer of each antibody that is used in the experiment and follow the laboratory staining procedure, incubation times, wash cycles, and resuspension media.
e. Follow the procedure for whatever compensation program your instrument is designed to use and save the calculated compensation settings.
f. Set up and run compensation verification controls. See next section.

4.8 Compensation Verification Procedures

New compensation settings should be verified for acceptability by running stained samples, preferably using the same fluorochromes used in the experiment. The most popular approach is using Fluoroescence-minus-one-controls (FMO). However, a less expensive and time-consuming method that may be just as informative and uses significantly fewer antibodies than the FMO method. In this panel, only one fluorochrome is used in each tube, representing all of the conjugates used in the panel. Stained cells are preferred, but comp beads will work too. This panel is referred to as the fluorescence-minus-all-but-one (FMABO) controls. This is a label-specific panel, not a lot-specific panel.

Example of the FMABO panel

a. Tube 1: CD8 FITC/FL2 blank/3-blank, 4-blank, 5-blank, etc.

b. Tube 2: FL1 blank /CD8 PE /3-blank, 4-blank, 5-blank, etc.

c. Tube 3: FL1 blank/FL2 blank/CD8 PerCP/FL4-blank, 5-blank, etc.

d. Tube 4, etc., until each color conjugate has been used once.

e. Last tube: a fully stained multicolor control tube. Each of the antibodies in the tube should react with a population in the sample as well as be negative for other populations. An example of the multicolor compensation verification tube:
CD20 FITC/CD2PE/CD5 PerCPcy5.5/CD8 PE-cy7/CD3 APC/CD7 or CD14 APC-H7/CD4 BV421/CD19 BV510*
*CD14 on monocytes is much brighter than CD7 on lymphocytes

f. Acquire the FMABO panel on the instrument using the previously calculated compensation settings. Using a bi-exponential/offset display, analyze each file one at a time and look at each positive result vs. the other fluorescence parameters used in the panel. This panel checks that the correct compensation is used between any two colors (see Fig. 27).

g. Analyze each FMABO control and change the X-axis to the positive parameter vs. the negative channel on the Y-axis for all subsequent tubes. Hint #1: for future use, save the display that includes each possible color combination so that it can be used

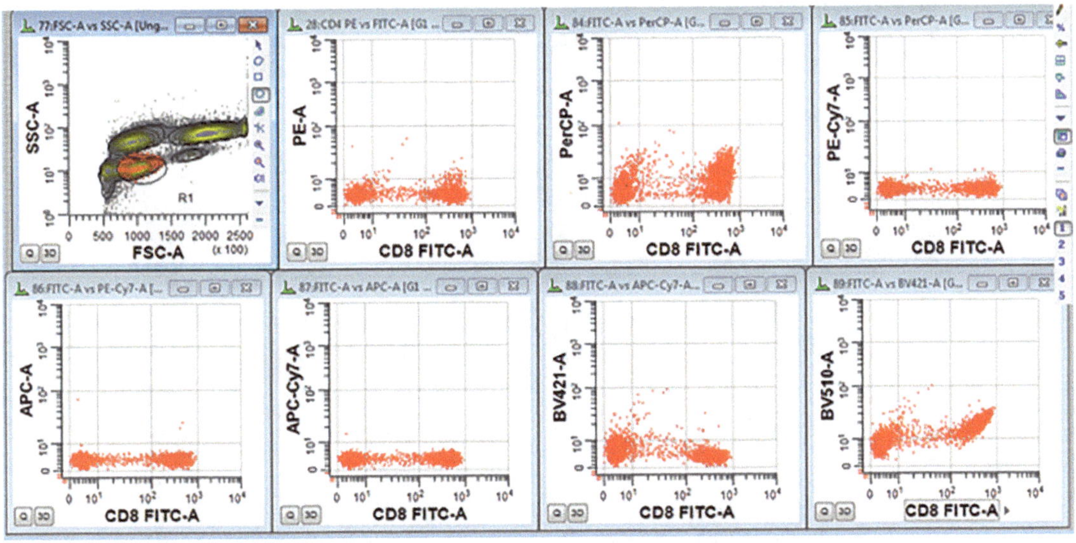

Fig. 27 EQA proficiency testing results over time. The percent CD19 over three different surveys shows results that fall both above and below the mean of the acceptable results. Conversely, the results of the percentage of NK Lymphocytes reported consistently run above the mean during the same period, creating a bias. This may be indicative of an analysis gating issue or an antibody that should be further investigated for suitability

again for checking the compensation of a fully stained sample in the experiment. The multicolor compensation control is described in the next section.

The Fully Stained Compensation Control Tube

The last compensation control tube should include all fluorochrome-conjugates used in the experiment. The 8-color cocktail used in the example includes several lymphocyte and monocyte antibodies found in normal peripheral blood. Some will co-express, and some will be exclusive. Use the results of this tube to do one last check of each combination to ensure no spillover or over-compensation exists. Keep a copy of the analysis worksheets showing all the fluorescent combinations for future reference, so that if questions arise about the compensation in an experiment, analyze the results in question and use the previously saved dot-plot combinations to quickly identify any compensation issues. By observing each conjugate vs. the others, you should be able to see if adjustments are necessary (see Fig. 28).

This example only shows half of the possible combinations, FITC vs. X and PE vs. X, for up to eight fluorochromes. More can be added or subtracted depending on the number of fluorochromes used in the experiment. Instrument compensation settings can now be monitored daily with this cocktail using a normal whole blood control. Verification of cocktail compensation should be monitored daily, with records of review kept for quality control purposes.

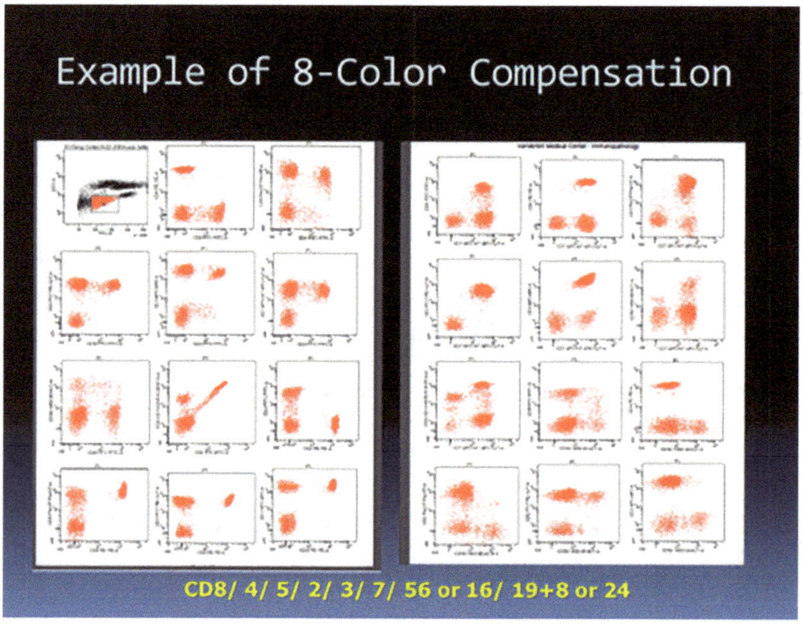

Fig. 28 Optimizing PMT settings using CD4-stained cells. CD4-Stained WBCs can be used to set up optimized PMT settings. Gating on both the lymphocytes and monocytes CD4 stains T-helper cells bright, monocytes mid-intensity positive, and is negative on CD8+ T-Cytotoxic Suppressor lymphocytes. Adjust the PMT voltage until all three populations can be clearly visualized with good resolution between the peaks

4.9 Immunophenotyping Quality Control Records

1. A Quality Assurance Plan—a statement that explains all aspects of quality control, including pre-analytic, analytic, and post-analytic procedures and results of those procedures.
2. Pre-Analytic Records
   a. Reagents: written procedure consistent with manufacturer's instructions OR records of method accuracy evaluation if alternative procedures are used, such as for a lab developed test (LDT). These records should be kept for at least 2 years.
   b. Antibody validation: all antibodies are validated on the cell sub-populations of interest in the context of the antibody combination used in an assay
   c. New reagent lot/confirmation testing for suitability
   d. Procedure validation: written records of any procedure validation, including signature approval. These records should be kept as long as the procedure and/or instrument stays in use.
   e. Written procedures should be reviewed and approved initially and every 2 years unless changes are made, which should be approved prior to use in the lab [33].
3. Analytic Records
   a. Instrument maintenance records of all maintenance procedures as well as any maintenance performed by a service engineer, including scheduled maintenance.
   b. Instrument performance records, including all daily QC results, and a signed monthly summary, should be kept for at least 2 years.
4. Post-Analytic Records
   a. Flow cytometry data for evaluation of hematolymphoid studies, PNH, and congenital immunodeficiency evaluations should be retained for at least 10 years.
   b. Routine lymphocyte subset and CD34+ enumeration data are retained for at least two years.
   c. Stored data must include raw list-mode data and final interpretation. Storage of gated data is encouraged but not required [33].

References

1. Department of Health and Human Services, Centers for Medicare and Medicaid Services (1992) Clinical laboratory improvement amendments of 1988; final rule. Fed Reg (Feb 28):7146 [42CFR493.1256]

2. Department of Health and Human Services, Centers for Medicare and Medicaid Services (2003) Clinical laboratory improvement amendments of 1988; final rule. Fed Reg (Jan 24):7166 [42CFR493.1256(f)]

3. CAP Flow Cytometry Checklist (2022) College of American Pathologists, 325 Waukegan Rd. Northfield, IL 60053. https://www.cap.org
4. CAP All Common Checklist (2022) COM.10050, College of American Pathologists, 325 Waukegan Rd. Northfield, IL 60053. www.cap.org
5. Clinical and Laboratory Standards Institute (CLSI) (2007) Clinical flow cytometric analysis of neoplastic hematolymphoid cells; approved guideline, 2nd edn. CLSI document H43-A2 (ISBN 1-56238-635-2). Clinical and Laboratory Standards Institute, 940 West Valley Road, Suite 1400, Wayne, Pennsylvania 19087-1898 USA
6. Banfi G, Salvagno GL, Lippi G (2007) The role of ethylenediamine tetraacetic acid (EDTA) as in vitro anticoagulant for diagnostic purposes. Clin Chem Lab Med 45(5):565–576
7. Bowen RAR, Remaley AT (2014) Interferences from blood collection tube components on clinical chemistry assays. Biochem Med 24:31–44. https://doi.org/10.11613/BM.2014.006
8. Novelli P, Savoia I, Cambieri R et al. (2000) Collagenase digestion and mechanical disaggregation as a method to extract and immunophenotype tumor lymphocytes in cutaneous T cell lymphomas. Clin Exp Dermatol 25:423–431
9. Becton Dickinson Medimachine™ brochure. Becton Dickinson biosciences, 2350 Qume Dr., San Jose, CA., 95131
10. Greig B, Stetler-Stevenson M, Lea J (2014) Stabilization media increases recovery in paucicellular cerebrospinal fluid specimens submitted for flow cytometry testing. Cytometry B Clin Cytom 86(2):135–138. https://doi.org/10.1002/cyto.b.21096
11. Davis BH, Dasgupta A, Kussick S, Han J-Y, Estrellado A, on behalf of ICSH/ICCS working group (2013) Validation of cell-based fluorescence assays: practice guidelines from the ICSH and ICCS-part II-preanalytical issues. Cytometry B 84B:286–290
12. Caldwell CW (1998) Analyte-specific reagents in the flow cytometry laboratory. Arch Pathol Lab Med 122:861–864
13. Wood BL, Arroz M, Barnett D, DiGiuseppe J, Greig B, Kussick SJ, Oldaker T, Shenkin M, Stone E, Wallace P (2007) 2006 Bethesda International Consensus recommendations on the immunophenotypic analysis of hematolymphoid neoplasia by flow cytometry: optimal reagents and reporting for the flow cytometric diagnosis of hematopoietic neoplasia. Cytometry B (Clin Cytom) 72B:S12–S22
14. CD-Chex Plus Immunology Control for Cluster Designation package insert, Streck Inc., 7002 S. 109 Street, La Vista, NE 68128
15. Hulspas R (2010) Curr Protoc Cytom Chapter 6:Unit 6.29. https://doi.org/10.1002/0471142956.cy0629s54
16. Clinical and Laboratory Standards Institute (CLSI) (2007) Clinical flow cytometric analysis of neoplastic hematolymphoid cells; approved guideline, 2nd edn. CLSI document H43-A2 (ISBN 1-56238-635-2). Clinical and Laboratory Standards Institute, 940 West Valley Road, Suite 1400, Wayne, Pennsylvania 19087-1898 USA
17. BD Biosciences (2009) Technical bulletin: an introduction to compensation for multicolor assays on digital flow cytometers. https://www.bdbiosciences.com/documents/Compensation_Multicolor_TechBulletin.pdf
18. Biolegend Blog. The stain index: what is it and what does it tell you. https://www.biolegend.com/en-us/blog/the-stain-index-what-is-it-and-what-does-it-tell-you
19. CAP All Common Checklist, COM.30600 (2022) College of American Pathologists, 325 Waukegan Rd. Northfield, IL 60053. https://www.cap.org
20. Heinlein C, Deppert W, Braithwaite AW, Speidel D (2010). A rapid and optimization-free procedure allows the in vivo detection of subtle cell cycle and ploidy alterations in tissues by flow cytometry. Cell Cycle. 9:3584–3590. https://doi.org/10.4161/cc.9.17.12831.
21. Darzynkiewicz Z, Halicka HD, Zhao H (2010) Analysis of cellular DNA content by flow and laser scanning cytometry. Adv Exp Med Biol 676:137–147
22. Nunez R (2001) DNA measurement and cell cycle analysis by flow cytometry. Curr Issues Mol Biol 3(3):67–70
23. Multicolor flow cytometry: beyond the basics, course workbook. BD Biosciences, San Jose, CA., 2014
24. Bagwell CB, Adams EG (1993) Fluorescence spectral overlap compensation for any number of flow cytometry parameters. Ann N Y Acad Sciences 677:167–184
25. Becton Dickinson Biosciences (2009) Procedure for manual setting of compensation for multicolor flow cytometric analysis. https://www.bdbiosciences.com/en-us/resources/protocols/setting-compensation-multicolor-flow
26. CAP All Common Checklist, COM.30570 (2023) College of American Pathologists, 325 Waukegan Rd. Northfield, IL 60053. https://www.cap.org
27. Maecker HT, Trotter J (2006) Flow cytometry controls, instrument setup, and the determination of positivity. Cytometry A, Cytometry 69A(9):1037–1042. https://doi.org/10.1002/cyto.a.20333
28. Landay A et al (1992) Clinical applications of flow cytometry: quality assurance and immunophenotyping of peripheral blood lymphocytes, Document H42-T, vol 12, No. 6, pp 6–8
29. Keeney M, Gratama JW, Chin-Yee IH, Sutherland DR (1998) Isotype controls in the analysis of lymphocytes and CD34+ stem and

progenitor cells by flow cytometry – time to let go!. Cytometry 343:280–283

30. Lindmo T, Fundingsrud K (1981) Measurement of the distribution of time intervals between cell passages in flow cytometry as a method for the evaluation of sample preparation procedures. Cytometry 2:151
31. Schmid I, Krall WJ, Uittenbogaart CH, Braun J, Giorgi JV (1992) Dead cell discrimination with 7-amino-actinomycin D in combination with dual color immunofluorescence in single laser flow cytometry. Cytometry 13(2):204–208
32. CAP All Common Checklist, COM.01300 (2023) College of American Pathologists, 325 Waukegan Rd. Northfield, IL 60053. www.cap.org
33. CAP Flow Cytometry Checklist, FLO.23706, (2022) College of American Pathologists, 325 Waukegan Rd. Northfield, IL 60053. https://www.cap.org

Check for updates

Chapter 16

High Immunophenotyping of Human PBMCs Using Lyophilized Antibodies

Kim Han

Abstract

Immunophenotyping of human peripheral blood mononuclear cells (PBMCs) is a critical technique for understanding immune responses in both health and disease. Traditionally, this process relies on the use of liquid antibody cocktails, which are time-consuming to prepare, prone to laboratory error, and have limited shelf lives. Recent advancements in lyophilization of antibodies and other reagents have addressed these challenges by enhancing reagent stability, extending shelf life, and simplifying transport and storage. Dried antibody cocktails have been shown to perform comparably to liquid cocktails in flow cytometry assays while reducing preparation time and minimizing handling errors. This review focuses on the use of lyophilized reagents for high-dimensional immunophenotyping of human PBMCs, discussing their advantages in large-scale studies, global health initiatives, and clinical research. By streamlining immune profiling workflows, lyophilized reagents have the potential to improve accessibility and facilitate the adoption of immunophenotyping in resource-limited settings.

Keywords Immunophenotyping, Peripheral blood mononuclear cells, Flow cytometry, Lyophilization, Dried antibody cocktail

1 Introduction

Determination of the phenotype of the major leukocyte subtypes and their state of activation is standard in clinical diagnosis and clinical research. Human peripheral blood mononuclear cells (PBMCs), which include a variety of immune cells such as T cells, B cells, monocytes, dendritic cells (DC), and natural killer (NK) cells, are central to the immune response against infections, cancers, autoimmune diseases, and immune deficiencies. The ability to identify and quantify these cells is essential for elucidating immune mechanisms and developing targeted therapies. High-dimensional flow cytometry has emerged as the gold standard for comprehensive leukocyte immunophenotyping, allowing the simultaneous measurement of multiple surface markers on individual cells, thus enabling the detailed study of immune cell populations and their functional states [1,2].

Angélique Biancotto and Pradeep K. Dagur (eds.), *Immunophenotyping*, Methods in Molecular Biology, vol. 3087,
https://doi.org/10.1007/978-1-0716-5607-5_16,

One of the significant challenges in flow cytometry-based immunophenotyping is the preparation and storage of antibody cocktails. Traditionally, these antibodies are prepared in liquid form, but this method poses difficulties in terms of reagent degradation due to aging, temperature variations, and light exposure. To overcome these challenges, the field has turned to lyophilization (freeze–drying) techniques, which provide a more stable and convenient method for antibody cocktail preparation. Lyophilized antibody cocktails offer a robust solution for high-dimensional immunophenotyping of PBMCs. These reagents can be stored for extended periods without degradation, making them ideal for large-scale studies, where consistency across different laboratories and research sites is critical [2–4]. For example, large-scale clinical studies, such as those conducted by the EuroFlow Consortium and the Human Immunophenotyping Consortium (HIPC), have relied on lyophilized antibody cocktails for standardized immunophenotyping of various leukocyte populations. The ability to store reagents in a stable, ready-to-use form reduces variability caused by differences in reagent handling and storage conditions, ensuring more reliable results [5–7]. Furthermore, the use of lyophilized reagents enables the efficient conduct of large-scale, longitudinal studies, such as those aimed at tracking immune responses in patients over time. These studies have important implications for monitoring therapeutic interventions and understanding the immune system's response to various diseases [8–11].

1.1 Benefits of Lyophilized Reagents

- **Stability and shelf life**: One of the primary advantages of lyophilized reagents is their improved stability. This makes them especially useful in large-scale studies that involve multicenter collaborations.
- **Convenience and cost-efficiency**: Lyophilized reagents simplify the workflow of immunophenotyping by eliminating the need for complex antibody preparation protocols. This ready-to-use antibody cocktail reduces the chance of user error, contributing to more consistent results and saving both time and labor.
- **Consistency across studies**: Lyophilized antibody cocktails are essential for ensuring consistency in large multicenter clinical trials. By using the same reagent batches across different locations, these reagents help minimize inter-laboratory variability, making it easier to compare results across various research sites and ensuring data harmonization.

1.2 Key Applications Using Lyophilized Reagents

- **EuroFlow Consortium**: The EuroFlow Consortium developed standardized immunophenotyping panels for the investigation of normal and malignant leukocytes, where lyophilized antibody cocktails played a crucial role.

- **Human Immunophenotyping Consortium (HIPC)**: In clinical trials, the use of lyophilized reagents for the study of T cells, B cells, NK cells, and other immune cell subsets has helped reduce variability, especially when investigating the immune status of cancer patients or studying immunodeficiency syndromes. The HIPC project helped standardize the markers used in flow cytometry, paving the way for reproducible, multicenter studies.
- **Companies**: Many manufacturers have developed lyophilized reagents for immunophenotyping. For instance, BD Biosciences (Lyoplate) has collaborated with various institutions to produce dried antibody formulations for the study of different leukocyte subsets. As well as Beckman Coulter (DuraClone), BioLegend (Veri-Cells), and Miltenyi Biotec (StainExpress), have implemented dried reagents for screening tubes focused on lymphocyte subsets in peripheral blood, bone marrow aspirates, and other fluids.

1.3 Advanced Technologies in Lyophilized Reagents

- **Comprehensive leukocyte immunophenotyping (CLIP)**: The development of CLIP panels represents a significant leap in immunophenotyping. CLIP panels aim to provide an in-depth, broad characterization of the immune system by using a combination of both surface and intracellular markers [1,12].

- **Spectral flow cytometry and mass cytometry**: Spectral flow cytometry (SFC) is a technique that uses a spectrograph and multichannel detector to analyze single cells or particles at a high resolution. The cytometry by time of flight (CyTOF) has expanded the capacity to analyze a greater number of immune cell markers, providing a more detailed understanding of the immune responses and functional states of immune cells and their roles in diseases [13–15).

Additionally, recent research has also shown that lyophilized antibody panels can provide valuable insights into the molecular mechanisms regulating immune function. For example, the study by Han et al. [16] demonstrated how fasting influences the responsiveness of human CD4$^+$ T helper cells, using lyophilized antibody panels for immunophenotyping. This work highlighted the importance of high-quality immune profiling in understanding the effects of metabolic states on immune cell function. Particularly, immune cell profiling in cancer immunotherapy is used to monitor patient responses to treatments, tracking the changes in immune cell populations over time and identifying the biomarkers for disease severity, progression, and therapeutic efficacy [17,18]. Such studies underscore the potential of lyophilized reagents to enhance our understanding of immune cell biology in both health and disease.

In conclusion, the integration of lyophilized reagents into immunophenotyping workflows represents a significant advancement in

96-well	1	2	3	4	5	6	7	8	9	10	11	12	Cocktails
A	CD45/CD3/HLA-DR/CD28/CD4/CCR7/CD45RA/CD38/CD8												T-cell
B	CD45/CD3/HLA-DR/CD39/CD4/CD25/CCR4/CD127/CD45RO												Treg
C	CD45/CD3/IgD/CD10/CD19/CD24/CD27/CD38/CD20												B-cell
D	CD45/CD14/HLA-DR/CD163/CD123/CD56/CD11c/CD16/CD3+CD19+CD20												DC/Mono/NK
E	CD45/CD3/HLA-DR/CXCR5/CD4/CXCR3/CCR6/CD38/CD8												T1/T2/T17/Tf
F													Empty wells
G													Empty wells
H													Empty wells

Fig. 1 Assay lyoplate (number 175) from BD Biosciences is a 96-well plate with rows A through F and columns 1 through 12 containing lyophilized antibody cocktail buttons. Each row includes 12 replicates of the complete cocktail as follows: T-cell (row A), Treg-cell (row B), B-cell (row C), mono/DC/NK-cell (row D), and T-cell subtypes, T1/T2/T17/Tf (row E)

both research and clinical settings. This approach offers numerous advantages in terms of reagent stability, cost, and ease of use, particularly in large-scale studies, disease phenotyping, and the emerging field of personalized medicine. As research continues to explore the complexities of immune responses in various diseases, lyophilized reagents will play an increasingly important role in providing high-quality, reproducible data that is critical for advancing our understanding of disease pathogenesis and therapeutic intervention.

2 Materials

2.1 Peripheral Blood Mononuclear Cells (PBMCs) Samples

PBMCs had been prepared using density gradient centrifugation and preservation in liquid nitrogen for long-term storage. The cell separation used in this procedure was not processed in the same laboratory that performed the staining; hence, the methodology is not described (*see* **Note 1**).

2.2 Assay Lyoplate (BD Biosciences, Designate 175)

Each of the five rows contains 12 replicates of the five cocktails (Fig. 1 and *see* **Note 2**). Ten-color antibodies include the utilization of the 488 nm laser-excited conjugates available for incorporation of an additional conjugate—either FITC (rows B and D) or BB515 (rows A, C, and E) in each cocktail (Table 1).

2.3 Compensation Lyoplate (BD Biosciences, Designate 177)

Each row contains wells with individual antibodies that are constituents of one of the five cocktails, as well as negative and positive compensation bead particles. Beads are not used to incorporate viability dyes into the compensation matrix (Fig. 2). Instead, cellular material is stained separately, in tubes with cells, and these are acquired on the cytometer at the same time as bead acquisition. Several strategies are available for calculating the compensation matrices for each cocktail. For example, a single plate was used that contained the compensation beads and the assay cocktails. Combining the assay and compensation

	1	2	3	4	5	6	7	8	9	10	11	12	Cocktails
A		CD45 BUV395	CD28 BB515	CD4 PCP Cy5.5	CD38 APC	CD8 APC-H7	CD3 V450	HLA-DR V500	CCR7 PE	CD45RA PE-Cy7			T-cell
B		CD45 BUV395	39 FITC	CD4 PCP Cy5.5	CD127 AF647	CD45RO APC-H7	CD3 V450	HLA-DR V500	CD25 PE	CCR4 PE-Cy7			Treg-cell
C		CD45 BUV395	CD10 BB515	CD19 PCP Cy5.5	CD38 APC	CD20 APC-H7	CD3 V450	IgD V500	CD24 PE	CD27 PE-Cy7			B-cell
D		CD45 BUV395	CD163 FITC	CD123 PCP Cy5.5	CD16 APC	CD3/19/20 APC-H7	CD14 V450	HLA-DR V500	CD56 PE	CD11c PE-Cy7			DC/Mono/ NK-cell
E		CD45 BUV395	CXCR5 BB515	CD4 PCP Cy5.5	CD38 APC	CD8 APC-H7	CD3 V450	HLA-DR V500	CXCR3 PE	CCR6 PE-Cy7			T1/T2/T17/ Tf-cell
F													Empty wells
G													Empty wells
H													Empty wells

Fig. 2 Compensation lyoplate (number 177) from BD Biosciences is a 96-well plate with rows A through F and columns 2 through 10 containing lyophilized buttons of a single antibody from each cocktail per well, along with both negative and positive beads that are included in BD Biosciences Anti-Mouse Ig Compensation Particles Kit (part number 552843). The antibody in well E3 is a rat monoclonal anti-CXCR5 conjugate; therefore, the compensation bead set for this well is derived from the BD Biosciences Anti-Rat Ig Compensation Particle Kit (catalog number 552844)

wells in one plate results in the compensation beads being washed more aggressively and incubating longer than necessary, leading to inconsistencies when calculating matrices. An alternative strategy is to use separate plates for compensation and assay. Using a separate compensation plate—with more wells available—allows the use of the full complement of the conjugates for each cocktail (*see* **Note 3**).

2.4 Thawing Medium

RPMI 1640 media (Gibco, catalog number 11875-093) supplemented with 10% fetal bovine serum and 50 U/ml benzonase nuclease (Sigma, catalog number E1014).

2.5 Staining Buffer

Phosphate-Buffered Saline (PBS), pH 7.4 (Gibco, catalog number 10010-023) supplemented with 0.25 mM EDTA, 1% Bovine serum albumin or 1% fetal bovine serum (*see* **Note 4**).

2.6 Fixation Medium

PBS supplemented with 1% formalin solution: Prepared fresh from a 10% methanol-free buffered formalin (Protocol, catalog number 245-684).

2.7 Viability Dye

Fixable Yellow Live/Dead Cell Stain Kit (ThermoFisher Scientific, catalog number L34968): Stock solutions are prepared by the addition of 50 μl of DMSO to each vial of dye. Aliquots of 10 μl are placed in screw-capped freezing vials, which are stored at −80°C. One of these vials is sufficient to stain cells from five different samples.

3 Methods

The methods outlined below refer to the overall protocol used to perform staining of cells using two different lyoplates: compensation and assay. The compensation plate is handled in a slightly different manner than the other plate, as described below.

An adherent foil wrapping covers the lyoplates beneath the lid of the plate. The foil is removed from the plate at the time of use, and buttons within the wells are inspected, with any anomalies noted, in the event unexpected results are obtained that might be associated with any defects. The compensation plate is used in its entirety once they are unpackaged. However, the assay plates may have unused wells if there are less than 12 samples to be assayed. In this case, the foil wrapping is only partially removed so that it remains adhered to unused plate wells. If the plate is used again for a subsequent staining procedure, the wells are evaluated for evidence of hydration of the remaining lyophilized buttons. If this is noted in any well, the plate is discarded.

Once the substrate is added to the lyophilized antibody cocktail button, the workflow for surface-marker immunophenotyping using dried reagents is not markedly different from that used with liquid reagents. Hence, this chapter does not proceed past the preparation of samples to be presented for data acquisition on a flow cytometer; these steps, as well as data analysis strategies, are particular to individual flow cytometry laboratories that utilize liquid antibody cocktails. Depending on the vendor of the dried cocktails, they may be manufactured in tubes or plates. Hence, the particulars for the procedure following admixture are contingent on the staining vessel. There are advantages and disadvantages inherent in each type of containment regarding staining versus cytometer data acquisition.

Given the importance of fluorescence-minus-one (FMO) in immunophenotyping, the use of dried reagents for this procedure is a significant addition to the protocol [19]. Therefore, a limited number of FMO lyoplates might be prepared during the production run. The use of dried FMO cocktails in plates is a significant time- and cost-saving step when compared with using liquid cocktails in tubes. That said, once the initial quadrant gate placements are derived from the initial FMO file analyses, subsequent FMO plates are only run following significant alterations following servicing of the cytometer. The strategy for confirming proper quadrant gate placement, after initial placement from analysis of FMO data files, involves the use of Boolean gating strategies to derive "NOT" gates in order to generate internal "negative" cell populations in the absence of markers that would definitively identify contextual leukocyte subpopulations (*see* **Note 5**).

3.1 Processing PBMCs Sample

1. Frozen PBMCs ($\approx 5–10 \times 10^6$) have been thawed in 10 ml of RPMI 1640 Thawing Medium.
2. The cells are centrifuged (ambient temperature, $400 \times g$, for 5 min) and removed supernatants.
3. Cell pellets are resuspended with 1 ml of staining buffer.

3.2 Processing the Compensation Lyoplate

1. Staining buffer is added to each well in 50 μl volumes, starting in well A2, proceeding down the column, then moving to the top of successive columns.
2. The compensation plate with lyophilized antibody is placed on a rotary plate mixer at low speed and incubated for 10 minutes at the ambient temperature, protected from light.
3. Cells from Sect. 3.1 are added into each well in 50 μl volumes, and starts pipetting of the cell sample.
4. The plate with cells is placed on a rotary plate mixer at low speed and incubated for 20 min at the ambient temperature, protected from light.
5. Cells with Viability Dye: Cells are placed in an empty well with 50 μl volumes. Before the addition of the viability dye, the stock solution of dye is tenfold diluted with staining buffer to make a working solution. After the addition of 5 μl of viability

working solution into the cell suspension, cells are incubated for 10 min at the ambient temperature, protected from light.

6. Cell with no dye (unstained): Cells are placed in an empty well with 50 μl volumes.
7. At the end of incubation, the plate is centrifuged (ambient temperature, 400 × *g*, for 5 min) and flicked to remove supernatants, followed by the addition of 100 μl of staining buffer.
8. The stained/unstained cells of each well are transferred into labeled polystyrene tubes and held at 4°C until cytometer acquisition, which is completed within 30 min of staining.

3.3 Processing the Assay Lyoplate

1. Once removed from its pouch and after peeling off the foil cover, 200 μl of cells from Sect. 3.1 is added into rows A–E in the column assigned to each sample. Pipetting of the sample proceeds using a new pipette tip for each well. Unstained cells are placed in an empty well.
2. Once all wells are filled, a 30 min incubation at the ambient temperature ensues, with the plate protected from light. After the first 5 min, the cells are mixed by gentle repetitive pipetting using a multipipettor. Following the resuspension of cells, there is an additional 25 min incubation, protected from light, at the ambient temperature on the rotary plate mixer.
3. Before the addition of the viability dye, the stock solution of dye is tenfold diluted with staining buffer to make a working solution. After the addition of 5 μl of viability working solution into the cell suspension, cells are incubated for 10 min at the ambient temperature, protected from light. Unstained cells are not added to the viability dye.
4. Lyoplate washing procedures are carried out at the ambient temperature, including centrifugations (400 × *g* for 5 min). The cells are washed with 200 μl of staining buffer per well and mixed by gentle repetitive pipetting, then centrifuged and flicked to remove supernatants. The washing step is processed twice (*see* **Note 6**).
5. The cells are fixed with 200 μl of 1% formalin in each well. The plate is then incubated at the ambient temperature, protected from light, for 30 min, then centrifuged and flicked to remove supernatants.
6. The cells are washed with 200 μl of staining buffer and mixed by gentle repetitive pipetting, then centrifuged and flicked to remove supernatants.
7. The final suspension volume of 250 μl of staining buffer is added, followed by gentle repetitive pipetting. Samples are collected on the cytometer immediately following the staining procedure. In the interval prior to acquisition on the cytometer,

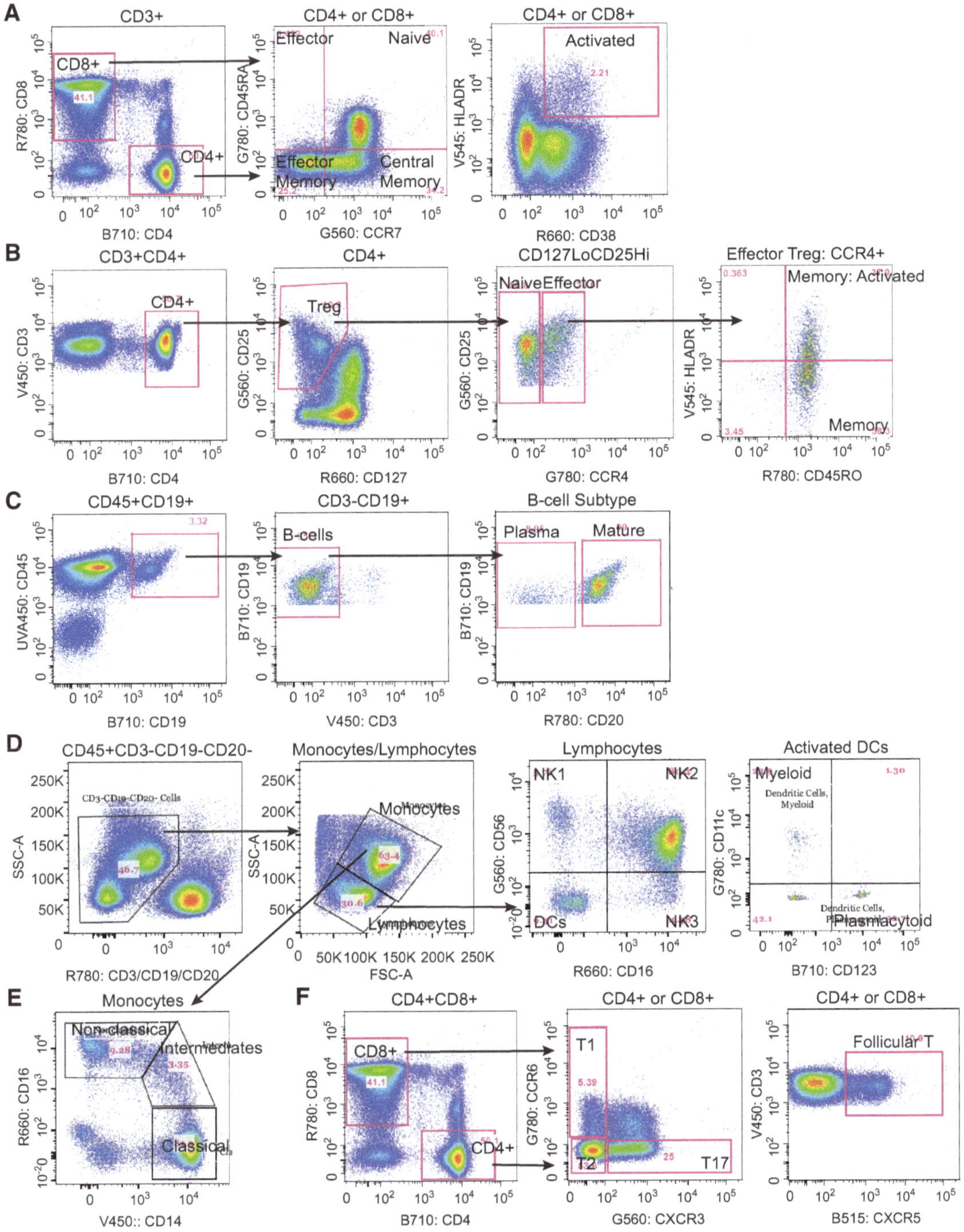

Fig. 3 High-dimensional flow cytometry shows the distinct cell type distribution of PBMCs. **A**, T-cells: Flow cytometry plots show naïve $CD4^+$ and $CD8^+$ positive T cells and activated T cells ($CD38^+$HLADR) gated on $CD45^+CD3^+$ cells. **B**, Treg-cells: Representative flow plot showing the expression of Treg cells ($CD127^{lo}CD25^{hi}$) gated on $CD3^+/CD4^+$ T cells. Treg-cells can be subtyped to naive ($CD25^+CCR4^-$) and effector cells ($CD25^+CCR4^+$) gated on Treg-cells. **C**, B-cells: Flow cytometric plots show immature ($CD19^+CD20^-$) and mature ($CD19^+CD20^+$) B-cells, gating on $CD45^+CD19^+$ cells. **D**, DC/NK-cells: Natural killer cells (NK1,

the plate is protected from light and held at 4°C until ready for acquisition.

3.4 Gating Strategy and Data Acquisition

Flow cytometry identifies differential activation of multiple lineages within PBMCs. The nine-color cell surface signatures on extracted PBMCs are shown by high-dimensional flow cytometry immunophenotyping (Fig. 3). Analysis excluded debris and doublets using light scatter measurements and dead cells by live/dead stain. The instrument configuration and gating strategy to identify immune cell subsets are provided in Tables 2 and 3; additional considerations for gating, including the use of FMO controls (*see* **Note 7**). Data analysis and gating were performed using standard flow cytometry analysis software (e.g., FlowJo). The cell populations stained with specific cell surface receptor antibodies used to designate cell lineage are shown with the corresponding flow plots. Following deep immunophenotyping and cytometric quantitation, the data shows that the most frequent $CD45^+$ cell subtype populations (Fig. 3A–F: A, T-cells; B, Treg-cells; C, B-cells; D, DC/NK-cells; E, Monocytes; F, T1/T2/T17/Tf-cells).

4 Notes

1. Although several cited studies report the use of dried antibody cocktails for bone marrow aspirates and other body fluids, peripheral blood remains the most commonly used sample type. This is typically prepared either by red blood cell lysis (performed before or after antibody staining) or by density gradient separation. In most cases, density gradient separation is followed by cryopreservation in liquid nitrogen for long-term storage. In this study, the samples were not processed in the same laboratory in which staining was performed; therefore, the processing methodology is not described here. The choice of sample preparation depends on the study design. For single-time-point analyses performed at a single site, lysed whole blood is generally preferred, as it minimizes cell loss and reduces the potential for phenotypic changes associated with cell separation or cryopreservation. In contrast, multicenter or

Fig.3 (continued) $CD16^-CD56^+$; NK2, $CD16^+CD56^+$; NK3, $CD16^+CD56^-$) and activated dendritic cells ($CD16^-CD56^-HLADR^+$) and dendrite cells ($CD16^-CD56^-$) show from lymphocytes without T and B cells ($CD45^+CD3^-CD19^-CD20^-$). Especially, DCs can be subtyped to myeloid DCs ($CD11c^+CD123^-$) and plasmacytoid cells ($CD11c^-CD123^+$) gated on activated DCs ($CD16^-CD56^-HLADR^+$). **E**, Monocytes: Flow plots of non-classical ($CD14^{low}CD16^{high}$), intermediate ($CD14^{med}CD16^{med}$), and classical monocytes ($CD14^{high}CD16^{low}$). **F**, T1/T2/T17/Tf-cells: Flow cytometric plots showing the expression of T follicular cells (Tf: $CXCR5^+$), T1 ($CXCR3^+CCR6^-$), T2 ($CXCR3^-CCR6^-$), and T17 ($CXCR3^-CCR6^+$) on $CD4^+$ or $CD8^+$ T cells

Table 2

Flow cytometry instrument (BD LSR Fortessa SORP) configuration with HTS

Laser line	Wave length (nm)	Laser type and source	Laser power (mW)	Detector name	Fluorochrome detected	DichroicLP filter	BP filter	Signal collected	Signal amplification	PMT
Blue	488	DPSS; coherent	100	FSC	Forward scatter	NA	NA	H, A, and W[b]	Linear	Photo diode
				B710	PerCP CY5.5	685	710/50	Area	Log	A
				B515	FITC	505	515/20	Area	Log	B
				SSC	Side Scatter	488		H, A, and W[b]	Linear	C
Green	532	DPSS; coherent	150	G780	PECy7	740	780/40	Area	Log	A
				G560	PE	empty	575/25	Area	Log	B
Red	637	DPSS; OBIS	140	R780	APC-H7	740	780/60	Area	Log	A
				R660	APC	empty	660/20Δ[a]	Area	Log	B
Violet	405	DPSS; OBIS	100	V565		557	560/40	Area	Log	A
				V545	Viability Stain	505	510/20	Area	Log	B
				V450	BV450	empty	450/50	Area	Log	C
Ultra violet	355	DPSS; genesis 355–100	45	UV450	BUV 395	empty	450/50	Area	Log	A

HTS equipped with high-throughput sampler system (BD), *DPSS* diode pumped solid state [a]Δ, Notch Filter [b]Height, area, and width

longitudinal studies, as described here, require the use of cryopreserved PBMCs to ensure consistency across sites and time points.

2. The original lyoplate design used in the HIPC project did not incorporate a 355 nm laser, as not all participating centers had this capability. In the present study, a CD45-BUV conjugate was included in each cocktail to take advantage of the 355 nm laser configuration. In addition, the original design utilized a viability dye excited by the 488 nm laser. This was modified to a 355 nm-excited viability dye (*see* Sect. 2.7), thereby freeing the 488 nm channel for additional fluorochrome use. This allowed incorporation of an additional marker using either FITC (rows B and D) or BB515 (rows A, C, and E) in each cocktail (see Fig. 2).
3. In the original lyoplate format, compensation beads and assay cocktails were included on a single plate. However, this approach can result in excessive washing and prolonged incubation of compensation beads, leading to variability in compensation matrix calculations. For this reason, it is preferable to use separate plates for compensation and assay. A dedicated compensation plate allows inclusion of all conjugates for each cocktail and provides greater flexibility during setup. Compensation matrices may be generated either by using all wells for each cocktail or by acquiring a reduced set of representative wells and generating cocktail-specific matrices within the cytometry software (e.g., FlowJo or BD FACSDiva). In this protocol, compensation matrices are generated only when significant changes to the cytometer occur. Minor day-to-day variation can be addressed by adjusting matrix values within the analysis software.
4. Different buffer conditions are used at various stages of the protocol. Some steps are performed using PBS alone, while others utilize Staining Buffer (PBS supplemented with serum and EDTA). PBS without protein is used prior to viability dye staining to avoid interference with dye uptake. It is also used during compensation staining, as serum components may interfere with antibody binding to compensation beads. PBS is further used during early wash steps prior to fixation. All other steps are performed using staining buffer to maintain cell integrity and minimize nonspecific binding.
5. The use of microtiter plates or 12 × 75 mm polystyrene tubes depends on the format provided by the manufacturer. Plate-based staining offers advantages in processing time, reagent use, and compatibility with high-throughput systems. In this protocol, staining is performed in plates, followed by transfer to tubes prior to cytometer acquisition. This approach is useful for samples with low cell numbers and for acquisition

at low flow rates. Because plate-based acquisition may require extended run times and does not allow temperature control, samples are maintained at 4°C until acquisition to prevent cell degradation.

6. All washing steps are performed at ambient temperature using centrifugation at 400×g for 5 min. Supernatants are removed by flicking the plate into a sink, which should be disinfected with 10% bleach after each wash step. For cell preparation steps performed in tubes, the same centrifugation conditions are used, with supernatants removed by decanting. Consistent handling across samples is important to minimize variability.

7. FMO controls are commonly used in high-dimensional immunophenotyping to establish accurate gating boundaries, particularly for dim or overlapping populations. The original lyoplate design did not include FMO capability; therefore, incorporation of dried FMO panels represents a useful extension of the protocol. A limited number of FMO lyoplates may be prepared and used during initial assay setup to define gating strategies. Once gates are established, FMO controls are typically not required for routine analysis unless substantial changes in staining conditions or instrument performance occur. Ongoing validation of gating may be achieved using Boolean gating approaches to generate internal negative populations, reducing the need for repeated FMO measurements.

References

1. Biancotto A, Fuchs JC, Williams A, Dagur PK, McCoy JP (2011) High dimensional flow cytometry for comprehensive leukocyte immunophenotyping (CLIP) in translational research. J Immunol Methods 363(2):245–261
2. Maecker HT, McCoy JP, Nussenblatt R (2012) Standardizing immunophenotyping for the human immunology project. Nat Rev Immunol 12:191–200
3. Chan RC, Kotner JS, Chuang CM, Gaur A (2017)) Stabilization of pre-optimized multicolor antibody cocktails for flow cytometry applications. Cytometry B Clin Cytom 92B:508–524
4. Langweiler M (2019) Immunophenotyping using dried and lyophilized reagents. Methods Mol Biol 2032:69–79
5. van der Velden VHJ, Flores-Montero J, Rawstron A, Asnafi V, Lecrevisse Q, Lucio P, Mejstrikova E, Szczepanski T, Kalina T, de Tute R, Bruggemann M, Sedek L, Cullen M, Langerak AW, Mendonca A, Macintyre E, Martin-Ayuso M, Hrusak O, Vidriales MD, Orfao A (2012) EuroFlow antibody panels for standardized n-dimensional flow cytometric immunophenotyping of normal, reactive and malignant leukocytes. Leukemia 26(9):1908–1975
6. Ivison S, Malek M, Garcia RV, Broady R, Halpin A, Richaud M, Brant RF, Wang S, Goupil M, Guan Q, Ashton P, Warren J, Rajab A, Urschel S, Kumar D, Streitz M, Sawitzki B, Schlickeiser S, Bijl JJ, Wall DA, Delisle J, West LJ, Brinkman RR, Levings MK (2018) A standardized immune phenotyping and automated data analysis platform for multicenter biomarker studies. JCI Insight 3(23):e12186
7. Finak G, Langweiler M, Jaimes M, Malek M, Taghiyar J, Korin Y, Raddassi K, Devine L, Obermoser G, Pekalski ML, Pontikos N, Diaz A, Heck S, Villanova F, Terrazzini N, Kern F, Qian Y, Stanton R, Wang K, Brandes A, Ramey J, Aghaeepour N, Mosmann T, Scheuermann RH, Reed E, Palucka K, Pascual V, Blomberg BB, Nestle F, Nussenblatt RB, Brinkman RR, Gottardo R, Maecker H, McCoy JP (2016) Standardizing flow cytometry immunophenotyping analysis from the human

immunophenotyping consortium. Sci Rep 6(1):20686
8. Kalina T (2020) Reproducibility of flow cytometry through standardization: opportunities and challenges. Cytometry A 97A:137–147
9. Maecker HT, McCoy JP, the FOCIS Human Immunophenotyping Consortium (2010) A model for harmonizing flow cytometry in clinical trials. Nat Rev Immunol 11:915–978
10. Villanova F, Meglio PD, Inokuma M, Aghaeepour N, Perucha E, Mollon J, Nomura L, Hernandez-Fuentes M, Cope A, Prevost AT, Heck S, Maino V, Lord G, Brinkman RR, Nestle FO (2020) Integration of lyoplate based flow cytometry and computational analysis for standardized immunological biomarker discovery. PLoS One 8(7):e65485
11. Newton HA, Dobrovolskaia MA (2022) Immunophenotyping: analytical approaches and role in preclinical development of nanomedicines. Adv Drug Deliv Rev 185:114281
12. Verschoor CP, Kohli V, Balion C (2018)) A comprehensive assessment of immunophenotyping performed in cryopreserved peripheral whole blood. Cytometry B Clin Cytom 94B:818–826
13. Nolan JP, Condello D (2013) Spectral flow cytometry. Curr Protoc Cytom 63:1.27.1–1.27.13.
14. Sharma S, Boyer J, Teyton L (2024) A practitioner's view of spectral flow cytometry. N Methods 21:740–743
15. Iyer A, Hamers AAJ, Pillai AB (2022) CyTOF® for the masses. Front Immunol 13:815828
16. Han K, Singh K, Rodman MJ, Hassanzadeh S, Wu K, Nguyen A, Huffstutler RD, Seifuddin F, Dagur PK, Saxena A, McCoy JP, Chen J, Biancotto A, Stagliano KER, Teague HL, Mehta NN, Pirooznia M, Sack MN (2021) Fasting-induced FOXO4 blunts human $CD4^+$ T helper cell responsiveness. Nat Metabol 3:318–326
17. Greenplate AR, Johnson DB, Ferrell PB, Irish JM (2016) Systems immune monitoring in cancer therapy. Eur J Cancer 61:77–84
18. Blache U, Weiss R, Boldt A, Kapinsky M, Blaudszun A, Quaiser A, Pohl A, Miloud T, Burgaud M, Vucinic V, Platzbecker U, Sack U, Fricke S, Koehl U (2021) Advanced flow cytometry assays for immune monitoring of CAR-T cell applications. Front Immunol 12:658314
19. Feher K, Kirsch J, Radbruch A, Chang H, Kaiser T (2014) Cell population identification using fluorescence minus-one controls with a one-class classifying algorithm. Bioinformatics 30(23):3372–3378

Chapter 17

Immunophenotyping by Single-Cell CITE-Seq

Wilson Dos-Santos Bele, Sandra Curras-Alonso and Fabien Delahaye

Abstract

Single-cell transcriptomics has revolutionized our understanding of cellular heterogeneity by enabling high-resolution gene expression profiling at the individual cell level. However, traditional single-cell RNA sequencing (scRNA-seq) lacks direct protein quantification, limiting comprehensive immunophenotyping. Cellular Indexing of Transcriptomes and Epitopes by sequencing (CITE-seq) overcomes this limitation by integrating antibody-derived tag (ADT) quantification with scRNA-seq, allowing simultaneous measurement of surface protein and gene expression from the same cell. This multimodal approach enhances immune cell characterization, revealing new functional states and rare subpopulations in complex biological systems. Here, we provide a detailed protocol for performing CITE-seq, from sample preparation to sequencing and data analysis. We highlight key experimental considerations, discuss challenges related to antibody selection and batch effects, and provide troubleshooting strategies to ensure robust and reproducible results. The integration of transcriptomic and proteomic data through CITE-seq provides unparalleled insights into cellular function, with broad applications in immunology, oncology, and systems biology.

Keywords Single-cell RNA sequencing, CITE-seq, Antibody-derived tags, Immunophenotyping, Multimodal profiling, Single-cell proteomics, Immune cell characterization, Transcriptomics, High-throughput sequencing, Cellular heterogeneity

1 Introduction

The immune system is composed of a diverse array of cell types that coordinate complex responses to infection, inflammation, and disease. Immunophenotyping is essential to understanding immune cell diversity, function, and interactions. Traditional approaches, such as flow cytometry and mass cytometry (CyTOF), provide high-dimensional protein profiling but lack transcriptomic resolution. Conversely, single-cell RNA sequencing (scRNA-seq) enables global transcriptome analysis but does not capture protein expression, which is crucial for defining immune cell states.

To bridge this gap, Cellular Indexing of Transcriptomes and Epitopes by sequencing (CITE-seq) [1] was developed as a multimodal single-cell technology that simultaneously quantifies gene

Angélique Biancotto and Pradeep K. Dagur (eds.), *Immunophenotyping*, Methods in Molecular Biology, vol. 3087, https://doi.org/10.1007/978-1-0716-5607-5_17,

expression and surface protein levels from the same cell. This is achieved by labeling proteins of interest with antibody-derived tags (ADTs) conjugated to oligonucleotides, which are then cosequenced alongside cellular mRNAs. By integrating transcriptomic and proteomic data, CITE-seq enables precise immune cell classification, identification of rare subpopulations, and characterization of functional states that are not discernible using transcriptomics or proteomics alone.

CITE-seq has found widespread applications in immunology, oncology, and systems biology, providing new insights into immune cell differentiation, tumor microenvironments, and disease progression [2]. However, this technique presents technical challenges, including batch effects, antibody specificity, and data integration complexities, which require careful experimental design and computational analysis [3].

In this chapter, we present a comprehensive CITE-seq workflow, from sample preparation and antibody staining to sequencing and data processing. We highlight key considerations for experimental optimization and provide troubleshooting strategies to ensure robust and reproducible results. By leveraging CITE-seq, researchers can achieve an unprecedented level of immune cell characterization, paving the way for deeper insights into health and disease.

1.1 General Introduction to Single-Cell Transcriptomics

Single-cell transcriptomics has revolutionized molecular biology by enabling the study of gene expression at single-cell resolution, providing an unprecedented level of detail in understanding cellular heterogeneity. Unlike bulk RNA sequencing, which captures the average gene expression profile of a mixed population of cells, single-cell RNA sequencing (scRNA-seq) allows researchers to dissect individual cellular states, identify rare subpopulations, and track dynamic processes such as differentiation, immune responses, and disease progression.

1.1.1 The Need for Single-Cell Resolution

Traditional transcriptomic techniques rely on bulk RNA sequencing, where RNA from millions of cells is pooled together before analysis. While powerful, bulk sequencing masks cellular heterogeneity, averaging out differences between individual cells and making it difficult to distinguish unique cellular states within complex tissues. For example, in the immune system, a bulk approach may obscure functionally distinct subsets of T cells or monocytes, which could have critical implications in understanding infection responses or tumor microenvironments.

Single-cell transcriptomics addresses this limitation by isolating individual cells and sequencing their RNA profiles separately. This allows researchers to:

- Identify new cell types and states, particularly in tissues with high heterogeneity.

- Map cellular differentiation trajectories, reconstructing how stem cells give rise to specialized cell types.
- Explore cellular plasticity and transitions, detecting intermediate states that are often missed in bulk approaches.
- Understand tissue microenvironments by uncovering interactions between different cell types in health and disease.

1.1.2 Key Technological Advances in Single-Cell RNA Sequencing

Several scRNA-seq technologies have been developed to enable high-throughput single-cell profiling. These approaches can be broadly classified into microfluidics-based, droplet-based, and plate-based methods:

Microfluidics-based approaches (e.g., Smart-seq, Fluidigm C1)

- Early single-cell methods relied on microfluidic platforms to capture individual cells and perform cDNA synthesis in small reaction volumes.
- Smart-seq2, an improved version of Smart-seq, provides full-length transcript coverage but is limited in throughput.

Droplet-based approaches (e.g., 10x Genomics Chromium, Drop-seq, inDrop)

- High-throughput platforms like 10x Genomics encapsulate individual cells in nanoliter-sized droplets along with barcoded beads [4].
- These barcodes allow mRNA transcripts from each cell to be uniquely indexed and later sequenced in bulk.
- Droplet-based approaches have enabled the sequencing of millions of cells in a single experiment, making them the dominant strategy for large-scale single-cell studies.

Plate-based approaches (e.g., MARS-seq, STRT-seq)

- These involve sorting single cells into microwell plates using fluorescence-activated cell sorting (FACS) or laser capture microdissection.
- They allow targeted selection of cells but are lower in throughput compared to droplet-based methods.

Each of these approaches has its own strengths and limitations. While microfluidic and plate-based methods offer high sensitivity and allow full-length transcript analysis, droplet-based methods provide the highest scalability, making them ideal for large datasets such as those used in single-cell immune profiling.

1.1.3 Challenges in Single-Cell Transcriptomics

Despite its transformative potential, scRNA-seq presents several technical challenges:

- Transcript dropouts: Because mRNA molecules are captured stochastically, lowly expressed genes may not be detected, leading to sparsity in the data.
- Cell dissociation bias: Single-cell isolation often requires enzymatic dissociation of tissues, which can alter gene expression profiles or selectively damage fragile cell types.
- Data integration across batches: Experimental variations across sequencing runs introduce batch effects that must be corrected using computational techniques [5,6].
- Computational complexity: Single-cell datasets contain millions of transcripts counts across thousands of cells, requiring sophisticated bioinformatics methods for normalization, clustering, and trajectory inference [7–10].

1.1.4 Recent Advances and Multimodal Approaches

To overcome some of these challenges and further expand the scope of single-cell analysis, recent innovations have focused on multimodal single-cell profiling [11,12]. These approaches aim to capture multiple layers of cellular information beyond just RNA sequencing, including:

- Simultaneous RNA and protein quantification (e.g., CITE-seq, REAP-seq).
- Epigenomic profiling at single-cell resolution (e.g., scATAC-seq, multiome assays).
- Spatial transcriptomics, preserving tissue context (e.g., 10x Visium, Xenium, MERFISH, NanoString GeoMx).

Among these multimodal strategies, CITE-seq has gained significant attention in immunology and cancer research, as it enables the integration of transcriptomic and proteomic data at the single-cell level. This method will be further explored in the next section.

1.2 CITE-seq: Integrating Transcriptomics and Proteomics in Single Cells

Single-cell RNA sequencing (scRNA-seq) has significantly advanced our ability to dissect cellular heterogeneity, yet it presents a major limitation: it only captures mRNA expression, which does not always correlate with protein abundance. In immunology, where cellular identity and function are often defined by surface protein markers, transcriptomic data alone may be insufficient for accurate immunophenotyping.

To overcome this gap, Cellular Indexing of Transcriptomes and Epitopes by sequencing (CITE-seq) was developed as a multimodal approach that combines single-cell transcriptomics and proteomics. By using oligonucleotide-labeled antibodies, CITE-seq enables the simultaneous quantification of surface protein expression and mRNA levels from the same single cell. This powerful technology bridges transcriptomics and protein-level characterization, providing a more comprehensive view of immune cell states, differentiation trajectories, and disease mechanisms.

1.2.1 Principles of CITE-seq Technology

CITE-seq relies on antibody-derived tags (ADTs), synthetic DNA barcodes conjugated to monoclonal antibodies, that specifically bind to cell surface proteins. The key steps of CITE-seq are as follows:

(A) **Antibody Staining with Oligonucleotide-Conjugated Antibodies**—Label cells with DNA-barcoded antibodies targeting surface proteins [13].

- A panel of antibody-derived tags (ADTs) is selected based on the desired surface markers.
- Cells are incubated with DNA-barcoded antibodies in a controlled environment to ensure optimal staining.

(B) **Single-Cell Isolation and Barcoding**—Capture individual cells in microdroplets or nanowells and assign them a unique barcode.

- Antibody-stained cells are loaded into a single-cell capture platform such as 10x Genomics Chromium, Drop-seq, or inDrop.

Each single cell is coencapsulated with a barcoded bead, enabling tracking of both RNA transcripts and antibody-derived tags (ADTs) from the same cell.

(C) **Cell Lysis and Reverse Transcription**—Capture and convert mRNA and ADT oligonucleotides into complementary DNA (cDNA).

- The encapsulated cells are lysed, releasing mRNA and DNA-barcoded antibody tags into the droplet.
- Reverse transcription (RT) occurs within each droplet, generating cDNA for both the mRNA and the antibody-derived tags (ADT).

The barcoded cDNA molecules are then amplified for downstream processing.

(D) Library Preparation and Sequencing

- ADTs and mRNA are amplified and prepared as separate sequencing libraries.
- Sequencing is performed using platforms such as **Illumina NovaSeq or NextSeq**, ensuring simultaneous readout of transcriptomic and protein data.

(E) Data Processing and Integration— Computational pipelines such as Seurat [3], Scanpy [14], or CITE-seq Count are used to:

- Demultiplex individual cells.
- Normalize ADT and RNA data.
- Integrate and cluster multimodal datasets to define cellular states.

By combining high-dimensional transcriptomic data with precise protein quantification, CITE-seq provides a richer and more robust cellular phenotype than scRNA-seq alone.

1.2.2 Key Advantages of CITE-seq

CITE-seq offers several **major advantages** over conventional single-cell approaches:

- Simultaneous transcriptomic and proteomic profiling to enhance cell type classification [15]: Unlike traditional scRNA-seq, which only provides mRNA data, CITE-seq enables a direct readout of surface protein expression, ensuring a more precise immunophenotyping.
- Overcoming RNA dropout issues: scRNA-seq often suffers from dropout events, where low-abundance transcripts are not detected. By capturing stable surface proteins, CITE-seq provides a more reliable marker profile.
- High-throughput compatibility: The method is scalable, allowing the simultaneous analysis of thousands to millions of cells in large studies.
- Reduced reliance on fluorescence-based flow cytometry: While flow cytometry and mass cytometry (CyTOF) are gold standards for protein detection, they are limited in transcriptomic resolution. CITE-seq overcomes this by integrating both molecular layers into a single experiment.

1.2.3 Challenges and Considerations in CITE-seq

Despite its advantages, CITE-seq presents several technical and computational challenges that must be addressed:

(A) Antibody Selection and Specificity

- Antibodies used in CITE-seq must be highly specific and well-validated to prevent nonspecific binding and background noise.
- Antibody titration is essential to optimize staining conditions and avoid saturation effects.
- The use of isotype controls and unstained samples can help assess background signal.

(B) Batch Effects and Technical Variability

- Differences in antibody labeling efficiency and sequencing depth can introduce batch effects.
- Proper experimental design, including biological and technical replicates, is crucial for reproducibility.
- Computational correction methods such as Harmony, Liger, or MNN-correct can be used to minimize batch effects [16].

(C) Data Integration and Normalization

- The dynamic range of protein expression is different from mRNA expression, requiring specialized normalization strategies (e.g., centered log-ratio transformation for ADTs).
- Integration of transcriptomic and proteomic layers requires multimodal analytical tools such as Seurat v4 WNN (weighted nearest neighbor) or MultiVI (Variational Inference for multimodal data).

1.2.4 Applications of CITE-seq in Immunology and Disease Research

CITE-seq has been widely adopted in immunology, oncology, and regenerative medicine [17–21], offering new insights into cellular dynamics and disease mechanisms. Key applications include:

(A) Immune Cell Profiling in Health and Disease.

- Characterization of immune subsets: CITE-seq allows precise distinction between closely related immune cell populations, such as regulatory T cells (Tregs) vs. effector T cells, or classical vs. non-classical monocytes.
- Single-cell immune responses: Studies have leveraged CITE-seq to analyze T and B cell responses to infections and vaccines, such as in COVID-19 and cancer immunotherapy research.

(B) Tumor Microenvironment and Cancer Research [22]

- Deciphering tumor immune landscapes: CITE-seq helps identify tumor-infiltrating lymphocytes (TILs) and their functional states, aiding in cancer prognosis and treatment strategies.
- Checkpoint inhibitor studies: By integrating protein-level detection of PD-1, CTLA-4, and PD-L1, CITE-seq enhances our understanding of immune evasion mechanisms in tumors.

(C) Hematopoietic and Stem Cell Research [23–25]

- CITE-seq has been used to track hematopoietic differentiation, mapping stem cell fates and lineage commitments with unprecedented resolution.
- The technique is also valuable for understanding stem cell plasticity and regenerative potential in various tissues.

CITE-seq represents a paradigm shift in single-cell immunophenotyping, offering an integrated view of transcriptomic and proteomic landscapes. By combining high-resolution mRNA sequencing with precise protein quantification, CITE-seq enables a deeper understanding of cellular identity, function, and dynamics. Despite some technical challenges, careful experimental design and computational integration allow researchers to leverage this powerful tool for immunology, oncology, and beyond.

2 Materials

2.1 Reagents

2.1.1 Cell Labeling

1. Phosphate-buffered saline (PBS): 137 mM NaCl, 2.7 mM KCl, 10 mM Na_2HPO_4, 1.8 mM KH_2PO_4, pH 7.4.
2. 10× PBS.
3. 1× PBS: add 5 ml 10× RNase-free PBS (pH 7.4) to 45 ml nuclease-free water in a 50 ml conical tube.
4. Ficoll-Paque solution: Density gradient medium for PBMC isolation.
5. Staining buffer: PBS, 0.04% bovine serum albumin (BSA).
6. Cell lysis buffer: 10 mM Tris-HCl, 10 mM NaCl, 0.2% Triton X-100, pH 7.5.
7. PBS with 1% FBS: PBS, 1% fetal bovine serum.
8. PBS with 10% FBS: PBS, 10% fetal bovine serum.
9. TotalSeq-C Universal Panel + customized ADT: Prevalidated antibody panel for cell surface protein labeling.
10. Cell washing buffer: PBS, 10% fetal bovine serum.
11. Cell resuspension buffer: PBS, 1% fetal bovine serum.
12. FACS staining buffer: PBS, 1% BSA.

2.1.2 Reagents for Single Cell Library Preparation

1. SPRIselect Reagent: Magnetic bead-based purification system for cDNA cleanup.
2. Dynabeads MyOne SILANE: Magnetic beads for post-GEM-RT cleanup.
3. RT Reagent E: Reverse transcription enzyme mix.
4. Poly-dT RT Primer B: Oligo-dT primer for cDNA synthesis.
5. Amp Mix: DNA polymerase mix for cDNA amplification.
6. cDNA Primers: Oligonucleotide primers for cDNA amplification.
7. Fragmentation Enzyme: For cDNA fragmentation prior to library construction.
8. Ligation Mix: For adapter ligation during library preparation.
9. DNA Ligase: Enzyme for joining DNA fragments.
10. Feature cDNA Primers 4: Primers used for cell surface protein library construction.

11. Reducing Agent B: Added during library preparation for optimal reaction conditions.
12. Cleanup Buffer: Used for sample purification steps.
13. Partitioning Oil B: Required for GEM generation in the Chromium system.
14. Recovery Agent: For phase separation post-GEM-RT cleanup.
15. Dual Index Plate: For sample multiplexing in sequencing.
16. Elution Solution I: Used in DNA elution from beads.
17. 80% Ethanol: Freshly prepared for bead washing during cleanup steps.
18. 10% Tween-20: Added in purification steps to prevent DNA loss.
19. PCR Tubes 0.2 ml 8-tube strips.
20. Molecular Grade Nuclease-free Water.
21. Glycerin (glycerol), 50% (v/v) aqueous solution.
22. 1× PBS phosphate-buffered saline, 1× without calcium and magnesium.
23. Qiagen Buffer EB.
24. Low TE Buffer (10 mM Tris-HCl pH 8.0, 0.1 mM EDTA).

2.2 Equipment

2.2.1 Equipment for Cell labeling

1. Biosafety Cabinet (Class II): For safe sample handling and preparation.
2. Centrifuge with Temperature Control: For cell pelleting and washing.
3. Automated Cell Counter: For determining cell viability and concentration.
4. Fluorescence-Activated Cell Sorter (FACS): For sorting and verifying labeled cells.
5. Incubator (37°C, 5% CO_2): For cell incubation steps.
6. Magnetic Separator Stand: For bead-based labeling and purification.
7. Vortex Mixer: For gentle resuspension of labeled cells.
8. Multichannel Pipettes: For efficient and precise reagent handling.
9. Hemocytometer or Counting Chamber: For manual cell counting if necessary.
10. Microcentrifuge: For quick spins during labeling protocols.
11. Cooling Block or Ice Bucket: To keep reagents and samples at appropriate temperatures.

2.2.2 Equipment for Single Cell Preparation

1. **Chromium X/iX Instrument 10X Genomics:** For GEM generation and single-cell partitioning.
2. **Chromium GEM-X Single Cell 5′Reagent Kits v3:** Specialized microfluidic chip for barcoding.
3. **Biosafety Cabinet (Class II):** For safe handling of biological samples.
4. **Thermal Cycler:** For cDNA synthesis, amplification, and library construction.
5. **10x Magnetic Separator B:** For magnetic bead-based sample purification.
6. **Automated Cell Counter:** For cell viability and concentration measurements.
7. **Qubit Fluorometer:** For DNA quantification.
8. **Bioanalyzer or TapeStation System:** For quality control of cDNA and libraries.
9. **Vortex Mixer and Plate Shaker:** For reagent mixing and bead resuspension.
10. **Multichannel Pipettes:** For precise liquid handling.
11. **Microcentrifuge:** For quick sample spins during cleanup steps.
12. **LabChip Traces or Agilent TapeStation:** For size selection and quality assessment of DNA fragments.
13. **Illumina Noaseq:** For high-throughput sequencing.
14. **10x Vortex Adapter:** For bead resuspension before use.

3 Methods

3.1 Step 0: Sample Preparation

3.1.1 Sample Preparation Before Staining

Overview of the library preparation and protocol is represented in Table 1.

For PBMC and other single-cell suspensions, standard preparation steps ensure high viability and efficient downstream processing.

For PBMC **samples** with 1 million cells in 1 ml PBS (*see* **Note 1**)

- Prepare single-cell suspension using standard dissociation protocols.
- Filter cells through a 40 μm strainer to remove debris and aggregates.

Table 1
Overview of CITE-seq library preparation and sequencing

Protocol steps and timing			
Step	**Description**	**Timing**	**Stop and store conditions**
Step 0: Sample preparation	Variable*	-	-
Refer to appropriate Demonstrated Protocols for details.			
0.1	Variable 1 Sample collection & processing	1 h to 8 h	Fresh samples preferred ≤4°C
0.2	Viability check and debris removal	30 min to 4 h	Proceed immediately
0.3	Oligonucleotide-conjugated antibody staining	30 min (4°C)	Stain in PBS + 2% FBS
0.4	Final wash & single-cell preparation	1 h–2 h	
Step 1: GEM Generation and Barcoding			
1.1	Prepare Master Mix	20 min	-
1.2	Load GEM-X Chip	10 min	-
1.3	Run the Chromium X/iX	6 min	-
1.4	Transfer GEMs	3 min	-
1.5	GEM-RT Incubation	55 min	4°C ≤ 72 h/–20°C ≤ 1 week
Step 2: Post GEM–RT Cleanup & cDNA Amplification			
2.1	Post GEM-RT Cleanup – Dynabeads	45 min	-
2.2	cDNA Amplification	40 min	4°C ≤ 72 h/–20°C ≤ 1 week
2.3	cDNA Cleanup – SPRIselect		
2.3A	Pellet Cleanup	30 min	4°C ≤ 72 h/–20°C ≤ 4 weeks
2.3B	Supernatant Cleanup	20 min	-
2.4	Post cDNA Amplification QC & Quantification	50 min	-

Table 1 (continued)

Protocol steps and timing			
Step	**Description**	**Timing**	**Stop and store conditions**
Step 3: V(D)J Amplification from cDNA (Optional)			
3.1	V(D)J Amplification 1	50 min	-
3.2	Post V(D)J Amplification 1 Cleanup (Double Sided – SPRIselect)	30 min	-
3.3	V(D)J Amplification 2	25 min	-
3.4	Post V(D)J Amplification 2 Cleanup (Double Sided – SPRIselect)	30 min	-
3.5	Post V(D)J Amplification QC & Quantification	40 min	4°C ≤ 72 h
Step 4: V(D)J Library Construction (Optional)			
4.1	Fragmentation, End Repair & A-tailing	50 min	-
4.2	Adaptor Ligation	30 min	-
4.3	Post Ligation Cleanup – SPRIselect	25 min	-
4.4	Sample Index PCR	40 min	4°C ≤ 72 h
4.5	Post Sample Index PCR Cleanup – SPRIselect	30 min	4°C ≤ 72 h/−20°C long term
4.6	Post Library Construction QC	50 min	
Step 5: Gene Expression Library Construction			
5.1	GEX Fragmentation, End Repair & A-tailing	50 min	
5.2	GEX Post Fragmentation, End Repair & A-tailing Double Sided – SPRIselect	30 min	
5.3	GEX Adaptor Ligation	25 min	
5.4	GEX Post Ligation Cleanup – SPRIselect	30 min	

Table 1 (continued)

Protocol steps and timing			
Step	**Description**	**Timing**	**Stop and store conditions**
5.5	GEX Sample Index PCR	40 min	4°C ≤ 72 h
5.6	Post Sample Index PCR Double Sided Size Selection – SPRIselect	30 min	4°C ≤ 72 h/−20°C long term
5.7	Post Library Construction QC	50 min	
Step 6: Cell Surface Protein Library Construction			
6.1	Sample Index PCR	40 min	
6.2	Post Sample Index PCR Size Selection – SPRIselect	20 min	4°C ≤ 72 h/−20°C long term
6.3	Post Library Construction QC	50 min	−20°C long term
Step 7: Sequencing			
7.1	Librarie Pooling & normalization	30/120 min	
7.2	Loading on sequencer	45 min	
7.3	Sequencing run	24 h to 48 h	
	Data processing & initial QC	6 h–24 h	

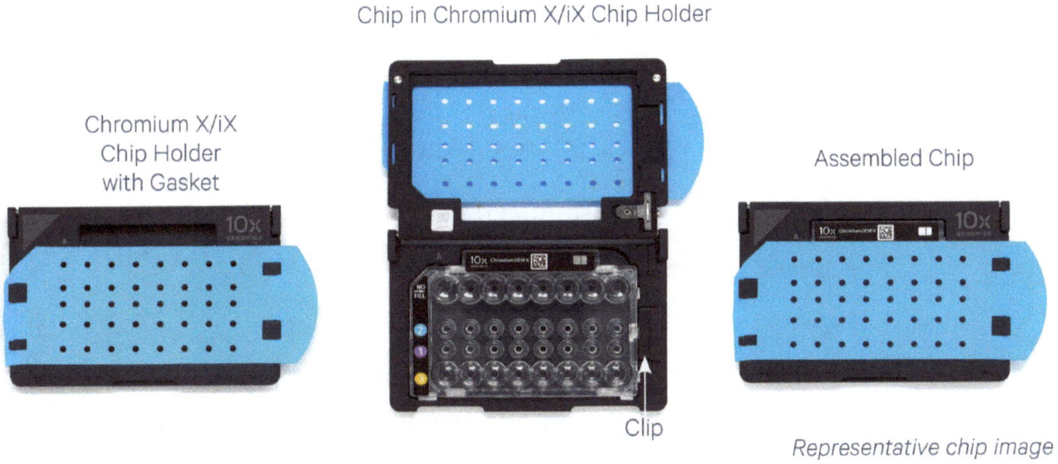

Fig. 1 GEM-X Chip Assembly. Schematic representation of the GEM-X 5′ Chip showing the chip holder, sealing gasket placement, and well layout for sample loading

- Count cells and ensure a concentration of 1×10^6 cells per sample.
- Assess viability (≥80% required).

3.1.2 Antibody Staining Protocol

Centrifuge samples **5 min at 400 × *g*, 4°C**

- Resuspend cell pellet in **50 µl PBS + 1% FBS.**
- Add **10 µl Human TruStain FcX™ BioLegend**, incubate 10 min at 4°C.
- Incubate **10 min on ice DO not wash proceed directly to staining.**

3.1.3 Oligonucleotide-Conjugated Antibody Staining (ADT Labeling)

Labels surface proteins with DNA-barcoded antibodies, allowing simultaneous proteomic and transcriptomic profiling.

- Prepare TotalSeq™-C Universal Mix (*see* **Note 2**).
- Equilibrate to **RT for 5 min**, spin **10,000 × *g* for 30 s at room temperature.**
- Resuspend in **27.5 µl PBS + 1% FBS**, vortex **10 s**, incubate **5 min at RT.**
- Vortex and spin **10,000 × *g* for 30 s at RT.**
- Transfer to **low protein binding tube**, pool all universal panel tubes.
- Centrifuge at **14,000 × *g* for 10 min at 4°C.**
- Centrifuge antibody pool (ADT add-on) at **14,000 × *g* for 10 min at 4°C.**
- Add **16.5 µl Universal Mix** to each sample.
- Add **6 µl ADT add-on.**
- Add **1 µl fluorophore-conjugated antibody for sorting** (e.g., **BCMA BioBright B515**).
- Add **PBS + 1% FBS up to 100 µl** (add **16 µl**).
- Incubate **30 min at 4°C.**
- Add **1 µl** LIVE/**DEAD™ Fixable Near IR (780)**, incubate **15 min at 4°C.**

3.1.4 Washing Steps (Critical for Background Reduction)

Removes unbound ADT antibodies to minimize background signal.

- Wash twice in **1.5 ml PBS + 1% FBS**, centrifuge **400 × *g*, 5 min at 4°C.**
- Resuspend pellet in **200 µl PBS + 1% FBS.**

  Key Considerations:

- Do not vortex or mix harshly to avoid cell damage.

- Use a low-retention pipette to prevent antibody loss.
- Ensure RNase-free conditions to protect transcriptomic integrity.

- Sorting cells before CITE-seq is essential for enriching specific populations, improving data quality, and removing debris or unwanted cells without traditional filtering.
- Resuspend sorted cells in **500 µl PBS + 10% FBS.**
- Verify cell concentration and viability (**95 µl cells + 5 µl DAPI**).
- Centrifuge sorted cells **10 min at 400 ×*g*, 4°C.**
- Resuspend pellet in **200 µl PBS + 1% FBS.**
- Count and check viability (**9 µl cells + 1 µl DAPI**) > 80%.

  Immediately proceed to single-cell encapsulation (Sect 3.2).

3.2 Step 1: GEM Generation and Barcoding

GEMs generation involves combining barcoded GEM-X Single Cell 5′ Gel Bead v3, a Master Mix with Feature Barcode-labeled cells, and Partitioning Oil B on a GEM-X 5′ Chip. To ensure single-cell resolution, cells are loaded at a limiting dilution, where 90%–99% of GEMs remain empty, and the remaining GEMs contain mostly one cell. Once GEMs are generated, Gel Beads dissolve, and any copartitioned cell is lysed. Released Gel Bead primers contain: Illumina TruSeq Read 1 (sequencing primer), 16 nt 10x Barcode (cell index), 12 nt unique molecular identifier (UMI), 13 nt template switch oligo (TSO).

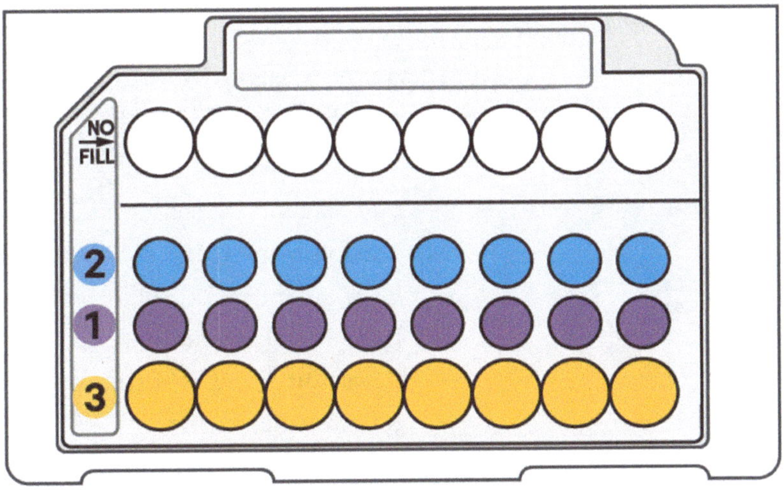

Fig. 2 GEM-X Chip Loading Layout. Diagram illustrating the row designations for loading Master Mix + Cell Suspension (Row 1), Gel Beads (Row 2), Partitioning Oil (Row 3), and the NO FILL row. Unused wells should receive 50% glycerol solution as indicated

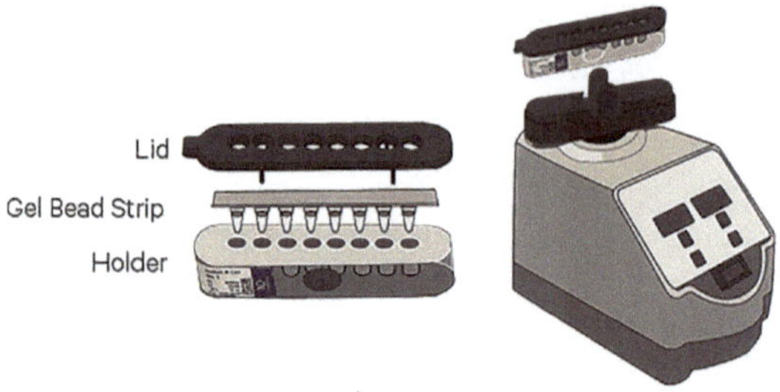

Fig. 3 Gel Bead Preparation. Illustration of the Gel Bead strip in the 10x Vortex Adapter, demonstrating the vortexing and centrifugation steps required before chip loading

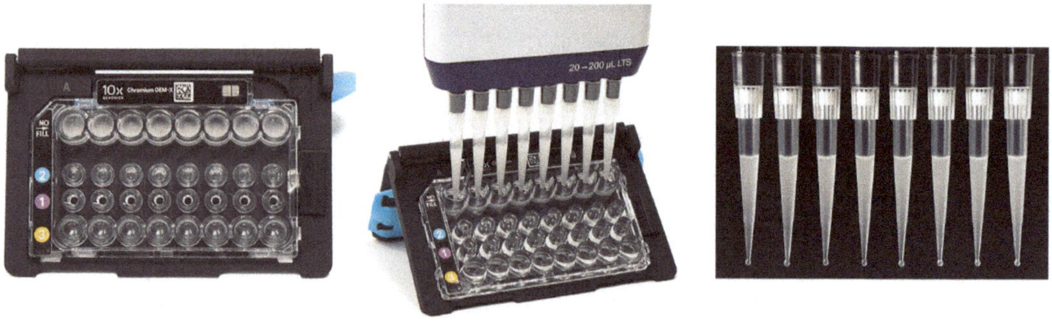

Fig. 4 GEM Transfer Procedure. Schematic showing the aspiration of GEMs from recovery wells and dispensing into tube strips on ice following Chromium X/iX run completion

During the reverse transcription (RT) a 10x Barcoded full-length cDNA is synthesized from polyadenylated mRNA using RT reagents in the Master Mix; feature Barcode-conjugated antibodies are captured, generating 10x Barcoded DNA corresponding to cell surface protein markers.

3.2.1 Prepare Master Mix

- Place all reagents on ice.
- Mix the following reagents thoroughly for 1Rx (*see* **Note 3**):
  - RT Reagent E: 16.3 µl.
  - Poly-dT RT Primer B: 2.2 µl.
  - Reducing Agent B: 2.0 µl.
  - RT Enzyme E: 7.0 µl.
- Pipette up and down 15 times to homogenize.
- Add 27.5 µl of Master Mix into each PCR tube strip.

3.2.2 Assemble GEM-X Chip (Fig. 1)

- Remove the GEM-X chip from its packaging and use within 24 h.
- Secure the chip into the Chromium X/iX chip holder.
- Attach the sealing gasket carefully, ensuring proper placement.
- Avoid touching the smooth surface of the gasket to prevent contamination.

3.2.3 Load GEM-X Chip

- Open the chip holder lid and lay the chip flat.
- After removing chip from the sealed bag, use in ≤24 h.
- Open the lid (gasket attached) of the assembled chip and lay flat for loading.
- Ensure that the Gel Beads are properly thawed and ready to use.
- When loading the chip, raising and depressing the pipette plunger should each take ~5 s. When dispensing, raise the pipette tips at the same rate as the liquid is rising, keeping the tips slightly submerged.

If loading less than eight samples/chip, add 50% glycerol solution to each unused well in row 1, 2, and 3 (Fig. 2)

- 60 μl in each unused well in row labeled 1.
- 60 μl in each unused well in row labeled 2.
- 250 μl in each unused well in row labeled 3.

DO NOT add 50% glycerol solution to the wells in top row labeled NO FILL.

DO NOT use any substitute for 50% glycerol solution.

Prepare Gel Beads (Fig. 3)

- Snap the tube strip holder with the Gel Bead strip into a 10x Vortex Adapter. Vortex 30 s.
- Centrifuge the Gel Bead strip for ~5 s.
- Confirm there are no bubbles at the bottom of the tubes & the liquid levels are even. Place the Gel Bead strip back in the holder. Secure the holder lid.

Gently pipette mix the cell suspension before adding to the Master Mix.

3.2.4 Prepare Master Mix + Cell suspension

Refer to the Cell Suspension Volume Calculator Table on **User Guide | CG000734** (*see* **Note 4**).

Add the appropriate volume of PBS to Master Mix.

Add corresponding volume of single cell suspension into the Master Mix. Total of 65 μl in each tube.

- Load the following reagents in the specified rows:
  Row 1: 60 μl of Master Mix + Cell Suspension.

Gently pipette mix the Master Mix + Cell Suspension. l Using the same pipette tip, dispense 60 µl Master Mix + Cell Suspension into the bottom center of each well in row labeled 1 without introducing bubbles. Wait 30 s.

Row 2: 60 µl of Gel Beads.

Puncture the foil seal of the Gel Bead tubes. l Slowly aspirate 60 µ l Gel Beads. l Dispense into the bottom center of each well in row labeled 2 without introducing bubbles. l Wait 30 s.

Row 3: 250 µl of Partitioning Oil. Dispense 250 µl Partitioning Oil B into the wells in row labeled 3 by pipetting two aliquots of 125 µl from a reagent reservoir.

- Ensure reagents are properly dispensed without introducing bubbles.
- Close the chip holder lid carefully without pressing on the gasket.

3.2.5 Run the Chromium X/iX

- Insert the loaded chip into the Chromium X/iX instrument tray.
- Select the "GEM-X 5'" program on the touchscreen.
- Start the run; the process takes approximately 6 min.
- Wait for the instrument to complete the process before proceeding.

3.2.6 Transfer GEMs (Fig. 4)

- After completion, eject and remove the chip from the holder. Place a tube strip on ice.
- Press the eject button of the Chromium X/iX and remove the chip.
- Discard the gasket. Open the chip holder. Fold the lid back until it clicks to expose the wells at 45°.
- Visually compare the remaining volume in rows labeled 1–2. Abnormally high volume in one well relative to other wells may indicate a clog.
- Slowly aspirate 100 µl GEMs from the lowest points of the recovery wells in the top NO FILL row without creating a seal between the tips and the bottom of the wells.

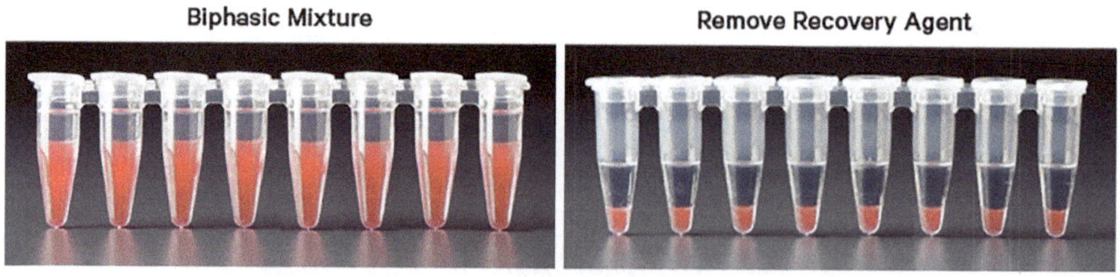

Fig. 5 Post GEM-RT Cleanup. Illustration of the biphasic separation following Recovery Agent addition, showing the Recovery Agent/Partitioning Oil (pink) layer and the aqueous phase (clear) layer

- Withdraw pipette tips from the wells. GEMs should appear opaque and uniform across all channels. Excess Partitioning Oil (clear) in the pipette tips indicates a potential clog.
- Over the course of ~20 s, dispense GEMs into the tube strip on ice with the pipette tips against the sidewalls of the tubes.
- It is recommended to start GEM incubation within 5 min. If multiple chips are run back-to-back, place the GEMs from the subsequent chips in separate thermal cyclers (*see* **Note 5**).

3.2.7 GEM-RT Incubation

- Incubate GEMs in a thermal cycler with the following program (*see* **Note 6**):
- 48°C for 45 min.
- 85°C for 5 min.
- Hold at 4°C.

Store at 4°C for up to 72 h or at −20°C for up to 1 week if not proceeding immediately.

3.3 Step 2: GEM-RT Cleanup and cDNA Amplification

3.3.1 Post GEM-RT Cleanup—Dynabeads

- Add 125 μl Recovery Agent to each sample (Post GEM-RT) at room temperature. If using sample stored at −20°C, equilibrate to room temperature before adding the Recovery Agent. DO NOT pipette mix or vortex the biphasic mixture. Wait 2 min.
  The resulting biphasic mixture contains Recovery Agent/Partitioning Oil (pink) and aqueous phase (clear), with no persisting emulsion (opaque).
  If biphasic separation is incomplete:
  Firmly secure the cap on the tube strip, ensuring that no liquid is trapped between the cap and the tube rim. Mix by inverting the capped tube strip 5×, centrifuge briefly, and proceed to Step b. DO NOT invert without firmly securing the caps. A smaller aqueous phase volume indicates a clog during GEM generation.

Table 2
Elution Solution I

Elution Solution I ***Add reagents in the order listed***		**PN**	**1× (μl)**	**10× (μl)**
	Buffer EB	—	98	980
	10% Tween 20	—	1	10
	Reducing Agent B	2000087	1	10
	Total		100	1000

- Slowly remove and discard 125 µl Recovery Agent/Partitioning Oil (pink) from the bottom of the tube. DO NOT aspirate any aqueous sample (Fig. 5).
- Prepare the Dynabeads Cleanup Mix: (*see* **Note 7**).
  - Cleanup Buffer: 182 µl.
  - Dynabeads MyOne SILANE: 8 µl.
  - Reducing Agent B: 5 µl.
  - Nuclease-free Water: 5 µl.
- Vortex and add 200 µl of the Cleanup Mix to each sample.
- Pipette mix ten times and incubate for 10 min at room temperature.
- Place tubes on a magnetic separator until the solution clears.
- Remove the supernatant without disturbing the beads.
- Wash the beads twice with 80% ethanol (300 µl each time) and remove the ethanol.
- Air dry the beads for 1 min.
- Resuspend the pellet in 35.5 µl of Elution Solution I and mix well.
- Incubate for 1 min, place on the magnetic separator, and transfer 35 µl of supernatant to a new tube.
- Prepare Elution Solution I (Table 2). Vortex and centrifuge briefly (*see* **Note 8**).

3.3.2 cDNA Amplification

- Prepare the cDNA Amplification Mix:
  - Amp Mix: 50 µl.
  - Feature cDNA Primers 4 Verify name & PN: 15 µl.
- Add 65 µl of the Amplification Mix to 35 µl of purified cDNA.
- Pipette mix 15 times and briefly centrifuge.
- Incubate in a thermal cycler with the following program:
- 98°C for 45 s.
- 15–18 cycles of (*see* **Note 9**):
- 98°C for 20 s.
- 63°C for 30 s.
- 72°C for 1 min.
- 72°C for 1 min.
- Hold at 4°C.
- Store at 4°C for up to 72 h or −20°C for ≤1 week or proceed to the next step.

3.3.3 cDNA Cleanup—SPRIselect

- Vortex to resuspend the SPRIselect reagent. Add 60 μl SPRIselect reagent (0.6×) to each sample and pipette mix 15× (pipette set to 140 μl).
- Incubate 5 min at room temperature.
- Place on the magnet•High until the solution clears.
- Transfer and save 75 μl supernatant in a new tube strip without disturbing the pellet. Maintain at room temperature. DO NOT discard the transferred supernatant (cleanup for Cell Surface Protein library construction).

Remove the remaining supernatant from the pellet without disturbing the pellet. DO NOT discard the supernatant (**Transferred Supernatant cleanup for 5′ Gene Expression library construction**). Immediately proceed to Pellet Cleanup.

3.3.4 Pellet Cleanup for 5′ Gene Expression Library

- Add 200 μl 80% ethanol to the pellet while still on the magnet• High. Wait 30 s.
- Remove the ethanol.
- Repeat Steps 63 and 64 for a total of two washes.
- Centrifuge briefly and place on the magnet•Low.
- Remove any remaining ethanol. Air dry for 2 min. DO NOT exceed 2 min as this will decrease elution efficiency.
- Remove from the magnet. Add 40.5 μl Buffer EB. Pipette mix 15×.

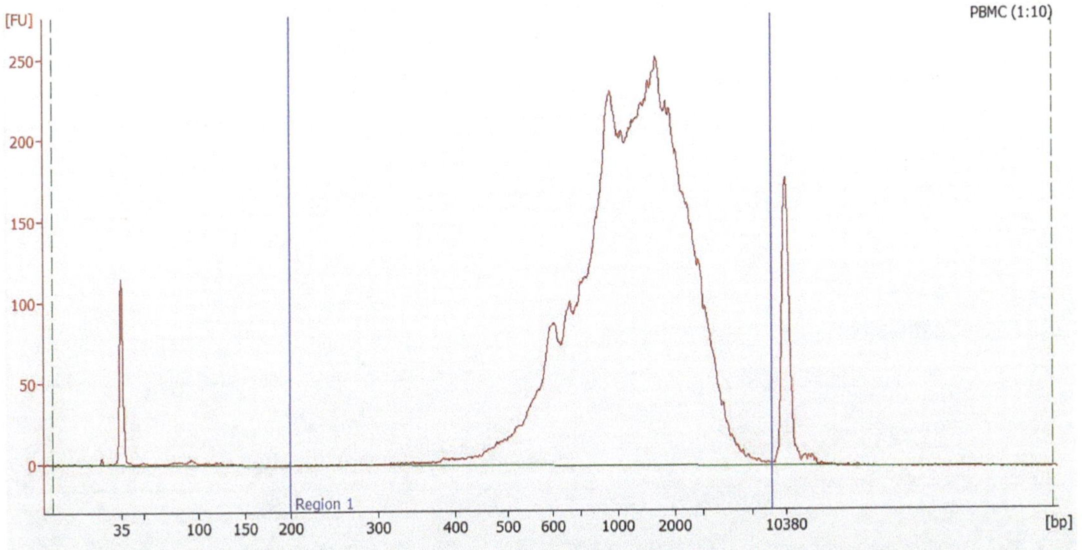

From [bp]	To [bp]	Corr. Area	% of Total	Average Size [bp]	Size distribution in CV [%]	Conc. [pg/μl]	Molarity [pmol/l]	Color
200	9,000	3,346.1	98	1,526	63.4	1,842.07	2,666.7	

Fig. 6 Representative Bioanalyzer Trace of cDNA after Amplification. Expected size distribution and quality metrics for postamplification cDNA

- Incubate 2 min at room temperature.
- Place on the magnet•High until the solution clears.
- Transfer 40 µl sample to a new tube strip.
- Store at 4°C for up to 72 h or at −20°C for up to 4 weeks or proceed to Post cDNA Amplification QC & Quantification for 5′ Gene Expression Library Construction.

3.3.5 Transferred Supernatant Cleanup for Cell Surface Protein

- Vortex to resuspend the SPRIselect reagent. Add 70 µl SPRIselect reagent (2.0×) to 75 µl of the transferred supernatant and pipette mix 15× (pipette set to 140 µl).
- Incubate for 5 min at room temperature.
- Place on the magnet•High until the solution clears.
- Remove supernatant.
- Add 200 µl 80% ethanol to the pellet. Wait 30 s.
- Remove the ethanol.
- Repeat Steps 77 and 78 for a total of two washes.
- Centrifuge briefly and place on the magnet•Low.
- Remove any remaining ethanol. Air dry for 2 min. DO NOT exceed 2 min as this will decrease elution efficiency.
- Remove from the magnet. Add 40.5 µl Buffer EB. Pipette mix 15×.
- Incubate 2 min at room temperature.
- Place the tube strip on the magnet•High until the solution clears.
- Transfer 40 µl sample to a new tube strip.
- Store at 4°C for up to 72 h or at −20°C for up to 4 weeks or proceed directly to Cell Surface Protein Library Construction.

Table 3
V(D)J Amplification mix

V(D)J Amplification 1 Rxn Mix *Add reagents in the order listed*	PN	1× (µl)	4× + 10% (µl)	8× + 10% (µl)
Amp Mix	2000047	50	220	440
T Cell Mix 1 v2 or B Cell Mix 1 v2	Human 2000242/ Mouse 2000256 or Human 2000254/ Mouse 2000258	48	211.2	422.4
Total		98	431.2	862.4

Table 4
PCR setting for V(D)J Amplification 1

Lid Temperature	Reaction Volume	Run Time
105°C	100μl	~20-30 min
Step	Temperature	Time *hh:mm:ss*
1	98°C	00:00:45
2	98°C	00:00:20
3	62°C	00:00:30
4	72°C	00:01:00
5 Different cycle numbers for T & B cells	T Cell: Go to Step 2, 11× (total 12 cycles) B Cell: Go to Step 2, 7× (total 8 cycles)	
6	72°C	00:01:00
7	4°C	Hold

3.3.6 Post cDNA Amplification QC & Quantification for 5′ Gene Expression Library

Use an Agilent Bioanalyzer, TapeStation, or qPCR to assess cDNA quality (Fig. 6).

Ensure the concentration meets the required range before proceeding to the next step.

Calculate: Multiply the diluted sample concentration [pg/μl] reported via Agilent 2100 Expert Software by the dilution factor and divide by 1000 to obtain the total cDNA yield in ng/μl.

Example of calculation:

Concentration: 1842.07 pg/μl Elution volume:40 μl; Dilution Factor:10

Total Yield = Conc'n (pg/μl) × Elution Vol. (μl) × Dilution Factor
1000 (pg/ng)

= 1842.07 × 40 × 10
1000 (pg/ng)

= 736.83 ng

Carrying Forward ONLY 25% of total cDNA yield for GEX Library (*see* **Note 10**)

= Total cDNA × 0.25 = 736.83 ng ×0.25 = 184.21 ng

3.4 Step 3: V(D) J Amplification from cDNA (Optional)

3.4.1 V(D)J Amplification 1

- Place a tube strip on ice and transfer 2 μl sample (post cDNA Amplification & QC, Step 2.3A) to the same tube.
- Prepare V(D)J Amplification 1 Reaction Mix on ice (Table 3). Vortex and centrifuge briefly.

Table 5
PCR setting for V(D)J Amplification 2

V(D)J Amplification 2 Rxn Mix *Add reagents in the order listed*	PN	1× (μl)	4× + 10% (μl)	8× + 10% (μl)
Amp Mix	2000047	50	220	440
T Cell Mix 2 v2 or B Cell Mix 2 v2	Human 2000243/ Mouse 2000257 or Human 2000255/ Mouse 2000259	15	66	132
Total		65	286	572

Table 6
PCR setting for V(D)J Amplification 2

Lid temperature	Reaction volume	Run time
105°C	100 μl	~25–30 min
Step	Temperature	Time *hh:mm:ss*
1	98°C	00:00:45
2	98°C	00:00:20
3	62°C	00:00:30
4	72°C	00:01:00
5 Different cycle numbers for T & B cells	T Cell: Go to Step 2, 9× (total 10 cycles) B Cell: Go to Step 2, 7× (total 8 cycles)	
6	72°C	00:01:00
7	4°C	Hold

- Add 98 μl V(D)J Amplification 1 Reaction Mix to each tube containing 2 μl sample.
- Pipette mix 5× (pipette set to 90 μl). Centrifuge briefly.
- Incubate in a thermal cycler with the protocol described in Table 4.

3.4.2 Post V(D)J Amplification 1 Cleanup—SPRIselect

- Vortex to resuspend SPRIselect Reagent. Add 50 μl SPRIselect Reagent (0.5×) to each sample. Pipette mix 15× (pipette set to 140 μl).
- Incubate 5 min at room temperature.
- Place tube strip on the magnet•High until the solution clears. DO NOT discard supernatant.
- Transfer 145 μl supernatant to a new tube strip.

- Vortex to resuspend SPRIselect Reagent. Add 30 µl SPRIselect Reagent (0.8×) to each sample. Pipette mix 15× (pipette set to 150 µl).
- Incubate 5 min at room temperature.
- Place on the magnet•High until the solution clears.
- Remove 170 µl supernatant. DO NOT discard any beads.
- Add 200 µl 80% ethanol. Wait 30 sec.
- Remove the ethanol.
- Repeat Steps 9 and 10 for a total of 2 washes.
- Centrifuge briefly. Place on the magnet•Low
- Remove remaining ethanol wash. DO NOT over-dry beads to ensure maximum elution efficiency.
- Remove from the magnet. Add 35.5 µl Buffer EB. Pipette mix 15×.
- Incubate 2 min at room temperature.
- Place on the magnet•Low until the solution clears.
- Transfer 35 µl sample to a new tube strip.
- Store at 4°C for up to 72 h or at −20°C for up to 1 week, or proceed to the next step.

3.4.3 V(D)J Amplification 2

- V(D)J Amplification 2 Reaction Mix (Table 5). Prepare on ice. Add reagents in the order listed. Vortex and centrifuge briefly before use.
- Add 65 µl V(D)J Amplification 2 Reaction Mix to each tube containing 35 µl sample.
- Pipette mix 5× (pipette set to 90 µl). Centrifuge briefly.
- Incubate in a thermal cycler with the protocol described in Table 6.

3.4.4 Post V(D)J Amplification 2 Cleanup—SPRIselect

- Vortex to resuspend SPRIselect Reagent. Add 50 µl SPRIselect Reagent (0.5×) to each sample. Pipette mix 15× (pipette set to 145 µl).
- Incubate 5 min at room temperature.
- Place tube strip on the magnet•High until the solution clears. DO NOT discard supernatant.
- Transfer 145 µl supernatant to a new tube strip.
- Vortex to resuspend SPRIselect Reagent. Add 30 µl SPRIselect Reagent (0.8×) to each sample. Pipette mix 15× (pipette set to 150 µl).
- Incubate 5 min at room temperature.
- Place on the magnet•High until the solution clears.

Table 7
PCR setting for End Repair & A-Tailing

Lid temperature	Reaction volume	Run time
65°C	50 µl	~35 min
Step	Temperature	Time hh:mm:ss
Precool block *Precool block prior to preparing the Fragmentation Mix*	4°C	Hold
Fragmentation	32°C	00:02:00
End Repair & A-Tailing	65°C	00:30:00
Hold	4°C	Hold

Table 8
Fragmentation mix

Fragmentation mix	PN	1× (µl)	4× + 10% (µl)	8× + 10% (µl)
Buffer EB	-	25	110	220
Fragmentation Buffer	2000091	5	22	44
Fragmentation Enzyme	2000090/			
	2000104	10	44	88
Total		40	176	352

Table 9
Adaptor Ligation Mix

Adaptor Ligation Mix	1× (µl)	4× + 10% (µl)	8× + 10% (µl)
Add reagents in the order PN *listed*			
Ligation Mix 2001109	40	176	352
DNA Ligase 220110/220131	10	44	88
Total	50	220	440

- Remove 170 µl supernatant. DO NOT discard any beads.
- Add 200 µl 80% ethanol. Wait 30 sec.
- Remove the ethanol.
- Repeat Steps 9 and 10 for a total of two washes.
- Centrifuge briefly. Place on the magnet•Low
- Remove remaining ethanol wash. DO NOT over-dry beads to ensure maximum elution efficiency.

- Remove from the magnet. Add 40.5 μl Buffer EB. Pipette mix 15×.
- Incubate 2 min at room temperature.
- Place on the magnet•Low until the solution clears.
- Transfer 40 μl sample to a new tube strip.
- Store at 4°C for up to 72 h or at −20°C for up to 1 week, or proceed to the next step.

3.4.5 Post V(D) J Amplification QC & Quantification

Use an Agilent Bioanalyzer, TapeStation, or qPCR to assess V(D) J cDNA quality.

Ensure the concentration meets the required range before proceeding to library construction.

3.5 Step 4: V(D)J Library Construction and Cleanup (Optional)

3.5.1 Fragmentation, End Repair & A-Tailing

- Prepare a thermal cycler with the incubation protocol described in Table 7.
- Vortex Fragmentation Buffer. Verify there is no precipitate.
- Prepare Fragmentation Mix on ice (Table 8). Pipette mix and centrifuge briefly.
- Transfer ONLY 10 μl purified V(D)J-amplified product from V(D)J Amplification (Step 3.4) to a tube strip. Note that only 10 μl (25%) V(D)J—amplified sample is sufficient for generating V(D)J library. The remaining sample can be stored at 4°C for up to 72 h or at −20°C for up to 4 weeks for generating additional V(D)J libraries (*see* **Note 10**).
- Add 40 μl Fragmentation Mix to each 10 μl sample.
- Pipette mix 15× (pipette set to 30 μl) on ice. Centrifuge briefly.
- Transfer into the precooled thermal cycler (4°C) and press "SKIP" to initiate the protocol.

Table 10
PCR setting for Adaptor Ligation

Lid temperature	Reaction volume	Run time
30°C (lid may be turned off if the instrument does not enable 30°C)	100 μl	15 min
Step	Temperature	Time *hh:mm:ss*
1	20°C	00:15:00
2	4°C	Hold

3.5.2 Adaptor Ligation

- Prepare Adaptor Ligation Mix (Table 9). Pipette mix and centrifuge briefly.
- Remove the sample from the thermal cycler.
- Add 50 μl Adaptor Ligation Mix to 50 μl sample. Pipette mix 15× (pipette set to 90 μl). Centrifuge briefly.
- Incubate in a thermal cycler with the protocol described in Table 10.

3.5.3 Post Ligation Cleanup—SPRIselect

- Vortex to resuspend SPRIselect Reagent. Add 80 μl SPRIselect Reagent (0.8×) to each sample. Pipette mix 15× (pipette set to 150 μl).
- Incubate 5 min at room temperature.
- Place on the magnet•High until the solution clears.
- Remove the supernatant.

Table 11
PCR setting for sample index PCR

Lid temperature	Reaction volume	Run time
105°C	100 μl	~30 min
Step	Temperature	Time *hh:mm:ss*
1	98°C	00:00:45
2	98°C	00:00:20
3	54°C	00:00:30
4	72°C	00:00:20
5	see Table 12 for # of cycles	
6	72°C	00:01:00
7	4°C	Hold

Table 12
PCR cycles recommendation based on cDNA input

cDNA input	Total cycles
<25 ng	9
25–150 ng	8
151–500 ng	7
501–1000 ng	6
>1500 ng	5

- Add 200 μl 80% ethanol to the pellet. Wait 30 s.
- Remove the ethanol.
- Repeat Steps 5 and 6 for a total of two washes.
- Centrifuge briefly. Place on the magnet•Low.
- Remove any remaining ethanol. Air dry for 2 min.
- Remove from the magnet. Add 30.5 μl Buffer EB. Pipette mix 15×. If beads still appear clumpy, continue pipette mixing until fully resuspended.
- Incubate 2 min at room temperature.
- Place on the magnet•Low until the solution clears.
- Transfer 30 μl sample to a new tube strip.

3.5.4 Sample Index PCR

- Choose the appropriate sample index sets to ensure that no sample indices overlap in a multiplexed sequencing run. Record the 10× sample index name (PN-3000431 Dual Index Plate TT Set A well ID) used.
- Add 50 μl Library Amp Mix (PN-2000531) or Amp Mix (PN-2000047/2000103) to 30 μl sample.
- Add 20 μl of an individual Dual Index TT Set A to each sample and record the well ID used. Pipette mix 5× (pipette set to 90 μl). Centrifuge briefly.
- Incubate in a thermal cycler with the protocol described in Tables 11 and 12.
- Store at 4°C for up to 72 h or proceed to the next step.

3.5.5 Post Sample Index PCR Cleanup—SPRIselect

- Vortex to resuspend the SPRIselect reagent. Add 80 μl SPRIselect Reagent (0.8×) to each sample. Pipette mix 15× (pipette set to 150 μl).
- Incubate 5 min at room temperature.
- Place the magnet•High until the solution clears.
- Remove the supernatant.
- Add 200 μl 80% ethanol to the pellet. Wait 30 s.
- Remove the ethanol.
- Repeat Steps 5 and 6 for a total of two washes.
- Centrifuge briefly. Place on the magnet•Low.
- Remove remaining ethanol. Air dry for 2 min.
- Remove from the magnet. Add 35.5 μl Buffer EB. Pipette mix 15×.
- Incubate 2 min at room temperature.
- Place on the magnet•Low until the solution clears.
- Transfer 35 μl to a new tube strip.

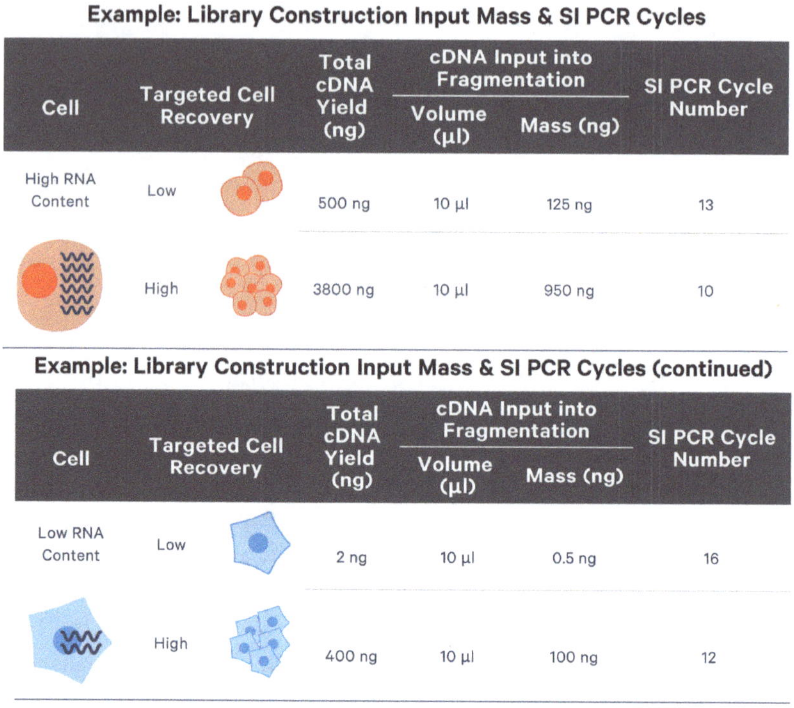

Fig. 7 (**a**) Library Construction Input Mass vs. SI PCR Cycles. Graphical representation of the relationship between input cDNA mass and recommended Sample Index PCR cycles for 5′ Gene Expression library construction. (**b**) Library Complexity Correlation. Graphical illustration showing that library complexity from 25% cDNA carry-forward is comparable to that from higher proportions

Table 13
PCR setting for Fragmentation, End Repair & A-Tailing

Lid temperature	**Reaction volume**	**Run time**
65°C	50 µl	~35 min
Step	Temperature	Time *hh:mm:ss*
Precool block *Precool block prior to preparing the Fragmentation Mix*	4°C	Hold
Fragmentation	32°C	00:05:00
End Repair & A-Tailing	65°C	00:30:00
Hold	4°C	Hold

Table 14
Fragmentation Mix

Fragmentation mix *Add reagents in the order listed*	PN	1× (μl)	4× +10% (μl)	8× + 10% (μl)
Buffer EB	-	25	110	220
Fragmentation buffer	2000091	5	22	44
Fragmentation enzyme	2000090/ 2000104	10	44	88
Total		40	176	352

- Store at 4°C for up to 72 h or at −20°C for long-term storage.

3.5.6 Post Library Construction QC

Use Agilent Bioanalyzer, Perkin Elmer LabChip, or Agilent Tapestation for QC.

Run 1 μl sample at 1:10 dilution on an Agilent Bioanalyzer High Sensitivity chip.

Select the region between 200 and 2000 bp to determine average size of the library. This will be used as the insert size for library quantification.

3.6 Step 5: 5′ Gene Expression Library Construction

3.6.1 Library Construction Input Mass & SI PCR Cycles

See Fig. 7a and 7b.

3.6.2 Fragmentation, End Repair and A-Tailing

- Prepare a thermal cycler with the incubation protocol described in Table 13.
- Vortex Fragmentation Buffer. Verify there is no precipitate.
- Prepare Fragmentation Mix on ice (Table 14). Add reagents in the order listed. Pipette mix and centrifuge briefly.
- Transfer ONLY 10 μl purified cDNA sample from Cleanup (Step 2.3 A) to a tube strip. The remaining cDNA sample can be stored at 4°C for up to 72 h or at −20°C for up to 4 weeks for generating additional Gene Expression libraries.
- Add 40 μl Fragmentation Mix to each 10 μl sample.
- Pipette mix 15× (pipette set to 30 μl) on ice. Centrifuge briefly.
- Transfer into the precooled thermal cycler (4°C) and press “SKIP” to initiate the protocol.

3.6.3 Post Fragmentation Cleanup—SPRIselect

- Vortex to resuspend SPRIselect reagent. Add 30 µl SPRIselect reagent (0.6×) to each sample. Pipette mix 15× (pipette set to 75 µl).
- Incubate 5 min at room temperature.
- Place on the magnet•High until the solution clears. DO NOT discard supernatant.
- Transfer 75 µl supernatant to a new tube strip.
- Vortex to resuspend SPRIselect reagent. Add 10 µl SPRIselect reagent (0.8×) to each sample. Pipette mix 15× (pipette set to 80 µl).
- Incubate 5 min at room temperature.
- Place on the magnet•High until the solution clears.
- Remove 80 µl supernatant. DO NOT discard any beads.
- With the tube strip still on the magnet, add 125 µl 80% ethanol to the pellet. Wait 30 s.
- Remove the ethanol.
- Repeat Steps 102 and 103 for a total of two washes.
- Centrifuge briefly. Place on the magnet•Low until the solution clears.
- Remove remaining ethanol. DO NOT over dry to ensure maximum elution efficiency.
- Remove from the magnet. Add 50.5 µl Buffer EB to each sample. Pipette mix 15× (pipette set to 45 µl).
- Incubate 2 min at room temperature.
- Place on the magnet•High until the solution clears.
- Transfer 50 µl sample to a new tube strip.

3.6.4 GEX Adaptor Ligation

- Prepare Adaptor Ligation Mix. Pipette mix and centrifuge briefly for 1 Rx
- Ligation Mix: 40 µl.
- DNA Ligase: 10 µl.

- Add 50 µl Adaptor Ligation Mix to 50 µl sample. Pipette mix 15× (pipette set to 90 µl). Centrifuge briefly.
- Incubate at 20°C for 15 min.

3.6.5 GEX Post Ligation Cleanup—SPRIselect

- Vortex to resuspend SPRIselect Reagent. Add 80 µl SPRIselect Reagent (0.8×) to each sample. Pipette mix 15× (pipette set to 150 µl).
- Incubate 5 min at room temperature.
- Place on the magnet•High until the solution clears.

Table 15
PCR cycles recommended based on cDNA input

cDNA input	Total cycles
0.25–50 ng	14–16
50–250 ng	12–14
250–600 ng	10–12
600–1100 ng	8–10
1100–1500 ng	6–8
>1500 ng	5

- Remove the supernatant.
- Add 200 μl 80% ethanol to the pellet. Wait 30 s.
- Remove the ethanol.
- Repeat Steps 117 and 118 for a total of two washes.
- Centrifuge briefly. Place on the magnet•Low.
- Remove any remaining ethanol. Air dry for 2 min. DO NOT exceed 2 min as this will decrease elution efficiency.
- Remove from the magnet. Add 30.5 μl Buffer EB. Pipette mix 15×.
- Incubate 2 min at room temperature.
- Place on the magnet•Low until the solution clears.
- Transfer 30 μl sample to a new tube strip.

3.6.6 GEX Sample Index PCR

- Choose the appropriate sample index sets to ensure that no sample indices overlap in a multiplexed sequencing run. Record the 10× sample index name (PN-3000431 Dual Index Plate TT Set A well ID) used.
- Add 50 μl Library Amp Mix (PN-2000531) or Amp Mix (PN-2000047/2000103) to 30 μl sample.
- Add 20 μl of an individual Dual Index TT Set A to each sample and record the well ID used. Pipette mix 5× (pipette set to 90 μl). Centrifuge briefly.
- Incubate in a thermal cycler with the following protocol.
  - 98°C for 45 s
  - 8–12 cycles of (Table 15)

    98°C for 20 s

    54°C for 30 s

    72°C for 20 s

- 72°C for 1 min
- Hold at 4°C

Table 15 recommends a starting point for optimization. The total cycles should be optimized based on 25% carry forward cDNA yield/input calculated during cDNA QC & Quantification.

3.6.7 Post Sample Index PCR Cleanup—SPRIselect

- Vortex to resuspend the SPRIselect reagent. Add 60 μl SPRIselect Reagent (0.6×) to each sample. Pipette mix 15× (pipette set to 150 μl).
- Incubate 5 min at room temperature.
- Place the magnet•High until the solution clears. DO NOT discard supernatant.
- Transfer 150 μl supernatant to a new tube strip.
- Vortex to resuspend the SPRIselect reagent. Add 20 μl SPRIselect Reagent (0.8×) to each transferred supernatant. Pipette mix 15× (pipette set to 150 μl).
- Incubate 5 min at room temperature.
- Place the magnet•High until the solution clears.
- Remove 165 μl supernatant. DO NOT discard any beads.
- With the tube still in the magnet, add 200 μl 80% ethanol to the pellet. Wait 30 s.
- Remove the ethanol.
- Repeat Steps 137 and 138 for a total of two washes.
- Centrifuge briefly. Place on the magnet•Low.
- Remove remaining ethanol. DO NOT over dry beads to ensure maximum elution efficiency.
- Remove from the magnet. Add 35.5 μl Buffer EB. Pipette mix 15× (pipette set to 35 μl).
- Incubate 2 min at room temperature.
- Place on the magnet•Low until the solution clears.

Table 16
Index PCR mixed

Sample Index PCR Mix	PN	1× (μl)	4× + 10% (μl)	8× + 10% (μl)	
	Amp Mix *Retrieve from Chromium GEM-X Single Cell 5' Feature Barcode Kit*	2000047	50	220	440
	Buffer EB	—	25	110	220
	Total		75	330	660

- Transfer 35 μl to a new tube strip.
- Store at 4°C for up to 72 h or at −20°C for long-term storage.

3.6.8 Post Library Construction QC

Use an Agilent Bioanalyzer, TapeStation, or qPCR to assess library quality.

Ensure the concentration meets the required range before proceeding to sequencing.

3.7 Step 6: Cell Surface Protein Library Construction

3.7.1 Sample Index PCR

- Choose the appropriate sample index sets to ensure that no sample indices overlap in a multiplexed sequencing run. Record the 10× sample index name (PN-3000510 Dual Index Plate TN Set A well ID) used.
- Prepare Sample Index PCR Mix. Add reagents in the order listed in Table 16.
- Transfer ONLY 5 μl from the Transferred Supernatant Cleanup (Step 86) to a new tube strip.
  The remaining sample can be stored at 4°C for up to 72 h or at −20°C
- Add 75 μl Sample Index PCR Mix to 5 μl Transferred Supernatant Cleanup sample.
- Add 20 μl of an individual Dual Index TN Set A to each sample and record the well ID used. Pipette mix 5× (pipette set to 90 μ l). Centrifuge briefly.
- Incubate in a thermal cycler with the following protocol:

- 98°C for 45 s.
- 12 cycles of
- 98°C for 45 s.
- 54°C for 30 s.
- 72°C for 20 s.

- 72°C for 1 min.
- Hold at 4°C.

3.7.2 Post Sample Index PCR Cleanup – SPRIselect

- Vortex to resuspend the SPRIselect reagent. Add 120 μl SPRIselect Reagent (1.2×) to each sample. Pipette mix 15× (pipette set to 150 μl).
- Incubate 5 min at room temperature.
- Place on the magnet•High until the solution clears. Remove the supernatant.

- Add 300 μl 80% ethanol to the pellet. Wait 30 s.
- Remove the ethanol.
- Add 200 μl 80% ethanol to the pellet. Wait 30 s.
- Remove the ethanol.
- Centrifuge briefly. Place on the magnet•Low. Remove remaining ethanol.
- Remove from the magnet. Add 35.5 μl Buffer EB. Pipette mix 15×.
- Incubate 2 min at room temperature.
- Place on the magnet•Low until the solution clears.
- Transfer 35 μl to a new tube strip.
- Store at 4°C for up to 72 h or at −20°C for long-term storage.

3.7.3 Post Library Construction QC

Use an Agilent Bioanalyzer, TapeStation, or qPCR to assess library quality (*see* **Note 11**).

Table 17
Library loading recommendations for Illumina Sequencing

Instrument	Loading concentration (pM)*	PhiX (%)
MiSeq	12	1
NextSeq 500/550	1.6	1
NextSeq 1000/2000	650	1
HiSeq 2500 (RR)	12	1
HiSeq 4000	240	1
NovaSeq 6000 Standard & Xp workflow	150	1

Table 18
Library pooling example

Libraries	Sequencing depth (read pairs per cell)	Library pooling ratio
Example 1		
V(D)J library 5' Gene Expression library Cell Surface Protein library	5,000 20,000 5,000	1 4 1
Example 2		
V(D)J library 5' Gene Expression library Cell Surface Protein library	5,000 50,000 5,000	1 10 1

Ensure the concentration meets the required range before proceeding to sequencing.

3.8 Step 7: Sequencing

3.8.1 Library Loading on Sequencer

Once quantified and normalized, libraries should be denatured and diluted as recommended for Illumina sequencing platforms. Refer to Illumina documentation for denaturing and diluting libraries. Refer to the 10x Genomics Support website for more information.

- **Dilution**: Adjust the pooled library to the optimal concentration for sequencing (Table 17).
- **Mixing**: Combine with sequencing reagents (e.g., loading buffer, control samples).
- **Flow Cell Loading**: Deposit the sample mix onto the sequencer's flow cell.
- **Prerun Calibration**: Perform an optional calibration check (if applicable) to ensure proper loading.

3.8.2 Library Pooling and Normalization (see **Note 12**)

- **Quantification**: Measure library concentration using a fluorometer (e.g., Qubit) or qPCR-based quantification.
- **Normalization**: Adjust all libraries to a uniform concentration.
- **Pooling**: Combine the libraries according to the sequencing platform's requirements (Table 18). *10% PhiX is recommended when pooling Gene Expression, V(D)J, and Cell Surface Protein libraries at a ratio of 1:1:1.
- **Quality check**: Verify library integrity using a Bioanalyzer or TapeStation.

Table 19
Library sequencing depth and run parameters

Parameter	Description
Sequencing depth	Minimum 5000 read pairs/cell for V(D)J library
	Minimum 20,000 read pairs/cell for 5' Gene Expression
	Minimum 5000 read pairs/cell for Cell Surface Protein
Sequencing type	Paired-end, dual indexing
Sequencing read	Recommended number of cycles
Read 1	28 cycles
i7 Index	10 cycles
i5 Index	10 cycles
Read 2	90 cycles

3.8.3 Sequencing Run (Variable: 24 h à 48 h+)

- **Start the Run:** Select the appropriate sequencing mode following recommendation in Table 19.
- **Monitor the Run:** Track run performance and error rates (e.g., Q-score).
- **Run duration:** Illumina NextSeq/NovaSeq: Typically 24 h to 48 h.

3.8.4 Data Analysis and Visualization

Sequencing data may be analyzed using Cell Ranger and visualized using Loupe Browser. Key features for these tools are listed below. For detailed product-specific information, visit the 10x Genomics Support website.

3.8.5 Cell Ranger

Cell Ranger is a set of analysis pipelines that processes Chromium Single Gene Expression data to align reads, generate Feature Barcode matrices and perform clustering and gene expression analysis.

- Input: Binary base call (BCL) and FASTQ
- Output: BAM, MEX, CSV, HDF5, Web Summary,.cloupe/.loupe
- Operating System: Linux

4 Notes

4.1 General considerations

Despite its advantages, CITE-seq presents several technical and computational challenges. Careful experimental design, including appropriate controls and replicates, is essential for generating high-quality and reproducible multimodal data.

Ensure all reagents and equipment are available and prepared. Fully thaw and thoroughly mix reagents before use.

Keep all enzymes and reagent mixes on ice during setup and use. Promptly move reagents back to the recommended storage

1. The following protocol is designed for one million cells per sample in 1 mL PBS, but quantities can be adjusted for lower cell numbers based on experimental needs. We have successfully performed the protocol with as few as 50,000 to 100,000 cells.

   Cell Viability Considerations: Ensure cell viability is above 80% before proceeding with staining and sequencing. Low viability affects data quality and increases background noise. If working with frozen PBMCs or dissociated tissue, thaw cells rapidly at 37°C, then transfer them immediately to cold media to prevent thermal shock. If cells are sticky or forming aggregates, filter through a 40 μm strainer before counting and staining.

If cells viability is lower than 80%, we have seen that PBS wash to remove dead cells, and debris may help to recover higher viability.

2. Titrate antibodies before use to determine the optimal concentration. Too much antibody leads to nonspecific binding, while too little reduces sensitivity. Include unstained, isotype controls, and single-stained controls to properly define background fluorescence and gating thresholds. If using a new antibody panel, run a pilot experiment with flow cytometry to confirm specificity before proceeding with CITE-seq.

   Keep ADT-conjugated antibodies at 4°C, and avoid repeated freeze-thaw cycles, which can degrade the oligonucleotide tags. Centrifuge antibody tubes before use (14,000*g*, 10 min at 4°C) to remove aggregates that could contribute to background noise. When working with a large panel of ADTs, consider pooling antibodies and premixing to ensure consistent staining.

3. Calculate reagent volumes with 10% excess of one reaction values.

4. The optimal input cell concentration depends upon the desired cell recovery target: we found 10X Genomics expected recovery (65%) optimistic, in our hand standard recovery is ~50% of loaded cells. Be aware that the presence of dead cells and debris in the suspension may reduce the recovery rate; thus, we always base our ratio on live cells.

   Avoid air bubbles when loading the Chromium chip, as this can disrupt droplet formation. If the sample appears clumpy, rerfilter with a 40 μm strainer before loading.

   For accurate single-cell partitioning, ensure cell concentration is within the optimal range (typically 700–1200 cells/μL for a recovery of 5000–10,000 cells). Do not overconcentrate cells (>1600 cells/μL) as this increases doublet formation and affects downstream analysis.

   Handle cells gently to avoid mechanical stress that could alter their gene expression profile. For immune cell profiling, avoid using strong trypsinization, as it can cleave surface markers and affect antibody binding. Keep cells on ice whenever possible to slow metabolic activity and RNA degradation. Limit the time between isolation and encapsulation to less than 30 min.

5. If GEM generation fails, check:
   - Bead suspension: Ensure the gel beads are fully resuspended before loading.
   - Partitioning oil: Confirm that partitioning oil has been added to each well.
   - Cell concentration: Overloaded cells can cause doublets or bead clogs.

6. Perform RT at 48°C for exactly 45 min, longer incubations do not improve efficiency and may degrade RNA.

7. Before using Dynabeads MyOne SILANE prepare the Dynabeads Cleanup Mix.

   - Vortex the Dynabeads thoroughly (≥30 s) immediately before adding to the mix.
   - Aspirate full liquid volume in the Dynabead tube with a pipette tip to verify that beads have not settled in the bottom of the tube. If clumps are present, pipette mix to resuspend completely. DO NOT centrifuge before adding to the mix.

8. If beads are lost during cleanup, check if:

   - Pipetting was too harsh, gently resuspend beads
   - SPRIselect beads were fully resuspended before use
   - If cDNA yield is low, ensure:
     - The cell recovery rate was optimal.
     - The reverse transcription enzyme mix was stored properly (−20°C).

   Avoiding Cross-Contamination: Use filtered tips and clean workspaces to prevent contamination between samples. Always process low RNA content samples first, then high-content ones, to prevent carryover.

9. The optimal cycle number is a trade-off between generating sufficient final mass for libraries & minimizing PCR amplification artifacts. The number of total cycles is based on the targeted cell recovery and is independent of the RNA content of the sample.

   - Targeted Cell Recovery <500 uses 17 total cycles.
   - Targeted Cell Recovery 501–2000 uses 15 total cycles.
   - Targeted Cell Recovery 2001–6000 uses 13 total cycles.
   - Targeted Cell Recovery >6000 uses 12 total cycles.

10. Correlation between input & library complexity

    A Single Cell Gene Expression library is generated using a fixed proportion (10 μl, 25%) of the total cDNA (40 μl) obtained. The complexity of this library will be comparable to one generated using a higher proportion (>25%) of the cDNA.

    The total number of SI PCR cycles (Step 128) should be optimized based on carrying forward a fixed proportion (10 μl, 25%) of the total cDNA yield calculated during Post cDNA Amplification QC & Quantification.

11. Assess cDNA and final libraries using TapeStation and Qubit:
    - Gene Expression Libraries: Peak at ~400 bp.
    - ADT Libraries: Peak at ~150 bp.
    - If excess adapter dimers are present (<200 bp fragments), perform a double-sided SPRIselect cleanup to remove them.

12. Different libraries maybe pooled for sequencing, taking into account the differences in cell number and per-cell read depth requirements between each library. Samples utilizing the same sample index should not be pooled together or run on the same flow cell lane, as this would not enable correct sample demultiplexing.

    The recommended sequencing depth per cell for Gene Expression Library is 20,000–50,000 reads/cell; for Antibody-Derived Tags (ADT) Library is 5000 reads/cell. If sequencing multiple libraries together, maintain the correct pooling ratio (e.g., 1:4:1 for V(D)J, Gene Expression, and ADT).

    Each sample index in the relevant Dual Index Kit contains a mix of one unique i7 and one unique i5 sample index. If multiple samples are pooled in a sequence lane, the sample index name (i.e. the Dual Index plate well ID) is needed in the sample sheet used for generating FASTQs with "cellranger mkfastq." If multiple libraries are pooled in a sequence lane, a separate sample index is needed with each library.

References

1. Stoeckius M, Hafemeister C, Stephenson W, Houck-Loomis B, Chattopadhyay PK, Swerdlow H, Satija R, Smibert P (2017) Simultaneous epitope and transcriptome measurement in single cells. Nat Methods 14(9):865–868. https://doi.org/10.1038/nmeth.4380
2. Cadot S, Valle C, Tosolini M, et al. (2020) Longitudinal CITE-Seq profiling of chronic lymphocytic leukemia during ibrutinib treatment: evolution of leukemic and immune cells at relapse. Biomark Res 8:72. https://doi.org/10.1186/s40364-020-00253-w10.1016/j.cell.2021.04.048
3. Hao Y, Hao S, Andersen-Nissen E, et al. (2021) Integrated analysis of multimodal single-cell data. Cell 184:3573–3587. https://doi.org/10.1016/j.cell.2021.04.048
4. Zheng GXY, Terry JM, Belgrader P, et al. (2017) Massively parallel digital transcriptional profiling of single cells. Nat Commun 8, 14049. https://doi.org/10.1038/ncomms1404910.1016/j.crmeth.2024.100938
5. Stuart T, Butler A, Hoffman P, et al. (2019) Comprehensive integration of single-cell data. Cell 177:1888–1902. https://doi.org/10.1016/j.cell.2019.05.031
6. Butler A, Hoffman P, Smibert P, Papalexi E, Satija R (2018) Integrating single-cell transcriptomic data across different conditions, technologies, and species. Nat Biotechnol 36:411–420. https://doi.org/10.1038/nbt.4096
7. McGinnis CS, Murrow LM, Gartner ZJ (2019) DoubletFinder: doublet detection in single-cell RNA sequencing data using artificial nearest neighbors. Cell Syst 8:329–337. https://doi.org/10.1016/j.cels.2019.03.003
8. Lun ATL, Riesenfeld S, Andrews T, et al. (2019) EmptyDrops: distinguishing cells from empty droplets in droplet-based single-cell RNA sequencing data. Genome Biol 20:63. https://doi.org/10.1186/s13059-019-1662-y
9. Lähnemann D, Köster J, Szczurek E, et al. (2020) Eleven grand challenges in single-cell data science. Genome Biol 21:31. https://doi.org/10.1186/s13059-020-1926-6
10. Svensson V, Natarajan KN, Ly LH, et al. (2017) Power analysis of single-cell RNA-sequencing experiments. Nat Methods 14:381–387. https://doi.org/10.1038/nmeth.4220

11. Stuart T, Satija R (2019) Integrative single-cell analysis. Nat Rev Genet 20:257–272. https://doi.org/10.1038/s41576-019-0093-7
12. Mimitou EP, Lareau CA, Lareau KY, et al. (2021) Scalable, multimodal profiling of chromatin accessibility, gene expression and protein levels in single cells. Nat Biotechnol 39:1246–1258. https://doi.org/10.1038/s41587-021-00927-210.1101/2021.04.01.437998
13. Stoeckius M, Zheng S, Houck-Loomis B, et al. (2018) Cell Hashing with barcoded antibodies enables multiplexing and doublet detection for single cell genomics. Genome Biol 19:224. https://doi.org/10.1186/s13059-018-1603-1
14. Wolf FA, Angerer P, Theis FJ (2018) SCANPY: large-scale single-cell gene expression data analysis. Genome Biol 19:15. https://doi.org/10.1186/s13059-017-1382-0
15. Caron DP, Specht WL, Chen D, et al. (2025) Multimodal hierarchical classification of CITE-seq data delineates immune cell states across lineages and tissues. Cell Rep Methods 5, 100938. https://doi.org/10.1016/j.crmeth.2024.100938
16. Korsunsky I, Millard N, Fan J, et al. (2019) Fast, sensitive and accurate integration of single-cell data with Harmony. Nat Methods 16:1289–1296. https://doi.org/10.1038/s41592-019-0619-0
17. Stephenson E, Reynolds G, Botting RA, et al. (2021) Single-cell multi-omics analysis of the immune response in COVID-19. Nat Med 27:904–916. https://doi.org/10.1038/s41591-021-01329-2
18. Good Z, Spiegel JY, Sahaf B, et al. (2022) Post-infusion CAR TReg cells identify patients resistant to CD19-CAR therapy. Nat Med 28:1860–1871. https://doi.org/10.1038/s41591-022-01960-7
19. Wells SB, Rainbow DB, Mark M, et al. (2025) Multimodal profiling reveals tissue-directed signatures of human immune cells altered with age. Nat Immunol 26:1612–1625. https://doi.org/10.1038/s41590-025-02241-4
20. Zhang B, Upadhyay R, Hao Y, et al. (2023) Multimodal single-cell datasets characterize antigen-specific CD8+ T cells across SARS-CoV-2 vaccination and infection. Nat Immunol 24:1725–1734. https://doi.org/10.1038/s41590-023-01608-9
21. Doolittle ML, Saul D, Kaur J, et al. (2023) Multiparametric senescent cell phenotyping reveals targets of senolytic therapy in the aged murine skeleton. Nat Commun 14:4587. https://doi.org/10.1038/s41467-023-40393-9
22. Leader AM, Grout JA, Maier BB, et al. (2021) Single-cell analysis of human non-small cell lung cancer lesions refines tumor classification and patient stratification. Cancer Cell 39:1594–1609. https://doi.org/10.1016/j.ccell.2021.10.009
23. Vanuytsel K, Villacorta-Martin C, Lindstrom-Vautrin J, et al. (2022) Multi-modal profiling of human fetal liver hematopoietic stem cells reveals the molecular signature of engraftment. Nat Commun 13:1103. https://doi.org/10.1038/s41467-022-28616-x
24. Zhang X, Song B, Carlino MJ, et al. (2024) An immunophenotype-coupled transcriptomic atlas of human hematopoietic progenitors. Nat Immunol 25:703–715. https://doi.org/10.1038/s41590-024-01782-4
25. Hanemaaijer ES, Müskens KF, Kal IJ, et al. (2026) Single-cell multiomic atlas of healthy pediatric bone marrow reveals age-dependent differences in lineage differentiation driven by stromal signaling. Nat Immunol 27:613–623. https://doi.org/10.1038/s41590-026-02422-9

Chapter 18

Immunophenotyping by Spatial Biology

Sandra Curras-Alonso, Yasmine Adimi, Clement Levin and Fabien Delahaye

Abstract

Studying the transcriptome and the proteome of cells is essential for gaining a detailed understanding of cellular behavior, development, drug action, and disease progression. Spatial biology emerges to advance our ability to study the expression of molecules within tissues while preserving their natural spatial context. These cutting-edge technologies enable the mapping of thousands of individual cells in their original environment by detecting the location and biological quantity of cellular contents, such as RNAs and proteins. Here, we present a protocol to combine spatial transcriptomics (Xenium) and spatial proteomics (PhenoCycler-Fusion) within 8 days on the same tissue section to successfully study the expression of hundreds of RNAs and tens of proteins simultaneously. The combination of these two technologies and consequent integration of the two data layers together with high-resolution H&E images allows for the extraction of a maximum of information from a single tissue section. Application of this protocol and the resulting integrated data will help researchers to understand complex biological processes and disease mechanisms, supporting more nuanced research in molecular biology and pathology.

Keywords Spatial biology, Spatial transcriptomics, Spatial proteomics, Immunophenotyping, RNA, Protein, Data integration

1 Introduction

1.1 General Introduction of Spatial Biology

Spatial tissue analysis represents a significant leap forward in understanding biological and pathological mechanisms. Unlike bulk techniques that average molecular information from dissociated samples, spatial approaches preserve tissue architecture and the topological relationships between cells, enabling a contextualized analysis of biological processes [1]. This spatial dimension is critical for understanding how cellular interactions and tissue organization influence biological functions, disease progression, and treatment response [2].

In pathology, spatial analysis plays an essential role in studying tumor microenvironments, immune infiltration patterns, and the interplay between tumor cells and surrounding stroma [3]. Recent

Angélique Biancotto and Pradeep K. Dagur (eds.), *Immunophenotyping*, Methods in Molecular Biology, vol. 3087, https://doi.org/10.1007/978-1-0716-5607-5_18,

technological advances have enriched these analyses, ranging from classical techniques such as hematoxylin and eosin (H&E) staining to highly multiplexed proteomic and transcriptomic approaches. These methods provide unparalleled resolution for mapping biomolecules at the cellular level while preserving spatial information [4].

The introduction of spatial technologies into biomedical sciences marks a turning point in tissue studies. These tools not only address fundamental questions about tissue biology but also improve diagnostics and personalized therapies, making them indispensable in both research and clinical practice [5].

1.2 Hematoxylin and Eosin (H&E): A Foundational Tool in Histology and Modern Pathology

Hematoxylin and eosin (H&E) staining remains a cornerstone of histological analysis, enabling the visualization of cellular and tissue organization with precise chromatic differentiation. Hematoxylin stains nuclei, rich in DNA, violet, while eosin marks cytoplasmic and extracellular structures pink. This technique provides critical insights into tissue morphology, allowing pathologists to identify specific cell types such as lymphocytes (dense, small nuclei), epithelial cells (oval nuclei with well-defined cytoplasm), and adipocytes (characterized by large lipid vacuoles). Additionally, H&E highlights morphological alterations associated with pathologies, including anisocytosis and nuclear abnormalities in poorly differentiated tumors or inflammatory infiltrates composed of granulocytes and macrophages [6].

Beyond its diagnostic utility, H&E serves as a crucial foundation for developing artificial intelligence (AI) algorithms in pathology. Its standardization and widespread availability enable the creation of large-scale datasets for training deep learning models. These models can automatically analyze histological images, extract complex features, and detect subtle anomalies that may be imperceptible to the human eye. For instance, AI-based classifiers can distinguish between healthy and tumor tissues, quantify intratumoral heterogeneity, and identify specific structures within the tumor microenvironment [7]. Advanced approaches, such as graph learning, further model the topological relationships between cells, offering a detailed representation of cellular interactions and growth patterns.

Thus, while simple in design, H&E remains indispensable in histology, bridging classical methodologies with modern digital technologies to enhance diagnostic accuracy and deepen the understanding of pathological mechanisms.

1.3 Spatial Transcriptomics (ST): Towards a Better Understanding of Tissue Architecture and Cellular Interactions

Spatial transcriptomics (ST) aims at describing the transcriptome landscape at the single-cell resolution in native histological tissue sections. Many different technologies have been developed in the last decade to map hundreds of RNAs in their spatial context, which can be divided into two main categories: in-situ sequencing-based approaches and hybridization-based approaches [8]. The former allows whole genome sequencing directly on the tissue, allowing de novo discovery of novel regions and cell types of interest within complex tissue structures. The latter, which is the primary focus of this chapter, detects expression of hundreds of targeted RNAs by hybridization of fluorescently labeled probes on tissue sections, allowing the mapping and measurement of RNA molecules with high sensitivity and specificity in a qualitative manner at the single-cell level.

1.3.1 Classical Strategies and Advancements in Hybridization-Based Spatial Transcriptomics

Classical approaches to spatially resolve gene expression data use computational single-cell RNAseq analysis to identify cell-type-specific markers as well as new cell subtypes and cell states. This allows the design of probes that are highly expressed and highly specific for each target RNA, which are then hybridized onto the tissue sections and detected either directly through fluorescent probes or indirectly via secondary fluorescently labeled probes using microscope imaging [9]. This approach has been used in complex and heterogenous tissues, such as the brain [10] or the lung [11].

To describe the vast heterogeneity of a tissue, it is essential to identify all the different cell types, which requires detecting hundreds of mRNA molecules within the same tissue section. One of the limitations to detect so many genes is that some are lowly expressed and therefore are difficult to detect. This issue has been overcome by amplifying RNA signal [12–15] or reducing tissue background and autofluorescence [16–18]. Another significant limitation of classical ST approaches is the number of spectrally resolved fluorophores, which are usually ranged from 3 to 4, as they limit the number of RNA targets that can be observed simultaneously. This limitation can be solved by doing sequential rounds of hybridization, by alternating rounds of probe hybridization, imaging, and signal removal [17,19–21], using, for example, microfluidic devices within the microscope [22]. This method enables a linear increase in the number of target mRNAs detected with each additional round of hybridization, which has required the development of complex segmentation algorithms to properly assign detected RNA transcripts to individual cells [23].

Other methods to increase the number of detectable transcripts include encoding strategies, in which throughout the different rounds of hybridization a unique barcode is built for each target mRNA. These methods allow the mapping of hundreds of target mRNAs in single cells through complex computational analysis, which typically include error-correction strategies [17,24–27]. Postprocessing of the data involves decoding of the barcodes, which allow the identification of each target, leading to the identification of hundreds to thousands RNAs.

1.3.2 A New Era with Xenium Technology

One of the latest technologies that has been developed includes RNA detection using padlock probes, highly specific and sensitive probes that allow direct detection of RNA transcripts of interest, which are ligated and amplified using rolling circle amplification (RCA) upon successful hybridization. RCA forms single stranded DNA repeats called rolling cycle products (RCPs), which contain barcodes that in the last hybridization round can be combinatorially decoded by hybridizing primary bridge probes and fluorescently labelled oligos over multiple cycles, generating a strong and specific signal that can be visualized using a microscope [28].

This is the technology in which the Xenium platform from 10x Genomics is based on. Xenium In Situ Gene Expression assays RNA at the subcellular level by using targeted probes in fresh-frozen (FF) or formalin-fixed and paraffin-embedded (FFPE) tissue sections. Through the FFPE protocol, which is the main focus of this chapter, FFPE tissue sections are placed on Xenium Slides, deparaffinized and decrosslinked. Then, the specific padlock probes (predesigned, add-on custom or standalone custom probe panels) are added to the tissue and hybridize the target RNA sequences. Each of these circularizable DNA probes contain two regions that hybridize to the target RNA and a third region that encodes a gene-specific barcode. When the two ends of the probes that bind the target RNA are ligated, a circular DNA probe is generated, which is enzymatically amplified, generating multiple copies of the barcodes. Then, for automated morphology-based cell segmentation analysis, cell segmentation reagents are added overnight to label cell nuclei, membrane, and interiors. Lastly, slides containing tissue sections are loaded on the Xenium Analyzer instrument, where cyclical rounds of hybridization of fluorescent probes to the amplified DNA probes, imaging, and removal occur, generating optical signatures specific for each barcode and converted into a gene identity.

Imaging is performed across multiple Z-planes with a 0.75 μm step size, covering the entire tissue thickness for each field of view (FOV) within the selected region, and through multiple fluorescent channels in every cycle. These images are then processed and stitched together to create a spatial map of the transcripts throughout the selected tissue section. The primary analysis to decode image data to transcripts and the secondary analysis to segment cells and assign transcripts occur automatically on the instrument. This high-throughput automated in situ analysis allows subcellular profiling of hundreds of RNA targets, uncovering novel aspects of cellular architecture and molecular function.

1.4 Spatial Proteomics

Spatial proteomics aims to map the protein landscape at single-cell resolution within native histological tissue sections. Over the past decade, various technologies have been developed to analyze protein expression in their spatial context, which can be broadly divided into two main categories: mass spectrometry-based approaches and antibody-based approaches. The former provides an untargeted and comprehensive view of the proteome, enabling the discovery of novel proteins and pathways across complex tissue structures. The latter, in which we will be focusing in this chapter, uses fluorescently labeled or barcoded antibodies to detect and quantify targeted proteins with high sensitivity and specificity, allowing precise spatial localization of proteins at the cellular level. Spatial proteomics offers invaluable insights into the organization and function of proteins in their native environment, driving advances in fields such as oncology, immunology, and regenerative medicine.

1.4.1 Immunohistochemistry (IHC): Single and Dual Staining for Targeted and Spatial Analysis

Immunohistochemistry (IHC) is a fundamental histological technique that localizes specific protein expressions within tissues. Single-marker IHC employs a single antibody directed against a target protein, conjugated to an enzymatic or fluorescent label. This approach is widely used to characterize specific biomarkers, such as HER2 in breast carcinomas, or to assess cellular proliferation through Ki-67 expression. It provides localized information on the abundance and distribution of the target protein within the tissue context [29].

Dual-marker IHC allows simultaneous detection of two distinct proteins on the same tissue section, using antibodies with different detection systems to ensure clear signal separation. This method enables analyses of colocalization or spatial relationships between proteins or cell types. For example, combining CD8 and PD-L1 staining can evaluate the proximity between cytotoxic T cells and tumor cells expressing immunosuppressive ligands, providing critical insights into immune evasion mechanisms in cancer. Such proximity analyses yield both qualitative and quantitative data on tissue microenvironments and their functional implications.

While limited by potential spectral overlap and the number of targets that can be simultaneously analyzed, dual-marker IHC represents a significant advancement over single-marker approaches. It allows more detailed investigations into the biological relationships between proteins and cells, paving the way for larger-scale multiplex methods.

1.4.2 Multiplex Immunofluorescence (IF): A High-Dimensional Approach for Tissue Analysis

Multiplex immunofluorescence (Multiplex IF) is an advanced histological method enabling the simultaneous detection of multiple proteins on a single tissue section. By employing antibodies conjugated to spectrally distinct fluorophores, this technique allows precise and quantitative visualization of multiple molecular targets while preserving spatial and topological tissue information [30].

This method is particularly suited for analyzing complex microenvironments, such as tumors, where multiple cell populations interact. For example, multiplex IF can simultaneously map immune markers (CD3, CD8, FOXP3) to characterize T-cell subtypes, alongside tumor-associated proteins like PD-L1 or Ki-67 [31]. These analyses provide a spatially resolved view of cellular heterogeneity and molecular interactions, offering critical insights into mechanisms such as immune surveillance and tumor escape.

A key advantage of multiplex IF is its ability to deliver high-resolution co-localization data. By analyzing multiple biomarkers concurrently, this technique quantifies cellular proximities and protein coexpression patterns, essential for mechanistic studies. For instance, in immuno-oncology, measuring the distance between CD8+ T cells and PD-L1-expressing tumor cells reveals essential information about immune responses and therapeutic targets [31].

However, this technique requires specialized equipment, such as multispectral fluorescence microscopes, and advanced computational tools for data analysis. By integrating multiplex IF with computational approaches, including machine learning, researchers can maximize the potential of multidimensional data generated by this technology. Multiplex IF thus emerges as a powerful tool for studies requiring simultaneous and contextualized characterization of multiple biomarkers, offering unprecedented resolution in normal and pathological tissue analysis.

1.4.3 Phenocycler Technology

One of the latest advancements in antibody-based spatial proteomics is the PhenoCycler (ex-CODEX) technology, which enables highly multiplexed protein detection and spatial mapping within tissue sections. PhenoCycler relies on iterative cycles of antibody-based labeling, signal amplification, and imaging to achieve single-cell resolution. The process begins with tissue preparation, where formalin-fixed paraffin-embedded (FFPE) or fresh frozen (FF) sections are stained using a panel of antibodies conjugated to DNA barcodes. These barcodes are subsequently hybridized with fluorescently labeled oligonucleotide reporters, enabling detection of specific proteins of interest. After imaging, the fluorophores are cleaved, allowing the same tissue section to be restained and imaged across multiple cycles. This iterative approach allows the detection of dozens to hundreds of proteins within the same tissue section, preserving spatial context while maintaining high sensitivity and specificity. Using PhenoCycler, researchers can generate spatially resolved proteomic maps, revealing intricate cellular interactions and molecular architectures within complex tissues. This platform is particularly powerful for exploring immune microenvironments, tumor heterogeneity, and tissue organization, driving new insights in fields such as oncology, immunology, and regenerative medicine [32].

1.5 Combination of Spatial Transcriptomics and Spatial Proteomics

Immunophenotyping has traditionally used flow cytometry or immunohistochemistry to study and characterize specific immune cell types based on their protein expression profiles. The ability to perform spatial transcriptomics and proteomics on the same tissue section extends this classical approach by preserving the spatial context of the cells within the tissue, enabling the simultaneous mapping of protein and gene expression, bringing a deep and precise characterization of immune cell subsets, states, and interactions within the tissue microenvironment. This spatially resolved data allows researchers to gain a comprehensive understanding of immune dynamics in health and disease, advancing fields like immune-oncology, inflammatory diseases, and personalized medicine.

In this protocol, we review how to perform spatial transcriptomics and spatial proteomics using Xenium In Situ with cell segmentation staining (10x Genomics) to measure gene expression in tissue sections that derive from formalin fixed and paraffin embedded (FFPE) tissue samples, followed by the PhenoCycler-Fusion (Akoya Biosciences), an ultrahigh-plex technique that allows the detection of multiple biomarkers at single-cell resolution.

2 Materials

2.1 Reagents

2.1.1 Spatial Transcriptomics (Xenium)

i. 1× PBS: add 5 ml 10× RNase-free PBS (pH 7.4) to 45 ml nuclease-free water in a 50 ml conical tube.

ii. PBS-Tween (PBS-T): add 175 μl 10% Tween-20 to 34.825 ml 1× PBS in a 50 ml conical tube.

iii. Decrosslinking buffer: mix 1017.5 μl Xenium FFPE Tissue Enhancer (10x Genomics), 68.8 μl Urea (8 M), and 13.8 μl Diluted Perm Enzyme B (10x Genomics).

iv. Probe Hybridization Mix: mix 630 μl Xenium Probe Hybridization Buffer, 288 μl TE buffer, 66 μl Xenium Pre-Designed Gene Expression Probes (10x Genomics), and 66 μ l Xenium Add-on Custom Probes (10x Genomics).

v. Ligation mix: mix 962.5 μl Xenium Ligation Buffer (10x Genomics), 27.5 μl Xenium Ligation Enzyme A (10x Genomics), and 110 μl Xenium Ligation Enzyme B (10x Genomics).

vi. Amplification Master Mix: mix 990 μl Xenium Amplification Mix (10x Genomics) and 110 μl Xenium Amplification Enzyme (10x Genomics).

vii. Autofluorescence solution: mix 1089 μl 100% ethanol and 11 μl Xenium Autofluorescence Mix (10x Genomics).

viii. Deionized Water/Xenium Instrument Wash Buffer: fill Reagent Bottle #1 (green bottle) with 500 ml of Nuclease-free Water/Nuclease-free Milli-Q Water and close the bottle using the standard cap.

ix. Xenium Sample Wash Buffer A: mix two packs of PBS-Tween (PPB005-20PAK) with 1 l Nuclease-free water in a 1 l glass bottle, close bottle and slowly invert 10× trying not to introduce bubbles. Pour the buffer into Reagent Bottle #2 (yellow bottle) and close the bottle using the standard cap.

x. Xenium Sample Wash Buffer B: fill Reagent Bottle #3 (red bottle) with 500 ml of Nuclease-free Water/Nuclease-free Milli-Q Water and close the bottle using the standard cap.

xi. Xenium Probe Removal Buffer: add 139.5 ml Nuclease-free water and 150 ml DMSO into a 500 ml glass bottle, close the bottle, and invert 10× to mix. Then add 7.5 ml KCl and 3 ml Tween-20, close the bottle, and invert 10× to mix avoiding the creation of bubbles.

2.1.2 Spatial Proteomics (Phenocycler-Fusion)

i. 1× PBS: add 100 ml of 10× PBS (Thermo Fisher Scientific) to 900 ml of Milli-Q water in a 1 l bottle and mix by shaking.

ii. 1× AR9 Buffer: add 10 ml of 10× AR9 Buffer (Akoya Biosciences) to 90 ml of Milli-Q water in a 100 ml bottle and mix by shaking.

iii. 1× Phenocycler-Fusion Buffer: add 100 ml of 10× Buffer for PhenoCycler (Akoya Biosciences) and 100 ml of Buffer Additive for PhenoCycler-Fusion (Akoya Biosciences) to 800 ml of Milli-Q water in a 1 l bottle and mix by pippeting up and down with pipet aid.

iv. Low DMSO: based on the number of marker cycles add the appropriate Volume of DMSO (Sigma-Aldrich) to the appropriate volume of 1× Phenocycler-Fusion Buffer. Stir with a magnetic stir bar and magnetic stirrer for 5 min.

v. High DMSO: based on the number of marker cycles, add the appropriate Volume of DMSO (Sigma-Aldrich) to the appropriate volume of 1× Phenocycler-Fusion Buffer. Stir with a magnetic stir bar and magnetic stirrer for 5 min (see Table 1).

vi. Storage Buffer: ready to use (Akoya Biosciences).

vii. Staining Buffer: ready to use (Akoya Biosciences).

viii. Hydratation Buffer: ready to use (Akoya Biosciences).

ix. Blocking Buffer: depending on the number of samples used, add the same volume of N/G/J/S Blocker to staining Buffer (Akoya Biosciences) in a tube and vortex gently. This solution should be prepared no earlier than 1 h before the staining (see Table 2).

x. Poststaining fixing solution: add 10 ml of 16% PFA (Electron Microscopy Sciences) to 90 ml of storage buffer (Akoya Biosciences) in 100 ml bottle and mix by shaking gently.

xi. Final Fixative Solution: add 20 μl of Phneocycler fixative reagent (Akoya Biosciences) in 1 ml of 1× PBS in 1.5 ml Eppendorf tube and vortex.

xii. Reporter stock Solution: depending on the number of cycles needed add *X* volume of Assay Reagent (Akoya Biosciences) and *X* volume of Nuclear Stain (Akoya Biosciences) to the

Table 1
Taken from PhenoCycler-Fusion User Guide_2.2.0

	Low DMSO			*High DMSO*			1×× PCF buffer cycles with additive (ml)
# of Marker cycles	Volume of pure DMSO (ml)	Volume of 1× PCF buffer with additive (ml)	Total (ml)	Volume of pure DMSO (ml)	Volume of 1× PCF buffer with additive (ml)	Total (ml)	
5	40	160	200	90	10	100	150
10	50	200	250	135	15	150	150
15	60	240	300	135	15	150	200
20	80	320	400	135	15	150	200
25	90	360	450	180	20	200	250
30	100	400	500	180	20	200	250
35	120	480	600	180	20	200	300
40	130	520	650	180	20	200	350
45	150	600	750	225	25	250	350
50	160	640	800	225	25	250	400
55	170	680	850	225	25	250	400
60	190	760	950	270	30	300	450

Table 2
Taken from PhenoCycler-Fusion User Guide_2.2.0

PhenoCycler reagent (μl)	2 Samples	4 Samples	6 Samples	8 Samples	10 Samples
Staining buffer	362	724	1086	1448	1810
N Blocker	9.5	19	28.5	38	47.5
G Blocker	9.5	19	28.5	38	47.5
J Blocker	9.5	19	28.5	38	47.5
S Blocker	9.5	19	28.5	38	47.5
Total volume	400	800	1200	1600	2000

corresponding volume to 1× PhenoCycler-Fusion Buffer in an opaque 1.5 ml tube or 15 ml tube covered with foil (see Table 3).

xiii. Methanol: ready to use (Sigma-Aldrich).

Table 3
Taken from PhenoCycler-Fusion User Guide_2.2.0

	Cycles/wells			
PhenoCycler 2.0 reporter stock solution (μl)	**5**	**10**	**15**	**20**
1X PhenoCycler-Fusion Buffer	1325	2650	3975	5300
Assay reagent	125	250	375	500
Nuclear stain	50	100	150	200
Total volume	1500	3000	4500	6000

2.2 Material

2.2.1 Spatial Transcriptomics (Xenium)

i. Xenium Analyzer (10x Genomics).
ii. 10x Genomics Reagent Kits & Consumables: Xenium De-coding Consumables, Xenium Decoding Reagents, Xenium Slides & Sample Prep Reagents, Xenium Cell Segmentation Add-on Kit, Gene Panels.
iii. Thermal cycler: Analytik Jena, Biometra TAdvanced 96 SG with 96-well block 846-x-070-241 and gradient function.
iv. Microtome and microtome blade, no specific vendor.
v. Water bath, no specific vendor.
vi. Section dryer oven, no specific vendor.
vii. Thermomixer with 2 ml adapter, no specific vendor.
viii. Vortex, no specific vendor.
ix. Fume hood, no specific vendor.
x. Mini centrifuge, no specific vendor.
xi. Refrigerated microcentrifuge, no specific vendor.
xii. Centrifuge with plate rotor, no specific vendor.

2.2.2 Spatial Proteomics (Phenocycler-Fusion)

i. Phenocycler Instrument (PhenoCycler-Fusion 2.1.0, Akoya Biosciences, S9043480).
ii. PhenoImager Fusion Microscope (PhenoImager Fusion 2.1.0, Akoya Biosciences, S9045103).
iii. Next-Gen Flow Cell Assembly Device (SF-000085, Akoya Biosciences, NA).
iv. Milli-Q water, no specific vendor.
v. PhenoImager Fusion Software (Akoya Biosciences).
vi. Phenocycler Experiment Designer Software (Akoya Biosciences).

vii. 96 Well Plates (Akoya Biosciences) and 96 Well Plates Seal (Akoya Biosciences).

viii. Amber 1.5 ml tubes, no specific vendor.

ix. Pressure cooker, no specific vendor.

2.3 Timing Table

This protocol planner provides all the steps and times to allow efficient experiment planning (see Table 4).

Table 4
Protocols and timing

Technology	Day	Step	Sub-steps	Timing	Dedicated protocol
Xenium	1	1. Tissue preparation	1.1 Face the block and rehydrate. 1.2 Sectioning 1.3 Section placement 1.4 Drying	~3 h Overnight	Tissue preparation guide – Section FFPE tissue onto Xenium Slides – Demonstrated protocol CG000578
	2	2. Fixation and permeabilization	2.1 Deparaffinization 2.2 Cassette assembly 2.3 Decrosslinking	3 h 10 min 1 h	Fixation and Permeabilization – Deparaffinize and decrosslink FFPE tissue sections – Demonstrated protocol CG000580
		3. Probe hybridization	3.1 Probe hybridization	10 min Overnight (16–24 h)	Xenium In Situ Gene Expression with Cell Segmentation Staining – For FF and FFPE samples on Xenium Slides – User Guide CG000749
	3	4. Posthybridization wash	4.1 Posthybridization wash	35 min	
		5. Ligation	5.1 Ligation	2 h	
		6. Amplification	6.1 Amplification 6.2 Postamplification wash	2 h 15 min	
		7. Cell segmentation staining	7.1 Block and stain	20 min 1 h + overnight (16–24 h)	
	4	8. Stain enhancement	8.1 Stain enhancement	1 h	
		9. Autofluorescence quenching	9.1 Autofluorescence quenching 9.2 Nuclei staining	30 min 10 min	
		10. Instrument run	10.1 Reagent plate preparation 10.2 Buffer preparation 10.3 Initialize instrument 10.4 Input experimental details 10.5 Load instrument 10.6 Sample scan 10.7 Region selection and initiate run	1 h ~10 min 5–10 min ~5 min 1 h ~10 min	Xenium Analyzer – Load Xenium Slide onto Xenium Analyzer – User Guide CG000584
	5	11. Run and post-run	11.1 Run 11.2 Postrun cleanup and unload 11.3 Data output 11.4. Long-term storage	~2–4 days 10 min	

(Continued)

Table 4 (Continued)

Technology	Day	Step	Sub-steps	Timing	Dedicated protocol
Phenocycler-Fusion	5	1. Instrument preparation	1.1. Maintenances 1.2. Design the Reporter Plate and calculate Buffer Volumes	1 h 30 min	PhenoCycler-Fusion User Guide_2.2.0 PhenoImager Fusion User Guide V2 _ 2 Flow Cell _ 2.2.0
		2. Buffer preparation	2.1. Prepare 1× PBS 2.2. Prepare 1XPCF 2.3. Prepare Low- and High-DMSO 2.4. Prepare AR9 Buffer		
	6	3. Morning setup: reagents and materials	3.1. Remove buffers from storage 3.2. Prepare work area and containers	10 min 15 min	PhenoCycler-Fusion User Guide_2.2.0
		4. Tissue preparation and Antigen Retrieval	4.1. Rinse steps if processing slide directly after a Xenium run or after short-term storage 4.2. Rinse steps if processing a slide after long-term storage 4.3. Antigen Retrieval 4.4. Cool and Rinse	20 min 20 min 30 min 1 h	
		5. Preparing the antibody cocktail solution	5.1. Prepare blocking buffer 5.2. Preparation of the First Antibody Cocktail	10 min 40 min	
		6. First Antibody Incubation	6.1. Dry the slide 6.2. Incubate with the First Antibody Cocktail	5 min 3 h	
		7. Preparation of the Second Antibody Cocktail	7.1. Prepare another batch of Blocking Buffer 7.2. Prepare the Second Antibody Cocktail	10 min 40 min	
		8. Wash between incubations		15 min	
		9. Second Antibody Incubation	9.1. Dry the slide 9.2. Add the Second Antibody Cocktail	5 min Overnight	
	7	10. Reporter Plate Preparation	10.1. Take out the reporters 10.2. Prepare Brown Eppendorf Tubes 10.3. Prepare the Reporter Stock Solution 10.4. Prepare Individual Reporter Tubes 10.5. Transfer 250 µl to the 96-well Plate 10.6. Protect the Plate	10 min 15 min 10 min 1 h 30 min 5 min	
		11. Tissue poststaining	11.1. Coplin Jar configuration 11.2. Tissue wash 11.3. Tissue fixation 11.4. Tissue wash 11.5. Ice-methanol incubation 11.6. Tissue wash 11.7. Tissue Fixation 11.8. Tissue wash 11.9. Store the tissue	10 min 5 min 15 min 2 min 5 min 2 min 25 min 2 min	PhenoImager Fusion User Guide V2 _ 2 Flow Cell _ 2.2.0 PhenoImager Fusion User Guide V2 _ 2 Flow Cell _ 2.2.0
		12. Flow cell preparation and assembly	12.1. Tissue wash 12.2. Assemble slide and flow cell 12.3. Buffer Addition	10 min 2 min 15 min	
		13. PhenoCycler and PhenoImager Setup	13.1. Power Up the Systems 13.2. Load the Slide 13.3. PhenoImager Setup	5 min 5 min 15 min	
		14. End of the Run and Flow cell removal	14.1. End of the Run 14.2. Flow cell removal	15 min 3 h or overnight	PhenoImager Fusion User Guide V2 _ 2 Flow Cell _ 2.2.0
H&E	8	1. PostInstrument Staining	1.1. Buffer preparation (quencher) 1.2. Quencher removal 1.3. Buffer preparation (H&E) 1.4. H&E staining 1.5. Coverslipping and imaging	15 min 20 min 15 min 1 h 35 min	H&E Staining – H&E stain post Xenium run – Demonstrated protocol CG000613

3 Methods

In this protocol, we describe how to combine Xenium protocol (10x Genomics) and Phenocycler-Fusion protocol (Akoya Biosciences) on the same tissue section to spatially map hundreds of RNAs and tens of proteins. On the one hand, we detail how to perform Xenium In Situ gene Expression with Cell Segmentation Staining to measure gene expression in FFPE tissue sections and spatially map hundreds of RNAs at a subcellular resolution by using targeted probes. The protocol describes all the different steps, from tissue preparation and predesigned and add-on custom probe hybridization to instrument run and final H&E staining. The complete protocol takes five lab workdays (plus instrument run time, which will depend on sample and panel size). On the other hand, we walk you through the process of using Phenocycler-Fusion to study and adjust the protein characteristics of cells within formalin-fixed, paraffin-embedded (FFPE) tissue samples. The Phenocycler technique allows for the visualization of multiple proteins simultaneously within tissue sections, retaining spatial information and revealing complex cellular interactions. This technique is based on the use of barcoded antibodies and reporter molecules, which systematically identify and quantify protein expressions within the tissue sections. The protocol encompasses all essential steps, including tissue preparation, phenocycler reagent application, controlled cycling conditions, instrument setup, and final tissue analysis. The entire workflow is designed to be completed within four laboratory workdays (excluding instrument run time, which varies based on sample complexity and experimental parameters).

3.1 Xenium Protocol

DAY 1: During the first day, tissue will be rehydrated, sectioned on the Xenium Slide and dried. Before proceeding to the experiment, consider verifying RNA and tissue quality on tissue sections placed on blank slides (*see* **Note 1**).

3.1.1 Tissue Preparation

Face the Block and Rehydrate

i. Prepare a water bath with Milli-Q water at 42°C with no bubble or particulates.

ii. Prepare an ice bath from filtered water.

iii. Remove FFPE tissue blocks from storage (recommended at 4°C) and set the microtome to 15 µm.

iv. Cut the tissue block at RT until the region of interest is exposed. This step is just necessary for new unused blocks.

v. Place the blocks on the ice bath, making sure that the exposed tissue is fully submerged.

vi. Incubate the block on the ice bath for 10–30 min. Monitor the tissue every 5–10 min during incubation until tissue surface is smooth and shiny and free of bumps.

vii. During this time, equilibrate Xenium Slides at RT for at least 30 min.

Sectioning

i. Set up the microtome at 5 μm and ensure that the clearance angle is set to 10°.
ii. After hydration, proceed to tissue sectioning (*see* **Note 2**). Remember to discard the first few sections.
iii. To collect sections, use a brush to pick the section and place it on the water at 42°C. Allow sections to float until they are flat and have no wrinkles or folds (*see* **Note 3**).

Section Placement

i. Insert the Xenium Slide into the water, and using the paintbrush or the probe, help the section to go on the Sample Area (10.45 mm × 22.45 mm) (*see* **Note 4**).
ii. Put the slide up and out of the water. Make sure there are no air bubbles.

Drying

i. Dry the tissue sections in a drying rack with the help of a fan at RT until no water remains. If no fan is used, the drying time may be extended to 30 min.
ii. Place the drying rack with the sections in an over or thermal cycler for 3 h at 42°C.
iii. Place the slides in a desiccator and keep the slides overnight at RT for proper drying.

END OF DAY 1
At this step there are two options:
Option 1—Proceed to deparaffinization and decrosslinking the day after.
Option 2—Slides with dry tissue can be stored in a desiccator at RT up to 4 weeks.

DAY 2: Before probe hybridization, tissue needs to be deparaffinized and descrosslinked. To do so, tissue that has been drying overnight in a desiccator is deparaffinized to remove the paraffin wax that has been used to preserve and stabilize the tissue, then rehydrated and finally descrosslinked to make the RNA from the tissue accessible. Then, the 10x predesigned panel and the add-on custom panel are added to the FFPE tissue sections and hybridized overnight. Each DNA probe contains one region that encodes a gene-specific barcode and two other regions that hybridize to the RNA of interest. During overnight hybridization, the two ends of the probes independently bind their complementary target RNA.

3.1.2 Fixation and Permeabilization

Deparaffinization

i. Take the slides from the desiccator after overnight drying.

ii. Incubate the slides in a Section Dryer Oven uncovered or on a Thermocycler Adaptor on a thermal cycler (with the tissue facing up), leaving the thermal cycler lid open for 2 h at 60°C.

During these 2 h you can prepare the next steps:

- Reagents
  - Preheat thermomixer or water bath to 37°C.
  - Around 30 min before the end of the 2 h, thaw Xenium FFPE Tissue Enhancer in the preheated thermomixer (or waterbath) for 30 min at 37°C, 300 rpm with shaking. Cool to RT for 5 min, vortex for 30 s, and centrifuge briefly.
  - Equilibrate Perm Enzyme B to RT and maintain at RT. Pipette mix (do not vortex) and centrifuge briefly before use.
- Buffers for deparaffinization:
  - 1× PBS. Invert to mix and maintain at RT. Keep it for later steps.

Nuclease-free water	45 ml
RNase-free PBS, pH 7.4 (10×)	5 ml
Total	50 ml

  - Jars for deparaffinization. Prepare in a fume hood.

Xylene (x2)
100% Ethanol (x2)
96% Ethanol (x2) (for 100 ml = 96 ml 100% ethanol + 4 ml nuclease-free water)
70% Ethanol (x1) (for 50 ml = 35 ml 100% ethanol + 15 ml nuclease-free water)
Nuclease-free water (x1)

- Buffers for decrosslinking:
  - PBS-Tween (PBS-T). Invert to mix and maintain at RT. Keep it for later steps.

1× PBS	34.825 ml
Tween-20 (10%)	175 µl
Total	35 ml

  - Diluted Perm Enzyme B. Mix with a pipet set to 600 µl and maintain at RT.

1× PBS	998 µl
Perm Enzyme B	2 µl
Total	1000 µl

- Decrosslinking buffer (for two slides). Pipet mix and maintain at RT in the dark.

Xenium FFPE Tissue Enhancer	1017.5 μl
Urea (8 M)	68.8 μl
Diluted Perm Enzyme B	13.8 μl
Total	1100 μl

i. After 2 h incubation, leave the slides at RT for 7 min to cool down.

ii. Immerse the slides in the Xylene jar 1 for 10 min and then in the Xylene jar 2 for another 10 min.

iii. Immerse the slides in the 100% Ethanol jar 1 for 3 min and then in the 100% Ethanol jar 2 for another 3 min.

iv. Immerse the slides in the 96% Ethanol jar 1 for 3 min and then in the 96% Ethanol jar 2 for another 3 min.

v. Immerse the slides in the 70% Ethanol jar for 3 min.

vi. Immerse the slides in the Nuclease-free water jar for 20 s.

Cassette Assembly

i. Use a lint-free laboratory wipe to remove any remaining water and place the slide in the cassettes (*see* **Note 5**).

ii. Add 500 μl of 1× PBS-T.

Decrosslinking

i. Place a Xenium Thermocycler Adaptor in the thermal cycler, prepare the incubation steps, and start the Decrosslinking Program. Lid temperature 80°C and reaction volume 100 μl.

Step	Temperature	Time
Hold	22°C	Hold
Decrosslinking	80°C	30 min
Reequilibrate	22°C	10 min
Hold	22°C	Hold

ii. Remove the 10× PBS-T.

iii. Add 500 μl of Decrosslinking Buffer. Tissue sections must be uniformly covered with no bubbles.

iv. Put a new Xenium Cassette Lid, place the Xenium Cassette on top of the Thermocycler Adaptor, close the thermal cycler lid, and skip Hold step.

v. During these 45 min you can prepare the next steps:

- Reagents
  - Preheat a heatblock or water bath to 95°C.
  - Thaw Xenium Probe Hybridization Buffer at RT for 15 min. Invert until clear and verify no precipitation. Maintain at RT.

- Equilibrate Xenium Pre-Designed Gene Expression Probes to RT and maintain at RT.
- Prepare Xenium Add-on Custom Probes. If already resuspended, equilibrate to RT. If not:
  - Briefly centrifuge custom probe panel tube.
  - Resuspend lyophilized custom probes in TE buffer: 140 µl if 4 reactions/kit and 700 µl if 16 reactions/kit.
  - Vortex the Add-on Custom Probes at RT for 5 min, centrifuge briefly an maintain at RT.

vi. Once the Decrosslinking Program has finished, remove the Xenium Cassette from the thermal cycler, throw away the lid and remove all Decrosslinking Buffer from the well.

vii. Wash the tissue three times by adding 500 µl of 1× PBS-T and incubating at RT for 1 min.

viii. Add 500 µl of 1× PBS-T one last time.

3.1.3 Probe Hybridization

Probe Hybridization

i. Prepare the incubation steps of the Hybridization Program and start it. Lid temperature 50°C and reaction volume 100 µl.

Step	Temperature	Time
Preequilibrate	50°C	Hold
Probe hybridization	50°C	Hold
Hold	50°C	Hold

ii. Centrifuge briefly the Xenium Pre-Designed Gene Expression Probes and the Xenium Add-on Custom Probes and save apart an aliquot of 68 µl (34 µl/slide) of each probe set.

iii. Preheat the probes at 95°C for 2 min in the preheated heat block or water bath and then incubate them 1 min on ice.

iv. Equilibrate the probes at RT, centrifuge briefly, and maintain at RT.

v. Prepare Probe Hybridization Mix (for two slides):

Xenium Probe Hybridization Buffer	630 µl
TE Buffer	288 µl
Xenium Pre-Designed Gene Expression Probes	66 µl
Xenium Add-on Custom Probes	66 µl
Total	1050 µl

vi. Remove all PBS-T from the well and add 500 µl of Probe Hybridization Mix. Tissue sections must be uniformly covered with no bubbles.

vii. Put a new Xenium Cassette Lid, place the Xenium Cassette on top of the Thermocycler Adaptor, close the thermal cycler lid and skip Preequilibrate step for overnight probe hybridization (16–24 h) (*see* **Note 6**).

END OF DAY 2

At this step there are no stopping options; protocol must be continued the day after.

DAY 3: After overnight hybridization, tissue needs to be washed to eliminate the excess of unbound probes. Then, the two ends of the probes that have bind the RNA of interest are ligated using a ligase to generate a circular DNA probe. This ligation is needed to ensure a unique level of probe specificity to the target region. Afterwards, the circularized probe is enzymatically amplified, which will generate hundreds of copies of the gene-specific barcode for each RNA of interest. After washing the excess of the amplified product and following a series of ethanol washes, blocking reagent is added to the tissue to reduce nonspecific binding during antibody staining. Cell segmentation antibodies are added for labelling of nuclei, membranes, and interiors during overnight incubation. This will serve as input for automated morphology-based cell segmentation analysis.

3.1.4 Post Hybridization Wash

Post Hybridization Wash

viii. Thaw at RT the Xenium Post Hybridization Wash Buffer for 30 min. Vortex, centrifuge briefly, and maintain at RT.

ix. During these 30 min, thaw reagents for the next day:

- Remove Xenium Decoding Reagent Module B from −20°C and thaw in a sealed mylar bag at 4°C overnight.
- Remove Xenium Cell Segmentation Detection Module from −20°C and thaw in sealed mylar bag at 4°C overnight.

x. Take the Xenium Cassette from the thermocycler, remove and discard the lid, and remove all Probe Hybridization Mix (*see* **Note 7**).

xi. Quickly add 500 μl of PBS-T to not let the sample dry and incubate for 1 min at RT. Tissue sections must be uniformly covered with no bubbles.

xii. Prepare the incubation steps of the Post Hybridization Wash Program and start it. Lid temperature 37°C and reaction volume 100 μl.

Step	Temperature	Time
Preequilibrate	37°C	Hold
Probe Hybridization	37°C	30 min
Hold	37°C	Hold

xiii. Remove all PBS-T and add 500 µl of PBS-T one last time. Incubate for 1 min at RT.

xiv. Remove all PBS-T and add 500 µl of Xenium Post Hybridization Wash Buffer.

xv. Put a new lid, place the cassette on the thermal cycler, close the lid and skip the Pre-Equilibrate step to start the Post Hybridization Wash. No stopping point, proceed to Ligation.

3.1.5 Ligation

Ligation

i. During the 30 min of Post Hybridization Wash, prepare the next steps:

- Equilibrate Xenium Ligation Buffer to RT for 15 min until completely thawed, vortex, centrifuge briefly, and maintain at RT.
- Pipette mix and centrifuge briefly the Xenium Ligation Enzyme A. Maintain on ice.
- Pipette mix and centrifuge briefly the Xenium Ligation Enzyme B. Maintain on ice.

ii. Prepare the Ligation Mix shortly before use (for two slides), pipette mix 10×, centrifuge briefly, and maintain on ice.

Xenium ligation buffer	962.5 µl
Xenium ligation enzyme A	27.5 µl
Xenium ligation enzyme B	110 µl
Total	1100 µl

iii. Once the Post Hybridization Wash Program has finished, remove the Xenium Cassette from the thermal cycler, remove and throw away the lid, and discard all Post Hybridization Wash Buffer.

iv. Quickly add 500 µl of PBS-T to avoid drying and incubate for 1 min at RT.

v. Prepare the incubation steps of the Ligation Program and start the program. Lid temperature 37°C and reaction volume 100 µl.

Step	Temperature	Time
Preequilibrate	37°C	Hold
Ligation	37°C	2 h
Hold	37°C	Hold

vi. Remove all PBS-T and add 500 µl of PBS-T again. Incubate for 1 min at RT.

vii. Remove all PBS-T and add 500 µl of PBS-T one last time. Incubate for 1 min at RT.

viii. Remove all PBS-T and add 500 µl of Ligation Mix.

ix. Put a new lid, place the cassette on the thermal cycler, close the lid and skip the pre-equilibrate step to start the Ligation. No stopping point, proceed to Amplification.

3.1.6 Amplification

Amplification

i. During Ligation Program prepare the next steps:

- Immediately after launching the Ligation Program, thaw the Xenium Amplification Mix on ice. Vortex and centrifuge briefly. Ensure no precipitate remains.
- Transfer Xenium Amplification Enzyme to ice just before use. Pipette mix, centrifuge briefly and maintain on ice.

ii. Prepare the Amplification Master Mix shortly before use (for two slides), pipette mix 10×, centrifuge briefly, and maintain on ice.

Xenium amplification mix	990 µl
Xenium amplification enzyme	110 µl
Total	1100 µl

iii. Once the Ligation Program has finished, remove the Xenium Cassette from the thermal cycler, remove and throw away the lid and discard all Ligation Mix.

iv. Quickly add 500 µl of PBS-T to avoid drying and incubate for 1 min at RT.

v. Prepare the incubation steps of the Amplification Program and start the program. Lid temperature 30°C and reaction volume 100 µl.

Step	Temperature	Time
Preequilibrate	30°C	Hold
Amplification	30°C	2 h
Hold	30°C	Hold

vi. Remove all PBS-T and add 500 µl of PBS-T again. Incubate for 1 min at RT.

vii. Remove all PBS-T and add 500 µl of PBS-T one last time. Incubate for 1 min at RT.

viii. Remove all PBS-T and add 500 µl of Amplification Master Mix.

ix. Put a new lid, place the cassette on the thermal cycler, close the lid and skip the Pre-Equilibrate step to start the Amplification. No stopping point, proceed to Wash.

3.1.7 Postamplification Wash

i. During Amplification Program prepare the next steps:
   - Reagents:
     - Thaw the Xenium Block and Stain Buffer at RT for 30 min. Vortex and centrifuge briefly. Maintain on ice.
     - Maintain Xenium Multi Tissue Stain Mix on ice.
     - Remove the Xenium Cassette Insert and the Slide Seals from −20°C and equilibrate to RT.
   - Buffers for Cell Segmentation Staining:
     - Ethanol preparation: once prepared, vortex and centrifuge briefly and maintain at RT.

100% Ethanol (for 15 ml = 15 ml 100% ethanol)
70% Ethanol (for 15 ml = 10.5 ml 100% ethanol + 4.5 ml nuclease-free water)

     - Prepare 1× Xenium Block and Stain Buffer. Vortex and centrifuge briefly and maintain at RT.

Nuclease-free water	990 µl
Xenium Block and Stain Buffer	330 µl
Total	1320 µl

ii. Once the Amplification Program has finished, remove the Xenium Cassette from the thermal cycler, remove and throw away the lid, and discard all Amplification Mix.

iii. Add 500 µl of TE Buffer and incubate for 1 min at RT.

iv. Remove the TE buffer and add 500 µl of TE buffer again. Incubate for 1 min at RT.

v. Remove all TE buffer and add 500 µl TE buffer one last time.

3.1.8 Cell Segmentation Staining

Block and Stain

vi. Remove all TE buffer and add 1 ml 70% ethanol and incubate for 2 min at RT.

vii. Remove the ethanol, add 1 ml 100% ethanol and incubate for 2 min at RT.

viii. Remove the ethanol, add 1 ml 100% ethanol and incubate for 2 min at RT.

ix. Remove the ethanol, add 1 ml 70% ethanol and incubate for 2 min at RT.

x. Remove the ethanol, quickly add 1 ml PBS-T and incubate for 1 min at RT. Important: do not let tissue dry.

xi. Remove all PBS-T and add 500 µl 1X Xenium Block and Stain Buffer for blocking. Important: do not discard remaining volume.

xii. Put a new lid and incubate for 1 h at RT.

xiii. During incubation, prepare Xenium Multi Tissue Stain Mix:

- Remove the sticker of the tube for better visibility, centrifuge for 5 s and make sure that the stain mix is at the bottom of the tube.
- Add 220 µl of previously prepared 1× Xenium Block and Stain Buffer to the Xenium Multi-Tissue Stain Mix. Pipette mix 15× avoiding bubbles and centrifuge briefly.
- Incubate resuspended Xenium Multi-Tissue Stain Mix for 30 min at RT.
- Centrifuge Xenium Multi-Tissue Stain Mix for 10 min at 14,000 rcf at 4°C and maintain on ice.

xiv. After 1 h incubation, remove and throw away the lid and discard the 1× Xenium Block and Stain Buffer. Important: do not let tissue dry.

xv. Place the Xenium Casette Insert onto the Xenium Casette using the forceps (*see* **Note 8**).

xvi. Add 100 µl Xenium Multi-Tissue Stain Mix under the Xenium Cassette Insert through the cut-out avoiding the introduction of bubbles (*see* **Note 9**).

xvii. Use the Slide Seal to seal the cassette. Never press down in the middle of the cassette; do it only on the edges or tissue will be damaged.

xviii. Incubate Xenium cassette overnight (16–24 h) at 4°C.

END OF DAY 3

At this step there are no stopping options; protocol must be continued the day after.

DAY 4: After overnight incubation of the tissue with the cell segmentation reagents, the excess of antibodies is washed away and the stain is enhanced. This is followed by autofluorescence quenching to reduce unwanted tissue autofluorescence and enhance signal-to-noise ratio. Then, nuclei are stained with DAPI and, to finish, the slides containing the tissue sections are loaded into the instrument for imaging and high throughput, automated in situ analysis. During the instrument run, which can take between 2 and 4 days, depending on the panel selected and the size of the tissue regions, fluorescently labeled oligos bind to the amplified DNA probes in continuous rounds of hybridization. These cyclical rounds of fluorescent probe hybridization, imaging, and removal of probes generate signatures that are specific for each barcode. The imaging data that are captured for multiple fluorescent channels in every cycle is then processed and stitched to build a spatial map of transcripts across the tissue sections.

3.1.9 Stain Enhancement

i. First thaw Xenium Stain Enhancer for 10 min at RT, centrifuge it for 5 s and make sure that the powder is at the bottom. Add 1100 μl of 1× PBS to the Xenium Stain Enhancer and pipette mix 5×. Maintain on ice.

ii. Preheat a thermomixer with a 2 ml thermoblock at 37°C for later.

iii. Remove Slide Seal from Xenium Cassette, add 200 μl PBS-T into Xenium Cassette Insert cut-out to float the insert and use the forceps to remove the Cassette Insert.

iv. Remove all Xenium Multi Tissue Stain Mix, add 500 μl PBS-T, and incubate 1 min at RT.

v. Remove all PBS-T, add 500 μl PBS-T for the second time, and incubate 1 min at RT.

vi. Remove all PBS-T, add 500 μl PBS-T for the third time, and incubate 1 min at RT.

vii. Remove all PBS-T, add 500 μl Xenium Stain Enhancer and incubate for 20 min at RT.

viii. During incubation, prepare the next steps:

- Reagents:
  - Thaw the Xenium Autofluorescence Mix in the preheated thermomixer with 2 ml thermoblock for 15 min at 37°C and 300 rpm with shaking. Cool to RT for 5 min, vortex for 30 s and centrifuge briefly.
  - Thaw Reducing Agent B at RT. Vortex and centrifuge briefly.
  - Thaw Xenium Nuclei Staining Buffer at RT. Vortex and centrifuge briefly and maintain in the dark until use.
- Buffers for Autofluorescence quenching:
  - Prepare Diluted Reducing Agent B (for two slides). Vortex and maintain at RT.

1× PBS	1089 μl
Reducing agent B	11 μl
Total	1100 μl

- Prepare 70% ethanol. Vortex and maintain at RT.

Nuclease-free water	660 µl
100% ethanol	1540 µl
Total	2200 µl

- Prepare Autofluorescence Solution. Vortex and maintain at RT in the dark until use.

100% ethanol	1089 µl
Xenium Autofluorescence Mix	11 µl
Total	1100 µl

- At this step you can start preparing Modules (10.1) and Buffers (10.2) for the Xenium instrument.

ix. After incubation, remove all Xenium Stain Enhancer, add 500 µl PBS-T, and incubate 1 min at RT.

x. Remove all PBS-T, add 500 µl PBS-T again, and incubate 1 min at RT.

xi. Remove all PBS-T and add 500 µl PBS-T one last time.

3.1.10 Autofluorescence Quenching

Autofluorescence Quenching

i. Remove all PBS-T and add 500 µl Diluted Reducing Agent B, put a new lid and incubate for 10 min at RT.

- At this step you can continue preparing Modules (10.1) and Buffers (10.2) for Xenium instrument.

ii. After incubation, remove and keep the lid, and discard Diluted Reducing Agent B.

iii. Add 1 ml of 70% ethanol and incubate for 1 min.

iv. Remove the ethanol, add 1 ml of 100% ethanol, and incubate for 1 min.

v. Remove the ethanol, add 1 ml of 100% ethanol one more time, incubate for 1 min, and remove all ethanol.

vi. Pipette the Autofluorescence Solution mix thoroughly, add 500 µl to each sample, close with the lid, and incubate for 10 min at RT in the dark.

- At this step you can finish preparing Modules (10.1) and Buffers (10.2) for the Xenium instrument.

vii. Prepare the incubation steps of the Drying Program and start the program. Lid temperature 37°C and reaction volume 100 µl.

Step	Temperature	Time
Preequilibrate	37°C	Hold
Drying	37°C	5 min
Hold	37°C	Hold

- This time can be used to initialize the instrument (10.3).

viii. Remove and throw away the lid and discard the Autofluorescence Solution.

ix. Add 1 ml of 100% ethanol and incubate for 2 min.

x. Remove the ethanol, add 1 ml of 100% ethanol and incubate for 2 min.

xi. For the last time, remove the ethanol, add 1 ml of 100% ethanol, and incubate for 1 min.

xii. Remove the ethanol and place the Xenium Casette without a lid on the thermal cycler, without closing the lid, and skip the Pre-Equilibrate step to start the Drying.

xiii. After 5 min remove the Xenium Cassette from the thermal cycler and quickly add 1 ml 1× PBS to rehydrate the tissue and incubate for 1 min at RT in the dark.

xiv. Remove all 1× PBS, add 1 ml PBS-T, and incubate for 2 min at RT in the dark.

At this step there are two options:
Option 1—Proceed directly to Nuclei Staining.
Option 2—Short-term storage: at 4°C in the dark up to 24 h with a lid.

Nuclei Staining

i. Remove all PBS-T, add 500 μl Xenium Nuclei Staining Buffer, and incubate 1 min at RT in the dark.

ii. Remove all Nuclei Staining Buffer, add 1 ml PBS-T, and incubate 1 min at RT in the dark.

iii. Remove all PBS-T, add 1 ml PBS-T again, and incubate 1 min at RT in the dark.

iv. Remove all PBS-T, add 1 ml PBS-T again, and incubate 1 min at RT in the dark.

v. Remove all PBS-T and add 1 ml PBS-T one last time.

At this step there are two options:
Option 1—Proceed directly to Xenium Run.
Option 2—Short-term storage: at 4°C in the dark up to 1 week with a lid.

3.1.11 Instrument Run

Reagent Plate Preparation

i. Prepare Decoding Reagent Module A:
   - Remove the plate from 4°C, open the mylar packaging and remove plate without removing the foil seal, mix gently by inverting the plate 20×. Try not to introduce bubbles and do not vortex. Maintain on ice.
   - Prepare a plate for counterbalancing the Decoding Reagent Module A plate with elastic and lid on, centrifuge 300 rcf for 1 min at RT, and leave at 4°C.

ii. Prepare Decoding Reagent Module B and Cell Segmentation Detection Module (C):
   - Remove plates from 4°C and equilibrate at RT for 30 min.
   - Open the mylar packaging and remove plates without removing the foil seal, mix gently by inverting the plates 20×. Try not to introduce bubbles and do not vortex. Maintain at RT.
   - Prepare two plates for counterbalancing the Decoding Reagent Module B plate and the Cell Segmentation Detection Module plate with elastic and lid on, centrifuge 300 rcf for 1 min at RT and leave at RT.

Buffer Preparation

i. Prepare Deionized Water/Xenium Instrument Wash Buffer:
   - Fill Reagent Bottle #1 (green bottle) with 500 ml of Nuclease-free Water/Nuclease-free Milli-Q Water to make Xenium Instrument Wash Buffer and close the bottle using the standard cap.

ii. Prepare Xenium Sample Wash Buffer A:
   - Mix two packs of PBS-Tween (PPB005-20PAK) with 1 l Nuclease-free water in a 1 L glass bottle, close bottle and slowly invert 10× trying not to introduce bubbles. Pour the buffer into Reagent Bottle #2 (yellow bottle) and close the bottle using the standard cap.

iii. Prepare Xenium Sample Wash Buffer B:
   - Fill Reagent Bottle #3 (red bottle) with 500 ml of Nuclease-free Water/Nuclease-free Milli-Q Water to make Xenium Sample Wash Buffer B and close the bottle using the standard cap.

iv. Prepare Xenium Probe Removal Buffer:
   - Prepare 300 ml 1× Xenium Probe Removal Buffer in a 500 ml glass bottle. First add the Nuclease-free water and the DMSO, close the bottle, and invert 10× to mix. Then

add the KCl and Tween 20, close the bottle, and invert 10× to mix avoiding the creation of bubbles.

Nuclease-free water	139.5 ml
DMSO	150 ml
KCl	7.5 ml
Tween 20	3 ml
Total	300 ml

- Pour the buffer into Reagent Bottle #4 (blue bottle), close the bottle using the standard cap, and maintain it at RT for 30 min.

Initialize Instrument

i. Turn on the instrument using the power switch at the right side of the instrument, launch the Xenium Analyzer Application by clicking the blue icon on the touchscreen and click "Start New Run" button on the instrument home screen.

ii. A series of automated system checks will be performed by the instrument to verify that all systems are working correctly (~3–5 min).

Input Experimental Details

i. Specify Run Name, which will be used in the Analysis Summary, to refer to this particular run.

ii. Select whether the sample was prepared with the Xenium Cell Segmentation Staining Reagents for multimodal cell segmentation (Xenium Analyzer Software version 2.0+ required for cell segmentation): in this case, YES. This step is very important for correct downstream functions.

iii. Enter the details of the run in the instrument screen:

- Cassette name: name/reference of the sample is recommended.
- Xenium Slide ID: 7 characters written on the Xenium Slide.
- Panel file: as we have used a predesigned panel + add on panel, we need to upload the custom panel file (.json) from a USB into the computer.

Load Instrument

i. After pressing Continue, open the front panel of the machine and follow the instructions of the screen to load the following items (*see* **Note 10**).

- Reagent bottles: change the standard cap for the Xenium buffer cap and load each bottle into the bottle carrier in the correct order (Bottle #1 green, Bottle #2 yellow, Bottle #3 red, Bottle #4 blue) and push the bottle carrier caps down to the top of the bottles to properly seal them. The Waste

Bottle needs to be empty and placed on the left side of the instrument, then push the bottle carrier cap down to seal the waste bottle (*see* **Note 11**).

- Reagent plates: do not forget to remove the elastic and the lid from the reagent plates before positioning them into their respective positions (Module A green, Module B red, Cell Segmentation Module blue) (*see* **Note 12**).
- Pipette tip rack: place the pipette tip rack into its position (yellow), push it down and remove the lid (*see* **Note 13**).
- Extraction tip: make sure you properly and firmly place the tip.
- Objective wetting consumable (OWC): place it behind the cassette carrier with the reagent priming reservoir on the left. If well placed, it will automatically be detected by the machine.
- Left and Right cassettes: open the cassette carriers, remove the lid of the cassettes, place them into the carriers and close them. If well placed, they will automatically be detected by the machine. Make sure the cassette carriers are clean before placing the samples; use 70% isopropanol and compressed air to properly clean the surface.
- Waste tip array: make sure that it is empty and clean.

ii. Once all the consumables are loaded correctly close the instrument front panel and click on continue (*see* **Note 14**).

Sample Scan

iii. After loading the instrument, the Sample Scan (~1 h) will begin, which takes an image of each slide so that region selection can be performed (*see* **Note 15**).

iv. Once finished, click on "Continue" and proceed to region selection.

Region Selection and Initiate Scan

v. For sample scanning, the different regions need to be selected. To do so, click on "Add Region" to select the area that needs to be imaged for each of the slides (*see* **Note 16**).

vi. Once finished, click "Start Run". Once the run starts it is possible to see the time remaining while the run is in progress.

END OF DAY 4

At this step there are no stopping options; protocol must be continued once the run is over.

DAY 5: During the run time it is possible to follow the progress in the touchscreen, which will show the estimated time remaining. In this process, the instrument performs real-time image processing, decoding, and secondary analysis. At the end, identified transcripts can be visualized using Xenium Explorer software. After the Xenium run, the tissue sections can be stained with H&E. To do so, first the autofluorescence solution needs to be removed in a destaining step, and then the sections can be stained and imaged to obtain histological data that can be combined with the gene expression data generated on the same tissue section.

Buh12 *Run and PostRun*

i. The duration of the run will depend on the panel selected and the size of the regions.

ii. Once the run is completed, the screen will show "Run complete." At this point, a button will appear to start the cleaning of the fluidic system.

Post Run Cleanup and Unload

i. Click on "Start Cleanup." The screen will show the progress of the cleanup and the estimated time remaining, which normally takes ~5 min. When finished, click on "Next."

ii. Open the front panel of the instrument and first open the cassette carrier, remove the cassettes containing the Xenium Slides, remove all liquid covering the slides and add 750 μl PBS-T to properly cover the tissue sections. Then reapply the lid and store at 4°C while doing the cleanup.

iii. Then, remove all consumables (reagent bottles, OWC, reagent plates, pipette tip rack, and extraction tip) and discard the liquid inside the waste bottle (keep the bottle and place it back in its position) and the tips inside the waste tip array.

iv. Once all consumables are removed, close the front door and click on "Continue."

Data Output

i. During the run, image processing, decoding, and secondary analysis are performed in real time on-instrument, generating a run-specific data output folder.

ii. To access the data, go to "Menu > Open Settings > Runs" and click on "View Analysis Summary" of the individual runs to open the summary of each sample.

iii. Do not forget to retrieve the data from the computer.

Long-Term Storage

i. For long-term storage, remove the PBS-T and add 1 ml PBS and incubate for 5 min at RT.

ii. Remove PBS and add 1 ml of storage solution (50% glycerol in PBS). Then reapply the lid and store at 4°C for up to 2 weeks.

At this step there are three options:
Option 1—Proceed directly to Phenocycler-Fusion protocol.
Option 2—Short-term storage at 4°C for up to 3 days.
Option 3—Long-term storage at 4°C for up to 2 weeks.

3.2 Phenocycler-Fusion Protocol

DAY 5: At the end of the Xenium run, we continue by preparing all necessary buffers and solutions required for the Phenocycler process. We perform routine maintenance on the Phenocycler instrument to ensure it is functioning optimally, which includes cleaning components and calibrating sensors. Using the Phenocycler Designer software, we calculate the precise volumes of reagents needed for our experiments.

3.2.1 Instrument Preparation

Maintenances

i. Maintenance Brightfield
   - Open PhenoIMAGER Fusion software.
   - Go on Fusion Dashboard.
   - Click on acquire for Brightfield References using the Reference Carrier.
   - Click View to see brightfield reference imagery and information.
   - Export for Diagnostics will save the set of reference images to aid in technical support.

ii. Maintenance wash
   - Open PhenoCycler Dashboard.
   - Load blank flow cells into the Phenocycler Flow cell slide carrier.
   - Fill a beaker with at least 200 mL of ddH_2O and place near the PhenoCycler.
   - Remove the tops of the Low DMSO Buffer bottle and the High DMSO Buffer bottle and place the fluidic lines into the beaker of water.
   - Click the Wash button.

Design the Reporter Plate and Calculate Buffer Volumes

i. Open the PhenoCycler Experiment Designer app and create your reporter plate layout tailored to your experimental requirements (*see* **Note 17**).

ii. Access the Run Buffer Calculator: Navigate to the Home screen of the PhenoCycler Experiment Designer software and use the Run Buffer Calculator to estimate the volumes (*see* **Note 18**).

3.2.2 Buffer Preparation

Prepare 1× PBS

i. Add 900 ml of Milli-Q water to a clean glass beaker.

ii. Add 100 ml of 10× PBS.

Prepare 1× PCF

i. Add 800 ml of double-distilled water (ddH_2O) to a clean glass beaker or a similar container.

ii. Using a pipette, add 100 ml of the 10× Buffer for PhenoCycler to the beaker.

iii. Pipette 100 ml of Buffer Additive for PhenoCycler-Fusion into the same container.

iv. With the same pipette, gently pipette the buffer additive solution up and down at least five times to ensure complete transfer.

v. Mix the solution thoroughly by pipetting up and down with a pipette aid or by stirring with a magnetic stir bar (*see* **Note 19**).

Prepare Low- and High-DMSO

i. Select and label two appropriate amber glass containers for preparing and storing the low DMSO and high DMSO solutions. If using a mixing bar, place a PTFE- or Teflon-coated mixing bar at the bottom of each container. Include the preparation date on each label to monitor solution usability.

ii. Refer to Table 1 and accurately measure the required volumes of DMSO, then transfer them into their designated low DMSO and high DMSO containers.

iii. Following Table 1, add the specified volumes of 1× PhenoCycler-Fusion Buffer to each corresponding low DMSO and high DMSO container.

iv. Combine the solutions by gently stirring with the mixing bar on a magnetic stir plate set to medium for 5 min, keeping the container lid off. Ensure the stirring is gentle to avoid vigorous agitation; adjust the stir plate speed as necessary. Alternatively, use a pipette aid to pipette the solutions up and down at least 15 times to achieve thorough mixing.

v. Allow both solutions to cool for at least 15 min.

vi. Remove the low DMSO and high DMSO bottles from the sidecar of the PhenoCycler instrument before filling to prevent spills.

vii. Carefully pour the low- and high-DMSO solutions into their respective bottles on the sidecar of the PhenoCycler instrument. Store any excess solution in a cool, dry, and dark place.

viii. Place the filled low- and high-DMSO bottles back into their designated positions on the sidecar of the PhenoCycler instrument.

Prepare AR9 Buffer

i. Add 90 ml of Milli-Q water to clean container.

ii. Add 10 ml of AR9.

END OF DAY 5

At this step there are no stopping options; protocol must be continued the day after.

DAY 6: During this day we carry out several critical steps of the experiment. First, performing antigen retrieval to improve the accessibility of the targeted epitopes. This is followed by the preparation and incubation of the two cocktail antibodies. Finally, we prepare the reporter plate.

3.2.3 Morning Setup: Reagents and Materials

Remove Buffers from Storage

i. Take the following buffers out of the refrigerator (+4°C) and place them on the bench RT:
   - Staining Buffer (4°C storage).
   - Hydration Buffer (4°C storage).
   - Storage Buffer (if performing both incubations on the same day).

ii. Remove Blockers G and S
   - Retrieve Blocker G and Blocker S from the freezer (−20°C).
   - Place them at +4°C (*see* **Note 20**).

Prepare Work Area and Containers

i. Under a fume hood, prepare Coplin jars for deparaffinization and rehydration.

ii. Ensure appropriate waste containers are available for xylene, ethanol, and other solvents.

3.2.4 Tissue Preparation and Antigen Retrieval

Rinse Steps if Processing Slides Directly After Xenium Run or After Short-Term Storage

iii. Remove PBS-T and remove slide from the Xenium cassette.

iv. Wash slides twice in PBS for 5 min at RT.

v. Transfer the slide into the first jar of water (H_2O 1) 3 min.

vi. Prepare a second jar of water (H_2O 2) for an additional rinse for 3 min.

vii. Proceed to Antigen Retrieval.

Rinse Steps if Processing Slide After Long-Term Storage

i. Remove storage solution (50% glycerol in PBS) and remove slide from Xenium cassette.

ii. Wash slides four times in PBS for 5 min at RT.

iii. Transfer the slide into the first jar of water (H_2O 1) 3 min.

iv. Prepare a second jar of water (H_2O 2) for an additional rinse for 3 min.

v. Proceed to Antigen Retrieval.

Antigen Retrieval

i. Prepare the pressure cooker
   - Add 1 L of water and insert a plastic jar filled with AR9 buffer.

ii. Perform antigen retrieval

- Place the slide in the plastic jar containing AR9 buffer.
- Cover the jar with aluminum foil and place it in the cooker.
- Close the cooker and run for 20 min under pressure (*see* **Notes 21 and 22**).

Cool and Rinse

i. After cooking, use forceps to carefully remove the jar and allow it to cool for 30 min at RT.

ii. Rinse the slide in a sequence of steps to thoroughly remove residual AR9 buffer: first, immerse the slide in distilled water for 2 × 3 min; then transfer it to 1× PhenoCycler Hydration Buffer for 2 × 3 min.

3.2.5 Preparing the Antibody Cocktail Solution

Prepare Blocking Buffer

iii. Gently tap and, if needed, briefly centrifuge them to ensure the liquid is at the bottom of the tubes.

iv. Prepare the Blocking Buffer (for two slides) (*see* **Note 23**).

Staining buffer	362 µl
Blocker N	9.5 µl
Blocker J	9.5 µl
Blocker G	9.5 µl
Blocker S	9.5 µl
Total	400 µl

v. Tap gently to mix (no vortex), then briefly centrifuge if needed.

Preparation of the First Antibody Cocktail

i. Take Out the Antibodies for the First Cocktail.

ii. Tap each antibody tube gently and briefly centrifuge.

iii. Prepare the antibody cocktail in a 2.5 ml Eppendorf tube according to your lab's protocol (volumes may vary) (*see* **Note 24**).

3.2.6 First Antibody Incubation

Dry the Slide

iv. Gently blot around the edges of the slide without touching the tissue section.

Incubate with the First Antibody Cocktail

i. Place the slide in a dark incubation chamber (to protect from light).

ii. Carefully add 190 µl of the first antibody cocktail onto the tissue, ensuring complete coverage.

iii. Cover the tissue with a piece of parafilm (do not stretch it; let it drop gently to avoid bubbles).
iv. Incubate for 3 h at RT (*see* **Note 25**).

3.2.7 Preparation of the Second Antibody Cocktail (see **Note 26***)*

Prepare Another Batch of Blocking Buffer

Staining buffer	362 μl
Blocker N	9.5 μl
Blocker J	9.5 μl
Blocker G	9.5 μl
Blocker S	9.5 μl
Total	400 μl

Prepare the Second Antibody Cocktail

As for the first cocktail:

i. Gently tap and briefly centrifuge each antibody tube.
ii. Mix in an Eppendorf tube per your standard protocol (*see* **Note 27**).

3.2.8 Wash Between Incubations

Wash after the 3 h Incubation:

i. Remove the parafilm and transfer the slide to a beaker/jar containing staining buffer.
ii. Soak for 2 min, discard the staining buffer, then repeat (2 × 2 min).

3.2.9 Second Antibody Incubation

Dry the Slide

Gently blot around the edges of the slide without touching the tissue section.

Incubate with the Second Antibody Cocktail

i. Add 190 μl of the second antibody cocktail onto the tissue.
ii. Cover with a fresh piece of parafilm.
iii. Incubate either:

- Overnight (16–24 h) at +4°C. Or 3 h at RT (depending on your protocol and schedule).

3.2.10 Reporter Plate Preparation (see Note 28)

Take Out the Reporters

i. Place the reporter tubes on ice in the order of their intended cycles (groups of 3 or 2).
ii. Also retrieve the Nuclear Stain and Assay Reagent.
iii. Tap each tube gently and briefly centrifuge to bring liquid to the bottom.

Prepare Brown Eppendorf Tubes

i. Label them clearly according to the corresponding cycles.
ii. Do not prepare tubes for the "blank cycles" (negative controls).

Prepare the Reporter Stock Solution

i. In the dark (since Nuclear Stain is light-sensitive), combine:

Nuclear stain	150 µl
Storage buffer	3975 µl
Assay reagent	375 µl
Total	4500 µl

ii. Mix gently without excessive vortexing.

Prepare Individual Reporter Tubes

For each cycle:

i. Pipette 235 µl of the Reporter Stock Solution into an Eppendorf (for a 3-reporter cycle); 240 µl of Reporter Stock for a 2-reporter cycle and 245 µl of Reporter Stock for a 1-reporter cycle.
ii. Then add 5 µl of each corresponding reporter to the tube.
iii. Gently tap, briefly spin down, and keep on ice/in the dark.

Transfer 250 µl to the 96-Well Plate

i. Add 250 µl of each prepared reporter solution to the designated well.
ii. Also add 250 µl of Reporter Stock Solution alone to wells H1 and H2 (blank controls).

Protect the Plate

i. Cover the 96-well plate with aluminum foil.
ii. Store it in a dark place (e.g., a drawer or incubator protected from light).

END OF DAY 6

At this step there are no stopping options; protocol must be continued the day after.

DAY 7: This day is crucial for preparing tissue samples for automated imaging and data acquisition. The day involves fixation, rinsing, flow cell assembly, and setup of the PhenoCycler system to ensure optimal conditions for multiplex imaging. By following these steps carefully, you will ensure sample integrity and reliable results.

3.2.11 Tissue PostStaiNing

Coplin Jar Configuration

i. Fill two Colpin jars with 40 ml of staining buffer.
ii. Fill three Coplin jars with 40 ml of 1× PBS.

Tissue Wash

i. Gently wick away the antibody cocktail solution and place the sample slide in the first Coplin jar containing Staining Buffer.
ii. Lift and immerse the sample slide two to three times.
iii. Incubate for 2 min.
iv. Place the sample slide in the second Coplin jar containing staining buffer.
v. Incubate for another 2 min for a total of two washes (*see* **Note 29**).

Tissue Fixation

i. Prepare the poststaining fixing solution:
   - Under the fume hood, prepare 100 ml of 4% PFA solution.
   - Use a clean plastic Coplin jar.

16% PFA	10 ml
Storage buffer	90 ml
Total	100 ml

ii. Fix the tissue:
   - Place the sample slide in the Coplin jar containing poststaining fixing solution.
   - Incubate for 10 min (*see* **Note 30**).

Tissue Wash

i. Place the sample slide in the first Coplin jar containing 1× PBS.
ii. Lift and immerse each sample slide two to three times to ensure the Fixing Solution is removed from the top and bottom of each slide.
iii. Immediately move the sample slide to the second Coplin jar containing 1× PBS. Lift and immerse the sample slide two to three times.
iv. Immediately move the sample slide to the third Coplin jar containing 1× PBS for a total of three washes. Lift and immerse the sample slide two to three times.

Ice-Methanol Incubation

i. Prepare the ice-methanol Coplin jar:
   - Retrieve methanol from 4°C storage.
   - Pour approximately 40 ml of methanol into the Coplin jar.

ii. Ice-methanol incubation:
   - Remove the sample slide from the third Coplin jar containing 1× PBS and place it in the coplin jar containing ice-cold methanol.
   - Incubate at 4°C for 5 min.

Tissue Wash

i. Quickly transfer the sample slide from methanol to the first corresponding 1× PBS Coplin jar (*see* **Note 31**).

ii. Lift and immerse the sample slide two to three times to ensure methanol is removed from the bottom of the slide as well as the top.

iii. Transfer the sample slide to the second 1× PBS Coplin jar. Lift and immerse the sample slide two to three times.

iv. Transfer the sample slide to the second 1× PBS Coplin jar. Lift and immerse the sample slide two to three times.

Tissue Fixation

Prepare the final fixative solution:

i. Retrieve one aliquot of Phenocycler fixative reagent tube from −20°C storage.

ii. Briefly spin down the fixative reagent.

iii. In an Eppendorf tube prepare the final fixative solution:

Fixative reagent	20 μl
1X PBS	1000 μl
Total	1020 μl

iv. Mix thoroughly or vortex the solution.

v. Draw up 200 μl of the Final Fixative solution with a pipette.

vi. Remove the sample slide from the Coplin jar.

vii. Use a kimwipe to carefully dry excess buffer solution around the sample.

viii. Place the slide on the tray of the slide staining box.

ix. Add 200 μl of Final Fixative Solution to the top corner of the sample slide.

x. Place the lid on the slide staining box and incubate for 20 min.

Tissue Wash

i. Remove the sample slide/s from the slide staining box and place in the first Coplin jar containing 1× PBS. Lift and immerse the sample slide two to three times to ensure Fixative Solution is removed.

ii. Move each sample slide/s to the second Coplin jar containing 1× PBS. Lift and immerse each sample slide two to three times.

iii. Move each sample slide/s to the third Coplin jar containing 1× PBS for a total of three washes. Lift and immerse each sample slides two to three times.

Store the Tissue

i. Label a new Coplin and pipet 40 ml of storage buffer.

ii. Place the sample slide in the Coplin jar keeping the tissue facing the same direction for all slides.

iii. Cap the Coplin jar and seal around the edges with parafilm (optional) for storage (*see* **Notes 32 and 33**).

3.2.12 Flow Cell Preparation and Assembly

Tissue Wash

Place the slide in PBS for 10 min to ensure all residual reagents are removed.

Assemble Slide and Flow Cell

i. Retrieve the flow cell:

- Take a flow cell from the storage drawer (to the right of the pressure cooker).
- Remove the plastic protective slide from the flow cell.

ii. Prepare the slide

- Gently blot the edges of the slide with Kimwipes, avoiding contact with the tissue.

iii. Secure the slide to the flow cell:

- Align the slide carefully with the flow cell.
- Press firmly for 30 s to ensure proper sealing.

Buffer Addition

i. Place the slide-flow cell assembly in a jar containing PhenoCycler Buffer with Additive.

ii. Fill the jar halfway, insert the assembly, and top off with buffer until the opening is covered.

iii. Incubate for 10 min.

iv. Clean the slide:

- Perform three strokes with Milli-Q water in one direction across the tissue.
- Follow with three strokes using ethanol in the same direction.

v. Dry the slide:

- Use Kimwipes to clean the edges, back, and bottom of the slide.

3.2.13 PhenoCycler and PhenoImager Setup

Power Up the Systems

i. Turn on the following:
   - The computer, PhenoCycler, and PhenoImager systems.

ii. Open the Fusion 2.0 application:
   - Select "Cat 2," create a new experiment, and load the protocol.
   - Enter sample names and configure imaging to 16-bit mode.

Load the Slide

Place the slide in the Flow Cell Slide Carrier:

- Follow software instructions to perform washes and priming cycles.

PhenoImager Setup

i. Prepare for imaging:
   - Ensure the black clamp is fully raised before placing the slide.

ii. Begin imaging and monitor the process via the software (*see* **Note 34**).

END OF DAY 8

At this step there are no stopping options; protocol must be continued the day after.

DAY 8.

3.2.14 End of the Run and Flow Cell Removal

End of the Run

i. Remove the slide from the carrier.

ii. Place blank slide and perform post experiment clean and maintenance.

Flow Cell Removal

i. Place the tissue sample slide with the flow cell in a staining dish or Coplin jar containing HistoClear®.

ii. Allow the slide to sit in HistoClear overnight or at least 2 h.

iii. After overnight incubation, the flow cell should separate from the tissue slide.

iv. Carefully remove the tissue sample slide from the HistoClear.

v. Rinse the tissue sample with DI water using at least 1 L of DI water to ensure thorough rinsing.

vi. Next, wash the tissue sample slide in 1× PhenoCycler™ Buffer without additive, 3 × 10 min.

vii. If you are not performing H&E immediately, transfer the tissue sample slide(s) to the PhenoCycler Storage buffer for storage.

viii. Remember to dispose of the dissociated flow cell and HistoClear following the institutional guidance and requirements.

ix. If short-term storage, put the slides in a Coplin jar containing a storage buffer and store for less than 5 days.

At this step there are two options:
Option 1—Proceed directly to post-run H&E staining.
Option 2—Short-term storage at 4°C for up to 5 days before H&E staining.

3.3 H&E STAINING: PostRun Staining

3.3.1 Buffer Preparation for Quenching

i. Prepare fresh Quencher Removal Solution in a fume hood shortly before use (needs to be used in the 10 min following preparation):

- Weight 87 mg of sodium hydrosulfite in a fume hood and add it to a conical tube containing 50 ml Milli-Q Water. Cap tube, vortex 10 s to dissolve and maintain at RT.

ii. Prepare five jars in a fume hood for quencher removal:

Quencher Removal Solution (x1)
Milli-Q water (x3)
1X PBS (x1)

3.3.2 Quencher Removal

i. Take the cassettes that were kept at 4°C, remove all the PBS-T, and remove the slide from the cassette.

ii. Immerse slides in the jar containing the Quencher Removal Solution and incubate for 10 min at RT.

iii. Immerse slides in the first jar containing Milli-Q water for 1 min at RT, then in the second jar containing Milli-Q water for 1 min at RT, and finally in the third jar containing Milli-Q water for 1 min at RT.

3.3.3 Buffer Preparation for H&E

i. Filter Hematoxylin & Eosin solutions using filter paper.

ii. Prepare 16 jars in a fume hood for H&E staining:

Hematoxylin Solution (x1)
Bluing Solution (x1)
70% Ethanol (x1)
95% Ethanol (x3)
Eosin Solution (x1)
100% Ethanol (x2)
Xylene (x2)
Milli-Q Water (x5)

3.3.4 H&E Staining

i. Immerse slides in Milli-Q jar 1 for 2 min at RT.

ii. Immerse slides in hematoxylin solution for 20 min at RT.

iii. Immerse slides in Milli-Q jar 2 for 1 min at RT, then in Milli-Q jar 3 for 1 min at RT, and finally in Milli-Q jar 4 for 1 min at RT.

iv. Immerse slides in Bluing Solution jar for 1 min at RT.

v. Immerse slides in Milli-Q jar 5 for 1 min at RT.

vi. Immerse slides in 70% Ethanol jar for 3 min at RT.

vii. Immerse slides in 95% Ethanol jar 1 for 3 min at RT.

viii. Immerse slides in Eosin Solution jar for 2 min at RT.

ix. Immerse slides in 95% Ethanol jar 2 for 30 s at RT and again in 95% Ethanol jar 3 for 30 s at RT.

x. Immerse slides in 100% Ethanol jar 1 for 30 s at RT and again in 100% Ethanol jar 2 for 30 s at RT.

xi. Immerse slides in Xylene jar 1 for 3 min at RT and to finish Immerse slides in Xylene jar 2 for 3 min at RT.

3.3.5 Coverslipping and Imaging

i. Before proceeding to coverslipping, remove any residual droplets with a lint-free laboratory wipe to make sure that the slide is properly dried.

ii. Add 150–200 μl mounting media to uniformly cover the tissue using a wide-bore pipette tip.

iii. Slowly apply a coverslip without introducing bubbles and allow mounting media to spread and settle.

iv. Let the coverslip dry for 30 min at RT before proceeding to imaging. If imaging is not done immediately, samples can be stored at 4°C for 3 days.

v. Proceed to imaging.

For long-term storage after H&E staining and imaging maintain slides at RT in the dark.

The combination of spatial transcriptomics and spatial proteomics on the same tissue section using Xenium In Situ followed by the PhenoCycler Fusion generates a highly multimodal imaging dataset. This dataset enables the study of hundreds of RNAs and tens of proteins, providing insights into the spatial organization and function of tissues. Following the combination of these two protocols, data are ready for integration and extensive analysis to elucidate immune cell phenotyping. Alternatively, PhenoCycler Fusion (Akoya) can also be performed followed by Visium (10×), as previously described in the Adimi et al. study [33]. Here, researchers demonstrated that RNA integrity was preserved throughout the PhenoCycler processing steps, confirming the compatibility of these two methodologies on a single tissue section. This approach enabled identification of specific biomarkers, including SOX9 and CD70, associated with poorly differentiated head and neck squamous cell carcinomas (HNSCC). The combination of spatial proteomics and transcriptomics on a single histological section represents a powerful multiomics approach. Integration of these two techniques generated a highly informative dataset from the same tissue section, providing complementary layers of molecular information, allowing to draw meaningful conclusions and to gain a deeper understanding of the tumor architecture, offering valuable insights into the spatial organization and complexity of head and neck squamous cell carcinomas.

4 Notes

1. Before starting the experiment, we recommend verifying tissue section quality. For that, after sectioning the tissue on blank slides, tissue should be DAPI and H&E stained to examine tissue morphology. This will allow us to look for signs of over- or under-fixation and artifacts. Having this information prior to the experiment will help to select the area of interest. DV200 asses of RNA integrity is also recommended to avoid a DV200 score of less than 30%, as this would correlate with low transcript density and, therefore, low-quality data. Moreover, it is strongly recommended to practice sectioning within the Sample Area of the Xenium Slides using standard blank slides.
2. Slow sectioning speed is recommended. Faster sectioning speed may lead to cracks or tears in the sections or damage to the tissue block. Sometimes tissue blocks need to be trimmed or scored to make them compatible in size to the Sample Area on Xenium slides.
3. 42°C is the recommended water bath temperature for most tissues, but sometimes bath temperature may need optimization based on the tissue type. In the event that the tissue is taking too long to expand, turn the water bath temperature up by 1° or 2° and let the section float for longer. On the contrary, if the tissue is expanding too quickly and dissociating, turn the water bath temperature down by 1° or 2 ° and shorten the floating time.
4. It is very important not to obstruct the fiducial frame. Tissue must be place within the Sample Area. It is possible to place as many tissue sections as you can fit within the sample area, as long as the tissue sections and the wax surrounding each tissue do not overlap. In addition, the sections on the slides should be uniform without any cracks, tears, or folds.
5. The Xenium slide needs to be placed in the bottom half of the cassette with the tissue facing upwards. The label needs to

match the bottom of the cassette. Then apply the top half of the cassette-securing clips of top half with tabs of bottom half on both sides of the cassette and apply some pressure until clips correctly click shut. Proper assembly of the cassette is very important because otherwise this can lead to poor assay performance.

6. Tighten the thermal cycler lid until an audible click is heard so that the lid is in direct contact with the Xenium Cassette Lid. The proper tightening of the thermal cycler lid is critical to avoid the evaporation of the Probe Hybridization Mix overnight and ensure assay performance. Nevertheless, ensure over-tightening to avoid breaking the slide.
7. If processing two slides at the same time, remove and add reagents from one slide first, and then proceed with the second slide. Verify that at least 400 μl are spired from each slide; a lower volume may suggest evaporation overnight, which can affect the results.
8. With the help of the forceps, slowly place the insert within the well of the cassette, which corresponds to the sample area. It is important to do it slowly, to avoid damaging the tissue.
9. Bubbles may impact the quality of staining and assay performance; therefore, if bubbles are introduced, do not aspirate or pop the bubbles; instead remove the insert and reapply it slowly.
10. You can click on each raw to view detailed help and instruction animations on how to load each item on the right side of the screen.
11. Push firmly downward on the bottle carrier caps of each bottle to ensure correct sealing of the bottles; otherwise, this will lead to an error of the system.
12. Make sure the label is facing you and that you have firmly pressed down the plates so that they are properly positioned; otherwise, this will lead to an error of the system.
13. Firmly press the pipette tip rack down so that it is properly positioned; otherwise, this will lead to an error of the system. Do not forget to remove the lid.
14. At this point the instrument just detects if the consumables are present or not; however, if they are not properly placed, this will be not detected at this step. Unproper loading of the consumables will be detected later, leading to a system error that forces the user to unload the machine, close the app, and restart the whole process again. We advise to verify that all the reagents have been properly placed to avoid this loss of time.
15. Check often the screen of the instrument, because it is at this time that the machine will show an error if any of the

consumables/reagents have not been properly loaded. If an error occurs, the Support team needs to be contacted, but most probably you will have to cancel the run, clean the machine, unload the instrument, and restart the whole process.

16. One single box on the grid corresponds to one field of view (FOV), and multiple FOVs confirm one region. If several sections of the same sample are placed on one single slide, select various FOVs as one single region, as regions do not need to be contiguous. If different samples are located in one single slide, select different FOVs as different regions. It is important to avoid selecting empty FOVs and overlapping regions.
17. Maximum Cycles: Single Slide (Position 1): Up to 64 cycles (including blank cycles). Dual Slide (Positions 1 and 2): Up to 63 cycles distributed between both positions.
18. The Run Buffer Calculator within the PhenoCycler Experiment Designer software assists in determining the required amounts of 1× PhenoCycler-Fusion (PCF) buffer, High DMSO, and Low DMSO buffers for single or dual-slide runs.
19. Ensure to scale this recipe according to the volume of 1× PhenoCycler-Fusion Buffer required for your experimental run. The prepared 1× PhenoCycler-Fusion Buffer should be stored at RT and remains stable for up to 2 weeks.
20. Do not vortex antibodies, blockers, or reporters. Gently tap to mix, and if necessary, briefly centrifuge to bring the liquid to the bottom of the tube.
21. Make sure the pressure cooker is set to pressure cooking mode (not steam mode).
22. Although the Xenium protocol includes a decrosslinking step, which is akin to antigen retrieval, a second round of antigen retrieval, as outlined in this protocol, is necessary for epitope unmasking and accurate protein detection by antibody-based methods [34].
23. Prepare the blocking buffer less than 1 h before the first antibody incubation.
24. Important: If an antibody is removed or changed in the panel, update the 96-well plate layout accordingly.
25. You may take a 2 h lunch break as part of this 3 h incubation, as long as the total antibody incubation remains 3 h.
26. Start ~1 h before the end of the first antibody incubation.
27. Update the 96-well plate if any antibody is removed or replaced.
28. Perform these steps in parallel or during the second antibody incubation.

29. Prepare the poststaining Tissue Fixation during the wash incubation.
30. During the 10 min incubation, prepare the ice-cold methanol Coplin jar and three other 1× PBS Coplin jars.
31. Methanol dries tissue faster than other buffers. Move quickly to prevent sample degradation.
32. Tissues can now be used directly to run a PhenoCycler-Fusion Experiment.
33. For best results, store at 4°C for <5 days.
34. Overnight run.

References

1. Lee S, Kim G, Lee J et al (2024) Mapping cancer biology in space: applications and perspectives on spatial omics for oncology. Mol Cancer 23: 26. https://doi.org/10.1186/s12943-024-01941-z
2. Hsieh W-C, Budiarto BR, Wang Y-F et al (2022) Spatial multi-omics analyses of the tumor immune microenvironment. J Biomed Sci 29:96. https://doi.org/10.1186/s12929-022-00879-y
3. Fu T, Dai L-J, Wu S-Y et al (2021) Spatial architecture of the immune microenvironment orchestrates tumor immunity and therapeutic response. J Hematol Oncol 14:98. https://doi.org/10.1186/s13045-021-01103-4
4. Lammi MJ, Qu C (2024) Spatial transcriptomics, proteomics, and epigenomics as tools in tissue engineering and regenerative medicine. Bioengineering (Basel) 11:1235 https://doi.org/10.3390/bioengineering11121235
5. Maciejewski K, Czerwinska P (2024) Scoping review: methods and applications of spatial transcriptomics in tumor research. Cancers (Basel) 16:3100 https://doi.org/10.3390/cancers16173100
6. Yoon C, Park E, Misra S et al (2024) Deep learning-based virtual staining, segmentation, and classification in label-free photoacoustic histology of human specimens. Light Sci Appl 13: 226. https://doi.org/10.1038/s41377-024-01554-7
7. Baxi V, Edwards R, Montalto M, Saha S (2022) Digital pathology and artificial intelligence in translational medicine and clinical practice. Mod Pathol 35:23–32. https://doi.org/10.1038/s41379-021-00919-2
8. Curras-Alonso S, Soulier J, Walter T et al (2021) Spatial transcriptomics for respiratory research and medicine. Eur Respir J 58:2004314. https://doi.org/10.1183/13993003.04314-2020
9. Halpern KB, Shenhav R, Matcovitch-Natan O et al (2017) Single-cell spatial reconstruction reveals global division of labour in the mammalian liver. Nature 542:352–356. https://doi.org/10.1038/nature21065
10. Codeluppi S, Borm LE, Zeisel A et al (2018) Spatial organization of the somatosensory cortex revealed by osmFISH. Nat Methods 15:932–935. https://doi.org/10.1038/s41592-018-0175-z
11. Curras-Alonso S, Soulier J, Defard T et al (2023) An interactive murine single-cell atlas of the lung responses to radiation injury. Nat Commun 14: 2445. https://doi.org/10.1038/s41467-023-38134-z
12. Choi HMT, Chang JY, Trinh LA et al (2010) Programmable in situ amplification for multiplexed imaging of mRNA expression. Nat Biotechnol 28:1208–1212. https://doi.org/10.1038/nbt.1692
13. Choi HMT, Beck VA, Pierce NA (2014) Next-generation in situ hybridization chain reaction: higher gain, lower cost, greater durability. ACS Nano 8:4284–4294. https://doi.org/10.1021/nn405717p
14. Saka SK, Wang Y, Kishi JY et al (2019) Immuno-SABER enables highly multiplexed and amplified protein imaging in tissues. Nat Biotechnol 37:1080–1090. https://doi.org/10.1038/s41587-019-0207-y
15. Rouhanifard SH, Mellis IA, Dunagin M et al (2018) ClampFISH detects individual nucleic acid molecules using click chemistry-based amplification. Nat Biotechnol https://doi.org/10.1038/nbt.4286
16. Goh JJL, Chou N, Seow WY et al (2020) Highly specific multiplexed RNA imaging in tissues with split-FISH. Nat Methods 17:689–693. https://doi.org/10.1038/s41592-020-0858-0

17. Moffitt JR, Hao J, Wang G et al (2016) High-throughput single-cell gene-expression profiling with multiplexed error-robust fluorescence in situ hybridization. Proc Natl Acad Sci U S A 113:11046–11051. https://doi.org/10.1073/pnas.1612826113
18. Shah S, Lubeck E, Schwarzkopf M et al (2016) Single-molecule RNA detection at depth by hybridization chain reaction and tissue hydrogel embedding and clearing. Development 143:2862–2867. https://doi.org/10.1242/dev.138560
19. Pichon X, Lagha M, Mueller F, Bertrand E (2018) A growing toolbox to image gene expression in single cells: sensitive approaches for demanding challenges. Mol Cell 71:468–480. https://doi.org/10.1016/j.molcel.2018.07.022
20. Lohoff T, Ghazanfar S, Missarova A et al (2022) Integration of spatial and single-cell transcriptomic data elucidates mouse organogenesis. Nat Biotechnol 40:74–85. https://doi.org/10.1038/s41587-021-01006-2
21. Mateo LJ, Sinnott-Armstrong N, Boettiger AN (2021) Tracing DNA paths and RNA profiles in cultured cells and tissues with ORCA. Nat Protoc 16:1647–1713. https://doi.org/10.1038/s41596-020-00478-x
22. Weber C, Defard T, Lelek M, et al (2026) autoFISH—a modular toolbox for sequential smFISH experiments. Commun Biol 9, 752. https://doi.org/10.1038/s42003-026-09979-z
23. Defard T, Laporte H, Ayan M et al (2024) A point cloud segmentation framework for image-based spatial transcriptomics. Commun Biol 7:823. https://doi.org/10.1038/s42003-024-06480-3
24. Lubeck E, Coskun AF, Zhiyentayev T et al (2014) Single-cell in situ RNA profiling by sequential hybridization. Nat Methods 11:360–361. https://doi.org/10.1038/nmeth.2892
25. Chen KH, Boettiger AN, Moffitt JR et al (2015) RNA imaging. Spatially resolved, highly multiplexed RNA profiling in single cells. Science 348:aaa6090 https://doi.org/10.1126/science.aaa6090
26. Eng C-HL, Lawson M, Zhu Q et al (2019) Transcriptome-scale super-resolved imaging in tissues by RNA seqFISH. Nature 568:235–239. https://doi.org/10.1038/s41586-019-1049-y
27. Zhang M, Eichhorn SW, Zingg B et al (2021) Spatially resolved cell atlas of the mouse primary motor cortex by MERFISH. Nature 598:137–143. https://doi.org/10.1038/s41586-021-03705-x
28. Lee H, Marco Salas S, Gyllborg D, Nilsson M (2022) Direct RNA targeted in situ sequencing for transcriptomic profiling in tissue. Sci Rep 12: 7976. https://doi.org/10.1038/s41598-022-11534-9
29. Hussaini HM, Seo B, Rich AM (2023) Immunohistochemistry and Immunofluorescence. Methods Mol Biol 2588:439–450. https://doi.org/10.1007/978-1-0716-2780-8_26
30. Harms PW, Frankel TL, Moutafi M et al (2023) Multiplex Immunohistochemistry and Immunofluorescence: a practical update for pathologists. Mod Pathol 36:100197. https://doi.org/10.1016/j.modpat.2023.100197
31. Tan WCC, Nerurkar SN, Cai HY et al (2020) Overview of multiplex immunohistochemistry/immunofluorescence techniques in the era of cancer immunotherapy. Cancer Commun (Lond) 40:135–153. https://doi.org/10.1002/cac2.12023
32. Black S, Phillips D, Hickey JW et al (2021) CODEX multiplexed tissue imaging with DNA-conjugated antibodies. Nat Protoc 16:3802–3835. https://doi.org/10.1038/s41596-021-00556-8
33. Adimi Y, Levin C, Lombardi EP et al (2024) A new spatial multi-omics approach to deeply characterize human cancer tissue using a single tissue section. 2024. https://doi.org/10.1101/2024.10.03.616510
34. Sivajothi S, Soja E, Bhargava S et al (2024) FFPE Tissue processing for multimodal imaging assays (Phenocycler-Fusion + H&E) following Xenium In Situ Gene Expression. protocols.io. https://doi.org/10.1016/j.modpat.2023.100197

Check for updates

Chapter 19

High-Dimensional Immunophenotyping of Human Leukocytes by Spectral Cytometry

Sarah Fidai, Shady Banyamin, Sahithi Thanikanti, Chih-Hsiung Chen and Ankit Saxena

Abstract

Spectral cytometry has rapidly evolved over the past two decades. The development of complete spectral analysis using the flow cytometer as a spectrometer has been actualized due to advances in optics, detectors, and electronics that have made single-cell flow spectroscopy a reality. Polychromatic flow cytometry is a complex process, including the design of antibody panels and instrument compensation. Spectral cytometry has contributed significantly to effectively dealing with the complexity of polychromatic flow cytometry. It has substantial potential to improve primary and clinical research. Here, we provide an overview of the methodology to develop a 15-color human polychromatic flow cytometry panel using spectral flow cytometry.

Keywords Spectral flow cytometry, Immunophenotyping, Whole blood, T cell, Tregs, Monocytes, Proliferation, Programmed death receptor -1, Dendritic cells

1 Introduction

Spectral flow cytometry measures the entire emission spectra for every fluorochrome across all laser lines [1]. Spectral fingerprints of fluorochromes can be identified uniquely using mathematical approaches even when their maximum emissions are very similar [2,3]. Leveraging this full-spectrum technology in a four-laser system, we demonstrate the ability to combine 14 fluorescently labeled antibodies and a fixable live dead dye to immunophenotype human peripheral blood mononuclear cells (PBMCs) in one tube. The panel can be used for studies aimed at characterizing the immune response in the context of various disease models such as infectious or autoimmune diseases, oncology and immunotherapy, and the discovery of novel biomarkers.

The panel presented in this chapter examines the frequency of T, B, NK, DCs, and various myeloid cell subsets. Dead cells were

Angélique Biancotto and Pradeep K. Dagur (eds.), *Immunophenotyping*, Methods in Molecular Biology, vol. 3087,
https://doi.org/10.1007/978-1-0716-5607-5_19,

excluded using a viability dye (LIVE/Dead Fixable dye). The following markers were used to characterize the indicated cell types: CD3, CD4, and CD8 for the main T cell populations; CD19 for B cells; CD56, NK cells; HLA-DR, CD11b, and CD11c defining myeloid cell subsets; CD14 and CD16 to identify classical, nonclassical, and intermediate monocyte populations. We included FoxP3 as a regulatory T cell (Treg) marker and PD-1, TIM-3, and LAG-3 as immune exhaustion markers.

Human peripheral blood mononuclear cells (PBMCs) were isolated [4], cryopreserved [5], and later used for flow cytometry staining. Single-color controls were prepared from UltraComp eBeads™ (ThermoFisher Scientific catalog: #01-2222-42), formalin-fixed PBMCs for fixable live dead control, and unstained cells were PBMCs. Cells were acquired on Cytek® Aurora cytometer equipped with 4 lasers and 38 detectors using Spectroflo® v3.0.3 software. Samples were acquired at low to medium speed. The flow cytometry results were analyzed using FlowJo™ v10.8 Software (BD Life Sciences).

2 Materials

2.1 Instrument and QC particle

- Cytek Aurora flow cytometer- (Cytek Biosciences, Fremont, CA) equipped with 4 LASERS (355, 405, 488, and 640 nm) and 64 detectors.
- Computer analysis software capable of reading .FCS files.
- QC beads—Spectroflow (Cytek Biosciences catalog: SKU B7-10001).

2.2 Reagents and Plastic Ware

- Fluorochrome-conjugated antibodies (Table 1; various vendors).
- Trypan Blue Stain 0.4% (catalog: T10282).
- Fc-blocking reagent (catalog: 422301).
- Fixable Live/dead reagent (catalog: 65-0866-14).
- Cell staining buffer (catalog: 420201-BL).
- Fixation buffer (catalog: 420801).
- 96-well plate (catalog: 650261-656).
- Benzonase Nuclease (catalog: E1014-25KU).
- Perm/Wash Buffer (catalog: 554723).
- RPMI 1640 Medium (catalog: 11875093).

Table 1
List of human antibodies

Antigen	Fluorochrome	AB clone	Vendor	Cat number
Live Dead	eFluor506	NA	eBiosciences	65-0866-14
CD3	Spark Blue 550	SK7	BioLegend	344852
CD4	BUV563	SK3	BD Biosciences	612912
CD8a	BUV661	RPA-T8	BD Biosciences	750699
CD19	BUV737	HIB19	BD Biosciences	741829
CD56	BUV805	QA17A16	BD Biosciences	748608
CD14	Pacific Blue	M5E2	BD Biosciences	982504
HLA-DR	PE-Dazzle 594	L243	BioLegend	980408
CD16	BV750	3G8	BioLegend	302082
CD11c	BB515	B-Iy6	BD Biosciences	564490
CD11b	BB700	ICRF44	BD Biosciences	742210
PD-1	Pe-Cy7	A17188B	BioLegend	621616
Tim-3	BV605	F38-2E2	BioLegend	345018
LAG-3	BV785	11C3C65	BioLegend	369322
FoxP3	APC	PCH101	Invitrogen	501128852

3 Methods

3.1 List of Biological Targets

In the present immune panel, we included the lineage markers for major immune cell subsets (such as CD3 for T cells and CD19 for B cells) along with their exhaustion markers and immune suppressor Treg cell marker transcription factor FoxP3 (Table 2).

3.2 Instrument QC

The instrument manufacturer provides bead setup and tracking features and includes the beads and accompanying analysis software that uses algorithms to differentiate the fluorescence signals from each bead type based on size and fluorescence intensity in each detector for a specific instrument. Linearity, detector efficiency (Qr), optical background (Br), electronic noise, and laser delays are all tracked. The median fluorescence intensity (MFI) and percent robust CV (%rCV) are measured for each bead intensity in all fluorescence detectors (PMT). PMT voltages are then adjusted to maximize population resolution in each detector; PMT voltages should not vary more than 10 V from their baseline, and %rCV should be <6%. Levy–Jennings plots (Gain or PMT voltage over time) are used to facilitate monitoring of performance (Fig. 1). CST beads are a set of fluorescent

Table 2
List of target antigens and their association with immune cell lineage and function

Antigen	Marker
Live Dead	Live dead
CD3	T cells
CD45	Leuckocytes
CD4	T helper
CD8a	T cytotoxic
CD19	B cells
CD56	NK cells
CD14	Monocytes
HLA-DR	Major histocompatibility class II marker
CD16	Monocyte-associated marker
CD11c	DC-associated marker
CD11b	Macrophage-associated marker
CD80	Antigen-presenting cells, B cells activation marker
PD-1	Programmed Death-1 receptor, exhaustion marker
Tim-3	T cell activation and exhaustion
LAG-3	T cell activation and exhaustion
FoxP3	Regulatory T cell marker.

beads that can be used to assess the performance of the lasers and detectors, determine optimal values for laser delay, and determine optimal baseline voltages for every PMT.

3.3 Characterization of Fluorochrome Brightness

Antibodies are often used in mixtures, often referred to as antibody cocktails. Each antibody is present in the cocktail at a concentration determined by titration. Antibodies have a range of concentrations in which they optimally bind to antigens. If too little is used, the fluorescence will produce an inaccurate amount of intensity, and, depending on the magnitude, a cell positive for the antibody may not be detected. However, if too much antibody is used, the sample's actual amount of the target antigen may be masked, and the fluorescence of the negative cells may increase.

During titration, serial dilutions of antibodies are used to stain identical aliquots of cells, and the dilution with the best separation between negative and positive cells determines the optimal titer. For titration, each staining can be viewed on the same axis using the concatenation feature in FlowJo software (Fig. 2a and b). The separating concentration can be determined by the difference between the two fluorescence intensities.

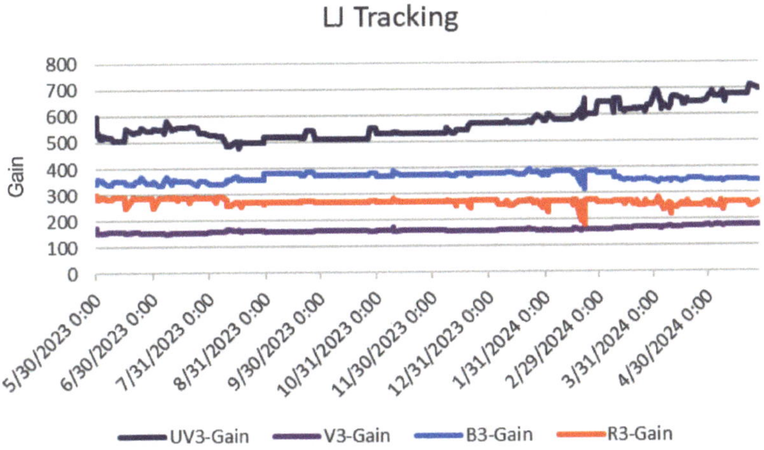

Fig. 1 Example of QC plots. On a daily basis, SpectroFlo® QC Beads are run on the cytometer. The median fluorescence intensity of each PMT for each laser is generated by SpectroFlo® software version 2.1. Levy–Jennings plots, which display the performance characteristics of an instrument over time, are useful for assessing instrument variations. Here, a Levy–Jennings plot of the Gain of bead standards at a constant voltage is shown for four PMTs across several months, with the performance of each detector shown in a different color

The performance of an antibody-fluorochrome conjugate is measured by the mean fluorescence intensity (MFI) and stain index (SI) values. MFI is the mean of the fluorescence intensity in the chosen fluorescence channel, and SI is the ratio of the difference between the positive signal and negative signal (D) and the width of the negative signal (W) (SI = D/W). Both MFI and SI should be obtained for each antibody-fluorochrome conjugate used in a staining combination.

3.4 Panel Design Using Cytek Spectral Viewer

Panel design for spectral polychromatic flow cytometry was performed in four steps. Three of these steps were performed using the Cytek spectral viewer online tool (https://cloud.cytekbio.com/spectrum/cloudspectrumviewer). First, the Similarity Index tool identified all fluorochromes with unique spectral signatures and minimum peak channel overlap (Fig. 3). Second, another feature available in the Cytek spectral viewer online tool called the Complexity Index (C.I.) was determined. C.I. measures the interference among a specific combination of fluorochromes and predicts the impact on the autofluorescence distribution of the spectrally unmixed results while also considering spillover. Third, staining index reduction was calculated (Figs. 2 and 4).

3.5 Titration of Reagents

For a typical titration experiment, it is recommended to perform an 8-twofold serial dilution. Typically, start the highest concentration at a dose and double the amount of antibody volume suggested by the vendor per staining reaction. For example, if the recommended volume for an antibody is 2 μg, then 4 μg should be used to begin

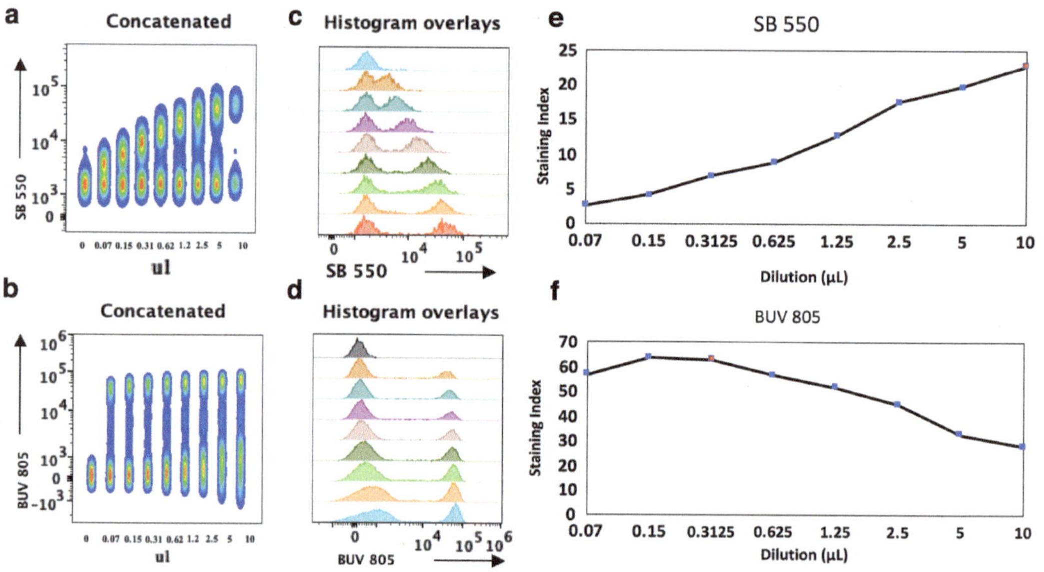

Fig. 2 Examples of an antibody titration experiment and a staining index chart. (**a**) and (**b**) Two representative titration experiments showing concatenated pseudocolor leukocyte populations acquired by flow at different serial dilutions (left) and Staining indices have been calculated as described in Sect. 3.8.4. A staggered histogram overlay shows the positive and negative populations for the same. For CD3 Spark Blue (SB) 550 and CD56 BUV 805, the plateau of the stain index is reached around 10 and 0.3125 µl (marked in red) of antibody volume (**e** and **f**), indicating that this is the minimal voltage required to deliver optimal resolution at the given concentration

with. If the volume of the antibody is suggested, then start with double the volume as the highest titration value. To obtain the saturating titration dose, plot a curve using dilution value vs staining index (Fig. 2e–f). The dilution at which S.I. is highest is the correct volume of antibody to be used per reaction.

3.6 Cell Thawing

Cryopreserved human PBMCs were used for this experiment.

1. Prepare thawing media mix: 10 ml RPMI with 5% FBS and 1 µl Benzonase nuclease (−20C).
2. Cells from the Liquid N2 tank are put in a dry ice container while transferring to the bench.
3. From the dry ice container, hold the cryovial over the surface of the water bath at 37°C with an occasional "flick" (~1 min)
4. Once the ice starts thawing, transfer cells to an empty 15 ml conical tube with a wide-bore pipette tip.
5. Add warm thawing media while gently swirling to reduce osmotic shock.
6. Centrifuge 300 × *g* for 5 min, aspirate supernatant and disrupt the pellet.
7. Resuspend the pellet in 5 ml RPMI with 5% FBS.

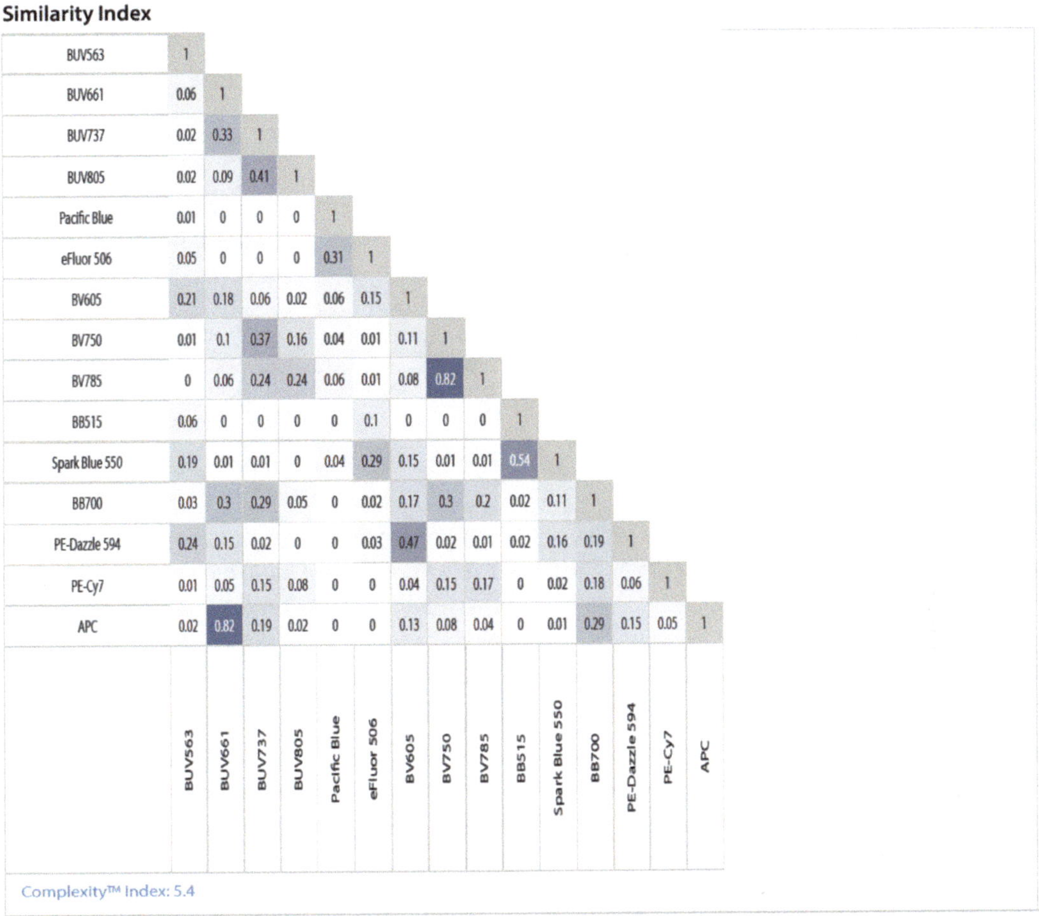

Similarity Index

	BUV563	BUV661	BUV737	BUV805	Pacific Blue	eFluor 506	BV605	BV750	BV785	BB515	Spark Blue 550	BB700	PE-Dazzle 594	PE-Cy7	APC
BUV563	1														
BUV661	0.06	1													
BUV737	0.02	0.33	1												
BUV805	0.02	0.09	0.41	1											
Pacific Blue	0.01	0	0	0	1										
eFluor 506	0.05	0	0	0	0.31	1									
BV605	0.21	0.18	0.06	0.02	0.06	0.15	1								
BV750	0.01	0.1	0.37	0.16	0.04	0.01	0.11	1							
BV785	0	0.06	0.24	0.24	0.06	0.01	0.08	0.82	1						
BB515	0.06	0	0	0	0	0.1	0	0	0	1					
Spark Blue 550	0.19	0.01	0.01	0	0.04	0.29	0.15	0.01	0.01	0.54	1				
BB700	0.03	0.3	0.29	0.05	0	0.02	0.17	0.3	0.2	0.02	0.11	1			
PE-Dazzle 594	0.24	0.15	0.02	0	0	0.03	0.47	0.02	0.01	0.02	0.16	0.19	1		
PE-Cy7	0.01	0.05	0.15	0.08	0	0	0.04	0.15	0.17	0	0.02	0.18	0.06	1	
APC	0.02	0.82	0.19	0.02	0	0	0.13	0.08	0.04	0	0.01	0.29	0.15	0.05	1

Complexity™ Index: 5.4

Fig. 3 Fluorochrome selection and compatibility—the Spectral signatures of 15 fluorochromes were visualized using the Cytek Aurora's full spectrum viewer weblink (https://cloud.cytekbio.com/spectrum/cloud-spectrumviewer) and the similarity and complexity indexes were generated. The Similarity Index Matrix (SIM) measures how similar two spectra are to each other. A similarity index of "0" indicates a Unique fluorochrome signature, and a value of "1" indicates virtually no difference between the fluorochromes. The chart displays the numerical value for each pair of fluorochromes identified for use in the panel

8. Centrifuge 300 × *g* for 5 min, aspirate supernatant and disrupt the pellet.
9. Resuspend in 1 ml RPMI with 5% FBS (or more, depending on how many cells) to obtain a proper cell concentration for cell counting.
10. Count cells using a hemocytometer.

3.7 Hemocytometer Cell Counting

1. 10 μl 4% Trypan Blue Dye
2. 10 μl of media containing cells diluted in 1 ml RPMI.

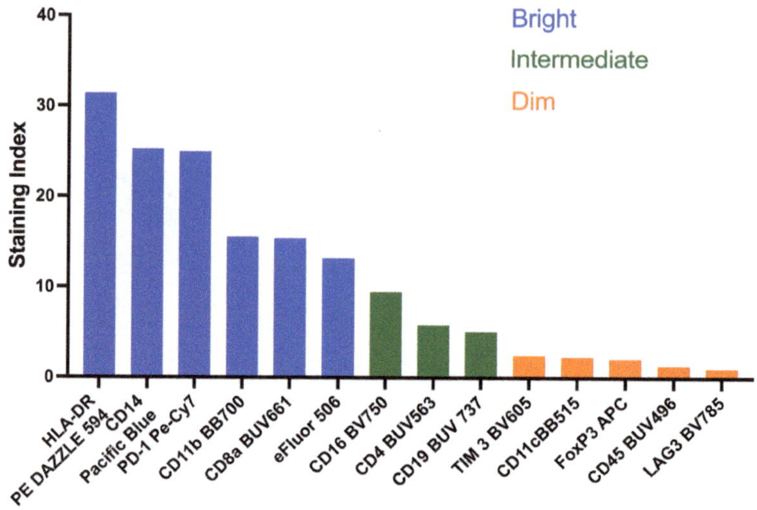

Fig. 4 Fluorochrome brightness charts assign fluorochromes to three broad categories to allow some generalization for relative brightness (e.g., BV785 at the lower end of the scale or PE Dazzle 594 dyes being at the upper end of the scale)

3. Place the hemocytometer under the microscope and count the cells.
4. Formula to count cells.

$$\text{Total Number of cells} = \text{Avg number of cells counted in all four large square chambers of hemocytometer} \times \text{Volume RPMI} \times 2 \times 10^{\wedge}4$$

3.8 Antibody Titration

Antibody titration can be performed in a 96-well U-bottom plate or 12 × 5 ml polystyrene FACS tubes. We have shown titration in a 96-well U-bottom plate to ease handling multiple antibodies.

3.8.1 Surface Staining

1. Obtain PBMCs from above.
2. Prepare **plate #1** for surface staining: make a cell concentration. of 2×10^6/ml as stock concentration, as we will need to pipette approximately 0.1×10^6/ml cells into each well of a 96-well plate. One column of the plate will correspond to one antibody titration. Prepare one well as your unstained control. See below:
3. Spin plate at 300 × *g* for 5 min at 4°C. Flick supernatant.
4. Prepare another 96-well U bottom plate **#2** for surface staining: Add 200 µl of cell staining buffer in wells A1–A11 of a 96-well plate, and add 100 µl of staining buffer to the rest of the wells B1–11, C1–C11, D1–D11, and so on.

5. Add twice the amount of antibody recommended by the vendor to well A1–A11. For example, if the technical detail sheet of the antibody recommends using 0.25 μg/ml or 5 μl of Ab, then add twice this amount as well, i.e., 0.5 μg/ml or 10 μl.
6. Mix the well using a multichannel pipette and transfer 100 μ l of the solution from the antibody-containing well A1–A11 into well B1–B11. Mix ten times. Then, 100 μl from wells B1–B11 is transferred into wells C1–C11, and so on. From well A-12, remove 100 μl and discard—these are your unstained cells. Alternatively, unstained cells can be prepared in a single tube to add one more column of antibody titration.
7. Using a P200 multichannel pipette, transfer the diluted antibody to the wells containing cells (**Plate #1**) and mix it four to five times.
8. Incubate this plate at 4°C for 15 min in the dark.
9. After incubation, 100 μl of staining buffer was added to each well-containing cell.
10. Centrifuge at 300 × *g* for 5 min at 4°C. Flick the supernatant.
11. Repeat Steps 6 and 7.
12. Resuspend cells in 200 μl of staining buffer and store at 4°C until the acquisition.

3.8.2 Intracellular Staining

Preparing cells in plate 1 for intracellular staining:

1. Determine cell concentration and prepare 2×10^6/ml as stock concentration.
2. Pipet approximately ~200 μl~ from the stock (approx. 0.2×10^6 cells) into each well of a 96-well plate
   a. One antibody titration per column
      i. For example, one antibody fills the well A1–H1 with cells + unstained in A12.
3. Add 100 μl Fix/Perm Buffer. Mix with multichannel pipette four to five times.
4. Incubate for 20 min at 4°C.
5. Centrifuge it at 300 × *g* for 5 min at 4°C. Discard supernatant.
6. Add 100 μl Perm/Wash Buffer. Mix with multichannel pipette four to five times.
7. Centrifuge it at 300 × *g* for 5 min at 4°C. Discard supernatant.

3.8.3 Titration Dilution in Plate 2 for Intracellular Staining

Take another 96-well plate. Make serial dilutions of your antibody in the Staining Buffer as follows:

1. Add 100 μl of Perm Wash Buffer to well A1 and 50 μl to the rest of the well B1–H1. Each column is a different antibody.

2. Add twice the amount of antibody recommended by the vendor to well A1. For example, if the technical detail sheet of the antibody recommends using 0.25 μg/ml or 5 μl of Ab, then twice the amount is added to the well A1.
3. Adjust a P200 pipette to 50 μl, and using this pipette, mix ten times the cell staining buffer containing antibody in well A1 and transfer this to well B1, mix ten times, and repeat it until you reach well H1. Discard the 50 μl cell staining buffer remaining in the pipette tip.
4. Using a P200 multichannel pipette, transfer the diluted 50 μl antibody to the wells containing cells (Plate 1) and mix it four to five times.
5. Incubate this plate at 4°C for 20 min.
6. After incubation, 100 μl of Perm/Wash was added to each well-containing cell, and the centrifuge was placed at 300 × *g* for 5 min at 4°C.
7. Discard the supernatant and resuspend cells in 200 μl of FACS buffer.
8. Cells are ready to be acquired on the Flow Cytometer.

Stained cells can be acquired using a plate reader per the defined procedure. The preferred method of plate orientation would be from unstained, followed by increasing antibody volume/concentration.

3.8.4 Titration Data Analysis

Optimal concentrations of all antibodies were determined through titration experiments. Twofold dilutions of antibodies were tested. As explained above. Files were concatenated for analysis using FlowJo™ v10.8 Software (BD Life Sciences). An antibody dose with the brightest staining index should be used on the multicolor panel.

Staining index calculation:

1. Separation of peaks between the positive and negative population divided by two times the standard deviation of the negative population.

   Stain Index = (MFI pos − MFI neg)/2× Standard Deviation of negative population.
2. Relative brightness and background shifting will provide information on the coupling between targets with fluorochromes.
3. Excess antibodies will bind at low affinity and create a background that reduces resolution. This may result in a false negative.
4. Through the staining index, we can find the optimal amount of antibody that will generate specific staining with the least amount of background.

3.9 15-Color Human Immunophenotyping Panel

1. Prepare cryopreserved PBMCs as described in Sect. 3.6.
2. Add fresh staining buffer and adjust cell suspension to a concentration of $1\text{–}2 \times 10^6$ cells/ml in ice-cold PBS, 10% FCS. Use a 40 μm mesh to remove clumps and obtain single-cell suspension. Cells are usually stained in polystyrene $12 \times 75\ \text{mm}^2$ Falcon tubes, microcentrifuge tubes, and 96-well round-bottomed microtiter plates.
3. Centrifuge cells at $300 \times g$ for 5 min at 4°C. Gently Disrupt the pellet after centrifugation.
4. Add live/dead aqua stain (manufacturer's recommended amount or titrated dose) along with FcR blocker (manufacturer's recommended amount) to the cells.
5. Incubate for 10 min at room temperature.
6. Add the titrated amount of fluorochrome-conjugated antibodies to the cells.
7. Incubate cells at 4°C for 15 min.
8. Add 1 ml of staining buffer (100 μl staining buffer in plate) and wash the cells twice ($300 \times g$ for 5 min).
9. Add 100 μl Fix/Perm Buffer to the cells. Mix four to five times. Incubate at 4°C for 20 min. Add 500 μl of 1× perm wash buffer (100 staining buffer in plate) and centrifuge ($300 \times g$ for 5 min). Discard the supernatant and break the pellet gently.
10. Add 100 μl Perm/Wash Buffer containing the intracellular antibodies (Ki-67 and FoxP3) to the cells.
11. Incubate for 20 min at 4°C.
12. Wash cells twice with perm wash buffer and once with staining buffer by centrifuging at $300 \times g$ for 5 min at 4°C. Discard supernatant and resuspend cells in 300–500 μl of cell staining buffer.
13. Cells should be stored at 4°C until acquired. Cells should be acquired immediately or within 24 h of staining.

3.10 Data Collection on Flow Cytometer

1. Specific methods depend on the available make and model of the flow cytometer. Here, we have used Cytek® Aurora (Cytek Biosciences, Fremont, CA) equipped with four lasers (355, 405, 488, and 640 nm) and 64 detectors using SpectroFlo® software as CytekAssaySetting.
2. Single-color controls for each fluorochrome are prepared using UltraComp eBeads™ Compensation Beads (cat no# 01-2222-41, Thermo Fisher Scientific and unstained cells are kept as a negative control). Note- Single-color controls can be prepared using cells or compensation beads as long as the bright signal is observed in each respective detector PMT.

3. Open SpectroFlo® software on the instrument controller and look for the green light appearing on the bottom right corner of the software to ensure proper instrument connection. Once the instrument is connected, select Create a New Experiment and add fluorochromes under the Fluorescent Tags tab.
4. Select Tabs Groups, define carrier type—Tube or plate, Markers—the name of markers associated with each fluorochrome, Keyword—any special details to the fluorochrome, and Under acquisition, define your gate to be used for data collection and a number of cells to be acquired. Users can select a Raw worksheet or a predefined unmixed worksheet for experiment groups. Single-color controls will be acquired by default on the Raw worksheet.
5. It is essential to establish forward (FSC-A) and side scatter (SSC-A) gates appropriately to exclude debris and cellular aggregates from the analysis. We recommend quickly checking the multicolor stained tube to check the intensities of all the dyes on the raw worksheet and adjusting intensities by putting a negative number in all channels box in case intensities are out of scale (Important note- do not change individual channel values as it will change the signature of the fluorochrome). Save the acquisition setting voltages with a specific name or experiment name.
6. Record data for the unstained and single-color controls approximately 5000 events for beads or cells.
7. On finishing single-color tube acquisition. Click on Unmix and follow the steps.
8. Adjust the gates on the histogram for the positive and negative population for each single-color control and select unstained as the negative control. Check that the correct detector is chosen for each fluorescent tag.
9. Click Live Unmix at the bottom right of the window.
10. A new Unmixed sheet will pop up. Gates should be created on this new unmixed worksheet. Save this sheet with a proper experiment name.
11. Click the edit button at the top of the experiment browser window and select the newly saved unmixed worksheet in front of the group where experimental tubes are collected. Click save and start acquiring samples.
12. Use menu commands or plot buttons to create dot plots and contour plots on the worksheet.
13. Acquire cells at an event rate of 3000–5000 events per second and record data.
14. Use a series of plots and establish a gating hierarchy to ensure the recording of debris and doublet-free live T cells (at least

250,000 single live T cells to ensure a sufficient number of rare subsets to analyze in the recorded data file, if needed).

15. The FCS files can be located under the C drive CytekBio export folder post-acquisition. Files can be analyzed on SpectroFlo® software or any third-party software.

3.11 Data Analysis

1. Open a new workspace on Flow-Jo software and add samples. fcs files in separate group folders.
2. Open the .fcs files and start creating gates as follows.
3. Define the gated populations as shown in Fig. 5.
4. Cleaning gates: gate on Time on the X-axis and forward scatter area (FSC-A) on Y- the Y-axis. A rectangular gate on uniformly acquired cells is used. Avoid any gaps, as they may be due to clumps or clogs during the run. This gating helps discard any irregularity during cell acquisition due to clumps or clogs.
5. Singlet gating: select gate, right-click, and drill down the bivariant polychromatic plot with FSC-A on the X-axis and FSC-H on the Y-axis for preselected lymphocytes. Make a gate around singlets. If the area scaling for the FSC detector and Blue LASER is properly adjusted while data recording, then singlets will be visible on a diagonal line when FSC-H and FSC-A parameters are plotted as the X and Y axes of the bivariate plots. Doublets will fall toward FSC-A but not FSC-H and can be gated out to identify single cells.
6. Live cell gating: Live cells were identified by gating on the cells negative for fixable live-dead dye against SSC-A.
7. Gated CD3 T cells, CD19 B cells, and double-negative monocyte, macrophage, and NK cell subsets were analyzed.
8. T cells were analyzed for CD4 and CD8.
9. CD4 cells were further analyzed for Treg cells using the FOXP3 intranuclear transcription factor.
10. CD4, CD8, and Treg cells were gated on CD8 and CD4 vs. PD-1, LAG-3, and TIM-3 exhaustion markers.
11. The CD3 negative and CD19 negative subsets were further gated on CD14 and CD16 to identify CD14 pos CD16 neg Classical monocytes, CD16 pos CD14 negative non-classical monocytes, and CD14 pos CD16 pos intermediate monocytes.
12. Two more subsets were gated out from the CD3 negative and CD19 negative subsets—these are Lin-DR+ and CD56 pos HLA-DR neg.
13. Lin-DR+ cells were analyzed for CD11c-positive dendritic cells (DCs).

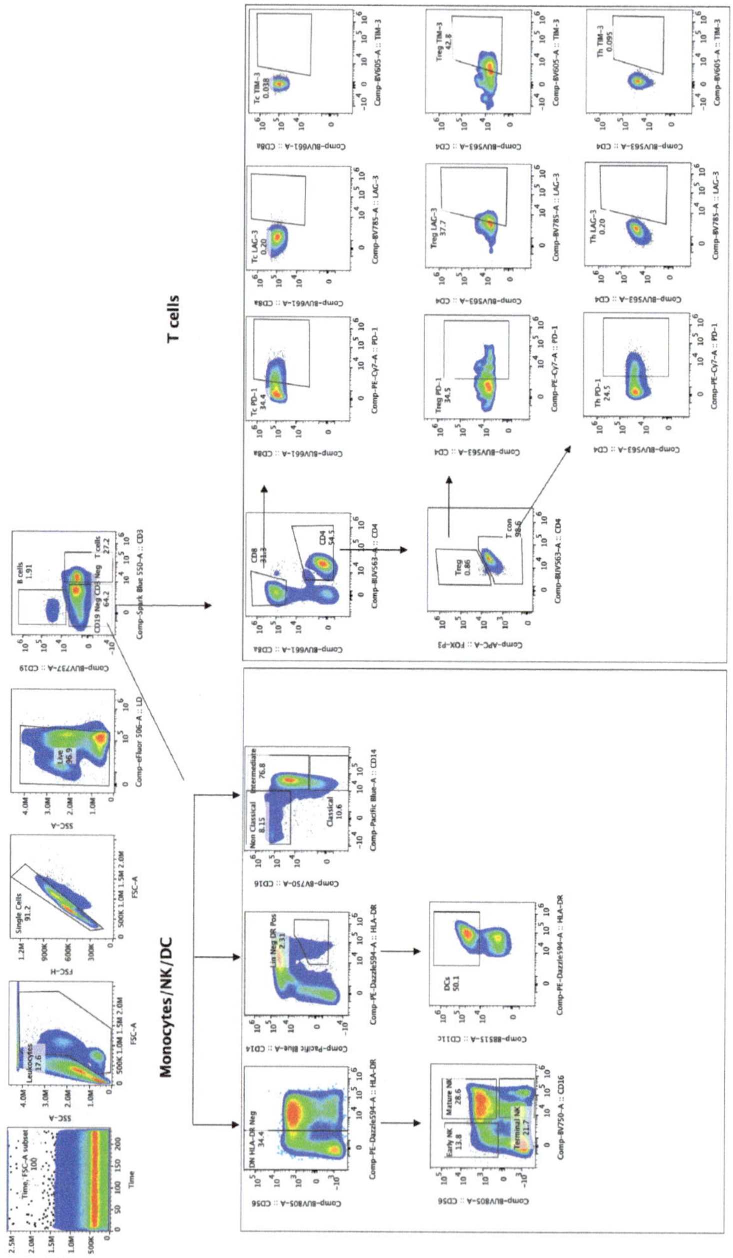

Fig. 5 Spectral cytometer 15 color panel Bivariant plots showing gating strategy. Bivariant plots the different cellular subsets characterized by manual gating. The main markers that showed different levels of expression across the various subsets of that population are presented for each population, showing T, B, NK, DCs, Tregs, and various myeloid cell subsets

14. CD56 pos HLA-DR neg from above were gated for CD56 vs CD16 and were defined as CD56 neg CD16 pos Terminal NK cells, CD56 pos CD16 neg Early NK cells, and CD56 pos CD16 pos Mature NK cells.

4 Notes

1. Spectral cytometry captures the emission of fluorochromes across the entire spectrum, allowing for the complete emission spectrum to be recorded and unmixed.
2. To assess the uniqueness of spectral signatures, use the similarity matrix tool. A matrix value greater than 0.9 suggests that the dyes are too similar and should not be used together in the same panel.
3. Utilize the complexity index to estimate potential spreading errors. A lower index indicates a higher probability of obtaining high-resolution data. Pair bright fluorophores with low-abundance markers and dim fluorophores with high-abundance markers to minimize spreading effects.
4. Ensure that antibody titration is conducted using the final, optimized staining protocol, which should include the same buffers, incubation times, temperatures, and fixation conditions.
5. Single-stain controls are essential. For the best results, use the same cell type as your sample for these controls to ensure matched autofluorescence. In cases where antigen abundance or signal-to-noise is low, leading to unclear separation from the background, consider using beads for single-color controls.
6. Always include an unstained cells as control. This allows the SpectroFlo software to identify and extract the intrinsic autofluorescence (AF) signature of your cells, enabling you to treat AF as an additional parameter and subtract it.
7. Employ a fixable viability dye to exclude dead cells, which can exhibit unique, high autofluorescence that interferes with the unmixing process.
8. Before proceeding to analysis, create an $N \times N$ plot (where all markers are plotted against each other) in SpectroFlo to identify any spilling of signal from one channel into another.
9. Finally, use the autofluorescence extraction feature in SpectroFlo to separate the autofluorescence spectrum from the fluorescent signal, which can significantly enhance the resolution of weak markers.

10. The use of azide in FACS buffer should be avoided, especially if cells are to be sorted and later used for functional studies. Azide inhibits various metabolic functions in cells, which could impede the downstream functional studies of sorted cells. Additionally, unfixed cells left in an azide buffer for an extended period may experience increased cell death compared to normal conditions.
11. Prolonged incubation (greater than 5 min) with ACK lysis buffer should also be avoided, as it can significantly impact lymphocyte viability. Supplementing the medium with 20 μg/ml DNase is beneficial for dissociating cell aggregates resulting from DNA released by dead cells. This is particularly helpful when working with cryopreserved cells, as the freeze–thaw process may cause cell lysis.
12. For cell staining, performing the procedure at 4°C and in the dark is recommended, as it prevents antibody capping and photobleaching. If there are any concerns regarding the pathogenicity of the cells involved, it is advisable to fix the cells after staining and before analyzing them with a flow cytometer. However, this should not be done if live cell sorting is intended.
13. After incubation, the fixative solution should be removed by washing the cells with FACS buffer. Prolonged fixation may alter the properties of fluorochromes and could lead to spillover in other channels.

References

1. Nolan JP, Condello D (2013) Spectral flow cytometry. Curr Protoc Cytom 63():1.27.1–1.27.13. https://doi.org/10.1002/0471142956.cy0127s63
2. Robinson JP, Ostafe R, Iyengar SN, Rajwa B, Fischer R (2023) Flow cytometry: the next revolution. Cells 12(14):1875. https://doi.org/10.3390/cells12141875
3. Park LM, Lannigan J, Jaimes MC (2020) OMIP-069: forty-color full spectrum flow cytometry panel for deep immunophenotyping of major cell subsets in human peripheral blood. Cytometry A 97(10):1044–1051. https://doi.org/10.1002/cyto.a.24213
4. Lim JF, Berger H, Su IH (2016) Isolation and activation of murine lymphocytes. J Vis Exp 116, 54596. https://doi.org/10.3791/54596
5. Nazarpour R, Zabihi E, Alijanpour E, Abedian Z, Mehdizadeh H, Rahimi F (2012) Optimization of human peripheral blood mononuclear cells (PBMCs) cryopreservation. Int J Mol Cell Med 1(2):88–93

Part IV

Translational and Analytical

Check for updates

Chapter 20

Build Heatmaps of Cell Frequencies and Marker Intensities for Correct Interpretations

Eugénie Lohmann, Laurent Gorvel, and Samuel Granjeaud

Abstract

High-content cytometry is an important technique in clinical research, producing data that is rich and complex to analyze. The resulting clusters of cells must be profiled to determine their role and function and analyzed to identify changes in cell frequency or marker intensity. The heat map is the most suitable tool for presenting this information in a synthetic way. However, its construction must be adapted to the information and the objective. Here we describe the step-by-step construction of heat maps of marker intensity and cell frequency using Excel, Phantasus, or R, including the conversion of numbers to colors and the organization of rows and columns. Researchers who master these steps will be able to correctly interpret heat maps and derive the most benefit from high-content cytometry.

Keywords Cytometry, Heatmap, Frequency, Intensity, Method, Build, Phantasus, Heat map, Pheatmap

1 Introduction

The rise of microarray technology in the late 1990s increased the need for visualizing large gene expression datasets to apprehend this data. The article by Eisen et al. [1] established heatmaps as a standard tool for visualization, exploration, and analysis. It significantly influenced the field and popularized the use of heatmaps in all omics, as well as the development of software with graphical user interfaces [2, 3].

In cytometry analyses, heatmaps are used to present in a compact graph the frequency changes of cell clusters (or populations) in the biological samples under study. The aim is to identify frequency changes related to the study question or, less interestingly, artefactual effects. Heatmaps are also used to present the expression profile of the cell clusters measured on the markers selected for the study in order to annotate these clusters and associate them with a cell population name.

Angélique Biancotto and Pradeep K. Dagur (eds.), *Immunophenotyping*, Methods in Molecular Biology, vol. 3087, https://doi.org/10.1007/978-1-0716-5607-5_20,

A heatmap is the representation of a matrix of numbers in an image, using a mapping table that associates a color with a number. A heatmap, therefore, comprises rows and columns, and the objects they represent must be defined and known. For cytometry, rows and columns may be clusters, markers, or samples. The numbers within the matrix are the result of a measurement and its processing, which must be identified in order to make authorized comparisons and avoid misinterpretation. As the color scale is used for all values in the matrix, it is important to know whether comparisons should be restricted within rows or within columns, or not at all. There is no single magic processing that applies to all types of data. Processing addresses a specific objective and derives from the nature of the measurements and their acquisition. Transforming and scaling are generic terms that compose the processing flow. They require a clear definition and a specific order, which is presented here. Workflows allow to build heatmaps, but their results can be used to apply statistical methods (principal components analysis, etc.) and machine learning techniques in order to explore the wealth of cytometry data deeply.

Frequency analysis is based on a matrix of cell counts with rows identifying clusters and columns identifying samples. The aim of the frequency heatmap is to highlight relevant clusters whose frequency changes between groups of samples. The statistical analysis of frequencies is called differential abundance (DA) analysis [4] and can be carried out with a dedicated framework such as egdeR [5, 6] or DESeq2 [7].

Cluster profiling is based on a matrix of median fluorescence intensities (MFIs), with rows identifying clusters and columns identifying markers. The term MFI is generic for any flow, mass, or spectral cytometry. The aim of the MFI heatmap is to link clusters to known cell populations and complement algorithms such as the MEM method [8].

The MFI analysis of a specified marker is based on a matrix of the MFI values measured per cluster and per sample, with rows identifying clusters and columns identifying samples. In practice, single-marker heatmaps derived from such matrices are rarely presented. However, they can highlight relevant clusters whose MFI changes between groups of samples. More broadly, the aim is to analyze the assemblage of MFI matrices for all markers of interest. This statistical analysis, known as differential state (DS) analysis [4], leads to the construction of a meta-heatmap, which integrates data across markers by applying adequate scaling to unify dynamic ranges and proper normalization to highlight changes in marker intensity.

Being a kind of image, a heatmap should have a carefully chosen color scale to be properly viewed and interpreted. Because cell frequency and marker intensity are on a continuous scale, gradients are the best choice. A single gradient color scale is usually the best option to cover the dynamic range from low to high values. If the midpoint between the minimum and maximum has a specific meaning, a better option is a color scale composed of two diverging gradients, one gradient covering the upper part of the range and the other the lower part. They meet at the midpoint, where they share the same color. When presenting changes that are symmetric by essence, the midpoint is the "no change" point, and the top and bottom must be symmetric, even if the effective data range is not symmetrical. Choosing the colors for the color scale is the final step. There are many options, but it's important to use common sense (e.g., avoid placing red at the lower end of a blue–white–red scale) and to consider color-blindness.

A heatmap converts numbers into a color matrix but does not order the rows or columns. When the order is defined by the experimenter or an annotation (such as sample groups), visualization is "supervised." When the order results from a data-driven algorithm, visualization is "exploratory." Adding annotations to the rows or columns of the heatmap is useful for visualizing and identifying associations between these annotations and the heatmap data. In the frequency heatmap, samples can be organized into sample groups to facilitate identification of a cluster whose frequency changes are associated with sample groups. But the sample order can also result from a hierarchical clustering, and this natural organization of the data allows us to check whether it correlates with the sample groups. The aim of hierarchical clustering (and other clustering algorithms) is to bring together elements that are similar. Elements are the rows and columns of the heatmap, and similarity is based on a distance (also called a metric), giving rise to a wide range of options.

There are many approaches to building heatmaps from data organized in a matrix. We use Microsoft Excel, as it is widely known and allows each step of the workflows to be easily built and clearly understood. But it does not allow for advanced and necessary final steps, such as hierarchical clustering. These final steps are performed using Phantasus [3], a web server with a graphical user interface for transcriptomics data analysis. In addition, we provide the same processing workflows using R and the pheatmap package [9].

2 Materials

2.1 Software

Requirements are the following:

1. Microsoft Excel, or any spreadsheet software (adaptations are left to the reader).
2. Access to the Phantasus website https://artyomovlab.wustl.edu/phantasus/.
3. R and the pheatmap package for the alternate R version.

Any recent version of this software will allow you to perform the steps we describe.

2.2 Experiment Results

The results are taken from a study of Michlmayr et al. [10]. They performed a 37-plex mass cytometry acquisition of peripheral blood mononuclear cells (PBMCs) of acute- and convalescent-phase samples obtained from 43 children naturally infected with chikungunya virus. Their analysis classified cells into 57 sub-communities of canonical leukocyte phenotypes and revealed a monocyte-driven response to acute infection, with the greatest expansions in "intermediate" CD14++CD16+ monocytes and an activated subpopulation of CD14+ monocytes. The FCS files and the attachment files are available on FlowRepository under ID FR-FCM-Z238. We focused on the smallest subset of cell populations called "Assignment."

We extracted counts and MFIs from the FCS files, although the frequency matrix is available. The extraction and preparation scripts are available in the GitHub repository accompanying this chapter. We used a heatmap of clusters versus samples to show counts lower than ten cells, which brings a fast overview. We removed clusters with a majority of low counts, reducing the dataset from 30 to 27 clusters. The cell populations differentially abundant in this experiment can be used to verify the heatmaps we obtained and compare them to the original article.

2.3 Input Data Files and Format

Cluster profiling is based on a matrix of untransformed MFIs. Rows identify clusters and columns identify markers. The MFI term is generic for any flow, mass, or spectral cytometry. Untransformed refers to the MFIs are expressed in the acquisition scale. It is possible to start the workflow with a matrix of transformed MFIs (which corresponds to the second step), and this is a better solution, as the transformed MFIs benefit from the transformation tuning performed by the experimenter.

Frequency analysis is based on a matrix of cell counts. Rows identify clusters, and columns identify samples (corresponding to FCS files). It is possible to start the workflow with a matrix of frequencies (which corresponds to the second step). We define frequency as the number of cells in a cluster (or a cell population) to the total number of cells in the whole sample (after cleaning steps) or to a selected population (e.g., T cells). Frequency is also called abundance.

An experiment usually compares samples from different groups. This information, as well as preparation and acquisition dates, is important for presenting and interpreting the results and the heatmaps. A metadata file is typically required to link samples to all this information.

We proposed preordered matrices in order to simplify the reading without a hierarchical clustering. Ready-to-use files

Table 1
Table of available files to reproduce the steps of each workflow

Data file name	Description
counts_Assignment_matrix.xlsx	An Excel sheet to start building the heatmap of frequencies. Also used for DESeq2 analysis.
counts_Assignment_matrix-single_sheet.xlsx	Final Excel sheet of processed counts leading to a heatmap.
counts_Assignment_matrix-multi_sheet.xlsx	Final Excel sheet of processed counts leading to a heatmap, where steps are presented in different sheets.
counts_Assignment_matrix-for_phantasus.xlsx	Final Excel sheet of processed counts leading to a log2 relative frequencies matrix for Phantasus.
mfis_Assignment_matrix.xlsx	Excel sheet to start building the profiling heatmap.
mfis_Assignment_matrix-single_sheet.xlsx	Final Excel sheet of processed MFIs leading to a heatmap.
mfis_Assignment_matrix-for_phantasus.xlsx	Final Excel sheet of processed MFIs in a matrix for Phantasus.
mfis_samples_Assignment_matrix.xlsx	Excel sheet to start building the meta-heatmap of marker intensity per cluster and per sample.
counts_Assignment.RData mfis_Assignment.RData mfis_samples_Assignment.RData	Data files for direct reading in R.

are listed in Table 1 and available at https://i-cyto.github.io/build-heatmaps-imp2/.

3 Methods

Workflows are summarized in Fig. 1 and apply to all cytometry technologies. Cluster profiling is based on the MFIs extracted for each marker of interest. The MFI must be transformed if it is not already. The arcsinh transformation is easy to implement (even in Excel), but the bi-exponential transformation is too complex. Next, the transformed MFIs are normalized to the maximum intensity by marker. Frequency analysis is based on the cell counts measured for each cluster in each sample. The counts are normalized per sample to the total count of the sample, leading to frequencies. The average frequency of each cluster is not interesting in itself, but changes in

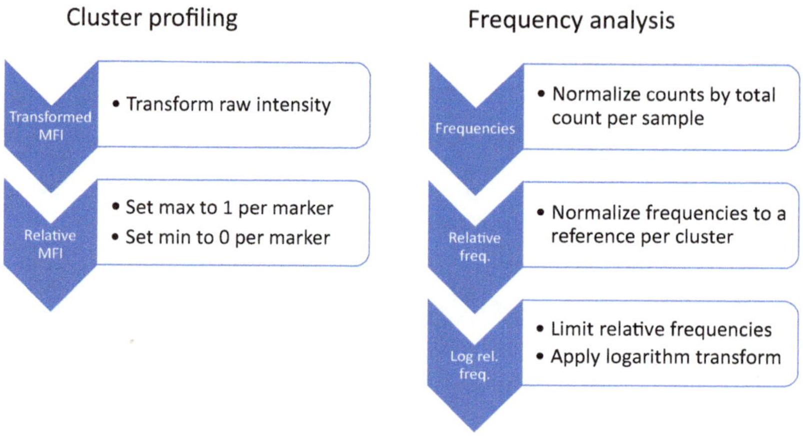

Fig. 1 Summary of the main steps of each workflow

frequencies within each cluster between samples are important. To capture these changes, the frequencies are normalized per cluster to a reference, usually the average frequency of the cluster. Finally, the logarithm transformation permits to express the increases and decreases in a symmetric way, but to avoid issues with zero frequencies (resulting from zero counts), a threshold can be applied beforehand. Other solutions exist to overcome the problem of the logarithmic transformation of zero (*see* **Note 1**).

Phantasus can carry out the complete workflow from raw MFIs to a heatmap, but with some cautions (*see* **Note 2**). It allows to perform statistical analyses (*see* **Note 3**). It can carry out the complete workflow from counts to heatmap, but with some cautions when some counts are zero (*see* **Note 4**). Adding a small value to counts is presented with Phantasus (for frequencies, *see* **Note 5**).

Here we report protocols built under Windows and an English version of Excel. *See* **Note 6** for Apple and Excel internationalization, and *see* **Note 7** for why the Excel file format is interesting.

3.1 Building the Heatmap of Cluster MFI Using Excel and Phantasus

3.1.1 Preparing

Make a copy of the MFI matrix Excel file and give it a name. Then open it.

Enlarge the clusters column. Reduce the width of the marker columns to fit them all into the screen.

3.1.2 Transforming MFI

The MFI are not transformed, which could be identified by looking at the minimum and maximum of all MFIs. Transformed MFIs are usually below ten.

Select and copy the entire region (Ctrl-* Ctrl-C). Go to A31, 3 lines below the copied region, and paste (Ctrl-V). This region will hold the transformed MFIs. Go to B32, the top left MFI, and enter the formula "`=ASINH(B2/5)`". This will transform all the MFIs with the asinh transformation and a co-factor of 5, which is adapted for mass cytometry. Copy the formula to the right up to the last marker and down to the last cluster.

3.1.3 Scaling MFI

We will compute the minimum and maximum of each marker (column). Go to A60 and enter "Maximum". Go to B60, enter the formula "`=MAX(B32:B58)`" and copy to the right up to the last marker. Go to A61 and enter "Minimum". Go to B61, enter the formula "`=MIN(B32:B58)`" and copy to the right.

Go to A31, select, and copy the entire region (Ctrl-* Ctrl-C). Go to A64, 3 lines below the min and max rows and paste. This region will hold the scaled MFIs. Go to B65, the top left MFI, and enter the formula "`=(B32-B$61)/(B$60-B$61)`". This will scale all the transformed MFIs per marker. Copy the formula to the right up to the last marker and down to the last cluster.

3.1.4 Coloring and Adjusting

Go to B65 and select the scaled MFIs region. In the "Conditional Formatting" menu, in the "Color Scales" sub-menu, select the "red–white–blue" color scale (red being the highest).

To hide the numbers in the cells, select the colored region, then click on the right button of the mouse to format the cells. In the cell format, select "Custom" and in the "Type" of the format, enter "`;;;`".

The final heatmap is presented in Fig. 2A. As the clusters are already annotated and organized in major cell populations, the heatmap permits to verify that the cluster annotations match the color profiles and that the workflow creates an informative and useful visualization.

We can modify the minima and maxima to adjust the coloring to the expected color. For example, the minimum of CD45 is nearly as high as the maximum because all cells were selected to be CD45+. However, the current coloring suggests that low CD45 clusters exist. The minimum intensity should be adjusted to zero, as zero is the expected minimum for CD45. This calibration ensures all clusters are rendered CD45+. A similar adjustment should be performed with maxima of lowly expressed markers, such as PD1, which appears too high in many clusters.

3.1.5 Exporting from Excel to Phantasus

Phantasus can import Excel files. *See* **Note 7** for reasons why using XLSX (Excel files) is a better format than CSV (Comma Separated Values files). We are going to copy the scaled MFI region to a new Excel file. As the colored heatmap region does not show values, Phantasus cannot find values, and we have to set up a region with the scaled values.

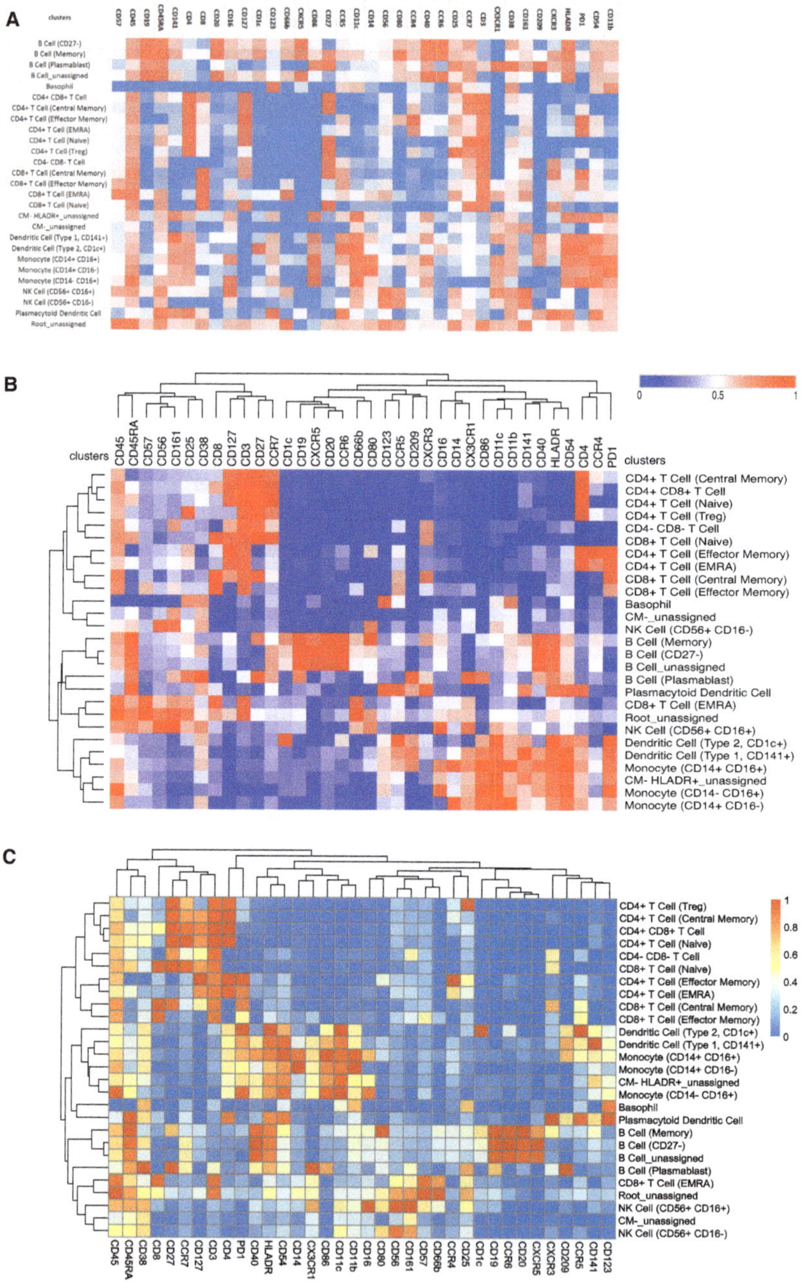

Fig. 2 Final heatmaps of transformed MFIs to profile the clusters. (**A**), Heatmap from Excel. (**B**), Heatmap from Excel and Phantasus with clustering of rows and columns to ease the interpretation by patterns. (**C**), Heatmap from pheatmap with clustering

Go to A64, and select the region of the heatmap (Ctrl-* Ctrl-C). Go to A94, 3 lines below the heatmap region, and paste. This region will hold the scaled MFI values. Go to B95, the first MFI value, and enter the formula "=B65". Copy it to the right and to the bottom of the region. This formula copies the scaled values. Remove the customized format and set it to "Number". Values are now visible inside the colored cells. Select the region (Ctrl-*), copy it (Ctrl-V), start a new Excel file, and paste only values (an example file is available).

Go to the Phantasus website, then select "My computer" in the "Choose a file" popup menu. If Phantasus is already open in your web browser, click on "File/Open" to import a file. Select the prepared Excel file on your computer. Phantasus presents the annotations and the numerical data matrix of the file. Row annotations are highlighted in light green and are on the left of the sheet. Column annotations are highlighted in light red and are at the top of the sheet. The data are highlighted in light blue. Click the first cell containing data if the MFIs region is not correctly identified. It is possible to transpose the matrix (i.e., swap rows and columns). Here we consider clusters are on the rows and markers are on the columns. Click "OK."

To tune the display to the effective scale of the data, open the wheel called "Options" on the top-right. Untick "Relative color scheme" that is using the minimum and maximum on each row to adjust the color scale per row. Verify that "Transform values" is set to "None" (more about on-the-fly transformations in **Note 8**), "Minimum" to "0," and "Maximum" to "1" (or to "−0.1" and "1.1" to lower the contrast of the color scale). This setting leads to minor changes in the display but it is the only way to show the scaled MFIs correctly. Before closing the window, it is possible to store this color scheme for reuse. Use the "Save color scheme" button to specify a name for this color scheme. It will be available in the "Saved color scheme" popup menu; select it and click "Load color scheme" to apply it.

3.1.6 Clustering with Phantasus

Clustering the scaled MFIs really eases the reading and interpretation of the heatmap in order to annotate. The ordering of clusters and markers will only be based on the scaled MFI values. To apply a clustering, click the "Tools" menu, then move the cursor down to "Clustering," then select "Hierarchical clustering" in the submenu. In the popup window, select "Euclidean" as "Metric" and "Complete" as "Linkage method" (this linkage makes clusters more homogeneous). Select the combination of rows and/or columns to which these settings will be applied to. Click "Close" to apply. The final heatmap is presented in Fig. 2B (*see* **Note 9** for exporting).

3.2 Building the Heatmap of Cluster Frequency Using Excel and Phantasus

3.2.1 Preparing

Make a copy of the count matrix Excel file and give it a name. Then open it.

Format the cluster names column, the sample names row, and the width of the count columns.

3.2.2 Computing Frequencies

Go to A30, two lines below the bottom of the count region. This line will hold the total counts. In A30, enter "`Total`". In B30, enter the formula "=SUM(B2:B28)" which computes the sum of counts for the first sample. Drag the formula to the right up to the last sample to copy it.

Go to A01, select all counts, including sample names and cluster names by typing keys Ctrl-*, then copy the region with Ctrl-C.

Go to A33, five lines after the end of the counts region, and paste by pressing Ctrl-V. This region will hold the frequency of clusters.

Go to B34, the top-left value of the pasted region. In B34, enter the formula "=B2/B$30", which divides each count by the total count of its sample. Copy the formula to the right up to the last sample and to the bottom up to the last cluster.

3.2.3 Computing Relative Frequencies

Go to CK33, two columns after the last sample of the frequency region. A new column will hold the average frequency of each cluster. Enter "Ave frequency". Go to the cell below at CK34. Enter the formula "=AVERAGE(B34:CI34)". Copy the formula down to the last cluster.

Go to A33, select all counts including sample names and cluster names, and copy the region using Ctrl-*, then Ctrl-C.

Go to A63, three lines below the frequencies region, and paste. This region will hold the relative frequencies. Go to B64. Enter the formula "=B34/$CK34" which divides the frequency by the average frequency of its cluster. Copy the formula to the right up to the last sample and to the bottom up to the last cluster.

3.2.4 Computing Logarithm of Limited Relative Frequencies

Go to A93, three lines below the relative frequency region. Enter "Threshold". Go to B93, and enter its value, i.e., the number "4". This value can be tuned to adjust the visual rendering.

Go to A63, select all relative frequencies, including sample names and cluster names, and copy the region (Ctrl-* Ctrl-C).

Go to A95, two lines below the threshold, and paste. The new region will hold the logarithm of threshold relative frequencies. In B95, enter the formula "=LOG(IF(B64>$B$93; $B$93; IF(B64< 1/$B$93; 1/$B$93; B64)))/LOG(2)". The external part of the formula computes the logarithm in base 2. The internal part of the first logarithm function aims to limit values higher than the threshold or lower than 1/threshold. Copy the formula to the right up to the last sample and to the bottom up to the last cluster.

3.2.5 Coloring

Go to A95, select all counts including sample names and cluster names, and copy the region (Ctrl-* Ctrl-C).

Go to A125, 3 lines below the ready-to-use values. The new region will hold the colorized version of these values. First, paste the copied region, then paste the values only. This will keep the formatting and the values, but not the formula.

In the "Conditional Formatting" menu, in the "Color Scales" sub-menu, select the "red–white–blue" color scale (red being the highest).

To hide the numbers in the cells, select the colored region, then click on the right button of the mouse to format the cells. In the cell format, select "Custom" and in the "Type" of the format, enter "`;;;`". The final heatmap is presented in Fig. 3A.

3.2.6 Exporting from Excel to Phantasus

Phantasus can import Excel files. *See* **Note 7** for reasons why using XLSX (Excel files) is a better format than CSV (Comma Separated Values files). Start a new Excel file and copy the log2 limited relative frequencies region with values (an example file is available). The colored heatmap region is not recognized as data by Phantasus.

Annotations to samples could be added as rows below the first row that contains the sample names. In the available example, we added rows for the group and date given in FCS file names.

Open the website, then select "My computer" in the "Choose a file" pop menu. Select the prepared Excel file on your computer. Phantasus presents the annotations and the numerical data matrix of the file. Row annotations are highlighted in light green and are on the left of the sheet. Column annotations are highlighted in light red and are at the top of the sheet. The data are highlighted in light blue. Click the first cell containing data if there are annotations. It is possible to transpose the matrix (i.e., swap rows and columns). Here, we consider clusters are on the rows and samples are on the columns. Click "OK."

To tune the display to the effective scale of the data, open the wheel called "Options" on the top-right. Untick "Relative color scheme" that is using the minimum and maximum on each row to adjust the color scale per row. Verify that "Transform values" is set to "None" (more about on-the-fly transformations in **Note 8**). Set the "Minimum" to "−2" and the "Maximum" to "+2". Before closing the window, it is possible to store this color scheme for reuse. Use the "Save color scheme" button to specify a name for this color scheme. It will be available in the "Saved color scheme" popup menu; select it and click "Load color scheme" to apply it.

3.2.7 Clustering with Phantasus

To apply a clustering, click the "Tools" menu, then move the cursor down to "Clustering," then select "Hierarchical clustering" in the sub-menu. In the popup window, select "Euclidean" as "Metric" and "Complete" as "Linkage method" (this linkage makes clusters more homogeneous). Select the combination of rows and/or columns to which these settings will be applied to. Click "Close" to apply.

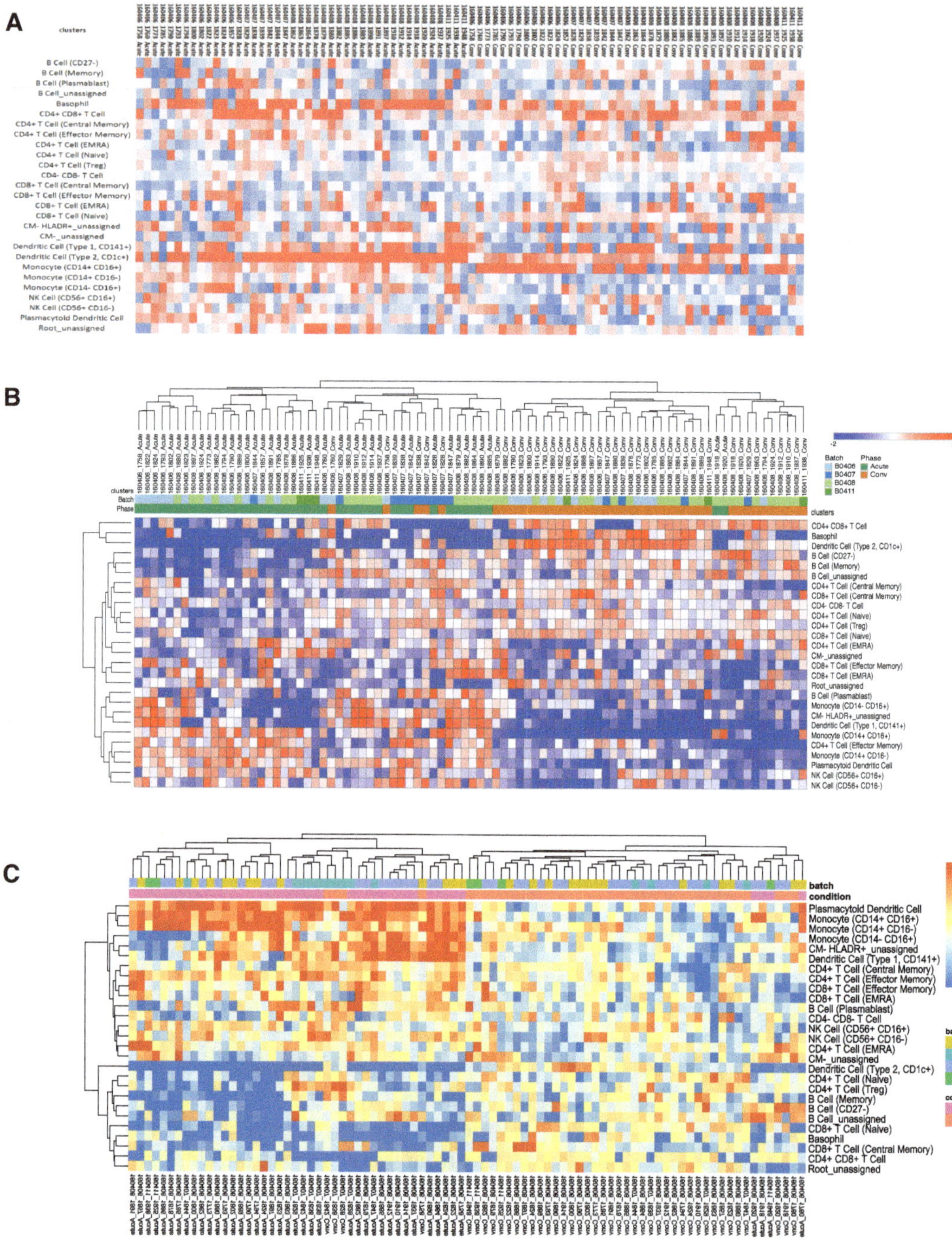

Fig. 3 Final heatmaps of relative frequencies in log scale. (**A**), Heatmap from Excel. (**B**), Heatmap from Excel and Phantasus with sample annotations and clustering of rows and columns. (**C**), Heatmap from pheatmap with sample annotations and clustering of rows and columns. Frequencies are relative to the "Conv" sample, making it easier to interpret changes and their significance and to assess the residual dispersion of the "Conv" sample group

The final heatmap is presented in Fig. 3B (*see* **Note 9** for exporting).

3.3 Building the Heatmaps with R and Pheatmap Package

We provide RMarkdown scripts and their corresponding HTML results to present the steps in the R programming language. The codes are accessible with a little knowledge of the R language. The aim is to clarify the steps and permit to build a reusable and reproducible workflow to create heatmaps. The pheatmap [9] function is particularly pleasant to create heatmaps and does not carry out on-the-fly transformation by default. It permits to define the limits of the color scale, which avoids the need of thresholding data. It also performs a hierarchical clustering by default without taking into account the limits applied for the color scale. In this section, we reduce the codes to the core formulas (*see* **Note 10** for loading prepared data files).

3.3.1 Heatmap of Scaled MFIs

Once the MFIs matrix is loaded (clusters on rows, markers on columns), the workflow is performed with the following code:

```
mfis = asinh(untransformed_mfis / 5) # cofactor for
mass cytometry
mfis_max = apply(mfis, 2, max) # max per column, 2nd
dimension
mfis_to_max = sweep(mfis, 2, mfis_max, "/") #
normalize per column
pheatmap(mfis_to_max)
```

To define a minimum per marker and to limit the color scale between 0 and 1, the code is below:

```
mfis_min = apply(mfis, 2, min)
mfis_to_min_max = sweep(mfis, 2, mfis_min, "-")
mfis_max = apply(mfis_to_min_max, 2, max)
mfis_to_min_max = sweep(mfis_to_min_max, 2, mfis_max,
"/")
pheatmap(mfis_to_min_max, breaks = seq(0, 1, length.
out = 101))
```

The final heatmap is presented in Fig. 2C.

3.3.2 Heatmap of Relative Frequencies

Once the count matrix is loaded (clusters on rows, samples on columns), the workflow is performed with the following code:

```
# build the frequency matrix
freq_raw = sweep(counts, 2, colSums(counts), "/")
# normalize: for each cluster, divide by its mean
frequency
freq_rel = sweep(freq_raw, 1, rowMeans(freq_raw),
"/")
# threshold relative frequency and convert to log2
thr = 4
```

```
freq_rel_lim = freq_rel # make a copy
freq_rel_lim[ freq_rel_lim > thr] = thr # replace
high values
freq_rel_lim[ freq_rel_lim < 1/thr] = 1/thr #
replace low values
log2_freq_rel_lim = log2(freq_rel_lim)
# heatmap
pheatmap(log2_freq_rel_lim, breaks = seq(-2, +2,
length.out = 101))
```

We performed a hard thresholding, although we can specify color scale limits to pheatmap, which gets rid of the logarithm of zero problem. However, pheatmap needs finite data in order to automatically determine the data range to match the color scale or to calculate clustering. Limiting the data range by hard thresholding removes the 0 from frequencies and also limits the impact of large differences on clustering. A few alternative solutions exist, but no consensus has been established to our knowledge (*see* **Note 1**).

In the differential analyses, one point is often overlooked: the reference group. When scaling frequencies to obtain relative frequencies, the average frequency of each cluster over all the samples is chosen. An interesting alternative is to select a group of samples in order to obtain a direct reading of changes with respect to this group and to evaluate the dispersion of this reference. Such a heatmap is presented in Fig. 3C and could be compared with Fig. 3A, B.

3.4 Building the Cluster Frequency Heatmap and Statistical Analysis with Phantasus

3.4.1 Importing Counts

Open the Phantasus website, then select "My computer" in the "Choose a file" popup menu. Select the Excel file of counts on your computer. Phantasus presents the annotations and the numerical data matrix of the file. Row annotations are highlighted in light green and are on the left of the sheet. Column annotations are highlighted in light red and are at the top of the sheet. The data are highlighted in light blue. Click the top left cell containing a count. The clusters must be on the rows and samples on the columns. Click "OK."

3.4.2 Performing the Statistical Analysis

Click "Tools/Differential expression/DESeq2" to perform the DESeq2 analysis via the menu. Enter "acute" in "Target level" and select all 43 entries. Enter "conv" in "Reference level" and select all 43 entries. When clicking "OK," the results appear on the right for each cluster, with log2FoldChange and padj (p-value adjusted for multi-testing) being the most important information for the interpretation. Don't rely on the colors in the heatmap (Fig. 4) as counts are normalized per row to the minimum and maximum.

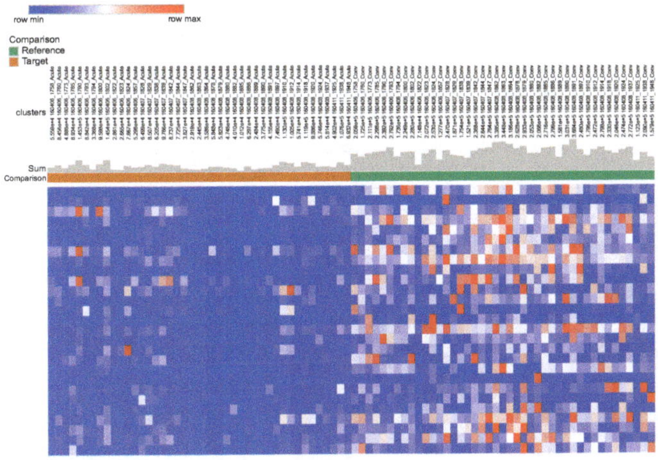

clusters	baseMean	log2FoldChange	lfcSE	stat	pvalue	padj
Dendritic Cell (Type 2, CD1c+)	293.89	-5.2424	0.53501	-10.049	9.2478e-24	2.4969e-22
Monocyte (CD14+ CD16+)	2406.6	2.3584	0.24141	9.9590	2.3035e-23	3.1097e-22
Plasmacytoid Dendritic Cell	796.77	1.2830	0.15890	8.1865	2.6885e-16	2.4197e-15
CD8+ T Cell (Naive)	9781.1	-1.1323	0.15389	-7.4752	7.7114e-14	5.2052e-13
B Cell (Memory)	2311.1	-1.1663	0.17495	-6.8319	8.3804e-12	4.5254e-11
CD4+ T Cell (Naive)	10650	-0.85401	0.15393	-5.6435	1.6661e-8	7.4975e-8
Monocyte (CD14+ CD16-)	21777	0.80838	0.15353	5.2825	1.2742e-7	4.9149e-7
CD4+ T Cell (Treg)	2339.7	-0.50020	0.10177	-4.9614	7.0003e-7	0.0000023626
CD8+ T Cell (Central Memory)	1523.0	-0.78457	0.16437	-4.8707	0.0000011122	0.0000033366
CD4+ T Cell (Effector Memory)	994.18	0.87258	0.20531	4.3534	0.000013402	0.000036184
CM- HLADR+_unassigned	1103.2	0.95454	0.22981	4.3104	0.000016298	0.000040003
B Cell (CD27-)	5365.8	-0.77397	0.19495	-4.1272	0.000036720	0.000080335
Dendritic Cell (Type 1, CD141+)	190.14	1.1788	0.30348	4.1152	0.000038680	0.000080335
Basophil	183.85	-1.6202	0.66812	-3.0100	0.0026125	0.0050383
CD4- CD8- T Cell	6925.0	-0.32579	0.11332	-2.9639	0.0030381	0.0054686
Root_unassigned	6025.5	-0.65441	0.26316	-2.6378	0.0083442	0.014081
Monocyte (CD14- CD16+)	788.71	0.56475	0.24994	2.3480	0.018875	0.029978
NK Cell (CD56+ CD16-)	2291.6	0.39601	0.19549	2.0405	0.041304	0.061955
CD4+ CD8+ T Cell	10670	-0.75474	0.59374	-1.6525	0.098432	0.13988
CD8+ T Cell (Effector Memory)	285.53	0.30372	0.21084	1.4761	0.13992	0.18890
CD4+ T Cell (EMRA)	263.02	0.24828	0.19439	1.3080	0.19086	0.23833
B Cell (Plasmablast)	528.30	0.26814	0.21422	1.2773	0.20148	0.23833
CD8+ T Cell (EMRA)	1454.8	0.28634	0.23136	1.2730	0.20302	0.23833
B Cell_unassigned	7046.2	-0.18003	0.18310	-0.97374	0.33019	0.37146
NK Cell (CD56+ CD16+)	7484.5	0.10346	0.17984	0.61991	0.53532	0.57815
CD4+ T Cell (Central Memory)	7168.6	-0.025370	0.14755	-0.18758	0.85121	0.88395
CM-_unassigned	2545.5	0.0054017	0.18953	-0.014649	0.98831	0.98831

Fig. 4 DESeq2 analysis with Phantasus. Counts were imported and directly subjected to a DESeq2 analysis via the menu "Tools/Differential expression/DESeq2." Clusters (identified cell populations) are sorted by increasing adjusted p-value. The colors in the heatmap are normalized per cluster (row) to the minimum and maximum by default. The coloring is biased by the fact that there are more cells in the "Conv" samples than in the "Acute" ones. The statistical analysis clearly takes this bias into account, as the "Monocyte (CD14+ CD16+)" cluster is statistically significant but not detectable in the heatmap. The sum of cells per sample (column) was added using "Tools/Created Calculated annotation" and displayed with four digits and as a bar chart

3.4.3 Computing the Relative Frequencies Heatmap

To avoid zero counts, we add 1 to all the counts after calculating the total counts. Next, we calculate frequencies and their logarithms and subtract a reference mean, which is the same as calculating the relative frequencies and then their logarithm. *See* **Note 5** to add a small value to frequency rather than counts. First, we calculate the total counts before adding 1. Select "Tools/Create Calculated Annotation," tick "Columns," click "Sum" in the "Operation" drop-down menu, and "OK." Then, we add 1 to the counts and calculate the frequencies. Select "Tools/Adjust," tick "One plus log 2," "Inverse log 2," and in the "Sweep" section, select "Divide" each "column" by field "Sum," and click "OK." Next, we calculate the logarithm. Select "Tools/Adjust," tick "Log 2" and click "OK." Now, the values are log2 frequencies. To define the "Conv" sample group as a reference for relative frequencies, select all its columns and then select "Tools/Create Calculated Annotation," tick "Rows," click "Mean" in the "Operation" drop-down menu, tick "Use selected rows and columns only," and click "OK." Then, select "Tools/Adjust," and in the "Sweep" section, select "Subtract" from each "row" field "Mean," and click "OK." To tune the display to the effective scale of the data, we select "View/Options," untick "Relative color scheme." Verify that "Transform values" is set to "None" (*see* **Note 8**), set the "Minimum" to "−3" and the "Maximum" to "+3" and click "Close."

3.4.4 Clustering with Phantasus

To apply a clustering, select "Tools/Clustering/Hierarchical clustering." In the popup window, select "Euclidean" as "Metric,"

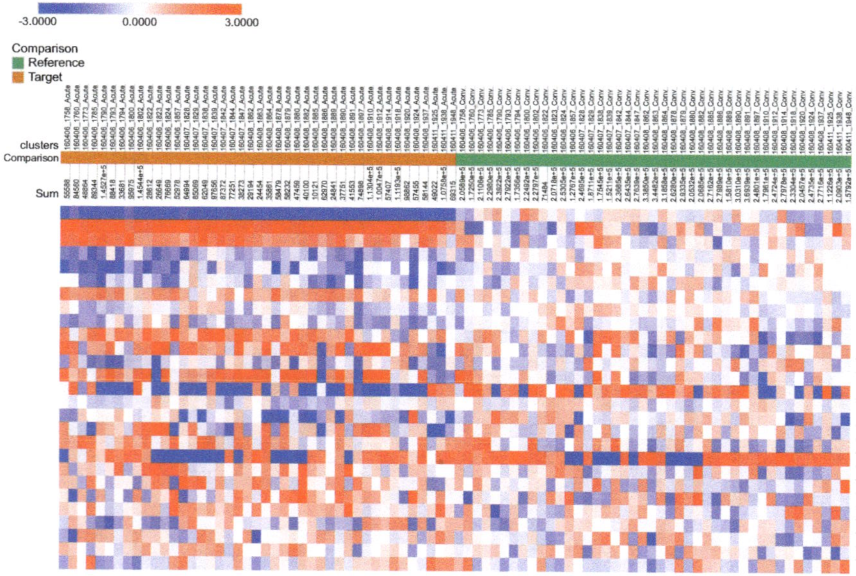

clusters	baseMean	log2FoldChange	lfcSE	stat	pvalue	padj	Mean
Dendritic Cell (Type 2, CD1c+)	293.89	-5.2424	0.53501	-10.049	9.2478e-24	2.4969e-22	-7.8731
Monocyte (CD14+ CD16+)	2406.6	2.3584	0.24141	9.9590	2.3035e-23	3.1097e-22	-7.9053
Plasmacytoid Dendritic Cell	796.77	1.2830	0.15890	8.1865	2.6885e-16	2.4197e-15	-8.1778
CD8+ T Cell (Naive)	9781.1	-1.1323	0.15389	-7.4752	7.7114e-14	5.2052e-13	-3.2631
B Cell (Memory)	2311.1	-1.1663	0.17495	-6.8319	8.3804e-12	4.5254e-11	-5.3314
CD4+ T Cell (Naive)	10650	-0.85401	0.15393	-5.6435	1.6661e-8	7.4975e-8	-3.2492
Monocyte (CD14+ CD16-)	21777	0.80838	0.15353	5.2825	1.2742e-7	4.9149e-7	-3.0295
CD4+ T Cell (Treg)	2339.7	-0.50020	0.10177	-4.9614	7.0003e-7	0.0000023626	-5.4884
CD8+ T Cell (Central Memory)	1523.0	-0.78457	0.16437	-4.8707	0.0000011122	0.0000033366	-6.1178
CD4+ T Cell (Effector Memory)	994.18	0.87258	0.20531	4.3534	0.000013402	0.000036184	-7.7745
CM- HLADR+_unassigned	1103.2	0.95454	0.22981	4.3104	0.000016298	0.000040003	-7.6375
B Cell (CD27-)	5365.8	-0.77397	0.19495	-4.1272	0.000036720	0.000080335	-4.3965
Dendritic Cell (Type 1, CD141+)	190.14	1.1788	0.30348	4.1152	0.000038680	0.000080335	-10.575
Basophil	183.85	-1.6202	0.66812	-3.0100	0.0026125	0.0050383	-10.530
CD4- CD8- T Cell	6925.0	-0.32579	0.11332	-2.9639	0.0030381	0.0054686	-4.0084
Root_unassigned	6025.5	-0.65441	0.26316	-2.6378	0.0083442	0.014081	-4.3652
Monocyte (CD14- CD16+)	788.71	0.56475	0.24994	2.3480	0.018875	0.029978	-7.7402
NK Cell (CD56+ CD16-)	2291.6	0.39601	0.19549	2.0405	0.041304	0.061955	-6.2420
CD4+ CD8+ T Cell	10670	-0.75474	0.59374	-1.6525	0.098432	0.13988	-4.9589
CD8+ T Cell (Effector Memory)	285.53	0.30372	0.21084	1.4761	0.13992	0.18890	-9.2277
B Cell (Plasmablast)	528.30	0.26814	0.21422	1.2773	0.20148	0.23833	-8.2937
CD8+ T Cell (EMRA)	1454.8	0.28634	0.23136	1.2730	0.20302	0.23833	-6.9068
CD4+ T Cell (EMRA)	263.02	0.24828	0.19439	1.3080	0.19086	0.23833	-9.2780
B Cell_unassigned	7046.2	-0.18003	0.18310	-0.97374	0.33019	0.37146	-4.1288
NK Cell (CD56+ CD16+)	7484.5	0.10346	0.17984	0.61991	0.53532	0.57815	-4.2297
CD4+ T Cell (Central Memory)	7168.6	-0.025370	0.14755	-0.18758	0.85121	0.88395	-4.2298
CM-_unassigned	2545.5	0.0054017	0.18953	-0.014649	0.98831	0.98831	-5.7811

Fig. 5 Final heatmap of frequencies relative to the "Conv" sample group along with the DESeq2 statistical analysis. As frequencies are relative to the "Conv" sample group, it is easier to interpret the changes and their strength by additionally relying on statistics

"Complete" as "Linkage method," "Rows" as "Cluster," and click "Close" to apply. We can also sort the clusters by adjusted p-values by clicking on the padj column heading. The final heatmap is presented in Fig. 5 (*see* **Note 9** for exporting).

3.5 Interpreting a Heatmap

A heatmap is a visualization tool for exploring and interpreting data. To read a heat map and draw correct conclusions, we need to understand which object is in the row and which is in the column, to know how the color scale corresponds to the numbers, and to understand why and how the numbers have been processed.

The heatmap of transformed and scaled MFIs of markers per cluster aims to link clusters to cell populations. Markers are scaled independently to their minimum and maximum. The minimum must be defined (automatically or manually) with care, as its value misleads the reading of positively selected cells such as CD45+ cells. The maximum must also be carefully defined, as its value may overestimate the intensity of markers that are weakly expressed either natively or as a result of negative selection (as in figure 3 by Liechti et al. [11]). These two limits need to be particularly scrutinized by analyzing the colors in the heatmap for each marker and comparing them with our knowledge. Applying a clustering algorithm to rows and columns is crucial to facilitate the annotation of clusters, and its parameters should also be evaluated.

The heatmap of log2 relative frequencies of clusters across samples aims to identify or point out changes between groups of samples, which is directly related to the study question. Changes are usually viewed and expressed as logarithm fold changes (logFC) and should be visualized with a symmetrical and limited color scale. When a group of samples is the reference in the experiment, it should be used to compute the relative frequencies instead of using all the groups. The heatmap then provides a direct reading of changes with respect to this group and of its dispersion. When the changes of relative frequencies are normalized to the standard deviation of the reference group (or the averaged within-group deviations), the unit of the color scale is no longer the logFC but a logFC relative to the dispersion and must be interpreted cluster by cluster. So, even if the color scale is presented with numbers, knowing the heatmap building is important to correctly interpret its units. Applying a clustering algorithm to clusters makes it easier to group them together based on similar changes over samples. Applying a clustering algorithm to samples makes it possible to verify the homogeneity of the groups studied or to detect subgroups.

We have presented the step-by-step construction of heatmaps used in cytometry. These workflows provide a clear understanding and enable the importance of steps and parameters to be assessed. However, R remains a more reproducible approach. A heatmap is not intended to perform statistical analyses but to complement them.

Statistical analysis of counts (aka DA) requires a dedicated framework [5–7], in which raw counts are not transformed (Fig. 4). A heatmap is a versatile visualization tool that covers many uses and can reveal or confirm interesting results of cytometry data, but understanding its building is mandatory.

4 Notes

1. We present a hard threshold of the relative frequency (to the chosen reference) before calculating the logarithm. An alternative well-known method is to add a prior of 1 to all the counts after calculating the total count per sample. Frequencies are then computed (with a denominator equal to the total count plus two times the prior in the empirical Bayesian framework). As long as the prior is small compared with the cell count the scientist considers convincing, and the total counts are much bigger, this method is a good approximation. Another method consists of adding a small value to the frequencies. In both methods, there is no rule for determining the additive value. Alternatively, all relative frequencies equal to zero can be replaced by the smallest non-zero value before applying the logarithmic transformation. There are probably other methods and transformations (square root, arcsine of the square root) that could be evaluated.
2. Phantasus can be used to carry out the whole workflow to obtain scaled transformed MFIs, but with some cautions. Transforming raw MFI is usually performed using a bi-exponential or arcsinh function. The log2(x+1) transformation available in Phantasus may look similar for mass cytometry intensities that are typically lower than a few thousand, but we cannot control the width of the linear region before moving to the log region. This should be avoided, and the safe solution is to start with transformed intensities. To scale markers (columns), click on "Tools/Create Calculated Annotation," then click "Columns" and select "Max" as "Operation." Columns can then be normalized to their own maximum. Select "Tools/Adjust," then at the bottom of the window in the "Sweep" section, select "Divide" each "Column" by field "Max." Phantasus proposes a heatmap ready for interpretation, bearing in mind that no fine-tuning of markers minima and maxima could be achieved, which means that maxima may be too high.
3. Phantasus allows to perform statistical analyses. DESeq2 should be performed on the raw counts, without any adjustment, even if the heatmap appears suboptimal. See Fig. 4.

4. When transforming values in the log2 scale, log2(0) is -Infinite by definition, but Phantasus chose to return 0 considering no value in the matrix would be lower than 1 as the matrix would represent counts. The log2 transformation of values between 0 and 1 will be negative and lower than the log2 result of 0, whereas they are higher than 0. For example, Phantasus will return log2(0.125) = log2(1/8) = log2(1/2^3) = -3, which is lower than the log2 returned value for 0, whereas 0.125 is higher than 0. As a result, any value below 1 becomes negative after transformation, even though it was originally greater than 0. This is a major problem when the values are frequencies (i.e., between 0 and 1 or between 0 and 100%). If there is no zero count, using Phantasus poses no problem for obtaining a correct heatmap. See Sect. 3.4 and **Note 5**.
5. Starting from the raw counts, we normalize these counts to a fixed total count for each sample and calculate the logarithm plus 1. Select "Tools/Adjust," tick "Scale column sum," enter a target total count, here "30000" (because many samples have a low cell count of the order of tens of thousands and only 1 is added to this target count), tick "One plus log2," and click "OK." Now, the values are log2 frequencies with a small value of 1/30,000 added. Next, we can follow the workflow applied to the counts and define the reference to obtain log2 relative frequencies (Fig. 6).
6. The shortcuts are valid for Excel under Windows. For Apple computer, the Ctrl key is replaced by the Apple key. As Excel translates the formula to your region, the words SUM, AVERAGE, and IF must be replaced with the corresponding words in your language. The list separator in formulas should also be replaced by "," instead of ";".
7. Excel files are interesting because the numbers are encoded in a format that does not depend on location. The decimal separator is displayed when numbers are displayed and adapted to localization, but their internal encoding is unique and portable. The same applies to date formats, but it is advisable to use the YYYY-MM-DD format, which is an unambiguous international standard.
8. Phantasus (and other software) transforms the data on-the-fly to derive a value that lies within the same range for the entire dataset. This leads to two datasets, the imported dataset itself (and possibly processed using "Tools/Adjust") and the dataset resulting from the on-the-fly transformation for coloring. Nevertheless, Phantasus still presents the actual data in the tooltip when the mouse is moved over the heatmap. This duality can be misleading, and we prefer to avoid it in order to visualize the result of a transformation on the data.

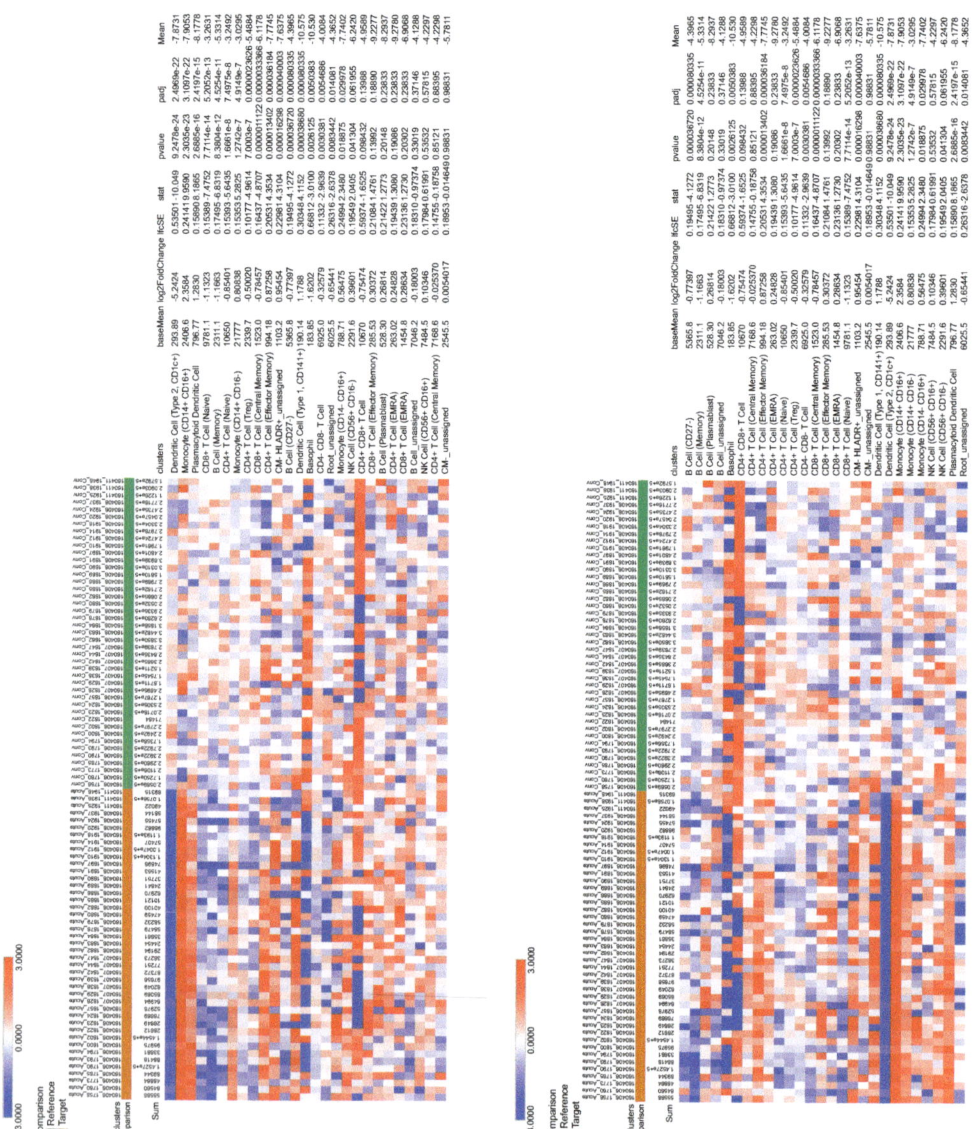

Fig. 6 Comparison of count processing: count + 1 workflow vs freq +1/30,000 workflow

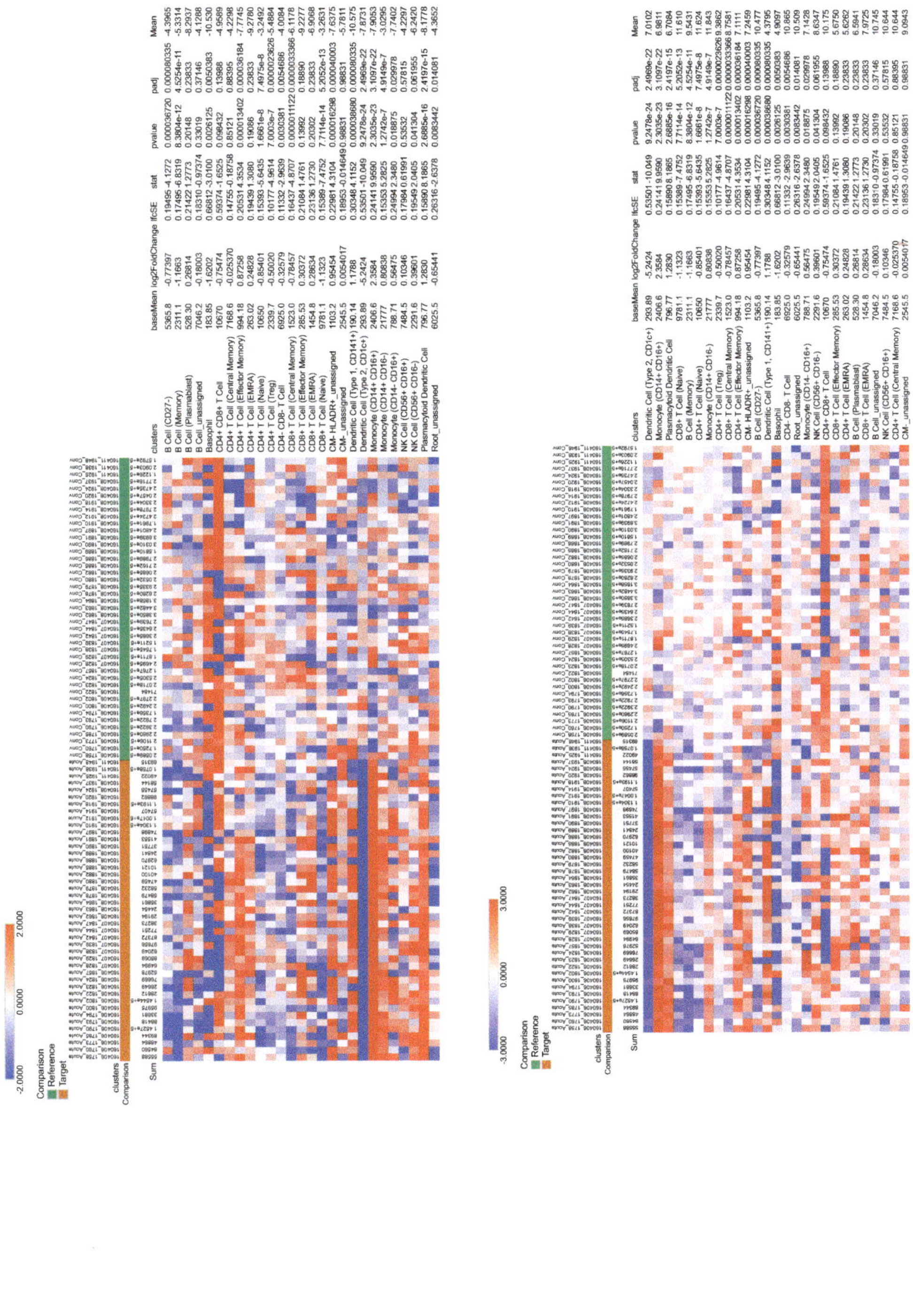

Fig. 6 (continued)

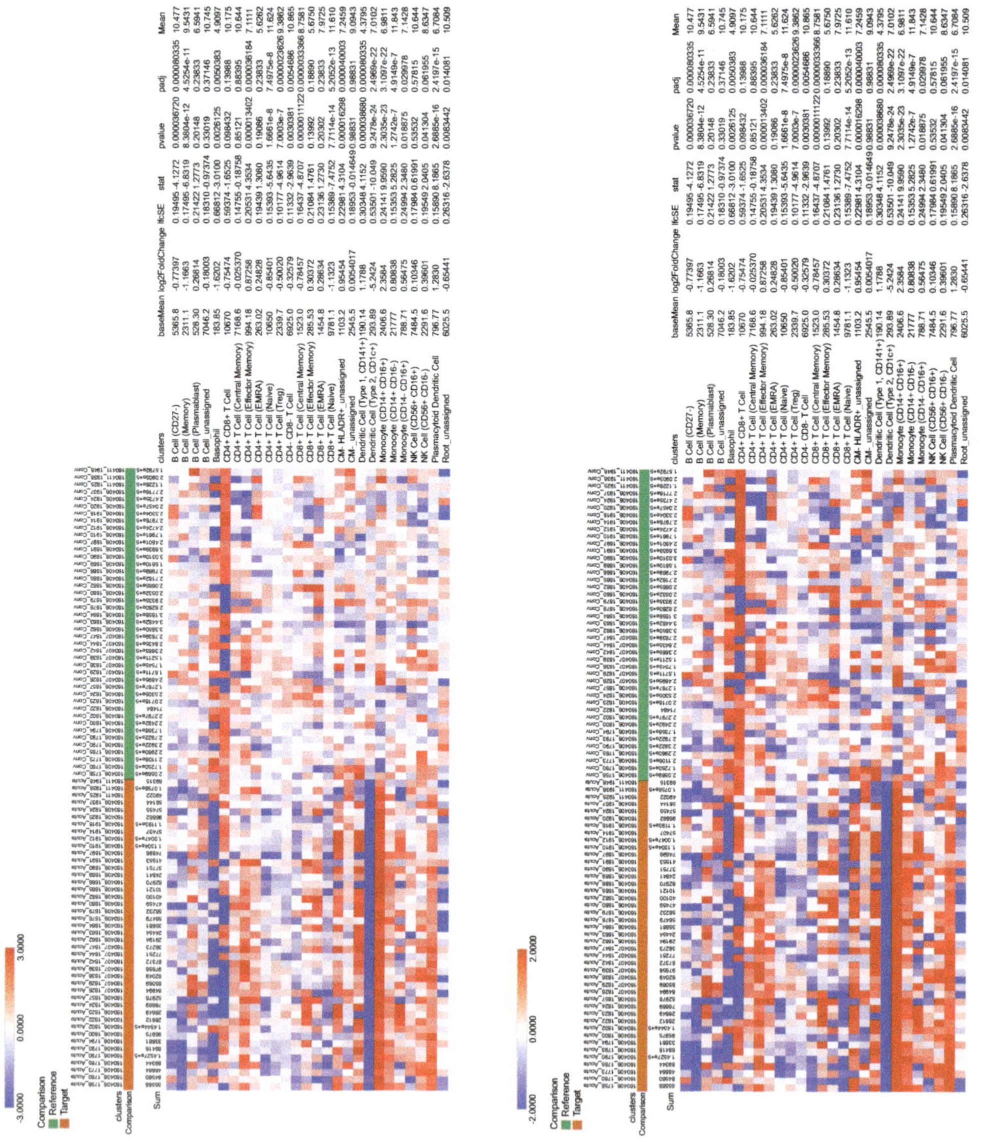

Fig. 6 (continued)

Phantasus offers on-the-fly transformations that apply to rows (typical of Omics analyses), which implies that the dataset is adequately prepared (transposition is available during the import). By default, the "Relative color scheme" is selected, using the minimum and maximum of each row to adjust the color scale per row. By unticking this option, two transformations are possible. "Subtract row mean, divide by row standard deviation" is the classical statistical standardization. "Subtract row median, divide by row median absolute deviation" is the alternative standardization robust to outliers. These two transformations could be interesting when we first discover the dataset, but scaling and color scaling are then specific to each row.

9. To export a nice figure from Phantasus, click on the icon showing a landscape (or use the menu "File/Save image"), then select either PNG for a screenshot or SVG (or PDF) for a scalable and editable graph. The exported figure will comprise the color scale and the legend of annotations. The color scale takes into account an on-the-fly transformation. If the default row min/max scaling is activated, the limits of the color scale are "row min" and "row max" instead of numerical values when the transformation is "None." The figure can be improved by removing the grid and adjusting the column width.

10. To load the prepared R data files, use the following commands: `load("mfis_Assignment.RData"); load("counts_Assignment.RData")`. To import the counts matrix from an Excel file, use the following code that requires the installation of the readxl package beforehand. Use a similar code for MFIs. `tmp = as.data.frame( readxl::read_excel["counts_Assignment_matrix.xlsx"]); counts = as.matrix(tmp[,-1]); rownames (counts) = tmp[,1]`.

References

1. Eisen MB, Spellman PT, Brown PO, Botstein D (1998) Cluster analysis and display of genome-wide expression patterns. Proc Natl Acad Sci U S A 95:14863–14868. https://doi.org/10.1073/pnas.95.25.14863
2. Saeed AI, Sharov V, White J et al. (2003) TM4: a free, open-source system for microarray data management and analysis. Biotechniques 34:374–378. https://doi.org/10.2144/03342mt01
3. Kleverov M, Zenkova D, Kamenev V et al. (2024) Phantasus, a web application for visual and interactive gene expression analysis. eLife 13:e85722. https://doi.org/10.7554/eLife.85722
4. Weber LM, Nowicka M, Soneson C, Robinson MD (2019) diffcyt: differential discovery in high-dimensional cytometry via high-resolution clustering. Commun Biol 2:1–11. https://doi.org/10.1038/s42003-019-0415-5
5. Robinson MD, McCarthy DJ, Smyth GK (2010) edgeR: a bioconductor package for differential expression analysis of digital gene expression data. Bioinformatics 26:139–140. https://doi.org/10.1093/bioinformatics/btp616

6. Chen Y, Chen L, Lun ATL et al. (2025) edgeR v4: powerful differential analysis of sequencing data with expanded functionality and improved support for small counts and larger datasets. Nucleic Acids Res 53:gkaf018. https://doi.org/10.1093/nar/gkaf018
7. Love MI, Huber W, Anders S (2014) Moderated estimation of fold change and dispersion for RNA-seq data with DESeq2. Genome Biol 15: 550. https://doi.org/10.1186/s13059-014-0550-8
8. Diggins KE, Gandelman JS, Roe CE, Irish JM (2018) Generating quantitative cell identity labels with marker enrichment modeling (MEM). Curr Protoc Cytom 83:10.21.1–10.21.28. https://doi.org/10.1002/cpcy.34
9. Kolde R (2019) pheatmap: pretty heatmaps. https://cran.r-project.org/package=pheatmap
10. Michlmayr D, Pak TR, Rahman AH et al. (2018) Comprehensive innate immune profiling of chikungunya virus infection in pediatric cases. Mol Syst Biol 14:e7862. https://doi.org/10.15252/msb.20177862
11. Liechti T, Weber LM, Ashhurst TM et al. (2021) An updated guide for the perplexed: cytometry in the high-dimensional era. Nat Immunol 22:1190–1197. https://doi.org/10.1038/s41590-021-01006-z

Chapter 21

High-Dimensional Multimodal Data Integration for Immune-Driven Diseases: A Practical Guide

Emiko Desvaux and Amazigh Mokhtari

Abstract

Recent advances in multiomics technologies have revolutionized the study of immune-driven diseases by enabling high-dimensional, single-cell resolution analyses. This chapter provides a practical guide for integrating CyTOF (mass cytometry) and single-cell RNA sequencing (scRNA-seq) data to address key challenges in this field. The integration of these modalities allows for consistent and reproducible cell type annotation, the transfer of annotations to assist in characterizing difficult-to-identify populations, and the transcriptional characterization of rare and heterogeneous subpopulations. Using tools such as OMIQ and R, the chapter outlines workflows for preprocessing, normalization, and scaling of CyTOF data, as well as dimensionality reduction and clustering techniques. The integration process involves creating Seurat objects, identifying common features, and using anchor-based methods to link CyTOF and scRNA-seq datasets. The chapter also discusses the use of multimodal deep learning techniques for rare subpopulation detection and emphasizes the importance of reproducibility and standardization in multiomics integration. By leveraging these methodologies, researchers can gain deeper insights into cellular heterogeneity and function, ultimately enhancing the understanding of immune-driven diseases. The chapter concludes by addressing integration challenges and proposing future directions for improving model interpretability and capturing nonlinear molecular interactions.

Keywords Multiomics, Data integration, CyToF scRNA-seq, Immuno-driven diseases

1 Introduction

Recent advances in multiomic technologies offer unprecedented opportunities to dissect the complex molecular mechanism underpinnings autoimmunity disease at single-cell resolution. With these rapidly evolving technologies, researchers generate an exponential amount of multimodal single-cell datasets on large cohort studies. However, despite the promise of multiomics, the complexity of high-dimensional datasets, coupled with the diversity of regulatory layers (genomic, transcriptomic, proteomic, epigenomic, and metabolic), presents an immense analytical challenge. To address these, emerging computational and machine learning methods are being

Angélique Biancotto and Pradeep K. Dagur (eds.), *Immunophenotyping*, Methods in Molecular Biology, vol. 3087,
https://doi.org/10.1007/978-1-0716-5607-5_21,

deployed to uncover the intricate molecular networks underpinning cellular functions and interactions.

In this chapter, we provide a practical guide to integrating CyTOF (mass cytometry) and single-cell RNA sequencing (scRNA-seq) data. The practical use case presented below specifically focuses on three core challenges and their solutions:

- Consistent and reproducible cell type annotation, leveraging tools such as OMIQ for CyTOF analysis.
- Transferring annotations from CyTOF to single-cell data to assist in annotating difficult-to-characterize populations using RNA data alone.
- Integrating multimodal datasets to transcriptionally characterize rare and heterogeneous subpopulations.

We use OMIQ and R as the primary tools for these tasks and provide concrete examples and workflows to enable readers to replicate these techniques in their own research.

2 Current Landscape of MultiOmic Integration Involving CyTOF

The integration of CyTOF and single-cell datasets has garnered increasing attention in immunological research, enabling a more comprehensive understanding of cellular heterogeneity and function. Researchers have developed various computational methods to align and interpret these diverse data types, addressing challenges such as data heterogeneity, rare subpopulation identification, reproducibility, and standardization. Several recent studies highlight these approaches:

2.1 Integration Challenges and Frameworks

2.1.1 Heterogeneity in MultiOmic Data

Integrating CyTOF data with other single-cell omics modalities, such as transcriptomics and genomics, presents significant challenges due to the inherent differences in data structures, measurement techniques, and feature spaces. To address this, computational frameworks like CyCombine or scMerge2 have been developed. These approaches are modular frameworks designed to integrate paired single-cell multiomics data, effectively bridging the gap between disparate datasets and facilitating a unified analysis [1, 2].

2.1.2 Rare Subpopulation Identification

Detecting rare cell subpopulations is crucial for understanding complex biological processes and disease mechanisms. Traditional methods often struggle with this due to the vast dimensionality and sparsity of single-cell data. Recent advancements have introduced multimodal deep learning techniques that enhance the sensitivity and accuracy of rare subpopulation detection. For instance, a study demonstrated the integration of single-cell RNA sequencing (scRNA-Seq) and CyTOF data to identify and transcriptionally characterize a rare subpopulation of CD11c-positive B cells. This methodology allowed for high-resolution analysis of this subpopulation without the need for prior cell sorting, revealing its heterogeneity in a manner not previously achievable [3].

These approaches leverage the combined information from multiple omics layers, improving the resolution at which these rare populations can be identified.

2.2 Reproducibility and Standardization

Ensuring reproducibility and standardization in multiomics integration is paramount for the validation and comparison of scientific findings. Automated and reproducible cell identification methods have been developed to address this need. For instance, Immunopred is a pretrained cell identification model that allows for the automated identification of immune cells without requiring user-defined prior knowledge of expected cell types. This tool enhances reproducibility by providing a consistent framework for cell type identification across different studies [4].

3 Methods: Data Preprocessing and Integration Workflows

In this chapter we took advantage of a rich dataset presented in Mennilo et al. generated a rich multimodal dataset to evaluate the effects of vedolizumab (VDZ) on ulcerative colitis (UC) across immune, stromal, and epithelial compartments [5]. The dataset includes single-cell RNA sequencing (scRNA-seq), where they profiled 93,900 cells from colonic biopsies and 20,130 peripheral blood leukocytes (PBLs). Cellular indexing of transcriptomes and epitopes by sequencing (CITE-seq) analyzed 93,900 mucosal biopsy cells to integrate transcriptomic and proteomic data. They performed cytometry by time-of-flight (CyTOF) on 684,249 cells from colonic biopsies, including samples from Healthy Controls (HC) ($n=4$), UC patients treated with 5-asathioprine (5-ASA) ($n=4$), and UC patients treated with vedolizumab (VDZ). Mennilo et al. underlined the added value of single-cell and spatial multiomics in identifying cellular dynamics and predictors of therapeutic response. Interestingly, they demonstrated that VDZ responders exhibited robust epithelial crypt base signatures and reduced activated stromal and MNP subsets, whereas nonresponders had persistent fibroblast and MNP activation despite treatment. Given the depth and value of this dataset, we propose to reanalyze the CyTOF data and integrate it with the scRNA-seq data to gain further insights.

Before integration, rigorous preprocessing is essential for ensuring that data from CyTOF and single-cell RNA-seq are of high quality and compatible for downstream analyses. The steps include:

3.1 Preprocessing and Normalization

CyTOF Data.

3.1.1 Scaling (or Transformation)

Scale settings may require adjustment if the default heuristic settings are not optimal. While this is rare for mass cytometry data, it is relatively common in fluorescent cytometry. The key parameter for *Arcsinh* scales is the cofactor—raising it compresses the region around zero, while lowering it expands this region. For biexponential scales, the corresponding parameter is the width basis. Scaling settings greatly influence data analysis outcomes and reproducibility, particularly when applying machine learning techniques.

3.1.2 Gating Live CD45+ Cells: Sequential Filters are Applied from the Root (100% events) as Follow (Fig. 1A)

1. *Time vs. Event Length* to remove acquisition artifacts and anomalies.
2. *Time vs. Residual* to eliminate signal outliers.
3. *Time vs. Width* and *Time vs. Offset* to further refine event quality.
4. *After ensuring high-quality events, apply Ir193Di* vs. *Ir191Di* to identify and retain single cells, excluding doublets and aggregates.
5. *Finally, gate 195Pt_Cisplatin* vs. *CD45* to isolate live $CD45^+$ cells, removing dead cells and nonimmune populations. This stepwise approach ensures a clean, high-quality dataset for downstream analysis.

3.1.3 Subsampling

Subsampling may be required when ensuring an equal number of events per file. This task can be achieved by applying the previously gated Live CD45+ cells filter to reduce the dataset accordingly. In this dataset, out of *1,261,759 total events, 121,390 Live $CD45^+$ cells* were selected for downstream analysis.

3.1.4 Dimensionality Reduction

Dimensionality reduction techniques aim to capture the variance and overall structure of high-dimensional data while representing its heterogeneity in a more concise, typically two-dimensional space. Methods like **UMAP** effectively preserve both local and global relationships in just two dimensions, whereas **PCA** often requires more dimensions to retain a comparable amount of variance. Techniques such as **t-SNE** and its optimized variant **opt-SNE** are particularly effective for visualizing complex structures by emphasizing local similarities, though they may distort global relationships. By projecting high-dimensional populations into a lower-dimensional space, these methods help identify distinct groups while maintaining meaningful biological interpretation. Although some information is inevitably lost during this transformation, the impact is often not practically significant. Visualization through dimensionality reduction provides a direct reference to variance, aiding in the identification of

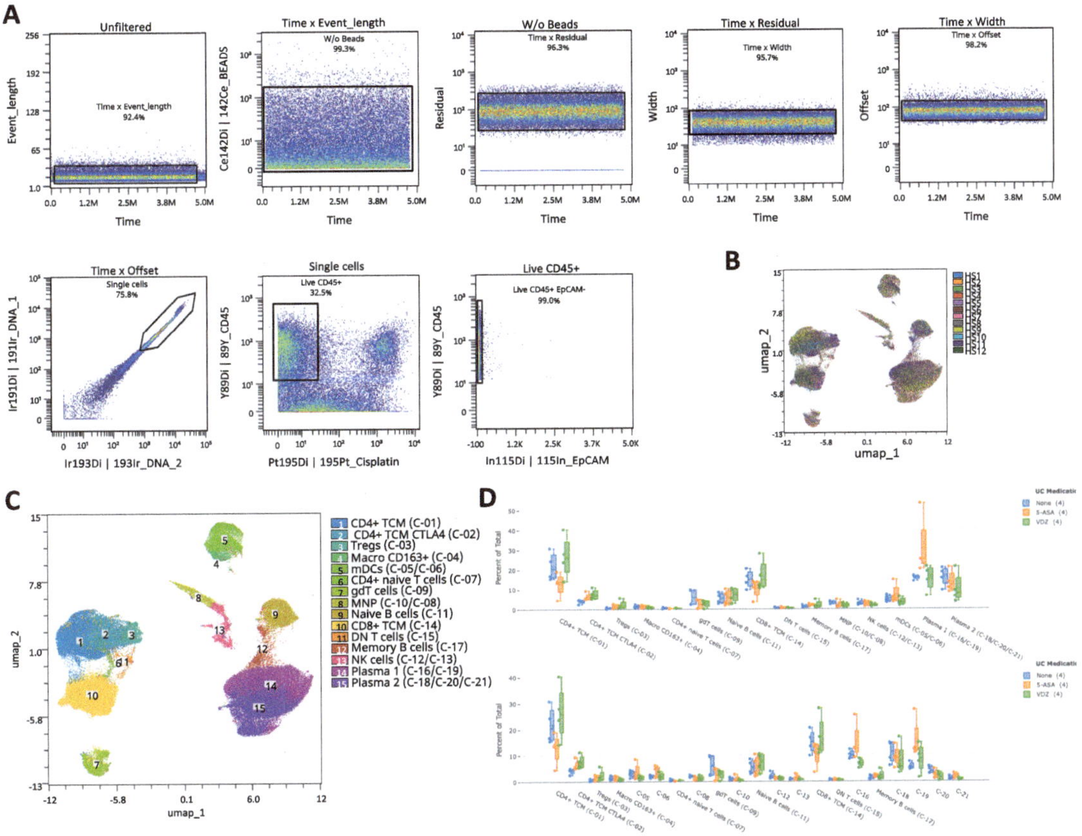

Fig. 1 CyTOF data processing and analysis workflow. (**A**) Gating strategy to isolate live $CD45^+$ cells: sequential gating filters applied from the root (100% of events) to remove acquisition artifacts, signal outliers, doublets, and dead cells, ensuring a high-quality dataset. (**B**) Clustering of single-cell data: FlowSOM clustering identifies 21 distinct immune cell clusters, with quality control measures applied to check for batch effects. (**C**) Cluster annotation: Identified clusters were consolidated into 15 immune cell populations using known phenotypic and canonical markers. (**D**) Cluster abundance representation: Immune cell populations were visualized and quantified according to clinical metadata

underlying patterns. The similarity between data points is computed using distance metrics, with Euclidean distance being the most used due to its efficiency and interpretability in high-dimensional spaces.

The following steps are suggested based on common analytical approaches. Given the high-dimensional nature of the data, we can explore expected hypotheses using a traditional supervised approach while also uncovering novel insights through unsupervised methods. Supervised analysis serves as an initial exploration layer, whereas unsupervised analysis provides an unbiased validation of findings, detects previously unaccounted populations (events that do not fit canonical classifications), and enables a more detailed investigation.

For this dataset, dimensionality reduction was performed using the following parameters:

1. *20 neighbors.*
2. *Minimum distance 0.3.*
3. *2 components.*
4. *Euclidean distance metric.*
5. *Learning rate 1.*
6. *300 Epochs.*
7. *Random seed 3107.*
8. *Spectral embedding initialization.*

3.1.5 Clustering

Clustering methods are essential for identifying distinct cell populations within high-dimensional cytometry data. **FlowSOM** is a widely used algorithm that employs self-organizing maps (SOMs) to create a structured representation of cellular phenotypes, allowing for rapid and efficient clustering while preserving hierarchical relationships. **SPADE** (Spanning-tree Progression Analysis of Density-normalized Events) constructs a minimum spanning tree to model cellular heterogeneity and identify rare populations by downsampling and density normalization. **Phenograph**, a graph-based clustering method, builds a k-nearest neighbor (k-NN) graph and applies the Louvain community detection algorithm to define clusters in an unbiased, data-driven manner. Each method has its strengths—FlowSOM excels in speed and visualization, SPADE is effective for rare population detection, and Phenograph provides robust and unsupervised clustering, making them valuable tools for single-cell analysis.

In this analysis workflow, FlowSOM was performed with the following settings:

1. *xdim 12.*
2. *ydim12.*
3. *rlen 10.*
4. *Euclidean distance metric.*
5. *Elbow metaclustering.*
6. *Random seed 3026.*

A total of *21* clusters were identified, potentially reflecting distinct immune populations or cell states. As a quality control measure, it is important to check that the clustering is not driven by individual samples corresponding to batch effects (Fig. 1B).

3.1.6 Cluster Annotation

Using prior research and well-established cellular markers, each cluster can be classified into specific immune cell types or states. This process involved associating the unsupervised clusters with known populations.

Identified clusters were further consolidated into 15 distinct populations (Fig. 1C). Cell populations from unsupervised clustering were annotated based on known phenotypical and canonical markers, including: **CD4+ TCM** (C-01), **CD4+ TCM CTLA4** (C-02), **Tregs** (C-03), **Macrophages CD163+** (C-04), **CD4+ naive T cells** (C-07), **gdT cells** (C-09), **Naive B cells** (C-11), **CD8+ TCM** (C-14), **DN T cells** (C-15), **Memory B cells** (C-17), **MNP** (C-10/C-08), **NK cells** (C-12/C-13), **mDCs** (C-05/C-06), **Plasma 1** (C-16/C-19), and **Plasma 2** (C-18/C-20/C-21). Clusters abundance has been represented according to the clinical metadata (Fig. 1D).

scRNA-seq Data: Normalize raw counts, data integration with harmony and scale data

Dimension Reduction:

Apply algorithms such as UMAP, leiden clustering, and Cell Annotation, then use the embedding for the integration.

3.2 Data Integration: A Practical Protocol for Integrating CyTOF and scRNA-seq Data

Multiomic data integration brings together diverse genomic layers to reveal underlying biological patterns and relationships, making it a key strategy for deciphering disease mechanisms and developing new therapies. Despite its power, integrating these datasets is challenging due to their multidimensional, noisy, and biased nature. To overcome these issues, recent methodologies have been developed that range from simple, step-wise overlap analyses to complex frameworks incorporating machine and deep learning algorithms. An ideal integration framework should extract relevant molecular and phenotypic features, capture the covariation among different omic layers, and accurately predict biological classes and phenotypes. While many approaches exist—categorized by experimental design, analytical strategy, or the underlying mathematical framework—no single gold standard has emerged, reflecting the field's ongoing evolution and the need for tailored integration strategies. In the following section we will explore two distinct integration strategies. First, a stepwise approach where we are enriching the annotation of scRNA data using nonmatching CyTOF measurements, and second, we demonstrate a more advanced unsupervised strategy to explore common variance of matching RNA and TotalSeq single-cell measurements.

Below, we outline a step-by-step workflow to integrate CyTOF (mass cytometry) data processed in OMIQ with a scRNA-seq dataset. This protocol details how to prepare both datasets, perform downstream analyses, and ultimately integrate protein (CyTOF) and transcript (scRNA-seq) information via anchor-based methods.

3.2.1 Prerequisites and Software Configuration

All steps in this tutorial were performed using R version 4.3. Before beginning, ensure that the following R libraries are installed and loaded:

```
# Install (if needed)
> install.packages(c(, "Seurat", "dplyr", "tidyr",
"ggplot2"))
```

```
> BiocManager::install("SingleCellExperiment")
# Load
> library(SingleCellExperiment)
> library(Seurat)
> library(dplyr)
> library(tidyr)
> library(ggplot2)
```

3.2.2 Data Preparation

Export Normalized Data and Its Respective Annotations

After completing gating, arcsinh transformation, and any normalization steps in OMIQ, export each sample's processed expression matrix. In addition, export also any annotation files (e.g., cell metadata) required for downstream analyses.

Data Import

Note: While this protocol outlines the essential steps for integrating CyTOF and scRNA-seq data, the deeper intricacies of single-cell analysis—such as dimensional reduction, clustering parameter selection, and advanced trajectory inference—lie outside its current scope. For comprehensive details on standard practices and methodological considerations in single-cell data processing, please refer to the original publication or existing literature that thoroughly describes these analytical workflows.

```
# import single cell RNA-seq data
> biopsy_scRNA <- readRDS("biopsy_scRNA.rds")
```

In R, gather your normalized samples in a list (e.g., NormFiles), then concatenate them:

```
# import CyTOF data
> NormFiles <- lapply(list.
files("OMIQ_CyTOF_Export/",
pattern = "_Normalized.csv",
full.names = TRUE),
read.csv, sep = ",")
#Concatenate the CyTOF data
> exprFcs <- do.call(rbind, NormFiles)
# import CyTOF data annotation
> AnnotationFiles <- list.
files("OMIQ_CyTOF_Export/",
pattern = "_Annotation.csv",
full.names = TRUE)
> annotationMetadataFiles
<- lapply(AnnotationFiles,
read.csv,
sep = "\t",
header = TRUE,
row.names = 1)
# Concatenate the CyTOF data annotation
> annotationMetadata <- do.call(rbind,
annotationMetadataFiles)
```

3.2.3 Convert CyTOF Marker Names to Gene Symbols

To effectively integrate the two datasets, it is crucial to unify their naming conventions. You may rename the scRNA-seq gene symbols to match the CyTOF protein markers or convert the CyTOF marker names into their corresponding gene symbols. In the following protocol, we adopt the latter approach to ensure consistent feature names across both datasets.

Create a Marker-Gene Mapping Data Frame:

With the help of resources like biomart (https://www.ensembl.org/info/data/biomart/index.html), match the names of the protein markers to their respective gene name and prepare a CSV file (e.g., **"Cytof_Convert.csv"**) with two columns:

- CytofMarkers: Original CyTOF marker names (e.g., CD3).
- GeneSymbol: The corresponding gene symbols (e.g., CD3E).

```
# Import the CyTOF marker conversion table
> cytofConvert <- read.csv("Cytof_Convert.csv",
sep=" ")
# Rename the columns to match the column names in
the CyTOF data
> for (i in seq_len(nrow(cytofConvert))) {
oldName <- cytofConvert$CytofMarkers[i]
newName <- cytofConvert$GeneSymbol[i]
if (oldName %in% colnames(exprFcs)) {
colnames(exprFcs)[colnames(exprFcs) == oldName] <-
newName }
}
```

3.2.4 Create a CyTOF Seurat Object

Seurat requires a matrix with features in rows and cells in columns, so transpose if necessary:

```
# Create a Seurat object from the CyTOF data
> CytofObj <- CreateSeuratObject(t(exprFcs))
# add the annotation metadata
> meta <- do.call(rbind, annotationMetadataFiles)
```

3.2.5 Seurat-Based Down Stream Analyses Workflow

After creating a Seurat object from CyTOF data (e.g., CytofObj), you can perform a standard single-cell analysis workflow using Seurat's built-in functions. Below is an example sequence of commands for variable feature selection, scaling, dimensional reduction, and clustering:

```
#Set Variable Features
## Use the entire marker list
> VariableFeatures(CytofObj) <- rownames(CytofObj)
#Scale Data
> CytofObj <- ScaleData(CytofObj, layer =
"counts")
#Principal Component Analysis (PCA)
```

```
> CytofObj <- RunPCA(CytofObj, verbose = TRUE,
npcs = 30)
# builds a nearest-neighbor graph using the first 20
principal components.
#Adjust k.param based on the expected complexity of
your data (e.g., number of clusters).
> CytofObj <- FindNeighbors(CytofObj, dims = 1:20,
verbose = TRUE, k.param = 20)
#Clustering
> CytofObj <- FindClusters(CytofObj, verbose =
TRUE)
#UMAP Embedding
CytofObj <- RunUMAP(CytofObj, dims = 1:20, verbose
= TRUE)
```

3.2.6 Identifying Common Features for Integration

To ensure robust alignment between the scRNA-seq and CyTOF datasets, first identify the features (genes) present in both datasets:

```
# Find the common features between the CyTOF and
scRNA-seq data
> common_features <- intersect(rownames(biopsy_
scRNA), rownames(CytofObj))
> common_features

[1] "TNFRSF9" "FCGR3A" "PTPRC" "CD8A" "CTLA4"
"ICOS" "PDCD1" "TIGIT" "CD38"
[10] "IL7R" "CD14" "HAVCR2" "HLA-DRA" "IL3RA"
"FOXP3" "CD274" "FUT4" "NCAM1"
[19] "CD3G" "IL2RA" "MKI67" "CD27" "CD4" "CD163"
"TRDC" "CD19" "ITGAX"
[28] "CCR7" "TBX21" "PECAM1" "CD7"
```

3.2.7 SCTransforming CyTOF and scRNA-seq Data for Integration

Seurat's SCTransform method is a robust normalization approach that uses regularized negative binomial regression to model and correct for technical noise in single-cell RNA sequencing data.

```
# Normalize the scRNA-seq data
> biopsy_scRNA <- SCTransform(biopsy_scRNA)
# Create a Seurat Cytof Assay
> CytofObj [["Cytof"]] <- CytofObj [["RNA"]]
> DefaultAssay(CytofObj) <- "Cytof"
CytofObj [["RNA"]] <- NULL
# Applies SCTransform to the newly created "Cytof"
assay.
> CytofObj <- SCTransform(ObjFcs, assay = "
CytofObj ")
# Prepares the object for marker-finding analyses
> CytofObj <- PrepSCTFindMarkers(CytofObj)
> DefaultAssay(CytofObj) <- "Cytof"
```

3.2.8 Creating a Data Layer Within the Seurat CyTOF Object

This step is necessary for layered data structures within Seurat, ensuring that both raw (counts) and normalized (data) values are retained and accessible for integration or downstream analyses.

```
# Normalize the 'Cytof' Assay
> CytofObj <- NormalizeData(CytofObj, verbose =
TRUE)
# Create a 'data' Layer
# Copies the normalized count data into a dedicated
'data' layer.
>CytofObj@assays[["Cytof"]]@layers[["data"]]
<-CytofObj @assays[["Cytof"]]@layers[["counts"]]
```

3.2.9 Finding Transfer Anchors for CyTOF-to-scRNA-seq Integration

Using anchor-based integration, link corresponding cell populations across the two modalities:

```
> transfer.anchors_cyt <-
FindTransferAnchors( reference = CytofObj,
query = biopsy_scRNA,
features = common_features,
reference.assay = "Cytof",
query.assay = "RNA",
reduction = "rpca",
dims = 1:30)
```

- **`reference`**: Your CyTOF dataset (CytofObj), using the "Cytof" assay.
- **`query`**: The scRNA-seq dataset (`biopsy_scRNA_conv`).
- **features**: Common genes across both modalities (`common_features`).
- **`reduction`**: Specifies the method ("rpca") used to project high-dimensional data into a lower-dimensional space for anchor identification, alternative methods could be also used, including: "pcaproject," "lsiproject," and "cca." However, the "rpca'" is less computationally burdensome while yielding reliable results.
- **`dims`**: The principal components or reduced dimensions (1:30) used for calculating anchors.

3.2.10 Performing CyTOF–scRNA-seq Integration

Impute CyTOF Signals in the scRNA-seq Space

```
> imputation_cyt <- TransferData(
anchorset = transfer.anchors_cyt,
refdata = GetAssayData( CytofObj, assay = "Cytof",
slot = "data")[common_features,],
weight.reduction= biopsy_scRNA [["pca"]],
 dims = 1:30)
```

- **`anchorset`**: The previously identified anchors linking CyTOF (reference) and scRNA-seq (query).
- **`refdata`**: The arcsinh-transformed CyTOF signals for the shared features.
- **`weight.reduction`**: Uses the PCA reduction from the scRNA-seq object to weight anchor-based imputation.
- **`dims`**: The principal components (1:30) guiding the projection.

Perform the Data Integration

```
# Add the imputed CyTOF data to the Seurat object
> biopsy_scRNA[["Cytof"]] <- imputation_cyt
# select the name of the variable where you stored
the Omiq exported annotation
> annotation_to_project <- "OmiqCytofAnnotation"
> celltype.predictions_cyt <- TransferData(
anchorset = transfer.anchors_cyt,
refdata = ObjFcs@meta.data[,
annotation_to_project],
weight.reduction = biopsy_scRNA_conv[["pca"]],
dims = 1:30)
# select the name of the variable where you stored
the Omiq exported annotation
> annotation_to_project <- "OmiqCytofAnnotation"
> celltype.predictions_cyt <- TransferData(
anchorset = transfer.anchors_cyt,
refdata = ObjFcs@meta.data[,
annotation_to_project],
weight.reduction = biopsy_scRNA[["pca"]],
dims = 1:30)
```

The resulting **imputation_cyt** object maps CyTOF-derived marker intensity values onto each scRNA-seq cell, enabling integrated downstream analyses of protein-level information in the transcriptional context.

After predicting cell types based on CyTOF-derived markers, you can incorporate these assignments into your scRNA-seq object's metadata:

```
> biopsy_scRNA <- AddMetaData( biopsy_scRNA,
metadata = celltype.predictions_cyt[, c("predicted.
id", "prediction.score.max")])
```

These annotations can then be used for downstream analysis, such as differential gene expression or cluster validation, linking the CyTOF-derived cell phenotypes with the transcriptional profiles of the scRNA-seq dataset.

3.2.11 Visualization

```
> P_fig_2A <- DimPlot( biopsy_scRNA,
group.by = "coarse_annotations_MK")+
ggtitle("scRNA-Based Annotation")
> P_fig_2B <- DimPlot( biopsy_scRNA,
```

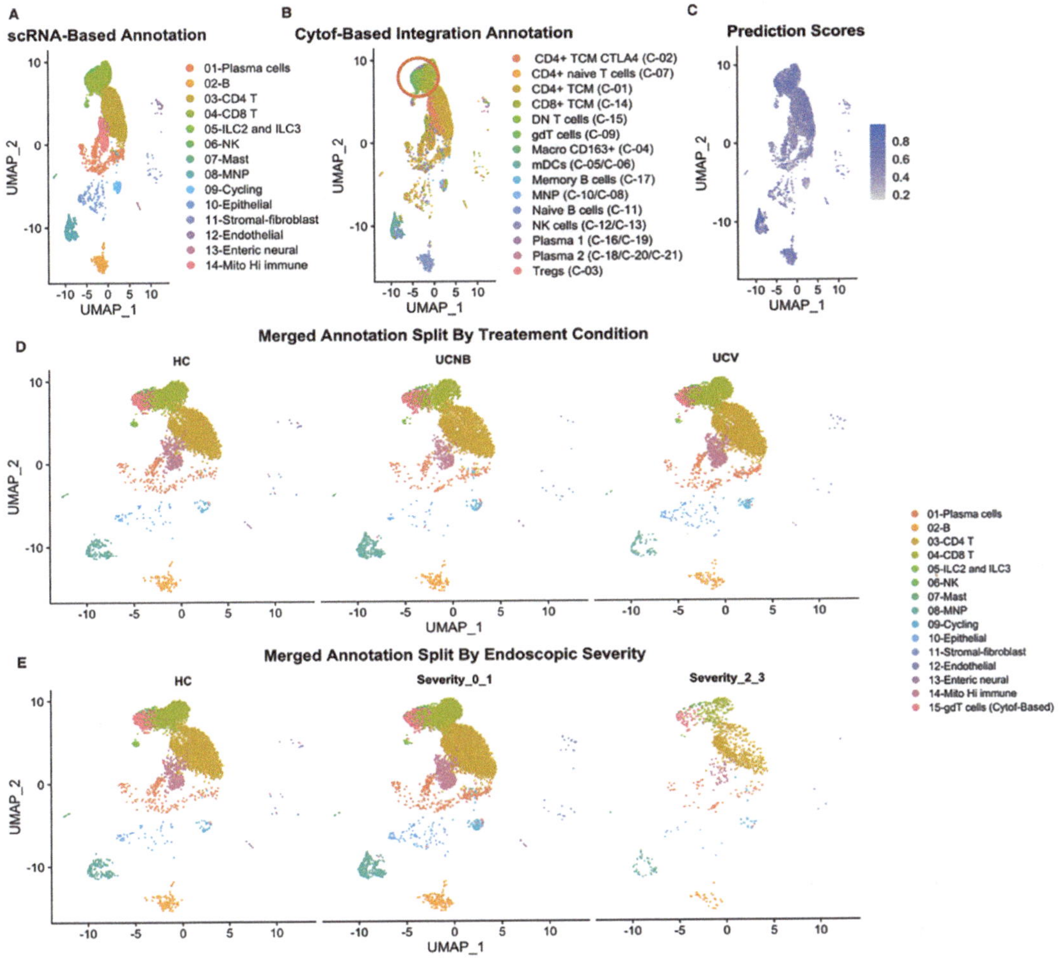

Fig. 2 scRNA and CyTOF based annotations. (**A, B**) Scatter plots—UMAP of the original scRNA and CyTOF merged annotations. (**C**) Scatter plots—UMAP of the prediction confidence scores. (**D**) Scatter plots—UMAP of the merged annotation split by treatment condition and endoscopic severity respectively

```
group.by = "predicted.id") +
ggtitle("Cytof-Based Integration Annotation")
> P_fig_2C <- FeaturePlot(biopsy_scRNA,
features = "prediction.score.max") +
ggtitle("Prediction Scores")
> P_fig_2D <- DimPlot( biopsy_scRNA, split.by =
'condition',
group.by = "finalAnnotation")+
ggtitle("Merged Annotation Split By Treatement
Condition")
> P_fig_2E <- DimPlot( biopsy_scRNA_conv,
split.by = 'Endoscopic_Severity_V1',
```

```
group.by = "finalAnnotation")+
ggtitle("Merged Annotation Split By Endoscopic
Severity")
```

In the panel of Fig. 2, the annotations derived from the CyTOF data in the previous section are projected onto the scRNA dataset. The comparison between the UMAP visualizations in Figs. 2A, B reveals a strong overall concordance between the two annotation strategies, a finding further supported by a high prediction score across the dataset as shown in Fig. 2C—except for the "Mito Hi immune" cluster. This cluster is characterized by cells exhibiting elevated mitochondrial gene expression, a hallmark of significant cellular mortality.

Notably, this integrated approach allowed for the clear identification of a gamma-delta T cell sub-cluster (cluster C-09, highlighted in red in Fig. 2B) among the CD8 T cells—a subset that was not apparent in the initial single-cell annotation. This demonstrates that data integration can uncover additional cell subtypes in cases where the granularity and specificity of RNA molecular signatures at the single-cell level may be insufficient.

Furthermore, we analyzed postmerged cell abundances by stratifying them based on treatment and endoscopic severity score (displayed in Fig. 2D and Fig. 2E, respectively) with the aim of revealing any subgroup-associated variations. However, the observed differences in cell abundance appear to be more attributable to variations in cell numbers across subgroups rather than to the treatment variable itself. While more in-depth scRNA analyses could be performed on this dataset, they fall outside the scope of this chapter.

3.3 A Practical Protocol for a Common Variance-Based Data Integration

Here we will demonstrate the main steps for the integration from the extraction of latent factors that capture key biological variations across modalities to the visualization of the results.

To do so, we downloaded PBMC data from [6]. Samples from eight volunteers enrolled in an HIV vaccine trial [7, 8], with ages ranging from 20 to 49 years. Each sample was profiled using the 10x Chromium 3′ platform coupled with 228 TotalSeq A antibodies, resulting in a dataset of 161,764 cells. However, due to computational resource limitations, we subsampled only 2000 cells for subsequent analyses.

For the purposes of this chapter, our analysis will focus on the treatment variable of interest, specifically comparing Treatment 1—an HIV-1 multiantigen plasmid DNA (HIV-MAG) combined with a Vesicular stomatitis virus (VSV) HIV Gag boost—with Treatment 2, which comprises HIV-MAG in combination with IL-12 and VSV HIV Gag.

3.3.1 Load the Library and Data

```
> library(MOFA2)
# Load the PBMC multimodal dataset
> pmcb_data
<- SeuratDisk::LoadH5Seurat("pbmc_multimodal.
h5seurat")
```

Make sure MOFA2 is installed, then load it for downstream multiomics integration steps; please refer to MOFA2 installation guidelines (https://biofam.github.io/MOFA2/installation.html).

```
# Extract common features (cell barcodes) between
the two modalities
> common <- intersect(
colnames(pmcb_data@assays$ADT@data),
colnames(pmcb_data@assays$SCT@data)
)
# Subset the data due to limitted computational
resources
> samplFeat <- sample(common, 2000)
```

3.3.2 Creating a MOFA Object from RNA and ADT Data

To initiate multiomics factor analysis, construct a **MOFA object** by combining the scRNA-seq (SCT assay) and ADT data matrices into a single input list. In this example, we select the top 1000 variable genes from the SCT assay and the shared subset of ADT features (samplFeat), ensuring consistent dimensions across assays for factorization.

```
# Create a MOFA object
> mofa <- create_mofa(list(
"RNA" = as.matrix(
pmcb_data@assays$SCT@data[
pmcb_data@assays[["SCT"]]@var.features[1:1000],
samplFeat
]
),
"ADT" = as.matrix(
pmcb_data@assays$ADT@data[, samplFeat]
)
))
> mofa
### Untrained MOFA model with the following
characteristics:
### Number of views: 2
### Views names: RNA ADT
### Number of features (per view): 5000 228
### Number of groups: 1
### Groups names: group1
### Number of samples (per group): 2000
```

3.3.3 Setting MOFA2 Model and Training Options

```
# Retrieve Default Model Options
> model_opts <- get_default_model_options(mofa)
# Retrieve Default Training Options
> train_opts <- get_default_training_options(mofa)
```

This function returns a set of parameters controlling how MOFA2 optimizes its factor model. These include settings for convergence criteria, learning rates, and regularization terms.

```
# Set the Number of Latent Factors
> model_opts$num_factors <- 4
```

A default of four factors is chosen here, but this can be adjusted based on dataset complexity and exploratory needs. Increasing the number of factors allows for finer granularity in data decomposition, while a smaller number may yield more interpretable results.

```
# Set a Fixed Random Seed for Reproducibility
> train_opts$seed <- 4321
# Prepare the MOFA object for training
> mofa <- prepare_mofa(
object = mofa,
model_options = model_opts,
training_options = train_opts)
# Train the MOFA model
> mofa <- run_mofa(mofa)
```

3.3.4 Integration Results

After training the MOFA2 model, it is crucial to evaluate how well the identified factors explain variability across different omics layers and whether they capture distinct or correlated signals. MOFA2 provides built-in functions for visualizing these properties.

Variance Decomposition

```
> P_fig3_A <-plot_factor_cor(mofa)
> P_fig3_B <-plot_variance_explained(mofa, max_
r2=50)[[1]]
> P_fig3_C <-plot_variance_explained(mofa,
plot_total = T)[[2]]
```

Unlike principal component analysis (PCA), MOFA2 does not enforce orthogonality constraints. Therefore, significant correlations among factors may indicate redundancy and suggest that the model

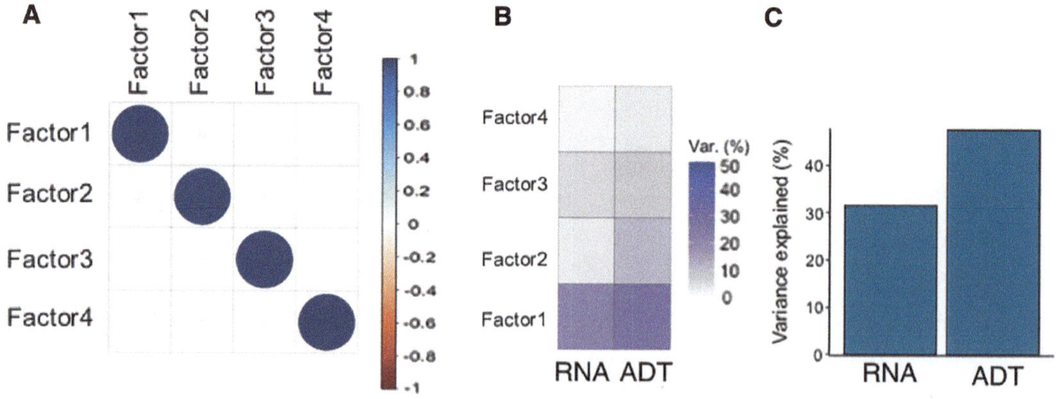

Fig. 3 Factor correlation and variance decomposition analysis. (**A**) Correlation heatmap of the extracted factors. (**B**) Heatmap of the factors shared explained variance with the ADT and RNA datasets. (**C**) Barplots of the total explained variance fraction in RNA and ADT

fit could be improved. Figure 3A does not demonstrate any correlation between the four extracted factors. In this analysis, the variance decomposition analysis illustrates the percentage of variance explained by each factor within each modality (RNA and ADT). As depicted in Fig. 3B, Factor 1 explains the largest portion of variance in both modalities. Furthermore, in Fig. 3C, the total variance explained across the four factors interestingly shows that even with a smaller number of features, the proteomics data exhibits higher percentage of total variance in the dataset compared to RNA with (~50% for ADT and 30% for RNA). This discrepancy may be attributed to the inherently higher signal-to-noise ratio and lower dimensionality of the proteomic measurements, which are designed to capture key cell surface markers. Additionally, the targeted nature of the ADT panel likely leads to a more focused dataset, where the dominant biological signals are more pronounced.

Factor's Characterization

```
> P_fig4_A <- correlate_factors_with_
covariates(mofa,
covariates = c("Treatment",
"CMV",
"time"))
> P_fig4_B <- plot_top_weights(mofa,
view = "RNA",
factors = 1,
sign = "positive",
nfeatures = 15)
> P_fig4_C <- plot_top_weights(mofa,
view = "ADT",
factors = 1,
sign = "positive",
nfeatures = 15)
```

The heatmap in Fig. 4A displays the correlations between the latent factors and a set of covariates, including our primary variable

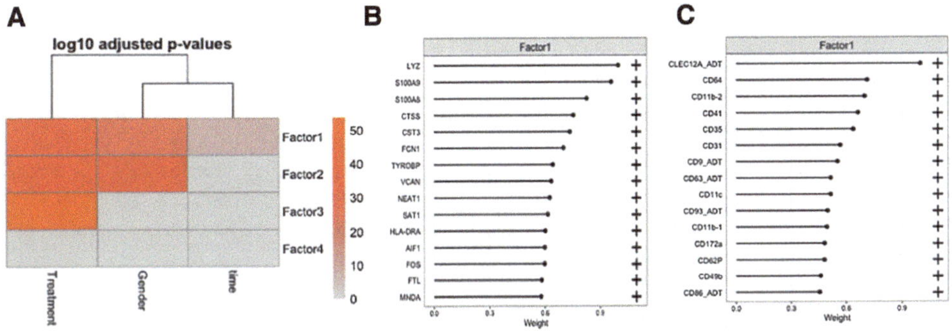

Fig. 4 Feature's contribution analysis. (**A**) Correlation heatmap of the extracted factors with variables of interest. (**B, C**) Barplots representing the weights of the extracted features from the first factor of the RNA and ADT datasets, respectively

of interest—Treatment—as well as potential confounders such as gender. This visualization serves two key purposes: it helps identify confounding variables that might need to be corrected for (notably, gender shows a strong correlation with Factors 1 and 2), and it highlights which factors are associated with the Treatment variable. While Factors 1, 2, and 3 all exhibit some degree of association with Treatment, prior results indicate that Factor 1 explains the highest proportion of variance, making it the most promising candidate for further characterization.

In addition, Fig. 4B, C illustrates the distribution of feature weights for each latent factor. These weights quantify the contribution of individual features to each factor: features with weights near zero have minimal association, whereas those with large absolute values contribute strongly to the factor. In this analysis, we focus on positive weights, which correspond to features with higher expression in Treatment 1 relative to Treatment 2. Notably, LYZ (Lysozyme) and CLEC12A (CD371) emerge as the top contributing features in Factor 1 for both RNA and ADT data, underscoring their potential biological relevance in differentiating the treatment groups.

Visualizations

```
> P_fig5_A <- plot_data_scatter(mofa,
view = "ADT",
factor = 1,
features = 1,
sign = "positive",
color_by = "Treatment"
) + labs(y="ADT")

> P_fig5_B <- plot_data_scatter(mofa,
view = "RNA",
factor = 1,
features = 1,
sign = "positive",
color_by = "Treatment"
) + labs(y="RNA")

> P_fig5_C <- plot_factor(mofa, factors=1, group_by
= " celltype.l1 ", color_by="LYZ") +
theme(caxis.text.x = element_text(color="black",
angle=40, vjust=1, hjust=1))

> P_fig5_D <- plot_factor(mofa, factors=1, group_by
= " celltype.l1 ", color_by="CLEC12A_ADT") +
theme(axis.text.x = element_text(color="black",
angle=40, vjust=1, hjust=1))
```

The scatter plots in Fig. 5A, B illustrate the association between the MOFA2-derived factor values and the expression levels of LYZ and CLEC12A, respectively. Each point represents an individual cell, with colors indicating the treatment administered to the donor. Both

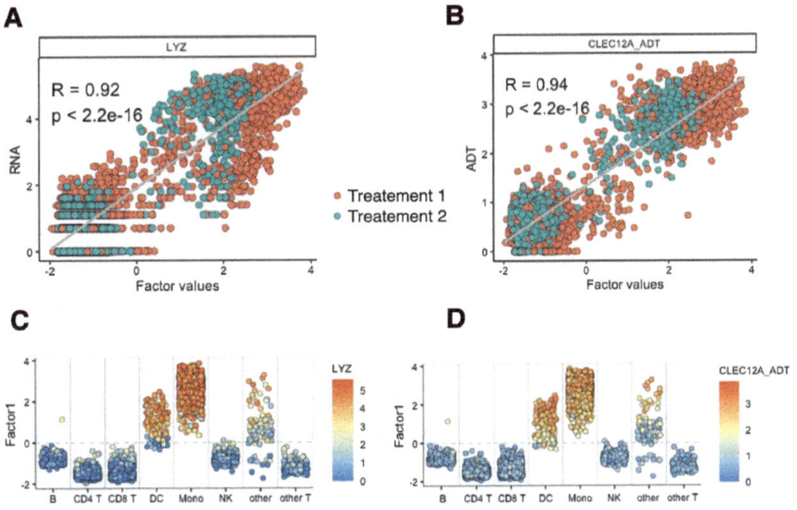

Fig. 5 Feature-trait association analysis. (**A, B**) Scatterplot of the correlations between levels of expressions and the factor values of the LYZ gene and CLEC12A protein, respectively, colored by the treatment conditions. (**C, D**) Factor value decomposition by cell type annotation for both LYZ gene and CLEC12A protein, respectively

plots reveal a strong positive correlation between factor values and expression levels, and notably, cells from patients receiving Treatment 1 exhibit higher expression and factor scores for these markers compared to those from Treatment 2.

In addition, Fig. 5C, D depicts the distribution of factor values across different annotated cell types, with each cell color-coded by its expression level of LYZ and CLEC12A. These visualizations confirm that both markers are predominantly expressed in monocytes and dendritic cells.

LYZ is known to serve as an innate activation marker—its gene product, lysozyme, is abundantly produced by phagocytes such as monocytes, macrophages, and neutrophils [9], while CLEC12A (CD371) is a C-type lectin receptor that modulates the activation status of myeloid cells.

The lower transcript levels of LYZ and reduced CLEC12A protein expression observed in Treatment 3 suggest that IL-12 exposure, combined with VSV HIV gag mo [6], may drive monocytes and dendritic cells toward a more inflammatory or activation-resistant state. This shift implies a distinct alteration in the activation profile of these immune cells, potentially modifying their role in the regulation of inflammatory responses.

4 Addressing Integration Challenges

Despite employing several unsupervised clustering methods for multiomic integration, our framework has several limitations. First, for both anchor-based methods using Seurat and cross-modality variance analysis with MOFA, we used parameters to enable an easier analysis, yet these defaults may not capture the full potential of each dataset; one should include parameter optimization, despite the increased computational demands. Additionally, in our protocol, approaches are entirely linear, which may overlook the nonlinear molecular interactions that underlie complex diseases. Exploring deep learning techniques—such as autoencoders, which have shown promise in oncology—could offer more nuanced insights, although these methods require large sample sizes and often suffer from reduced interpretability. Finally, incorporating sparsity constraints, as in methods such as PathME [10], might improve model interpretability by focusing on the most relevant features, thereby better reflecting the underlying biology.

5 Conclusion

By leveraging tools like OMIQ and R, we were able to annotate cell types that were previously absent from the initial analysis, transfer annotations across modalities, and uncover insights into rare populations. These tools enabled a more refined characterization of cellular diversity, enhancing our ability to detect subtle phenotypic differences and functional states within complex datasets. Furthermore, by integrating multiomics data, we improved the resolution of rare subpopulation identification, shedding light on their potential biological significance.

The methodologies described in this chapter provide a robust framework to guide researchers in overcoming analytical challenges and realizing the full potential of multimodal data integration. Standardized workflows and computational advancements, including machine learning-driven classification and automated annotation pipelines, further streamline the interpretation of high-dimensional datasets.

References

1. Pedersen CB, Dam SH, Barnkob MB et al (2022) cyCombine allows for robust integration of single-cell cytometry datasets within and across technologies. Nat Commun 13:1698. https://doi.org/10.1038/s41467-022-29383-5
2. Lin Y, Cao Y, Willie E, Patrick E, Yang JYH (2023) Atlas-scale single-cell multi-sample multi-condition data integration using scMerge2. Nat Commun 14(1), 4272. https://doi.org/10.1038/s41467-023-39923-2
3. Repapi E, Agarwal D, Napolitani G et al (2023) Integration of single-cell RNA-Seq and CyTOF data characterises heterogeneity of rare cell subpopulations. F1000Research 11:560. https://doi.org/10.12688/f1000research.121829.3

4. Saihi H, Bessant C, Alazawi W (2023) Automated and reproducible cell identification in mass cytometry using neural networks. Brief Bioinform 24(6):bbad392. https://doi.org/10.1093/bib/bbad392
5. Mennillo E, Kim YJ, Lee G et al (2024) Single-cell and spatial multi-omics highlight effects of anti-integrin therapy across cellular compartments in ulcerative colitis. Nat Commun 15:1493. https://doi.org/10.1038/s41467-024-45665-6
6. Hao Y, Hao S, Andersen-Nissen E, Mauck WM 3rd , Zheng S, Butler A, Lee MJ, Wilk AJ, Darby C, Zager M, Hoffman P, Stoeckius M, Papalexi E, Mimitou EP, Jain J, Srivastava A, Stuart T, Fleming LM, Yeung B, Rogers AJ, McElrath JM, Blish CA, Gottardo R, Smibert P, Satija R (2021) Integrated analysis of multimodal single-cell data. Cell 184(13):3573–3587. https://doi.org/10.1016/j.cell.2021.04.048.
7. Elizaga ML, Li SS, Kochar NK, Wilson GJ, Allen MA, Tieu HVN, Frank I, Sobieszczyk ME, Cohen KW, Sanchez B, Latham TE, Clarke DK, Egan MA, Eldridge JH, Hannaman D, Xu R, Ota-Setlik A, McElrath MJ, Hay CM, NIAID HIV Vaccine Trials Network (HVTN) 087 Study Team (2018) Safety and tolerability of HIV-1 multiantigen pDNA vaccine given with IL-12 plasmid DNA via electroporation, boosted with a recombinant vesicular stomatitis virus HIV Gag vaccine in healthy volunteers in a randomized, controlled clinical trial. PLoS One 13(9):e0202753. https://doi.org/10.1371/journal.pone.0202753.
8. Li SS, Kochar NK, Elizaga M, Hay CM, Wilson GJ, Cohen KW, De Rosa SC, Xu R, Ota-Setlik A, Morris D, Finak G, Allen M, Tieu HV, Frank I, Sobieszczyk ME, Hannaman D, Gottardo R, Gilbert PB, Tomaras GD, Corey L, Clarke DK, Egan MA, Eldridge JH, McElrath MJ, Frahm N, NIAID HIV (2017) Vaccine Trials Network. DNA priming increases frequency of T-cell responses to a vesicular stomatitis virus HIV vaccine with specific enhancement of $CD8^+$ T-cell responses by interleukin-12 plasmid DNA. Clin Vaccine Immunol 24(11):e00263-17. https://doi.org/10.1128/CVI.00263-17
9. Bonnardel J, Da Silva C, Henri S, Tamoutounour S, Chasson L, Montañana-Sanchis F, Gorvel JP, Lelouard H (2015) Innate and adaptive immune functions of peyer's patch monocyte-derived cells. Cell Rep 11(5):770–84. https://doi.org/10.1016/j.celrep.2015.03.067
10. Lemsara A, Ouadfel S, Fröhlich H (2020) *PathME*: pathway based multi-modal sparse autoencoders for clustering of patient-level multi-omics data. BMC Bioinform 21:146. https://doi.org/10.1186/s12859-020-3465-2

Kolibri 360 GmbH
Ferdinand-Jühlke-Straße 7
99095 Erfurt, Deutschland
produktsicherheit@kolibri360.de